17th International Conference Diagnostics of Machines and Vehicles 2018

MATEC Web of Conferences Volume 182 (2018)

Bydgoszcz, Poland
25 - 26 September 2018

Editors:

T. Kalaczynski
M. Zoltowski

ISBN: 978-1-5108-6805-2

Printed from e-media with permission by:

Curran Associates, Inc.
57 Morehouse Lane
Red Hook, NY 12571

Some format issues inherent in the e-media version may also appear in this print version.

Printed by Curran Associates, Inc. (2018)

For additional information, please contact EDP Sciences – Web of Conferences
at the address below.

EDP Sciences – Web of Conferences
17, Avenue du Hoggar
Parc d'Activité de Courtabœuf
BP 112
F-91944 Les Ulis Cedex A
France

Phone: +33 (0) 1 69 18 75 75
Fax: +33 (0) 1 69 28 84 91

contact-edps@webofconferences.org

Additional copies of this publication are available from:

Curran Associates, Inc.
57 Morehouse Lane
Red Hook, NY 12571 USA
Phone: 845-758-0400
Fax: 845-758-2633
Email: curran@proceedings.com
Web: www.proceedings.com

TABLE OF CONTENTS

17th International Conference
DIAGNOSTICS OF MACHINES AND VEHICLES

ORGANIZING COMMITTEE:

Dr inż. Tomasz Kałaczyński	**chairman**
Dr inż. Bolesław Przybyliński	
Dr inż. Marcin Łukasiewicz	
Dr inż. Sylwester Borowski	
Dr inż. Mariusz Żółtowski	
Dr inż. Joanna Wilczarska	**conference office**
Dr inż. Robert Kostek	
Dr inż. Michał Liss	

edp sciences

Web of Conferences

Statement of Peer review

In submitting conference proceedings to *Web of Conferences*, the editors of the proceedings certify to the Publisher that

1. They adhere to its Policy on Publishing Integrity in order to safeguard good scientific practice in publishing.
2. All articles have been subjected to peer review administered by the proceedings editors.
3. Reviews have been conducted by expert referees, who have been requested to provide unbiased and constructive comments aimed, whenever possible, at improving the work.
4. Proceedings editors have taken all reasonable steps to ensure the quality of the materials they publish and their decision to accept or reject a paper for publication has been based only on the merits of the work and the relevance to the journal.

Title, date and place of the conference:

17th International Conference DIAGNOSTICS OF MACHINES AND VEHICLES 25-26 September 2018. in Pieczyska near BYDGOSZCZ , Poland.

Proceedings editor(s): Tomasz Kałaczynski, Mariusz Żółtowski, conference science committee

Date and editor's signature

29.06.2018

Electric public transport vehicle with composite material body diagnostic - user safety

Marcela Escobar[1] , *Daniel* Cadena[1] , and *Leonel Francisco* Castañeda[1, **]

[1]Grupo de Estudios en Mantenimiento Industrial GEMI, departamento de ingeniería mecánica, Universidad EAFIT, Antioquia - COLOMBIA.

Abstract. The project contemplated the elaboration of modules of composite material, with which a bus body of less weight was built, compared with the traditional kind, and adapted for the chassis that are currently being used in the city. The structure was prepared for the evaluation of the security space of the future users. The panels were instrumented with different sensors and configured to the conditions of the test by means of a data acquisition system. The test was carried out following the procedure of regulation 66 of the UNECE. The results obtained demonstrated the feasibility of using this type of bodywork in public transport.

1 Description

Latin American countries are traditionally consumers of European and American transport technology. Due to its topographic conditions, public bus transport leads the mobility of big cities. In the last two decades, the city of Medellín has become a benchmark for social, economic and environmental transformation, this is possible in large part because of the integration of its inhabitants through its environmentally friendly public transport network; metro, tram, transport by cable, bicycles and buses with low emissions.

The feeder vehicles of this public transport network are an automobile fleet mainly made of metal body and uses fossil fuels; Diesel and gasoline. In order to generate alternatives for environmental improvements, the EAFIT university and the ICOLFIBRA company established an alliance to develop a body and a motor module in feeder microbuses. This article presents part of the results of the research project with code 67597475630 of the Administrative Department of Science and Technology - COLCIENCIAS, focused on the development of a vehicle with a body made of composite materials that meets the conditions of safety, accessibility and comfort.

1.1 CONTEXT

The Ministry of Transport of Colombia has published different standards, resolutions and technical regulations in order to ensure the safety, accessibility and comfort of users. In 1995 Resolution 7126 was issued, where some parameters are required for the manufacture and / or assembly of vehicles for public transport of passengers. Under this resolution, the chassis and bodies that meet the established requirements are approved by the Ministry. In 2009, the standards NTC 5206: 2009 and NTC 5701: 2009 are published, which cite global standards, and some sections of these become technical regulations in 2016 through resolution 3753. Comfort is evaluated from requirements of the NTC5206: 2009 [1] and is validated through the metrology of position of chairs, location of handholds, distances of corridors, height of the vehicle, width of doors, height of the steps, and the verification of the measures and quantities of emergency protocols, certificates of glasses and materials. Accessibility refers to sections of NTC 5701: 2009 that are addressed to the access of people with reduced mobility or PRM to the vehicle, the PRM space, travel distances,

* Leonel Francisco Castañeda Heredia: lcasta@eafit.edu.co

access ramps, PRM zones, entry platform are measured. Vehicle safety is verified through the structural strength of the body under the requirements of two different standards: the standard NTC5206: 2009 with the load test on roof and lateral load, applies 50% of the gross vehicle weight on the vehicle for 5 minutes and the roof should not suffer a deflection of more than 70 mm, and the ECE R66 [2] standard by means of the rollover test, which consists of dumping a complete vehicle under established conditions and the survival space should not be affected.

Numerous investigations have been carried out in which the properties of composite materials are evaluated in comparison with conventional materials in real applications, in order to create composites as an alternative that may have additional benefits in specific functions. [3, 4]

The norm states that the superstructure of the bus must support the potential energy transformed into kinetic energy in the following way:

$$E_T = 0.75 \cdot M \cdot g \cdot \Delta h \quad (1)$$

(M) represents the mass of the loaded vehicle, (g) states the standard earth gravity and (Δh) refers to the distance of the center of mass from the floor.

This article will show all the necessary conditions to perform a rollover test on a vehicle with a body made of composite materials.

2 Vehicle description

The vehicle to which the rollover test will be performed, is a minibus that transports 19 passengers plus driver. It is made up of a metallic chassis and a body made with 100% composite materials. The object of study is characterized by the complexity of its geometry and its body made of composite materials. A Class II vehicle (Vehicles to provide the collective public service within the national scope of action) will be used, which will be used to verify the requirements of annex 5 of Regulation 66 of the United Nations. The complete vehicle dimensions are shown in table 1.

Fig. 1. Bus used on test.

Table 1. Bus dimensions.

Vehicle	
No. Of passengers	19
Type of chairs	Plastic fixed chairs
Chassis serial number	304
Front overhang	1330 mm
Rear overhang	2177 mm
Total lenght	6857 mm
Total width	2296 mm
Total height from the floor	2888 mm
First step height	400 mm
Internal free space height	1800 mm
Bodywork weight	1780 kg
Chassis weight	2470 kg
Total weight	4250 kg
Number of operative doors	1 de 650 mm free space
Driver door	No
Hatch	1

The chassis brand is MITSUBISHI FUSO; its characteristics are shown in the following table.

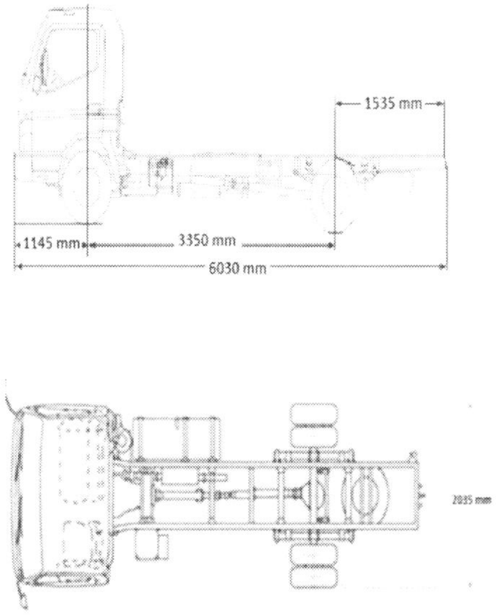

Fig. 2. Bus schematic.

Table 2. Bus technic description

Chasis	
Brand	MITSUBISHI FUSO
Reference	FE85DE6SLGP
Homologation	FTH-TPP 16165
Motor placement	Forward
Motor type	Diesel turbo charged 4 cilinders in line 4899 cm^3
Max power	177 HP@ 2700 rpm
Torque	54 kg.m @ 1600 rpm
Emission standard	Euro IV
Distance between axis	3350 mm
Gauge width	2035 mm
Front overhang	1145 mm
Rear overhang	1535 mm
Front axle weight	1485 kg
Back axle weight	885 kg
Total unoccupan weight	2370 kg
Permissible weight front axle	2800 kg
Permissible weight back axle	5100 kg
Gross Vehicle Weight	7500 kg
Tire	215/75 R17,5,

3 Rollover test description

The test is to tilt the vehicle, which is suspended on a tilting platform with the fixed suspension, it is slowly raised until the vehicle loses its equilibrium position, according to ECE R66 [2] the tipping conditions must meet the conditions shown in fig 3. The rollover test begins at the unstable position of the vehicle with zero angular velocity and ends at the moment of impact of the vehicle with the concrete. The test is satisfactory if the survival space is not affected. The survival space is defined as the space that must remain in the compartment or compartments of the driver, travelers and staff so that the driver, travelers and staff are more likely to survive in the event of rollover [5].

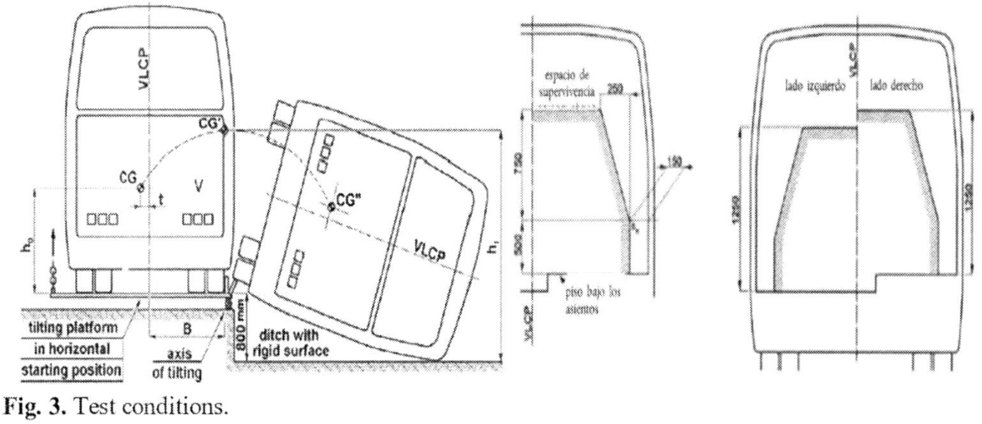

Fig. 3. Test conditions.

Table 2. Measurement elements used

MEASUREMENT ELEMENT	NOMENCLATURE	QUANTITY
Strain gauges	G	10
Displacement sensors	L, E	4
Accelerometers	A.D, A.T, A.C	3
Gyroscope	N/A	1

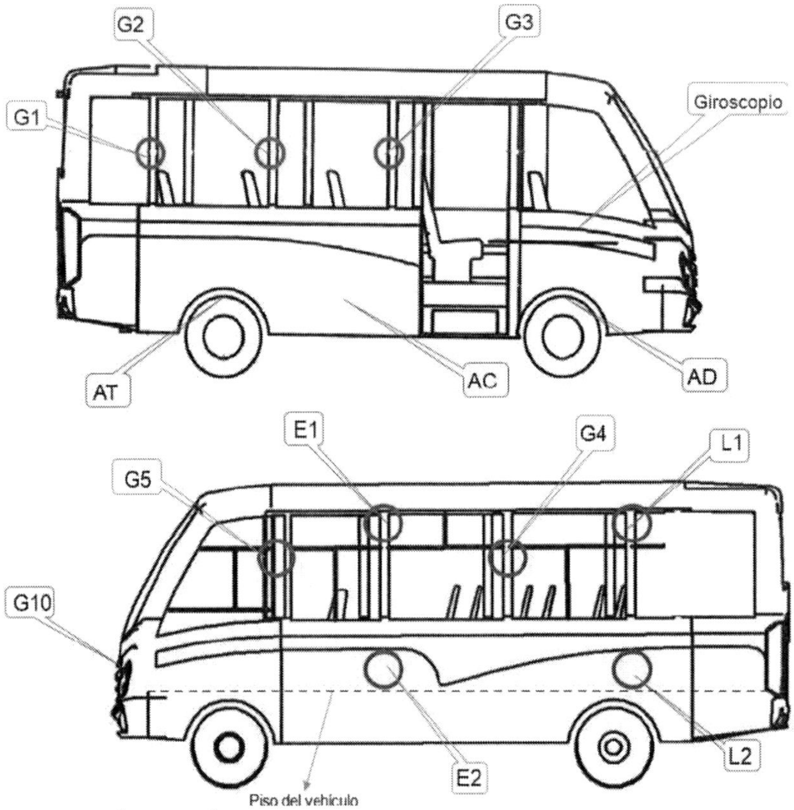

Fig. 4. Measurement elements placement.

4 Results and discussions

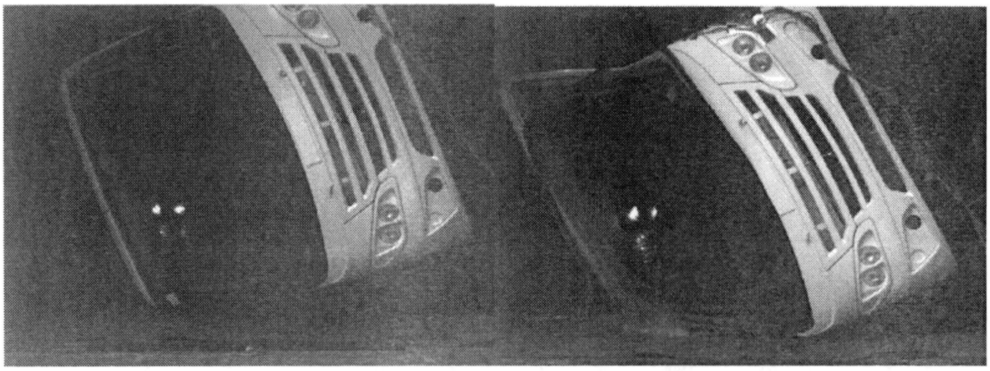

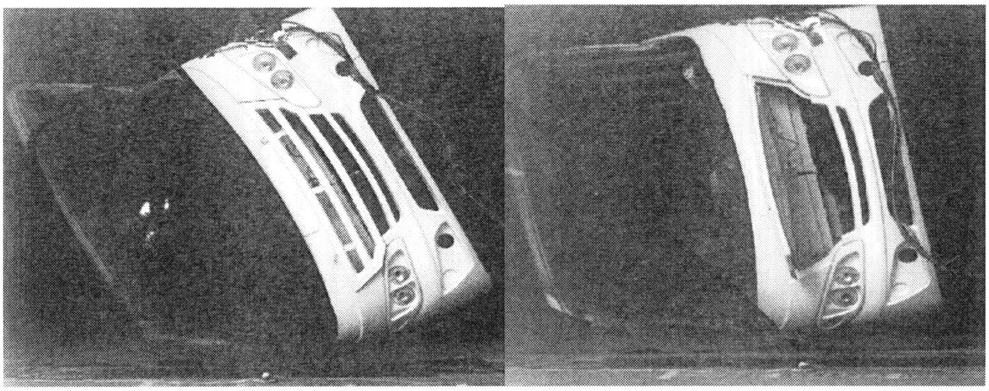

Fig. 5. Rollover test of a body made of composite materials.

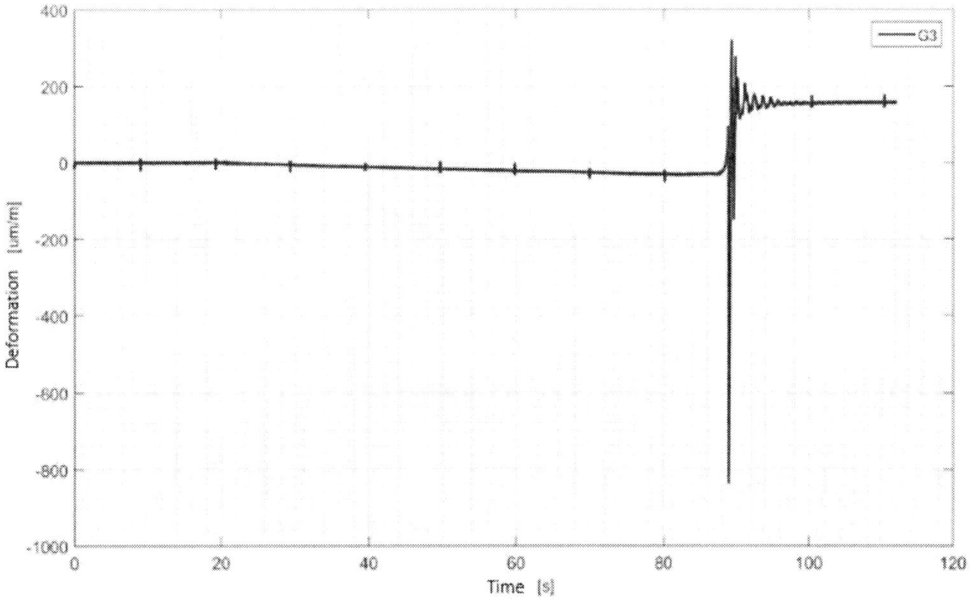

Fig. 6. G3 extensiometric gauge for critical pillar.

The graph shows the displacements registered at the moment of impact by the G3 gauge located in one of the most critical pillar, which corresponds to the one located on the side where the entrance door is located, a geometry that can structurally represent a greater challenge for the integrity of the vehicle. During the test it is observed that this element is the one that is subjected to the highest recorded deformation, however, the piece tends to recover the original shape and position.

It's remarkable to mention that in order to generate alternatives for environmental improvements, the EAFIT university and the company ICOLFIBRA established an alliance to develop a body and a motor module in the feeder microbuses, framed in the context of the Administrative Department of Science and Technology - COLCIENCIAS.

References

1. Norma Técnica Colombiana NTC – icontec, «Vehículos para el transporte terrestre público colectivo y especial de pasajeros. Requisitos y métodos de ensayo,» 2009
2. Official Journal of the European Union, «Large Passenger Vehicles with Regard To The Strength Of Their Superstructure, ECE 66-02, » 2010.

3. A. G. K. a. J. W. Kaczmar, «Application of Polymer Based Composite Materials in Transportation, » Progress in Rubber, Plastics and Recycling Technology, vol. 32, 2016.
4. E. F. A. P. G. M. A. Zinno, «Multiscale approach for the design of composite sandwich structures for train application, » Composite Structures, p. 2208–2219, 2010.
5. L. V. C. Colombo, «Experimental and numerical analysis of a bus component in composite material, » Composite Structures, pp. 1706-1715, 2010.

The method of diagnosing glued joints subjected to static loads

Anna Jasińska[1,*], *Bogdan* Ligaj[1] and *Mirosław* Miciak[2]

[1]Faculty of Mechanical Engineering, UTP University of Science and Technology in Bydgoszcz, al. Prof. S. Kaliskiego 7, 85-796 Bydgoszcz, Poland
[2]Faculty of Telecommunications, Computer Science and Electrical Engineering, UTP University of Science and Technology in Bydgoszcz, al. Prof. S. Kaliskiego 7, 85-796 Bydgoszcz, Poland

Abstract. The article presents a new type of sample developed for the purpose of assessing the strength of glued joints. The construction and the idea of testing glued joints were described in detail. In the article put a question: what geometric features of the sample significantly affect the change in the stress values at the measuring points S1, S2 and S3 under the influence of the acting bending moment? Analysis of the results of numerical calculations allowed to state that the width of the glued joint h, spacing of connectors L3 and the thickness of the upper beam g_1 significantly affect the value of stresses in points S1, S2 and S3.

1 Introduction

Designing of machine construction elements is based on criteria formulated by end users. Criteria enforce the search for cheaper and simple manufacturing technologies of elements. A commonly used method of joining structural elements is gluing. It is used in the production of vehicles (cars, buses, trains, trams etc.) as assembly connections, among others windshield with supporting structure. Such application requires from connections to maintain high strength properties over the assumed service life , to ensure a high level of safety for vehicle users. Conducting experimental tests of this type of connections, in terms of strength and durability, allows you to choose their geometric features that ensure compliance with normative and utility requirements.

At the initial stages of constructing the examination of structural elements of machines, they are carried out on samples. Their shape and dimensions are adopted on the basis of standards, examples of which are studies [1, 2, 3]. The results of tests carried out on standard samples are presented in the paper [4]. There are also many examples of tests carried out using non-standard samples [5, 6] from a wide range of scientific disciplines.

In the literature are presented the results of research on combining elements made of different materials. Standard samples or parts of structural elements are used in the tests. On their base, the strength and durability of connections is assessed. Presented at work [7, 8] the results of the glued joints tests were carried out on samples under the monotonic conditions of the increasing bending moment. The presented samples allow to assess the strength of glued joints in relation to the variable stiffness of the elements and the dimensions of the joints.

The purpose of the work is to determine, which of the geometrical features of a new type of sample significantly affects the value of stresses in points S1, S2 and S3 under constant load conditions, the value of the bending moment.

The scope of work includes the presentation of: construction of a new type of sample for testing glued joints, the idea of assessing the strength of joints in four-point bending

* Corresponding author: jasinska.an@wp.pl

conditions, results of numerical calculations for variable geometric features, analysis of results.

2 Innovative sample for testing glued joints

2.1 Sample structure

Strength testing of adhesive joints can be carried out using samples of shape shown on Figure 1. The construction of the sample was submitted as an invention to the Polish Patent Office, in which it was registered under number P.422884 [9].

The test sample is made of: an upper beam, a bottom beam, two glued joints and strain gauges The geometric features of the sample elements are following (Fig.1a):

- top beam(M1): L1 – length, H1 width, g_1 – thickness,
- bottom beam(M2): L2 – length, H2 – width, g_2 – thickness,
- glued joint (AJ): L3 – length, h – width, g_3 – thickness,
- strain gauges (T): unified elements e.g.. TEN-TFS5/120-P [10].

Geometrical relations between the elements of the sample are shown in the figure 1a and 1b. Glued joints (AJ) they are symmetrical in relation to the length L1 upper beam (M1). Distance between connectors AJ defines the dimension L4. Bottom beam (M2) with a length L2 it is located symmetrically to the beam M1 and connectors AJ. Beam widths: M1 and M2 and the length of the connector AJ they have the same values: H1 = H2 = L3. On beam M1 and beam M2 measuring points were determined (S), in which the measurement of deformations is carried out at the moment of testing. Measurement points S they are symmetrically oriented with respect to the axis of the sample in the width range H1 and H2, as well as length L1. In the symmetry axis of the sample, measuring points are located: S2 on a beam M1 and S3 on a beam M2. The axis of the sample is distant from the edge of the beam M1 by value 0.5L1. Position of the measuring point S1 in relation to the axis of symmetry determines the dimension F. Its value should be within the range $L4 < 2F < Y$. At each measuring point S1, S2 and S3 two strain gauges were mounted (T_0), located on both sides of the beam. The task of strain gauges T_0 is the measurement of deformations on the surface of the sample elements, i.e. the upper beam M1 and bottom beam M2.

This sample is designed to assess the strength of adhesive joints in conditions of controlled deformation of machine construction elements. The most advantageous form of loading the sample is a four-point bending with a monotonically increasing value of the bending moment. The structure of the sample also allows the implementation of tests in cyclic loading conditions with cycle asymmetry coefficient $R \geq 0$. The method of applying the load is shown on the Figure 1c.

During the tests, the sample rests on supports A remote from each other by value X. Forces loading the sample are applied in points B, remote from each other by value Y. Points of load application B they are symmetrical with respect to supports A. Position of the symmetry axis of the sample relative to the supports A defines the dimension X/2. The value of the bending moment loading the sample depends on the value of forces acting, as well as on the ratio of dimensions X/Y.

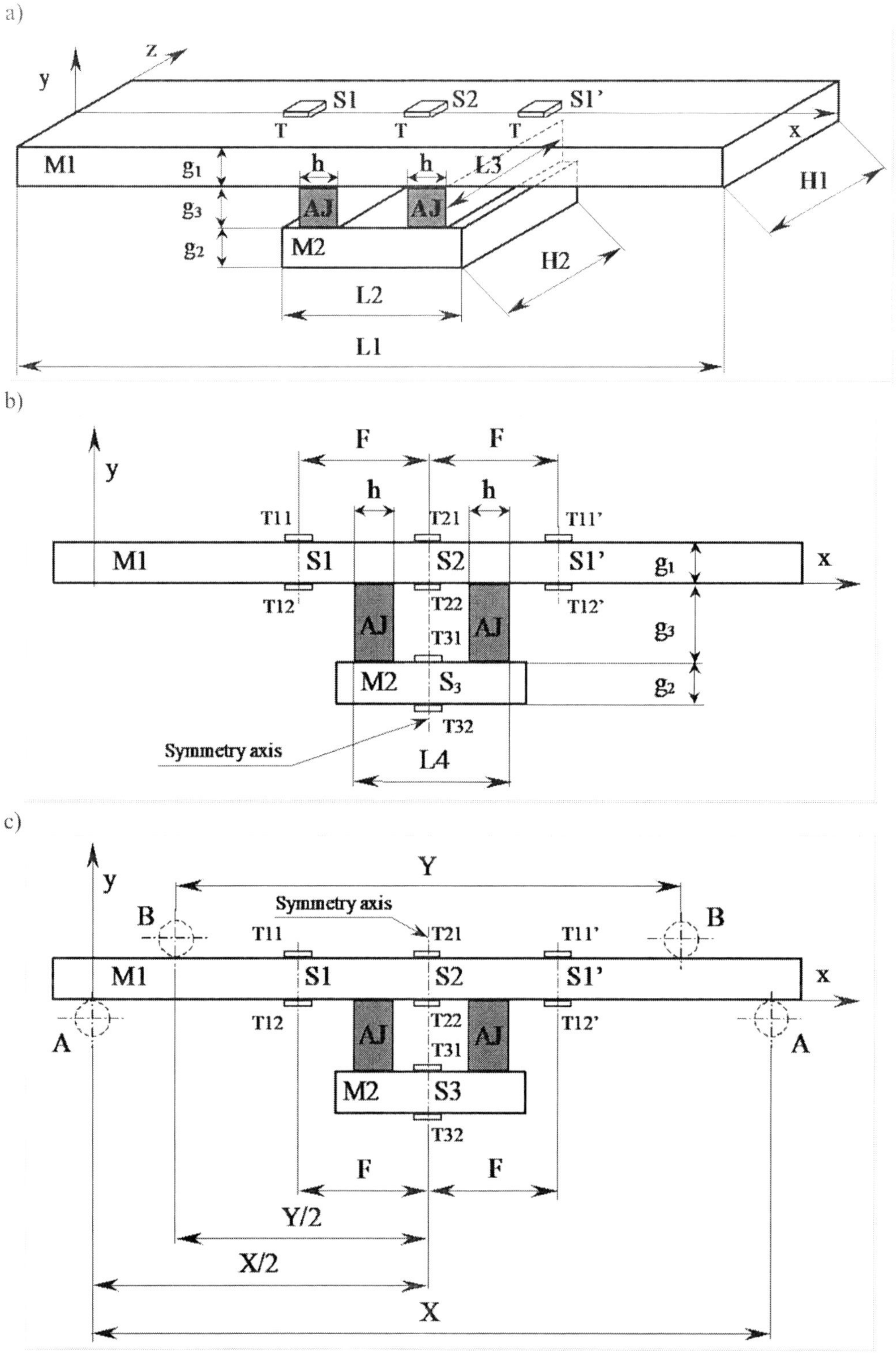

Fig. 1. Schematic of specimen for testing the strength of adhesive joint.

Strength and stiffness of the top beam M1, as well as the bottom beam M2 is shaped by the selection of: cross-section, geometric dimensions and material type. The cross-section of the beams is able to take the shape of simple figures (eg rectangle, square) or complex figures (eg "C" section, "L" section and others). The shape of the cross-section and its dimensions affect the value of the static moment of inertia I_z. Presented on a Figure 1 the sample is made of a beam M1 and beam M2 having a rectangular cross-section. Dimensions of the cross-section of the beam M1 are $H1 \cdot g_1$, but beam M2 are $H2 \cdot g_2$. The value of the moment of inertia of each beam is determined by the thickness $g_{()}$ ($I_{z()} = (H() \cdot g_{()}^3)/12$), because values H1 = H2. Coefficient of stiffness of beams M1 and M2 is the product of the static moment of inertia I_z and modulus of longitudinal elasticity E (modulus Younga). Modulus Younga E is a characteristic parameter for the material.

2.2 The idea of testing glued joints

Examination of a sample of an innovative shape in four-point bending conditions with a monotonically increasing torque value consists in measuring the deformation values at the measuring points: S1 (S1'), S2 and S3. On the basis of strain measurements, the value of stresses is determined using prawa Hooke'a. Measurement points S1 and S2 they are in places where the value of bending moment has a constant value. Point S1 it is located outside the glued joints, and the point S2 between glued joints. The change in the stress value at S1 is proportional to the change in the bending moment value (Fig. 3).

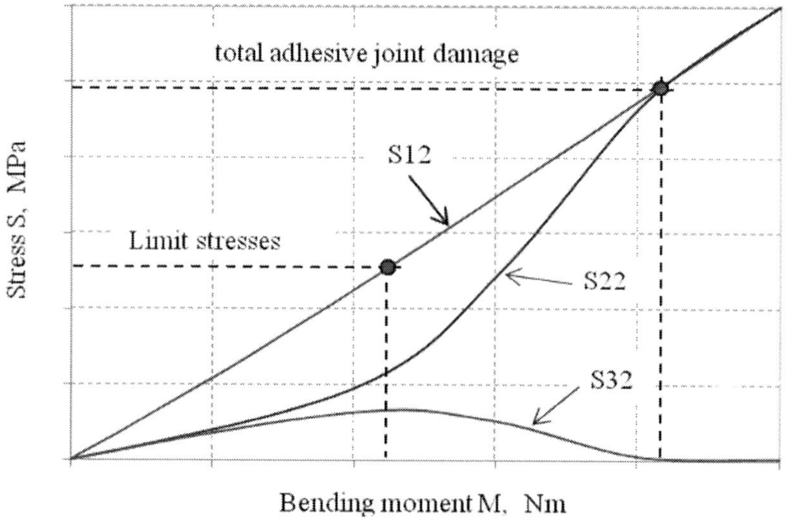

Fig. 3. The idea of assessing the strength of a glued joint on the basis of sample tests in four-point bending conditions.

Relationship between stresses in points S1 and S2 should be considered in three intervals (Fig. 3):

- section I – proportionally changing values of stresses S1 and S2 against the monotonically increasing value of the bending moment; this dependence indicates the lack of joint damage and work in the area of elastic stresses,
- section II - progressive increase of stresses S2 against stress values S1; This indicates the loss of load bearing capacity through the glued joint together with the increase of the bending moment, in the final stage the total failure of the joint,
- section III – equalization of stress values S1 and S2; applies to the situation of total damage to the glued joint.

The boundary between section I and section II should be considered as the value of allowable stresses, which do not damage the glued joint. Stress value S1 = S2 indicates that the joint is completely damaged. Stresses at S3 refer to load changes in the lower beam M2, which change in opposition to the value S2.

3 Numerical calculations

The purpose of calculations is to get the answer to the question: which geometrical features of the sample significantly affect the change in stress values at the measuring points S1, S2 and S3 under the influence of the acting bending moment?

3.1 Conditions for numerical calculations

Numerical calculations of stress values in elements of the test sample were carried out using the program Ansys. Dimensions of the model 3D adopted according to the signs presented on Figure 1. The following parameters were adopted for the purposes of calculations:

- constantly:
  - geometric:
    - spacing of supports X = 340 mm,
    - distance between the points of load application Y = 260 mm,
    - the width of the sample H1 = H2 = L3 = 40 mm,
    - length of the top beam L1 = 400 mm,
    - length of the bottom beam L2 = 90 mm,
    - distance of the measurement site S1 relative to the symmetry axis of the sample F = 80 mm,
  - material:
    - top bar M1 – structural steel $E = 2.1 \cdot 10^5$ MPa,
    - bottom bar M2 – structural steel $E = 2.1 \cdot 10^5$ MPa,
    - glued joint AJ – glue PLEXUS MA300 (E = 1610 MPa, S_u = 23.7 MPa) [11],
- variables:
  - geometric:
    - width of the glued joint h = 3 mm, 6 mm, 9 mm,
    - distance of glued joints L3 = 20 mm, 35 mm, 50 mm,
    - thickness of the top beam M1 g1 = 2 mm, 4 mm, 6 mm,
    - thickness of the bottom beam M2 g2 = 2 mm, 4 mm, 6 mm,
    - height (thickness) of the glued joint AJ g3 = 2 mm, 4 mm, 6 mm.

3.2 Calculation results

The results of numerical calculations allowed to determine the values of stresses in points S1, S2 and S3, which are presented in the tables 1-3.

Table 1. Stress value at point S1, S2 and S3 for the width of the glued joint h = 3 mm.

Lp.	Parametry zmienne					Wynik			Lp.	Parametry zmienne					Wynik		
	h	L3	g3	g1	g2	S1	S2	S3		h	L3	g3	g1	g2	S1	S2	S3
	mm	mm	mm	mm	mm	MPa	MPa	MPa		mm	mm	mm	mm	mm	MPa	MPa	MPa
1	2	3	4	5	6	7	8	9	1	2	3	4	5	6	7	8	9
1	3	20	2	2	2	61.4	44.4	2.4	42	3	35	4	4	6	22.8	21.2	0.5
2	3	20	2	2	4	61.6	42.5	2.2	43	3	35	4	6	2	17.5	17.4	0.5
3	3	20	2	2	6	61.6	42.1	1.4	44	3	35	4	6	4	17.5	17.3	0.4
4	3	20	2	4	2	22.7	21.3	0.8	45	3	35	4	6	6	17.5	17.3	0.2
5	3	20	2	4	4	22.8	21.0	1.0	46	3	35	6	2	2	62.0	46.7	3.9
6	3	20	2	4	6	22.8	20.7	0.7	47	3	35	6	2	4	62.1	45.5	2.3
7	3	20	2	6	2	15.5	14.8	0.4	48	3	35	6	2	6	62.2	45.2	1.5
8	3	20	2	6	4	15.5	14.7	0.5	49	3	35	6	4	2	22.7	21.7	1.0
9	3	20	2	6	6	15.5	14.7	0.5	50	3	35	6	4	4	22.8	21.6	0.6
10	3	20	4	2	2	61.1	48.3	3.5	51	3	35	6	4	6	22.8	21.5	0.4
11	3	20	4	2	4	61.2	47.1	2.2	52	3	35	6	6	2	17.5	17.4	0.5
12	3	20	4	2	6	61.2	46.8	13.0	53	3	35	6	6	4	17.5	17.4	0.3
13	3	20	4	4	2	22.7	21.8	1.0	54	3	35	6	6	6	17.5	17.4	0.2
14	3	20	4	4	4	22.7	21.6	0.7	55	3	50	2	2	2	65.2	36.6	4.6
15	3	20	4	4	6	22.7	21.5	0.4	56	3	50	2	2	4	65.4	34.8	1.7

16	3	20	4	6	2	15.5	14.9	0.5	57	3	50	2	2	6	65.4	34.4	0.7
17	3	20	4	6	4	15.5	14.8	0.4	58	3	50	2	4	2	23.1	20.3	0.8
18	3	20	4	6	6	15.5	14.8	0.3	59	3	50	2	4	4	23.1	20.1	0.8
19	3	20	6	2	2	60.9	50.0	3.7	60	3	50	2	4	6	23.1	20.0	0.7
20	3	20	6	2	4	61.0	49.0	1.9	61	3	50	2	6	2	48.0	47.0	1.0
21	3	20	6	2	6	61.0	48.8	1.1	62	3	50	2	6	4	48.0	47.0	1.0
22	3	20	6	4	2	22.7	21.9	1.0	63	3	50	2	6	6	48.0	47.0	1.0
23	3	20	6	4	4	22.7	21.1	0.5	64	3	50	4	2	2	64.0	41.8	3.4
24	3	20	6	4	6	22.7	21.8	0.3	65	3	50	4	2	4	64.1	40.9	2.7
25	3	20	6	6	2	15.5	14.9	0.4	66	3	50	4	2	6	64.1	40.7	2.1
26	3	20	6	6	4	15.5	14.9	0.3	67	3	50	4	4	2	22.9	21.2	1.2
27	3	20	6	6	6	15.5	14.9	0.2	68	3	50	4	4	4	22.9	21.0	0.9
28	3	35	2	2	2	63.1	39.4	1.8	69	3	50	4	4	6	22.9	21.0	0.7
29	3	35	2	2	4	63.3	37.4	1.6	70	3	50	4	6	2	48.0	47.0	1.0
30	3	35	2	2	6	63.3	36.9	1.3	71	3	50	4	6	4	48.0	47.0	1.0
31	3	35	2	4	2	22.9	20.8	0.8	72	3	50	4	6	6	48.0	47.0	1.0
32	3	35	2	4	4	22.9	20.4	0.8	73	3	50	6	2	2	63.4	44.6	2.5
33	3	35	2	4	6	22.9	20.2	0.6	74	3	50	6	2	4	63.6	43.6	2.9
34	3	35	2	6	2	17.5	17.2	0.4	75	3	50	6	2	6	63.6	43.2	2.0
35	3	35	2	6	4	17.5	17.1	0.4	76	3	50	6	4	2	22.8	21.5	1.2
36	3	35	2	6	6	17.5	17.1	0.3	77	3	50	6	4	4	22.9	21.4	0.8
37	3	35	4	2	2	62.4	44.2	3.3	78	3	50	6	4	6	22.9	21.3	0.6
38	3	35	4	2	4	62.5	43.1	2.3	79	3	50	6	6	2	48.0	48.0	1.0
39	3	35	4	2	6	62.5	42.8	1.6	80	3	50	6	6	4	48.0	48.0	1.0
40	3	35	4	4	2	22.8	21.4	1.0	81	3	50	6	6	6	48.0	48.0	1.0
41	3	35	4	4	4	22.8	21.2	0.7									

Table 2. Stress value at point S1, S2 and S3 for the width of the glued joint h = 6 mm.

Lp.	Parametry zmienne					Wynik			Lp.	Parametry zmienne					Wynik		
	h	L3	g3	g1	g2	S1	S2	S3		h	L3	g3	g1	g2	S1	S2	S3
	mm	mm	mm	mm	mm	MPa	MPa	MPa		mm	mm	mm	mm	mm	MPa	MPa	MPa
1	2	3	4	5	6	7	8	9	1	2	3	4	5	6	7	8	9
1	6	20	2	2	2	63.4	27.8	2.3	42	6	35	4	4	6	23.3	18.0	0.4
2	6	20	2	2	4	64.1	20.5	3.1	43	6	35	4	6	2	17.6	16.6	0.2
3	6	20	2	2	6	64.3	18.0	1.8	44	6	35	4	6	4	17.6	16.5	0.2
4	6	20	2	4	2	23.0	19.0	0.4	45	6	35	4	6	6	17.6	16.5	0.2
5	6	20	2	4	4	23.1	17.6	1.1	46	6	35	6	2	2	65.6	26.3	1.3
6	6	20	2	4	6	23.2	16.9	0.9	47	6	35	6	2	4	65.7	25.4	1.1
7	6	20	2	6	2	13.2	12.8	0.1	48	6	35	6	2	6	65.7	25.2	1.1
8	6	20	2	6	4	13.2	12.5	0.5	49	6	35	6	4	2	23.2	19.1	0.7
9	6	20	2	6	6	13.3	12.6	0.5	50	6	35	6	4	4	23.2	18.9	0.6
10	6	20	4	2	2	63.2	29.6	1.3	51	6	35	6	4	6	23.2	18.9	0.5
11	6	20	4	2	4	63.4	27.1	1.4	52	6	35	6	6	2	17.5	16.8	0.4
12	6	20	4	2	6	63.5	26.5	0.8	53	6	35	6	6	4	17.5	16.7	0.3
13	6	20	4	4	2	22.9	19.6	0.4	54	6	35	6	6	6	17.5	16.7	0.3
14	6	20	4	4	4	23.0	19.2	0.7	55	6	50	2	2	2	69.0	22.7	0.3
15	6	20	4	4	6	23.0	19.0	0.5	56	6	50	2	2	4	70.3	16.2	2.2
16	6	20	4	6	2	13.2	12.9	0.2	57	6	50	2	2	6	70.8	13.6	1.6
17	6	20	4	6	4	13.2	12.9	0.3	58	6	50	2	4	2	23.8	17.3	0.5
18	6	20	4	6	6	13.2	12.8	0.3	59	6	50	2	4	4	24.1	15.6	0.7
19	6	20	6	2	2	63.0	31.3	1.5	60	6	50	2	4	6	24.3	14.7	0.6
20	6	20	6	2	4	63.1	30.4	1.4	61	6	50	2	6	2	14.2	12.8	0.4
21	6	20	6	2	6	63.1	30.2	1.0	62	6	50	2	6	4	14.3	12.3	0.3
22	6	20	6	4	2	22.9	20.0	0.6	63	6	50	2	6	6	14.3	12.1	0.3
23	6	20	6	4	4	22.9	19.8	0.6	64	6	50	4	2	2	68.9	22.6	0.7
24	6	20	6	4	6	22.9	19.8	0.5	65	6	50	4	2	4	69.3	20.3	0.6
25	6	20	6	6	2	13.2	13.0	0.3	66	6	50	4	2	6	69.4	19.6	0.3
26	6	20	6	6	4	13.2	13.0	0.3	67	6	50	4	4	2	23.7	17.9	0.6
27	6	20	6	6	6	13.2	13.0	0.2	68	6	50	4	4	4	23.7	17.4	0.3
28	6	35	2	2	2	66.1	24.4	1.1	69	6	50	4	4	6	23.8	17.3	0.4
29	6	35	2	2	4	67.0	17.5	2.6	70	6	50	4	6	2	14.1	13.0	0.4
30	6	35	2	2	6	67.4	14.9	1.7	71	6	50	4	6	4	14.1	12.9	0.3

31	6	35	2	4	2	23.4	18.0	0.1	72	6	50	4	6	6	14.1	12.9	0.3
32	6	35	2	4	4	23.6	16.4	0.9	73	6	50	6	2	2	68.6	23.6	1.4
33	6	35	2	4	6	23.7	15.5	0.7	74	6	50	6	2	4	68.7	22.8	1.1
34	6	35	2	6	2	17.6	16.3	0.1	75	6	50	6	2	6	68.7	22.7	1.2
35	6	35	2	6	4	17.6	15.9	0.5	76	6	50	6	4	2	23.5	18.4	0.8
36	6	35	2	6	6	17.7	15.6	0.4	77	6	50	6	4	4	23.6	18.3	0.7
37	6	35	4	2	2	65.9	25.0	0.7	78	6	50	6	4	6	23.6	18.2	0.7
38	6	35	4	2	4	66.2	22.5	1.0	79	6	50	6	6	2	14.1	13.2	0.5
39	6	35	4	2	6	66.2	21.8	0.5	80	6	50	6	6	4	14.1	13.2	0.4
40	6	35	4	4	2	23.3	18.6	0.4	81	6	50	6	6	6	14.1	13.2	0.4
41	6	35	4	4	4	23.3	18.2	0.4									

Table 3. Stress value at point S1, S2 and S3 for the width of the glued joint h = 9 mm.

Lp.	Parametry zmienne					Wynik			Lp.	Parametry zmienne					Wynik		
	h	L3	g3	g1	g2	S1	S2	S3		h	L3	g3	g1	g2	S1	S2	S3
	mm	mm	mm	mm	mm	MPa	MPa	MPa		mm	mm	mm	mm	mm	MPa	MPa	MPa
1	2	3	4	5	6	7	8	9	1	2	3	4	5	6	7	8	9
1	9	20	2	2	2	64.8	21.4	1.2	42	9	35	4	4	6	24.0	14.0	0.7
2	9	20	2	2	4	65.7	12.6	3.5	43	9	35	4	6	2	17.7	15.8	0.2
3	9	20	2	2	6	66.1	8.3	2.8	44	9	35	4	6	4	17.7	15.3	0.5
4	9	20	2	4	2	23.3	17.2	0.2	45	9	35	4	6	6	17.8	15.1	0.4
5	9	20	2	4	4	23.5	14.5	1.7	46	9	35	6	2	2	68.0	16.2	0.4
6	9	20	2	4	6	23.7	12.6	1.7	47	9	35	6	2	4	68.3	14.0	1.1
7	9	20	2	6	2	17.5	16.1	0.1	48	9	35	6	2	6	68.4	13.1	0.7
8	9	20	2	6	4	17.6	15.3	1.0	49	9	35	6	4	2	23.7	16.2	0.4
9	9	20	2	6	6	17.7	14.6	1.1	50	9	35	6	4	4	23.8	15.6	0.3
10	9	20	4	2	2	64.9	19.6	1.2	51	9	35	6	4	6	23.8	15.3	0.2
11	9	20	4	2	4	65.3	14.8	2.5	52	9	35	6	6	2	17.7	15.8	0.3
12	9	20	4	2	6	65.5	12.7	1.7	53	9	35	6	6	4	17.7	15.7	0.2
13	9	20	4	4	2	23.3	17.3	0.2	54	9	35	6	6	6	17.7	15.6	0.1
14	9	20	4	4	4	23.3	16.0	1.1	55	9	50	2	2	2	70.8	18.5	0.9
15	9	20	4	4	6	23.4	15.7	0.8	56	9	50	2	2	4	72.3	11.6	2.2
16	9	20	4	6	2	17.5	16.2	0.1	57	9	50	2	2	6	73.1	8.2	2.2
17	9	20	4	6	4	17.5	15.9	0.6	58	9	50	2	4	2	24.3	15.7	0.9
18	9	20	4	6	6	17.5	15.7	0.5	59	9	50	2	4	4	24.9	12.9	1.0
19	9	20	6	2	2	64.9	19.1	0.6	60	9	50	2	4	6	25.3	10.9	1.3
20	9	20	6	2	4	65.1	16.7	1.4	61	9	50	2	6	2	17.8	15.3	0.7
21	9	20	6	2	6	62.2	15.8	0.8	62	9	50	2	6	4	18.1	14.1	0.6
22	9	20	6	4	2	23.2	17.5	0.2	63	9	50	2	6	6	18.3	13.1	0.9
23	9	20	6	4	4	23.2	16.9	0.5	64	9	50	4	2	2	71.3	15.7	0.6
24	9	20	6	4	6	23.3	16.7	0.4	65	9	50	4	2	4	72.1	12.0	1.6
25	9	20	6	6	2	17.5	16.3	0.1	66	9	50	4	2	6	72.5	10.2	1.4
26	9	20	6	6	4	17.5	16.2	0.3	67	9	50	4	4	2	24.3	15.3	0.7
27	9	20	6	6	6	17.5	16.1	0.2	68	9	50	4	4	4	24.6	13.9	0.6
28	9	35	2	2	2	67.6	19.5	0.1	69	9	50	4	4	6	24.8	13.1	0.6
29	9	35	2	2	4	68.9	11.9	2.7	70	9	50	4	6	2	17.8	15.2	0.5
30	9	35	2	2	6	69.4	8.2	2.4	71	9	50	4	6	4	17.9	14.7	0.3
31	9	35	2	4	2	23.7	16.2	0.5	72	9	50	4	6	6	18.0	14.4	0.4
32	9	35	2	4	4	24.2	13.5	1.3	73	9	50	6	2	2	71.5	14.7	0.8
33	9	35	2	4	6	24.4	11.5	1.5	74	9	50	6	2	4	71.9	12.8	0.7
34	9	35	2	6	2	17.7	15.7	0.4	75	9	50	6	2	6	72.0	11.9	0.6
35	9	35	2	6	4	17.9	14.7	0.8	76	9	50	6	4	2	24.3	15.3	0.7
36	9	35	2	6	6	18.0	13.8	1.0	77	9	50	6	4	4	24.4	14.6	0.2
37	9	35	4	2	2	67.9	17.0	0.2	78	9	50	6	4	6	24.5	14.3	0.1
38	9	35	4	2	4	68.5	12.9	2.0	79	9	50	6	6	2	17.8	15.3	0.5
39	9	35	4	2	6	68.8	10.9	1.6	80	9	50	6	6	4	17.8	15.1	0.1
40	9	35	4	4	2	23.7	16.1	0.3	81	9	50	6	6	6	17.9	15.0	0.1
41	9	35	4	4	4	23.9	14.7	0.8									

3.3 Selection of most important attributes

A set of attributes undergoes (in the form of parameter columns) initial data processing to enable their comparison and further analysis. Due to the type of processed data, standardization was used, in result of which, the variable mean expected value assumes zero and the mean variance is equal to one. The most commonly used standardization method can be expressed as:

$$z = \frac{x - \mu}{\sigma} \tag{1}$$

where: x – value of parameter, μ - mean value of parameter, σ – standard deviation of parameter.

There can be many parameters from one or more measurement but most of them are significantly interfered or independent of the monitored phenomenon. The proposed method involves choosing parameters with the use of rough set theory. This technique provides methods for determination of the most important attributes of a computer system without losing classification ability [12]. The rough set theory was developed by Zdzisław Pawlak [13, 14, 15] and is based on the concept of upper and lower approximation of the set, approximating space and models of sets. Measurement data is presented in the form of a table, whose columns correspond to the attributes and rows corresponds to particular objects or states. Such a record is called an information system [16]. Based on rough set theory is a reduct, which is the main part of the information system. As a result of this, it is possible to distinguish all the distinguishable objects from the original set of attributes, hence, reducts are minimal subsets of attributes maintaining characteristics of the whole set of attributes. Literature studies provides many algorithms for determination of sets of information system reducts. For the needs of our solution, the authors have taken into consideration the methods of: Johnson, global, local [17]. The algorithm of determining the set of reducts is as follows:

Input: Information system $\boldsymbol{A} = (U, A)$

Output: Set $RED_A(A)$ of all reduction A

1. Determine the discrimination matrix $M(A) = (C_{ij})$
2. Reduce M using the extraction laws: d number of non-empty fields of the M
3. Build the family of sets $R_0, R_1, ..., R_d$ as follows:
4. **begin**
5. $R_0 = 0$
6. for $i = 1$ to d
7. $R_i = S_i \cup T_i$, *where* $S_i = \{R \in R_{i-1}: R \cap C_i \neq 0\}$ and $T_i = (R \cup \{a\})$
8. **end**
9. Remove any unnecessary elements for each element of the R_d family
10. For each R_d family element, remove the repeating elements
11. **return** $RED_A(A) = R_d$

Significance of the attributes is described on the basis of the determined set of reducts, where the attributes are ordered according to the number of occurrence in the set. An exemplary set of reducts is presented in Table 4. Parameters ordered according to significance are presented in Table 5.

Table 4. The set of reducts for all attributes (S1,S2 and S3).

	Number	REDUCTS (no. of attributes)
1	5	{ h, L3, g3, g1, g2 }
2	5	{ L3, g3, g1, g2, S2 }
3	4	{ h, g3, g2, S1 }
4	4	{ g3, g2, S1, S2 }

Table 5. Parameters ordered according to significance.

Attribute	g3	g2	h	L3	g1
Number	4	4	2	2	2

The set of reducts for S1 is presented in Table 6. Parameters ordered according to significance are presented in Table 7.

Table 6. The set of reducts for S1.

	Number	REDUCTS (no. of attributes)
1	1	{ h }
2	4	{ h, L3, g3, g1 }
3	4	{ h, L3, g1, g2 }
4	5	{ h, L3, g3, g1, g2 }
5	4	{ h, L3, g1 }

Table 7. Parameters ordered according to significance.

Attribute	h	L3	g1	g3	g2
Number	5	4	4	2	2

The set of reducts for S2 is presented in Table 8. Parameters ordered according to significance are presented in Table 9.

Table 8. The set of reducts for S2.

	Number	REDUCTS (no. of attributes)
1	5	{ h, L3, g3, g1, g2 }
2	4	{ h, L3, g3, g1 }

Table 9. Parameters ordered according to significance.

Attribute	h	L3	g3	g1	g2
Number	2	2	2	2	1

The set of reducts for S3 is presented in Table 10. Parameters ordered according to significance are presented in Table 11.

Table 10. The set of reducts for S3.

	Number	REDUCTS (no. of attributes)
1	5	{ h, L3, g3, g1, g2 }
2	4	{ h, L3, g3, g1 }
3	4	{ h, g3, g1, g2 }
4	3	{ h, L3, g1 }
5	4	{ h, L3, g1, g2 }
6	4	{ L3, g3, g1, g2 }

Table 11. Parameters ordered according to significance.

Attribute	g1	h	L3	g3	g2
Number	6	5	5	4	4

Conclusions

Diagnostic possibilities of glued joints are related to the analysis of stress values at points S1, S2 and S3. The analysis of the results of numerical calculations shows that all indicated geometric parameters, i.e. h, L3, g_1, g_2 and g_3 they affect the value of stresses in points S1, S2 and S3. Considering the set of results S1, S2 and S3 as a whole, the most important parameters are: g_2 and g_3. Analyzing the results of stresses at individual measurement points, it can be indicated that the most significant impact on the value:

- S1 have dimensions:h, L3, g_1,
- S2 have dimensions:h, L3, g_3, g_1, g_2,

- S3 have dimensions: g_1, h, L3, g_3, g_2.

Further work will be related to the determination of the relationship between individual geometric parameters taking into account the variable material properties of individual sample elements.

References

1. European standard PN-EN 1465:2009, 2009
2. International standard ISO 5173, 2009
3. International standard ISO 6892-1, 2016
4. G. Szala, B. Ligaj, International Journal of Fatigue 91 (2016), pp. 39–49
5. G. Jain, A. Narad, L. C. Boruah, B. Rajkumar, . J. Conserv. Dent. 2015 Jul-Aug; 18(4), pp. 337–341, doi: 10.4103/0972-0707.159754
6. M. Wirwicki, T. Topolinski, Book Series: Advanced Materials Research, vol. 845, 2014, pp. 153-157, DOI: 10.4028/www.scientific.net/AMR.845.153
7. A. Jasiń ska, B. Ligaj, A. Mazurkiewicz, IOP Conference Series: Materials Science and Engineering, 2018
8. B. Ligaj, M. Wirwicki, A. Jasiń ska, K. Karolewska, IOP Conference Series: Materials Science and Engineering 344 (2018) 012013, doi:10.1088/1757-899X/344/1/012013
9. B. Ligaj, A. Jasiń ska and K. Karolewska, Patent application no. P.422884, 2017
10. Website:https://www.tme.eu/html/PL/tensometry-foliowe/ramka_2261_PL_pelny.html
11. P. Mać kowiak, B. Ligaj, Engineering Mechanics 2018, pp. ...., published: 2018
12. E. Szydłowska, XIII Konferencja uż ytkowników i deweloperów ORACLE, 262-267. Zakopane (2007)
13. J. Bazan, M. Szczuka, A. Wojna, M. Wojnarski, Lecture Notesin Artificial Intelligence 3366, 592–601,Springer-Verlag (2004)
14. J. Bazan, M. Szczuka, J. Wróblewski, Lecture Notes in Artificial Intelligence 2475, 397–404, Springer-Verlag (2002)
15. Z. Pawlak, Information Systems - theoretical foundations. WNT, Warszawa (1983)
16. B. Walczak, D.L. Massart, Gemoemetrics and Intelligent Laboratory Systems 47, 1-16, (1999)
17. M. Salamo, E. Golobardes, Recent Advances in Artificial Intelligence Research and Development. 217-234, (2004)

The choice of parameters hydrodynamic retarder brake with increased power capacity

Alexander Pavarekha[1], *Denis* Kapski[2], *Andrey* Voytik[1] and Marcin Łukasiewicz[3]

[1]Belarusian National Technical University, Department "Tractors", 220013 Minsk, Nezavisimosti, 65, Belarus
[2]Belarusian National Technical University, Faculty of automotive engineering, 220013 Minsk, Nezavisimosti, 65, Belarus
[3]Faculty of Mechanical Engineering UTP in Bydgoszcz, Department of Vehicle Engineering, Al. Prof. S.Kaliskiego 7, 85-796 Bydgoszcz, Poland

Abstract. The current trend of increasing the total weight and speed of vehicles, tougher requirements for road safety requires constant improvement of their braking systems. In this regard, for the main tractors and a number of other machines with a large operating weight is relevant to the use of brakes–retarders. This allows to increase the average speed, reduce the thermal stress of the main (working) brake mechanisms and increase their durability, which increases the level of active safety of vehicles. This article presents the results of studies of working processes in hydrodynamic brakes of retarders, determined the dependence of their braking efficiency on the design features. The studies were conducted using finite element models.

1 Introduction

Reducing the thermal stress of the working brake mechanisms in order to improve traffic safety on long or steep descents and at high speeds can be achieved with the help of a brake-retarder. The retarder serves to slow down and maintain the necessary speed of the car by absorbing part of its kinetic energy.

When braking with a decelerator, the wheels are less likely to lock, which reduces the possibility of lateral sliding, thereby increasing the lateral stability of the car on slippery roads. In addition to improving the safety of the brake, the retarder helps to reduce the wear of friction pairs of wheel, brakes, and the use of transmission brakes leads to an increase in the durability of engine parts.

Electromagnetic and hydrodynamic brakes are the most effective at present. The main disadvantages of electrodynamic brakes are large weight and size, high cost, loss of efficiency when heated.

Hydrodynamic brakes are devoid of the disadvantages of electromagnetic brakes and have a number of advantages, which has led to their widespread use in vehicles. However, there is low efficiency in reducing the angular velocity of the rotor.

Hydrodynamic brakes are devoid of the disadvantages of electromagnetic brakes and have a number of advantages, which has led to their widespread use in vehicles. However, there is low efficiency in reducing the angular velocity of the rotor.

Since the value of the brake torque of hydrodynamic retarder depends mainly on its geometric parameters (the active diameter of the cavity shape, the diameter of flow area, the type of blade system) and the frequency of rotation of the rotor, then the solution of the problem should be sought in the selection of these parameters.

2 Main part

Structurally, the hydrodynamic retarder brake is a hydraulic clutch, one of the wheels of which is connected to the rotating element of the transmission, and the second is connected to the fixed body. When the rotor rotates on the shaft, a braking torque is created due to the hydrodynamic resistance to the rotation of the rotor blades.

The installation location of the rotating fluid coupling element depends on the type of transmission. For mechanical transmissions, the installation of the rotor fluid coupling is carried out: on the shaft of the internal combustion engine; on the shaft of the gearbox; on the main transmission shaft; on the intermediate shaft of the cardan transmission. For hydro-mechanical transmission, the rotor is most often installed: on the turbine shaft of the torque Converter; on the primary shaft of the hydro-mechanical transmission (GMT); on the output shaft of the GMT; on the intermediate shaft of the GMT.

Depending on the shape of the working space hydraulic brakes-retarders can be open or closed rotor blades, with a symmetrical or asymmetric working cavity, and can also be made in the form of single or double fluid coupling.

In turn, the wheels of the brake-retarder can have straight, curved, as well as straight or curved inclined blades.

Adjustment of the braking torque can be carried out by changing the level of the working fluid in the cavity, by turning the blades of the fixed wheel, as well as by means of special damper-type dampers, partially covering the working cavity of the hydraulic retarder.

As the prototype was considered a wheeled chassis 8к8 with CMT [1]. In order to reduce the load of frictions and gears of the gearbox, to exclude the operation of the engine in brake mode, the retarder is installed on the output shaft of the GMT. However, in this case, the retarder should be more efficient than the one installed inside the gearbox. In this regard, to obtain the desired efficiency, the retarder should have a greater energy intensity. The purpose of the work is to select the parameters that provide a high level of power consumption of the hydraulic brake-retarder.

According to the results of the analysis of literature sources [2-3], it was found that the mechanism in the form of a double fluid coupling has the greatest efficiency. Therefore, as an initial version of the hydrodynamic brake-retarder mechanism was adopted in the form of a dual fluid coupling.

Figure 1 shows a three-dimensional model of the hydrodynamic retarder brake, made in CAD program Solid Edge ST9.

Increasing the energy consumption of the brake-retarder is possible by changing its design parameters and increasing the speed. In this paper, we investigate the dependence of the energy intensity of the hydrodynamic brake-retarder on the type of its blade system, with constant other design parameters. The study was conducted using Computational Fluid Dynamics (CFD) of software product FlowVision [4].

CFD is a computational method to simulate the dynamics and flows of liquids and gases. In addition, this method can be used to calculate heat and mass transfer, phase changes, chemical reactions, mechanical motion, stress and deformation of solid materials.

The arguments in favor of the calculation using the QFD methodology are:

- advanced analysis tools- for a large number of projects it is difficult and expensive to create an appropriate experimental model; in addition, CFD analysis allows to obtain data, which measurement in a physical experiment is impossible;
- the possibility of multivariate analysis - allows you to test a large number of project variants in a short time;
- increase productivity, design quality and reduce costs - CFD is a tool that ultimately contributes to the rapid appearance of the product on the market.

Three types of open blade systems were selected for comparative analysis:

- straight radial blades (basic version);
- circular blades;
- inclined blades (45^0).

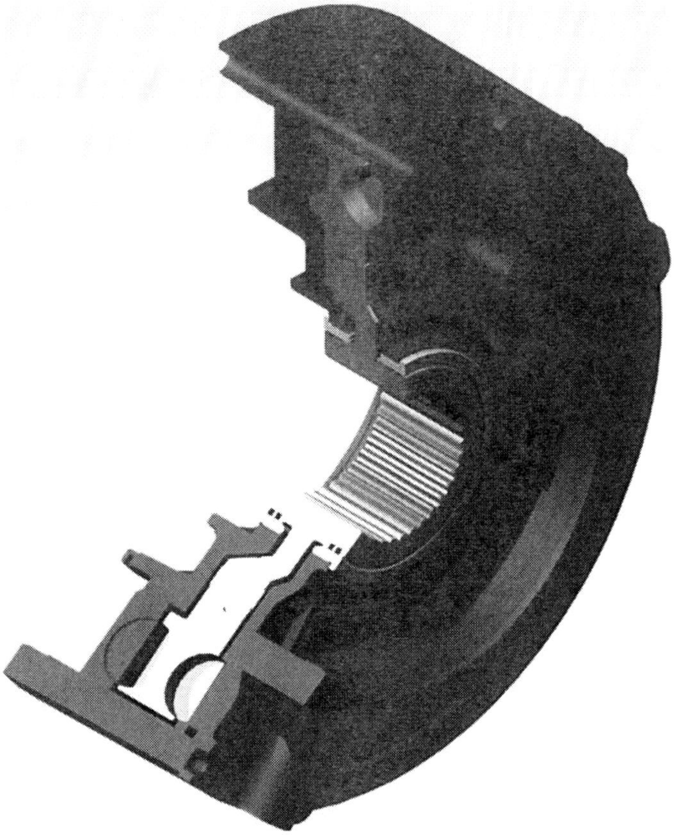

Fig. 1. 3-D model of hydrodynamic brake-retarder.

Figure 2 shows the basic geometric parameters of the three brake-retarder variants under study.

The figure adopted the designation: δ - the thickness of the blade; D - the active diameter of the brake-retarder; d - the inner diameter of the flow of the brake-retarder; R_0 - radius of curvature of the flow part of the retarder; R_l - the radius of curvature of the blades.

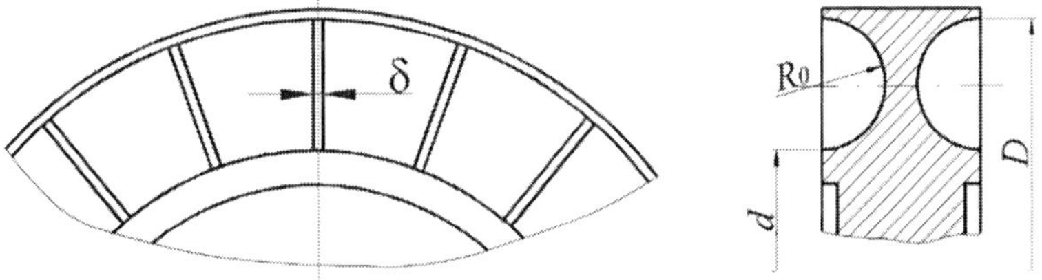

a)

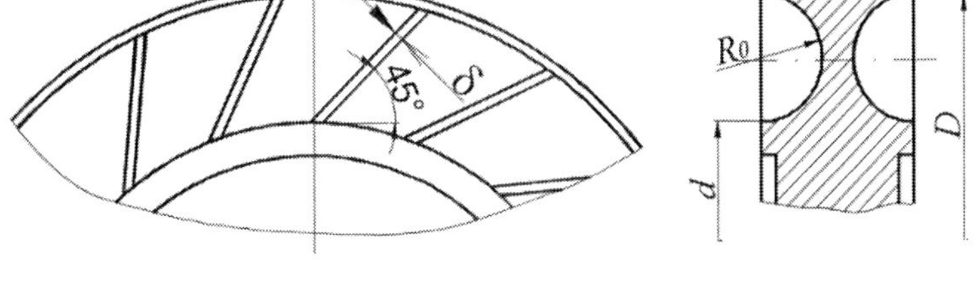

b)

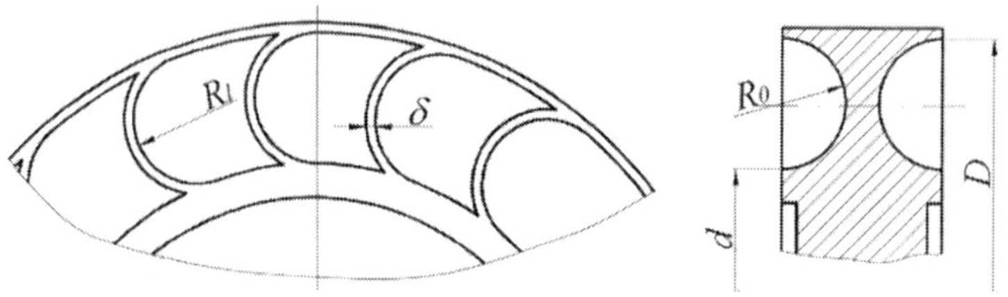

c)

Fig. 2. Geometric parameters of the brakes: a) basic version; b) with circular blade system; c) with inclined blades.

Simulation of oil flow in the flow part of the brake-retarder is performed within the model of turbulent fluid flow [5]. The equations describing the change in velocity, pressure, turbulent energy, and dissipation have the form:

- Navier-Stokes equation

$$\frac{\partial \boldsymbol{V}}{\partial t} + \nabla(\boldsymbol{V} \otimes \boldsymbol{V}) = -\frac{\nabla p}{\rho} + \frac{1}{\rho} \cdot \nabla(\mu + \mu_t) \cdot (\nabla \boldsymbol{V} + (\nabla \boldsymbol{V})^T) + g,$$

- continuity equation

$$\nabla(\boldsymbol{V}) = 0,$$

- equations for turbulent energy and dissipation rate

$$\mu_t = c_\mu \cdot \rho \cdot \frac{k^2}{\varepsilon},$$

$$\frac{\partial(\rho \cdot k)}{\partial t} + \nabla(\rho \cdot \boldsymbol{V} \cdot k) = \nabla\left(\left(\mu + \frac{\mu_t}{\sigma_k}\right) \cdot \nabla k\right) + \mu_t \cdot G - \rho \cdot \varepsilon,$$

$$\frac{\partial(\rho \cdot \varepsilon)}{\partial t} + \nabla(\rho \cdot \boldsymbol{V} \cdot \varepsilon) = \nabla\left(\left(\mu + \frac{\mu_t}{\sigma_\varepsilon}\right) \cdot \nabla \varepsilon\right) + c_1 \cdot \frac{\varepsilon}{k} \cdot \mu_t \cdot G - c_2 \cdot \rho \cdot \frac{\varepsilon^2}{k},$$

where $\boldsymbol{V}$ – the velocity vector; t – time; p – pressure; ρ – the fluid density; μ – the molecular dynamic viscosity; μ_t – the turbulent dynamic viscosity; g – the vector of gravitational acceleration; k – turbulent energy; ε – the rate of dissipation of turbulent energy.

The equation of motion of the wheel has the form:

$$J \cdot \dot{\omega} = \oint [r \cdot (p\mathbf{n} + S\mathbf{n})]\, ds + M,$$

where S – the stress tensor on the surface of a solid body; $\mathbf{n}$ – surface normal; J – moment of inertia tensor; ω – angular velocity of rotation of the body; M - external torque.

As the physical parameters of the working fluid in the simulation were taken: density ρ = 840 кг/м^3; molecular viscosity μ = 0.0071 кг/(м·с); the flow of the working fluid – 0.1 kg/s; working fluid temperature in the working area – no more than 120 °C; pressure at the inlet of the retarder – 5.5 MPa.

The algorithm of finite element analysis was used to solve the given equations [6].
The simulation used a uniform finite element design grid in the plane passing through the wheel rotation axis (figure 3a) and a uniform one in the perpendicular plane (figure 3b).

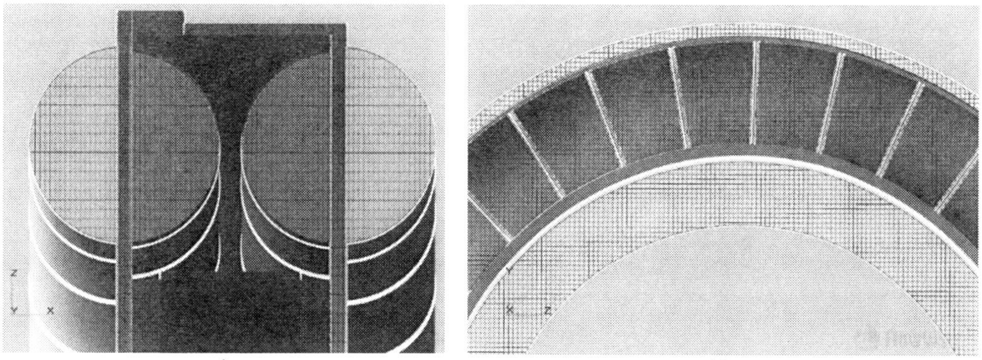

a) b)

Fig. 3. View of the calculated grid in the longitudinal and transverse planes.

Table 1 presents the results of the calculation of the braking performance of the retarder with straight radial blades.

Table 1. The results of the calculation of the basic option.

The rotor's rotation frequency, rpm	0	300	600	900	1200	1500	1800
The braking torque on the rotor, N·m	0,0	70,0	279,9	629,9	1119,8	1749,7	2519,5

Figure 4 shows the visualization of velocity fields in a plane passing through the axis of rotation of the wheels when n_{rot}=1500 rpm. As can be seen from the figure for the working fluid is turbulent in nature. The meridional component of the liquid flow rate is 53 m/s.

Figure 5 shows a visualization of the pressure distribution in the plane passing through the axis of rotation of the wheels at n_{rot} =1500 rpm. As can be seen from the figure, the flow of the working fluid has a turbulent character. The area of maximum pressure is located on the rotor blades. The region of minimum pressure is located in the region of formation of vortex flows.

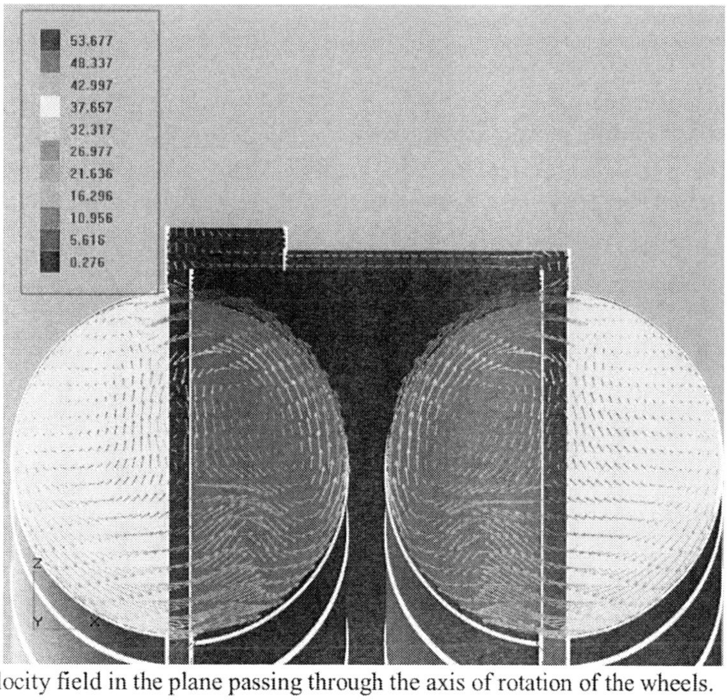

Fig 4. The velocity field in the plane passing through the axis of rotation of the wheels.

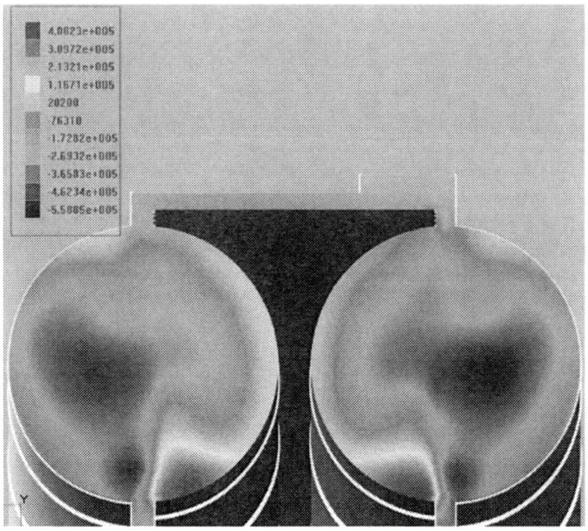

Fig. 5. Distribution of pressure in the plane passing through the axis of rotation of the wheels.

Table 2 presents the results of the calculation of the brake characteristics of the decelerator brake with circular blades.

Table 2. Results of calculation of the variant with circular blades.

The rotor's rotation frequency, rpm	0	300	600	900	1200	1500	1800
The braking torque on the rotor, N·m	0,0	235,4	941,5	2118,5	3766,1	5884,6	8473,8

Table 3 presents the results of the calculation of the brake characteristics of the retarder with inclined blades.

Table 3. Results of calculation of the variant with circular blades.

The rotor's rotation frequency, rpm	0	300	600	900	1200	1500	1800
The braking torque on the rotor, N·m	0,0	65,2	245,3	542,6	964,6	1507,2	2170,4

The dependences constructed according to tables 1 - 3 are shown in figure 6.

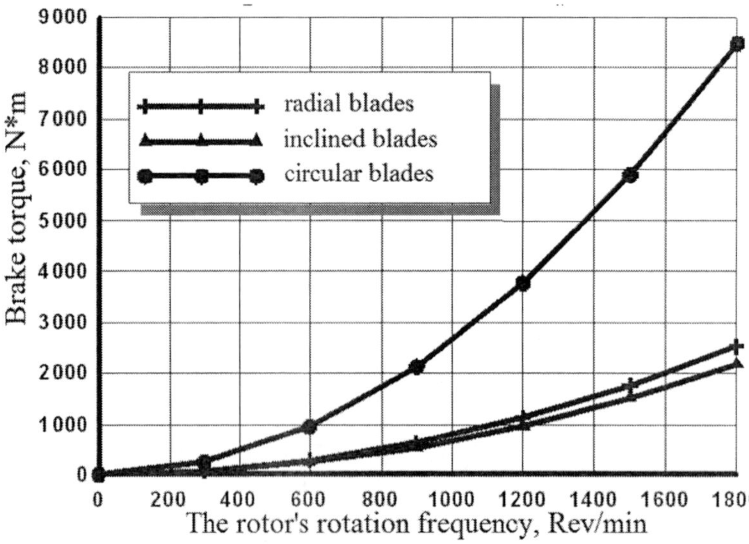

Fig. 6. The dependence of the brake torque on the rotor speed.

As can be seen from the dependencies shown in figure 6, the most effective is the retarder with circular blades. The brake torque of the brake with circular blades is 3.37 times higher than that of the brake with straight blades.

Further, additional studies were carried out to determine the effect of the radius of curvature of the circular blade system on the braking efficiency of the hydrodynamic retarder [7].

Taking as the basic version of the blade system with the radius of curvature of the blade equal to the radius of the flow part of the brake-retarder, research was performed in several variants with different radii of curvature of the blades R_l. Other geometric parameters remained unchanged.

For each of the considered variants, the fields of velocity and pressure distribution in the flow cavity of the retarder are obtained, similar to those shown in figures 4 and 5. The analysis of the obtained characteristics allowed to choose the geometry of circular blades, providing maximum efficiency of the brake-retarder.

Conclusion

In this work, the dependence of the braking torque on the rotor of the hydrodynamic retarder on the geometric parameters of the blade system is investigated.

The CFD method of calculation of the hydrodynamic brake-retarder is developed, which has the following properties:

- allows to calculate the braking characteristics of hydrodynamic brakes with different options of geometric parameters and types of blade systems;
- CFD calculation of the hydrodynamic retarder adequately replaces the physical experiment, which allows to reduce the time and money for the product development;
- various variants of performance of hydrodynamic brakes-retarders were investigated, in which the geometric parameters of the blade system were selected to increase the braking efficiency (energy intensity).

References

1. *Ground traction and transport systems. Encyclopedia*, **1**, (Moscow: Mashinostroenie, 2003)
2. S.P. Steshin, E. A. Yakovenko, *Hydrodynamic transmission,* (Moscow: Mashinostroenie, 1973)
3. *Automotive blade hidrosadenitis. Review.*/ A.T. Gapoyan, N.K. Dyachkov. (Moscow, NIIAVTOPROM, 1968)
4. A. A. Aksenov, Yu. I. Babiy, P. I. Karasev, K. V. Kuznetsov, I. V. Moskalev, Software systems FlowVision and IOSO for computer-aided design of technical systems // Cross-sectoral information service. Scientific and methodical journal, 4, 50–59 (2013)
5. L. D. Landau, E. M. Lifshitz, *Course of Theoretical physics,* **6**, (Moscow, FIZMATLIT, 2001)
6. John. Connor, K. Brebbia, *Finite element Method in fluid mechanics*, (Leningrad, Shipbuilding, 1979)
7. A.S. Pavarekha, V.S.Kovzun, SNTK-73. Materials of the 73th Scientific and Technical Conference, 24 (Minsk, BNTU, 2017)

Thermal imaging diagnostics locomotives

Andriy Sumtsov [1,*], *Anatoliy* Falendysh[1], and *Olha* Kletska[1]

[1]Ukrainian State University of Railway Transport, Department of Thermal Engineering and Heat Engines, Square Feuerbach 7, 61050, Kharkov, Ukraine

Abstract. The work considers the application of thermal imaging control in the operation of diesel locomotives. In particular, the diagnosis of electrical systems, cooling systems and compressors is considered. The results of the thermal imaging control of locomotives of the locomotive series CHME3 are presented. The directions of further researches for increase of efficiency of application of thermal imagers at diagnostics of a condition of systems and units of diesel locomotives are formulated.

1 Introduction

Continuous improvement of locomotives and increasing the degree of automation of internal control processes put the problem of optimal organization of technical operation at the forefront. An important role in this case is given to the technical condition of the locomotive, which, under the influence of external and internal factors, changes over time. Timely and complete information on the state of the ship together with the forecasting of its change will allow the locomotive to be use more efficiently, taking into account the requirements of traffic safety and the timely delivery of the cargo.

Determination of the technical condition of the locomotive carried out by various means of technical diagnostics. These include stationary systems, onboard built-in systems and portable diagnostic tools. The latter allow for diagnostic operations on locomotive by trained and qualified personnel.

The active development of science and electronics over the past decades has enabled the introduction of new types of diagnostic equipment to a wide extent into the production process of operation and repair of locomotives. Among these varieties of diagnostic devices and complexes are the means of thermal and thermal imaging control. All of them represent a group of means of non-destructive control of the state of nodes and systems that use as a diagnostic feature the temperature and its change.

Let's consider more detailed directions of application of thermal imaging control for diagnostics of knots and systems of locomotives.

Historically, the first units of the rolling stock, which were massively subjected to thermal control, were the rolling units of rolling stock. Various stationary and portable instruments for measuring infrared radiation were using for this purpose. With the advent of thermal imagers, it was possible to measure not only individual points, but also the entire surface of the boot knot. Due to this it is possible simultaneously to obtain an image of the temperature gradient of the entire surface to determine the place of damage or overload. In addition to the boot knots, methods of thermal imaging of other chassis units are actively implemented. In particular, brake discs, wheel band bands, motor-axial bearings for sliding and rolling. Active work in this direction is taking place in the USA, Russia. So in [1, 2] various constructions of equipment for diagnostics of chassis are considered. In [2], the technology of machine vision is considered, which allows to quickly and automatically control the state of the running gear of the rolling stock. Using a different spectral image

* Corresponding author: gurao@ukr.net

(visible and infrared) allows you to record the physical and thermal state and the correlation between them.

Recently, the use of thermal imaging control for the assessment of the state of electric circuits and electromechanics of rolling stock is actively being carried out. In electrical circuits, the state of the connections is checked. Comparison of individual compounds operating on equal terms allows rapid detection and removal of overheating places without allowing melting, burn-out or damage to current-carrying parts and isolation. This is due to the application of two approaches:

- determination of the absolute temperature of parts and units;
- definition of thermal anomalies in circles and nodes.

The first approach is aimed at obtaining surface temperature data and subsequent calculations for obtaining temperature data in the middle of the node. This approach is used to check the insulation tolerance and the state of the bearing units. In the last significant increase in temperature leads to the leakage of lubricants and, as a consequence, damage to the bearing.

The second approach is used to identify weak and problem areas in electric circuits. It is based on the comparison of the surface temperatures of the electric circuit or the comparison of temperatures of the same type of elements that operate under equivalent conditions. Thus, for example, in [3] the results of the study of the heating of collector plates of traction electric motors are presented. The thermal imaging diagnostics allows to reveal the uneven heating of separate plates and uneven heating of cockroaches compared with the working surface. In this way it is possible to detect low-quality connection of conductors with cockroaches and leakage of solder in the place of soldering. A similar way is the thermal diagnosis of terminal rails and other open connections. As a result, places of high load and low-quality contact, including low-quality soldered connections [3-5], are detected. The latter is a very important factor in preventing failures during operation.

Figure 1 shows the thermogram of the train contactors (a) and the terminal rail (b) CHME3 locomotive. It is clearly visible from the overload of the second train contactor compared to others. In accordance with the existing scheme of electric circuits diesel locomotive CHME3, the distribution of currents on all train contactors are uniform and, as a consequence, their heating should be uniform. The revealed overheating of the second contactor indicates a poor quality contact. In this way, a weak spot identified that requires intervention and regulatory work. As a result, the possibility of a contactor failure due to overheating of the insulation and contact gaps warned. Similarly, the analysis of the condition of the contacts and connections of the terminal rails carried out (Figure 1b and Figure 2). It is necessary to take into account the distribution of the load. So on the terminal blocks of diesel locomotives CHME3, the difference in currents provided by the construction can be 50%. Figure 2 shows defects as a contact of terminal block connections due to the insufficient effort of tightening the bolted connection. Thus, the use of test control during the conduct of TO-2 locomotives allows us to assess the state of the contact connections and formulate an operational plan for the regulation and adjustment of electrical contactors and connections.

Relatively new direction of application of thermal imagers is the diagnostics of the fuel system diesel locomotives. As shown by studies [3, 7], the use of thermal imagers allows us to detect malfunctions in the fuel system of high pressure. At the present stage, such a diagnostic can provide advice on the need to repair high-pressure fuel pumps or nozzles depending on the thermal state.

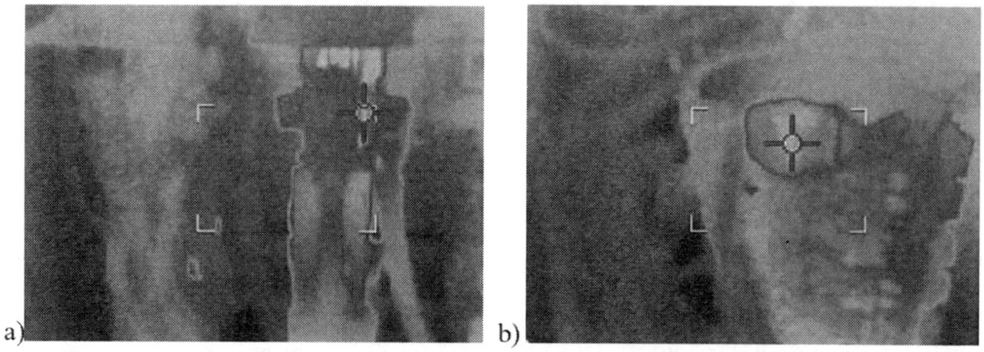

Fig. 1. Termograms of train contactors (a) and terminal rail (b) CHME3 locomotive.

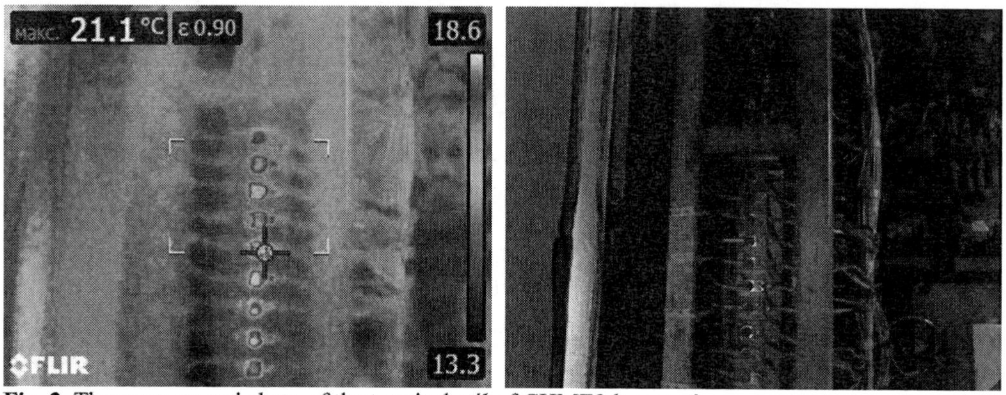

Fig. 2. Thermogram and photo of the terminal rail of CHME3 locomotive.

Thermal diagnostics is actively used to diagnose cooling systems. The main elements of the cooling system of the locomotive, which determine the efficiency of its operation, are the sections of the refrigerator. During operation, the ability of individual sections of the refrigerator locomotives to dissipate heat reduced. Thanks to the use of the thermal imager an express diagnosis of the sections of the refrigerators [3] is carried out. This allows you to adjust the volume of work when cleaning the sections and check the quality of the repairs. Studies [7, 8] have proven the feasibility of using thermal sights for the entrance control of the diesel locomotive cooling system when performing repairs and maintenance. Conducting thermo-controlled control of the refrigerator allows the timely detection of sections that do not provide the necessary cooling, and thereby prevent the increase in power take-off on the fan drive. This in turn reduces fuel consumption by 1 - 1.5% and does not allow for possible overheating of diesel in operation. This increases the reliability of the locomotive and its fuel efficiency, which makes it possible to reduce costs for the current maintenance of the locomotive as a whole.

Let's consider in more detail the application of thermal imaging diagnostics for technical maintenance and repairs of the cooling system of the locomotive. The thermal imaging technology based on measuring the temperature on the surface of the engine's refrigerating sections with subsequent processing of the results. During processing, temperature, uniformity of section heating and local overheating points are determined. Measurement carried out on a heated diesel locomotive with a stable mode of operation of the cooling system (without changing the temperature of the coolant). Figure 3 shows a thermogram and a photograph of fully functioning sections of the CHME3 diesel engine cooling system. The thermogram clearly demonstrates the uniform heating of all sections and throughout their length.

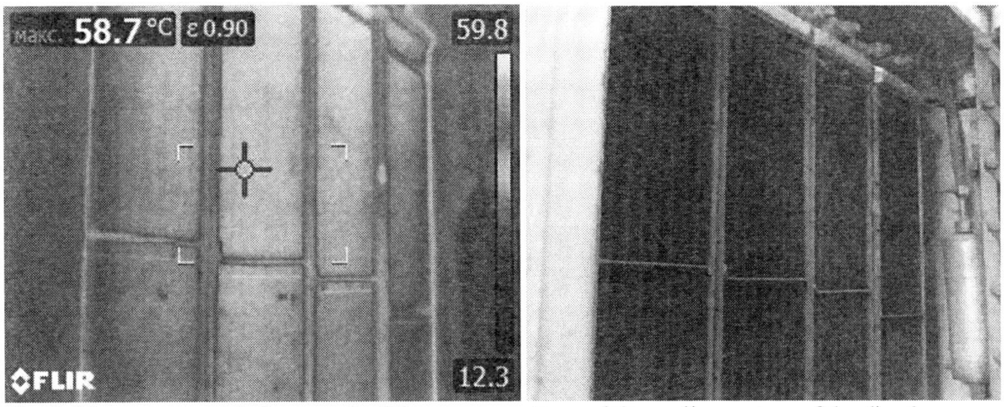

Fig. 3. Thermogram and photograph of the proper sections of the cooling system of the diesel locomotive CHME3.

In the event of worsening of the throughput of the section (due to the contamination of pipes or their drooping), heat transfer is greatly deteriorated. Figure 4 shows sections with uneven heating. So the most distant section is completely muffled and does not pass water. In it only the heating of the upper connecting flange occurs due to heat transfer from the surface of the collector. The two cooling sections located in the middle are dirty tubes. The second section of the arrangement has a slight contamination and the temperature of the drop in the section at 15°C. The third section has a low passage of tubes and, as a consequence, a drop of 35°C. Thus, from the four sections, two are required to rinse, one is repair and the first section is operational.

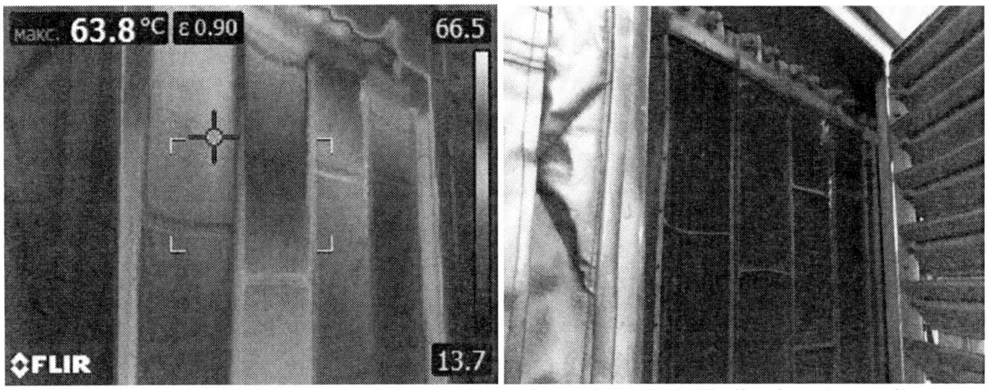

Fig. 4. Thermogram and photograph of the proper sections of the CHME3 diesel engine cooling system.

When using thermal sensors to check the status of sections of refrigerators with a two-row arrangement, it is necessary to check both the internal and external series of sections. Particularly relevant is checking the status of sections of refrigerators when switched to operation in winter conditions. When the cooling fluid passes through the tube, at low temperatures, there is the possibility of freezing sections with subsequent damage due to the expansion of water in the tubes. In order to avoid this phenomenon, locomotives are increasingly using cooling liquids with a low freezing point and a heat storage system.

In addition to the considered method of use of thermal imagers, they can be used for complex diagnostics of the efficiency of the refrigeration unit. Such an approach is proposed in [9]. As a diagnostic parameter, the temperature pressure of the tube wall of the section Δtn, °C (between the coolant and the outer surface of the refrigerator sections) is used:

$$\Delta t_j = t - t_{nj}, \qquad (1)$$

where:
t - water temperature at the entrance to the radiator section, [° C],
tpj - temperature of the tube surface section, [° C],
j - section number.

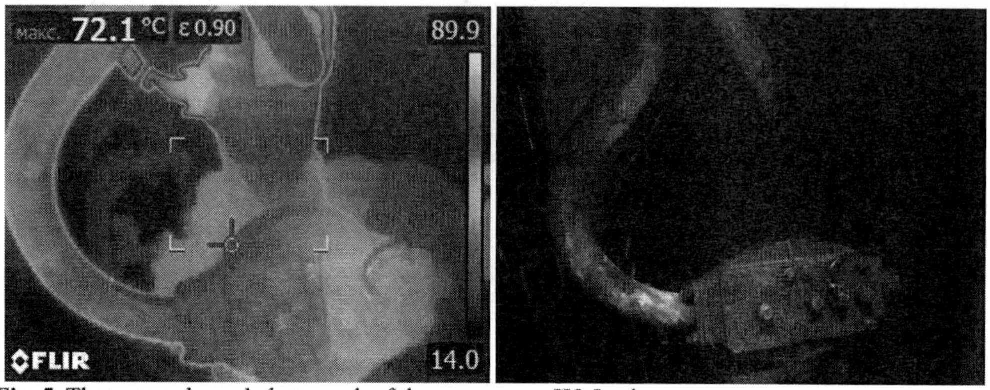

Fig. 5. Thermography and photograph of the compressor K2-Lock.

Refrigerator of the cooling system is subject to inspection in the process of carrying out of rheostats tests of diesel locomotive. This is due to the need for a long measure of temperature at the nominal (full power) mode, in which the greatest amount of heat is allocated to the cooling system. The thermal image of the sections of the sections is removed using the thermal imager. Each thermogram is treated with ten sections for typical sections and the surface temperature tj in each section is determined. Then, by the expression (1), the temperature difference is calculated in each section of the section, the average value of Δtj is determined and compared with the minimum allowable value). In this way, all sections are checked and a general conclusion is drawn about the possibility of cooling the system to dissipate heat from the diesel engine. This way the decision is made on the need to repair the refrigerator locomotive.

An important aspect of using thermal vision is the diagnosis of the state of compressor locomotives equipment. Due to the use of heat shocks, check the quality and uniformity of the valve mechanism, the condition of the bearing units and the radiator of cooling air. Figure 5 shows the work flow diagram of the compressor K2-LOK diesel locomotive CHME3.

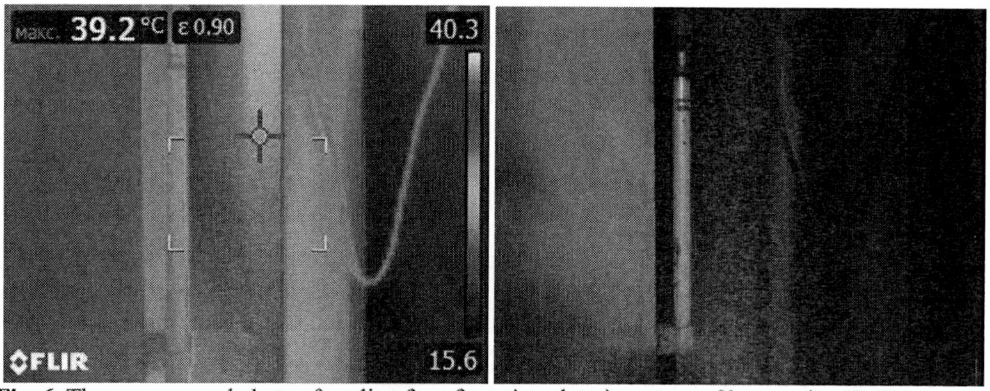

Fig. 6. Thermogram and photo of cooling fan of traction electric motors of locomotive ЧМЕ3.

The temperature difference in the first degree of compression of air is about 30°C. This parameter serves as an indirect proof of the normal operation of the compressor valve mechanism and the presence of compression.

In the same way as thermal inspection of sections of the refrigerator, an intermediate compressor cooler is checked. Figure 6 shows the results of this check. At the same time, the main difference is checking that I will not turn the tubes, but I will get a sharpening and, as a consequence, cooling quality by the flow of air.

The use of thermal imaging controls to diagnose the condition of bearing units has significant prospects. So this type of diagnosis is subject to all electric and auxiliary mannina. For an example, let's consider the results of diagnostics of cooling fan of traction electric motors of diesel locomotive CHME3 shown in Figures 7 and 8.

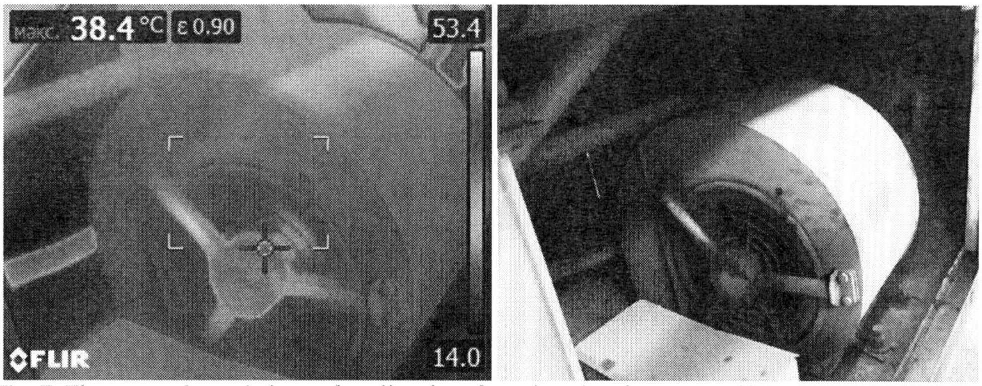

Fig. 7. Thermography and photo of cooling fan of traction electric motors of locomotive ЧМЕ3.

Figure 7 shows the thermal graph of the back cover of the fan. Located in the center of this cover, the roller bearing is a source of heat. Due to the friction that occurs as a result of the fan's operation, a tempo is created. With insufficient lubrication there is a rapid increase in temperature. With too high values there is a leakage of grease lubrication and there is a possibility of jamming of the fan. The given diagram shows a good bearing unit (the heating temperature does not exceed 40°C).

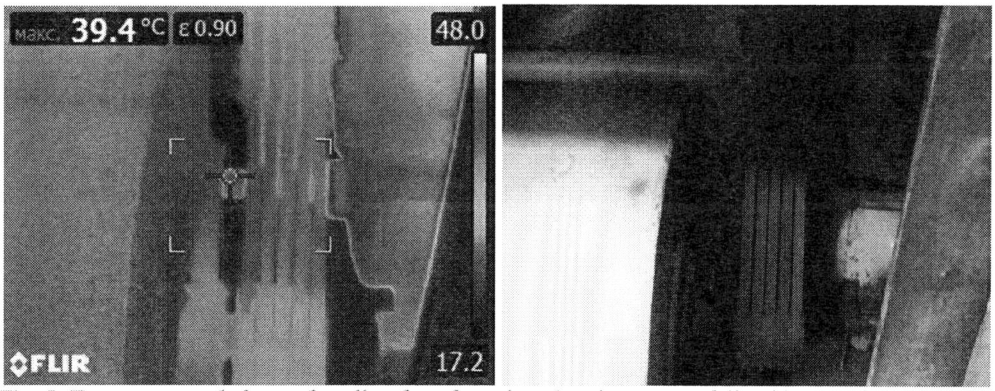

Fig. 8. Termogram and photo of cooling fan of traction electric motors of diesel locomotive CHME3.

Figure 8 represents the front cover of the cooling fan for traction motors. The condition of the bearing unit is similar to the rear. The heating temperature is lower by 10°C due to better air cooling.

On the thermogram of Figure 8 it is clearly visible to heat the belts and pulleys of the fan drive. The difference in the temperature of the microfibers of the pulleys clearly demonstrates the intensity of the use of a smaller pulley compared to a larger pulley, which explained by the higher frequency of rotation of the first one. The uniformity of the heating of the belts of the cavalry transmission indicates the correctness and uniformity of their tension. Proceeding from this, the presented thermograms prove the serviceability of the cooling fan of traction electric motors and its drive.

Another direction of thermal imaging control is to check the amount of fuel and its temperature in the fuel tanks. This direction on locomotives little used because of the presence of standard systems for determining the amount of fuel that has a higher accuracy of measurement compared with the thermal method. An example of such a measurement shown in Figure 9.

Fig. 9. Thermogram and photo of fuel tank diesel locomotive CHME3.

Thus, the use of thermal imaging control allows operative diagnostics of most nodes and locomotive systems with the detection of places with atypical thermal load. Depending on the equipment tested, after the diagnosis, the site is adjusted or repaired.

The use of thermal imaging diagnostics allows us to implement maintenance and repair on the actual condition of the nodes, which is difficult to verify by other types of diagnostics, for example, the quality of soldering of collector plates, contact in the places of connection of current-carrying parts of electric circuits, etc.

References

1. Багаутдинов В.Р., Ураксеев М.А. Методы контроля ходовой части железнодорожного транспорта. Тепловизионный метод. [Текст]./ В.Р. Багаутдинов, М.А. Ураксеев / ИЗВЕСТИЯ ВолгГТУ, 2012. – №10 (97). – С. 5 – 10
2. Machine Vision Using Multi-Spectral Imaging for Undercarriage Inspection of Railroad Equipment / J.M. Hart, E. Resendiz, B. Freid, S. Sawadisavi, C.P.L. Barkan, N. Ahuja
3. Овчаренко С.М., Балагин О.В., Балагин Д.В. Реализация комплексной системы бесконтактноготеплового контроля узлов тепловозов. / С.М. Овчаренко, О.В. Балагин, Д.В. Балагин // ИЗВЕСТИЯ Транссиба, 2014. – №4(20). – С. 35 – 40
4. Хачкинаян С.А., Губарев П.В., Рубан В.Г., Матва А.М. Методы тепловизионного контроля электрического оборудования электровозов. – Труды международной научно-практической конференции «ТРАНСПОРТ-2013». – С. 265-265
5. Рубан В.Г. Комплексный подход к организации тепловой диагностики узлов локомотивов в условиях депо / В.Г. Рубан, О.А. Ворон // Труды Всероссийской научно-практической конференции «Транспорт-2011», часть 2. Ростов-на-Дону, 2011.– с. 421-423
6. Боряк К. Ф. Перспективы внедрения температурного мониторинга для диагностики технического состояния механических узлов подвижного тягового железнодорожного состава / К. Ф. Боряк, С. Л. Волков, Е. С. Шпат // Техническая диагностика и неразрушающий контроль. - 2014. - № 2. - С. 50-53
7. Лобанов И. И. Повышение эксплуатационной эффективности тепловозных дизелей применением средств оперативной диагностики. Диссертация на соискание ученой степени кандидата технических наук по специальности 05.22.07 – Подвижной состав железных дорог, тяга поездов и электрификация – Москва, 2017. – 209 с

8. Горин А. В. Контроль технического состояния систем охлаждения тепловозных дизелей – Вестник ВНИИЖТ 3/2015. – С. 23 – 30
9. Анисимов А.С., Балагин О.В., Чулков А.В. Технология тепловизионного контроля технического состояния секций холодильников тепловозных дизелей. – Омский научный вестник, 2006. – №3(36). – С. 111 – 113

Operation of car turbochargers – selected problems

Hubert Wojciechowski[1,*1], Czesław Żdanowicz [2], Tony A. Hope[3]

[1]UTP University of Science and Technology in Bydgoszcz, Faculty of Mechanical Engineering, Poland
[2] BNTU University of Minsk, Faculty of Mechanical Engineering, Belarus
[3]Southampton Institute, U.K

Abstract. In order to increase the filling of the cylinder in the engine, a so-called charge is applied, which consists in increasing the amount of air input under increased pressure and at an unchanged temperature. This increases the density of that air and increases the mass in the engine cylinder. In most cars, the charging is carried out by an exhaust gas turbocharger or a mechanical compressor. This paper presents the issues that concern the failure of turbochargers resulting from their balance by weight reduction method. Selected aspects concerning consequences of incorrect balancing and methods of reducing unbalanced rotor mass. A key part of the work includes a proposal to balance turbochargers of automotive vehicles using the "weight-adjustment"method. In this scope, preliminary studies were presented, and conclusions observed during their implementation were described.

1 Introduction

Modern internal combustion engines are characterized by the downsizing (decrease in engine capacity, while maintaining high engine power) and aim to achieve the best overall efficiency [4].

These trends can be met by using mainly a turbo-charging system in the drive unit. The principle of the engine charging system is relatively simple and consists in injecting more air into the cylinder in order to burn the petrol-blend more efficiently. The turbocharger or compressor is responsible for this process. In this way, the engine produces more power, which improves vehicle performance.

Car companies use mechanical turbo-charging in their models - compressors, as well as turbocharging. The second option is much more commonly used because of its universality. They are installed in compression ignition engines, spark ignition engines and hybrid cars. The turbocharger is the basic and at the same time the main element in the system of turbo-charging [1,3,8]. It consists of three basic elements [3]: center hull, in which there is a bearing turbocharger rotor, turbine hull and compressor hull.

The basis of the efficiency of the turbines is the fact that they are able to pump a very large amount of air into the cylinder, and in addition they are powered by exhaust gas energy, which is not used in any way in naturally aspirated engines. The turbo-charging system has a significant influence on performance, so a noticeable power decrease can be one of the symptoms of turbocharger damage, especially if the problem is mainly related to the rotational range in which the turbine is most active. The technical sources are usually rotor damages, bearing damage and, in the case of turbines with variable geometry, hydraulic steering wheels.

* Corresponding author: hubert.wojciechowki@utp.edu.pl

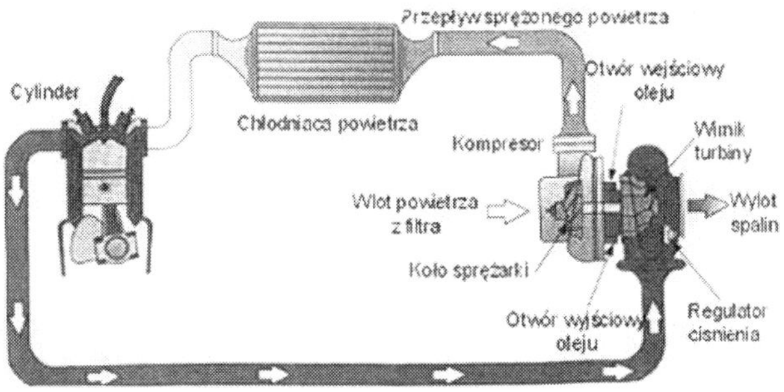

Fig. 1. Working principle of the turbocharger [2].

The turbocharger principle is based on the exhaust gas drive of the turbine rotor. The turbine is connected by a common shaft with a compression wheel, on which the air is aspirated. The compressed air is supplied to the combustion chamber of the engine cylinder under pressure - Fig. 1. To increase the efficiency of the turbocharger, an air cooler (intercooler) is used. Cooled air increases its density, which results an increase in the mass delivered to the combustion chamber.

Publications and scientific articles in this area of research discuss outside the typical operational defects of turbochargers do not focus on their faults, which are result from the methods of turbo-charging equipment regeneration [10,12]. Despite the huge popularity of engine turbo-charging systems for modern motor vehicles, manufacturers of turbochargers do not reveal the materials from which given components are constructed and do not provide information about theirs mechanical and physical properties[4]. Defects of the turbochargers components have a negative impact on the emission of harmful exhaust gases to the environment [10,11]. Description and modelling of the dynamic state, methods of diagnosing the state as well as the application of modern information techniques in turbochargers' research is so far hardly recognized and requires immediate improvement [8 ,11,12]. Literature describing the dependence of the method of regeneration and mainly balancing the turbocharger components -on the operation of the turbo-charging system is still very insignificant in this field. Existing publications so far describe mainly the balancing of turbochargers by means of a reduction method - **taking off unbalanced mass** and focusing mainly on damage to turbochargers resulting from improper operation of the turbo-charging device. Examples of this are the following articles: Balancing of turbocharger rotor [10], Damage analysis of turbocharger compressors [12], Analysis of the impact of operating conditions on the technical condition of turbocharger turbochargers of combustion engines [11]. Under heading [1] engine turbo-charging, general information can be found on, inter alia, balancing turbochargers using the' mass capturing' method.

This article presents the main causes of turbocharger damage, emissions in motor vehicles. The present methods of verifying the condition of charging devices have also been described and an **innovative method of balancing turbochargers** by **"adding mass"** has been proposed.

For the turbochargers regeneration, apart from the correct replacement of worn parts and the cleaning of other parts, the most important thing is the precise dynamic balancing of the rotating parts of the turbine.

2 Diagnostic tester – verification of actual parameters

In order to eliminate the undesirable condition of the turbocharger such as its failure, it is necessary to correctly identify and evaluate in detail the technical condition of the turbo-charger by available ways. The simplest and at the same time lowest effective method of diagnosing turbocharger is organoleptic. Turbo-charging devices are evaluated with human senses. An oiled turbocharger may indicate a broken housing, damaged seals or a worn out suction section of the turbocharger. The organoleptics make it possible to hear disturbing sounds from the device. The damaged turbocharger is characterized by rustling, whistle. This can be indicated damaged, broken turbocharger rotor and a clogged cable between the inlet and intake manifold [1,3,7].

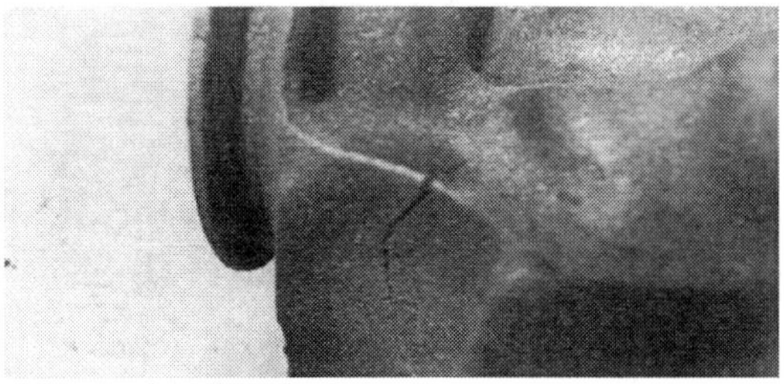

Fig. 2. Fracture of the outer part of the housing.

The color of the exhaust gas coming out of the exhaust system may also indicate the condition of turbo- charging device. The blue or black colour of the exhaust gases can indicate a turbocharger failure. This defect may indicate a cracked compression wheel or a defective oil system in the turbocharger. The presented method is one of the simplest, but in case of its credibility it becomes questionable because it does not ensure accuracy and insight in its assessment [10].

The parameter that characterizes turbo- charge system and can indicate its condition is the pressure of turbo-charging. The vehicle directive that has been drawn up stipulates that cars manufactured after 2000 must have 16-pin diagnostic connector. This makes it easier to diagnose any drivers emissions in a vehicle and the reading of operating parameters in any time unit.

The key information in the turbo-charging circuit is the actual data, which is stored by signals from sensors in the circuit. The evaluation of the above parameters is possible through additional functions, which are used as applications in diagnostic testers. An excellent example of this solution is the dashboard function in IDC5 software in the TEXA TXT's Tester. The turbo-charge pressure is checked here depending on [5]:

- rotor speed,
- compressed air pressure,
- exhaust gas temperatures,
- the blade angle of inclination in the variable geometry system.

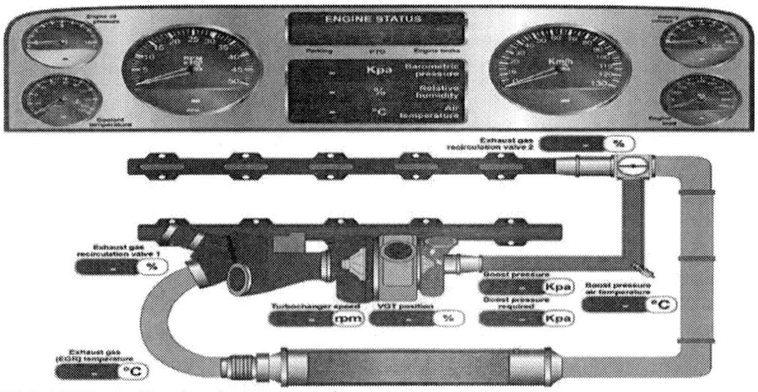

Fig. 3: DASHBOARD application in IDC5 software [5].

One of the main reasons of failure of turbochargers is unbalance of its rotating parts. To measure the balancing quality of the turbocharger's rotor we use two types of devices:

- quick-rotating turbocharger core correction machines (fig.4),

- slow rotating devices for balancing the turbocharger shaft (Fig. 5).

Fig.4. Core correction machine.

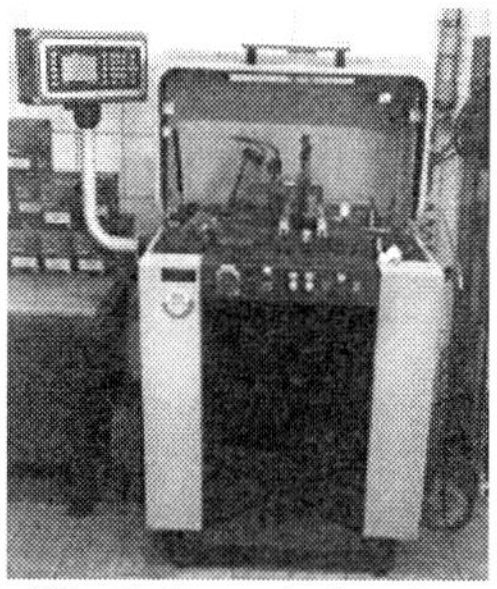

Fig.5. Rotor balancing machine.

The presented devices verify turbochargers in the context of unbalance measurements of rotor masses and correct the balance of turbine wheel and compression wheel with turbocharger shaft. Unbalanced weight is reduced with an accuracy of 0,01 mg/mm2 and the unbalance tolerance value is assigned to a given type of turbocharger and stored in the internal memory of the device. Unquestionably, the above method is one of the most accurate, precise methods of verifying turbo-charging devices.

3 Methods of identyfying the consition of turbochargers

Due to higher and higher operating speeds of turbochargers, even a minor unbalance causes an increase of vibrations, which in consequences may lead to a quicker damage of the whole component. The effect of its use is subsequent vibrations of rotors, causing much faster wear of bearings, rings and other parts of turbochargers.

Balancing is the process of checking the distribution of the rotor mass and correcting it, if necessary, until residual unbalance is obtained within the set tolerances. The aim is to achieve an even weight distribution of the rotor with respect to its geometric axis of rotation. In theory, the unbalance of rotating elements can be divided into three basic types: static, momentary and dynamic. In practice, we are dealing with all of them simultaneously, but the division applied makes it easier to understand the essence of the phenomena taking place and to choose a rational method of limiting their negative results.

3.1. Dynamic balancing of the rotor by the 'mass reduction' method

The main reason for damage and failure of turbo-charging devices is unbalance of turbocharger's rotor. Rotor unbalance creates excessive vibrations which are transmitted to the machine body, which leads to damage to the roller bearings. A characteristic feature of the rotor unbalance is a regular increase in bearing vibration amplitude, which depends on the rotational speed of the rotor of the turbo-charging device. These are always harmonic vibrations, the frequency of which is equal to the frequency of rotor rotation. Rotational speed of turbocharger rotor oscillates around 200 000 rpm, so the problem of natural frequency is particularly important. The absolute condition of the turbocharger's usefulness is not showing a tendency to any vibrations, whether the drive unit rotates or not [2,6,12].

Unbalance of basic components of the turbocharger may lead to various damages and even serious failures [3,9]:

- damage to the turbine wheel blades and compression,
- cracks in the turbocharger shaft,
- damage to radial cotact ball bearings and thrust bearings,
- jamming and damage of turbocharger's rotor,
- variable geometry system failure.

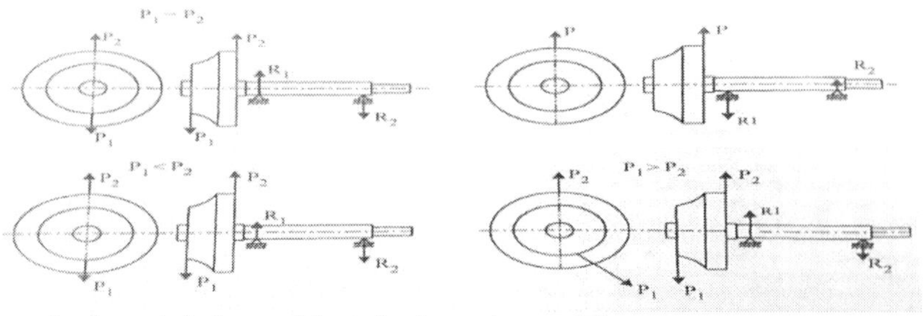

Fig.6. Illustrates the dynamic balance of the turbocharger's rotor [1].

The force pair P1 and P2 as shown in Fig. 6 is in the same mode of working as the axis of the rotor. On a bearing-mounted balancing station, the reactions R1 and R2 are generated, which are in the same position as the above forces. They have equal values but opposite directions (R1=-R2). The location of the rotor, as well as the value of unbalanced mass, is used to verify the place and mass of the material, which has to be removed for proper dynamic balancing of turbocharger rotor [1].

The turbocharger of vehicle is balanced in two stages:

- stage I - Balancing the shaft with the turbine wheel and the compression wheel with nuts and spacers and seals (Fig. 7),
- stage II - balancing of the bearing-mounted turbocharger rotor in the compressor housing with oil system connected (Fig. 8).

Fig. 7. Balancing the turbocharger shaft.

Fig. 8. Balancing the core of the turbocharger.

This type of balancing is the result of the recommended stricter requirements imposed by companies manufacturing turbo-charging devices. Correction of unbalance of turbochargers by "reduction of mass" certainly affects the structure of the balanced element, because excessive removal of material from the compression wheel or turbine wheel make them weak, which often causes cracking of these elements.

The following figures (9 and 10) show the points on the compression wheel from which the mass hass been reduced and the resulting wheel crack at the point where the unbalanced mass was gathered.

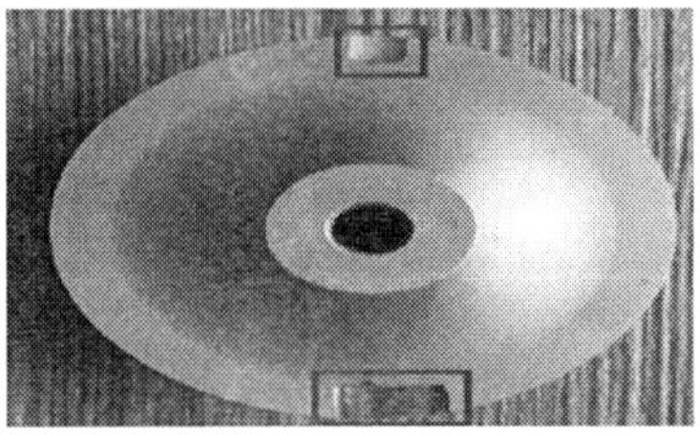

Fig. 9. Mass reduction points on the compression wheel.

Fig.10. Cracked compression wheel.

3.2 Dynamic balancing of the rotor by the 'mass addition' method

Most of the damage to turbochargers is caused by breakage of the compression wheel; turbine; and also by breakage of the turbocharger shaft. Such defects are caused by improper exploitation of turbocharged combustion engines and incorrect regeneration of turbocharger components. Incorrect balancing of basic components has a negative impact on the construction and mechanical properties of turbocharger components. In order to prevent damage to materials which result from improper regeneration of turbocharger components, innovative tests are carried out in order to minimize the above problem. The method of 'mass addition' is carried out in two stages:

- first stage concerns the correction of the unbalanced mass of the turbocharger rotor including sealants and spacer nuts,
- second stage is based on balancing the rotating element (core of turbocharger) with bearings in the body of the device.

Two types of devices were used to perform these tests:

- Schenck TB Comfort slow-running balancing machine for the first stage of balancing,
- the second stage was carried out on a high-speed Schenck MBRS 110 machine.

Disassembly of balanced components and input of data of turbochargers into the system is the identical as in the traditional method using balancers for rotors and cores of turbochargers. The difference is in the way of correcting unbalanced masses of rotating parts. The process of unbalanced mass correction was carried out by using a special thermo-resistant mass, which is designed, among other things, for the removal of exhaust gases and liquids whose temperature exceeds 1000°C. Characteristic feature of this substance is a high adhesion coefficient and short time of mass fixation. During the reading of the value of the unbalanced mass of the rotating element, the heat-resistant mass is applied to the outside of the turbine wheel and compressed via an insulin syringe. Before application, the rotor surfaces were thoroughly cleaned and the amount of mass applied was verified on an electronic scale with an accuracy of 0.01mg.

When the adhesive mass was bound, the turbocharger rotor was mounted in the core, which was subjected to strength tests on a device allowing to accelerate the rotating element to the speed of more than 200 000 1/min.

The above method has been used for turbocharger number 038145702E, which is used in vehicles of Volkswagen brand with engine capacity of 1900cm3 and after removal of main components of turbocharger was verified its rotor in terms of balance.

In the first stage, the basic dimensions of the rotor, which are necessary for the test to be carried out correctly, were entered into the TB COMFORT balancing system. The following dimensions have been set:

- diameter of the compression wheel,
- height of the compression wheel,
- turbine wheel diameter,
- turbine wheel height,
- turbocharger shaft length,
- distance between turbocharger shaft and balancing sensors placed in supports.

Then the rotor of the turbocharger was rotated in order to verify its level of unbalance. The obtained value of unbalance correction was reduced by application of thermo-resistant mass with the use of insulin syringe (Fig. 11 and 12).

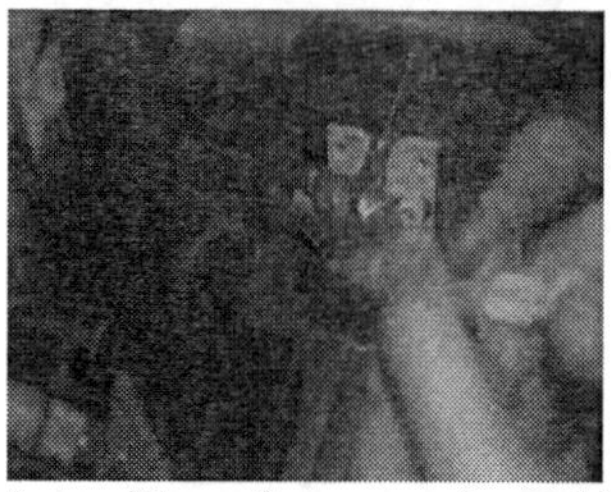

Fig. 11. Imposition of mass on the turbine.

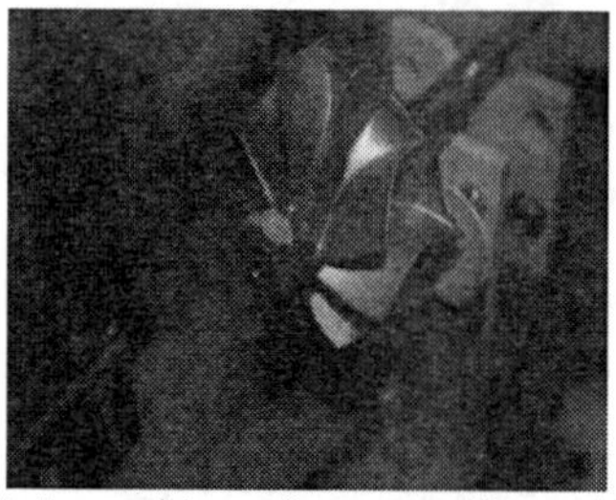

Fig. 12. Imposition of mass on the turbine wheel.

After the process of balancing the turbine wheel had been completed, the compression wheel was verified. The procedure was the same as in the case of the turbine wheel. The unbalance of the compression wheel was reduced by overlapping the same mass as that of the turbine wheel. The procedure of balancing the turbocharger rotor was carried out in a correct way, the size of unbalance was within the tolerance limits.

After the process of balancing the turbocharger's rotor, the balanced element was placed in the core of the turbocharger and mounted in the high speed Schenck MBRS110 balancing machine. During this process, the rotational speed of the rotating component exceeds 200 000 rpm. At this stage of the test, the high rotational speed strength of the mass of the rotor, which is common during the daily operation of the turbocharged drive unit, has been verified. During this test, it was observed that at a rotational speed of 200 000 rpm, the heat-resistant mass remained on the turbine wheel (Fig. 13), but broke out of the compression wheel (Fig. 14).

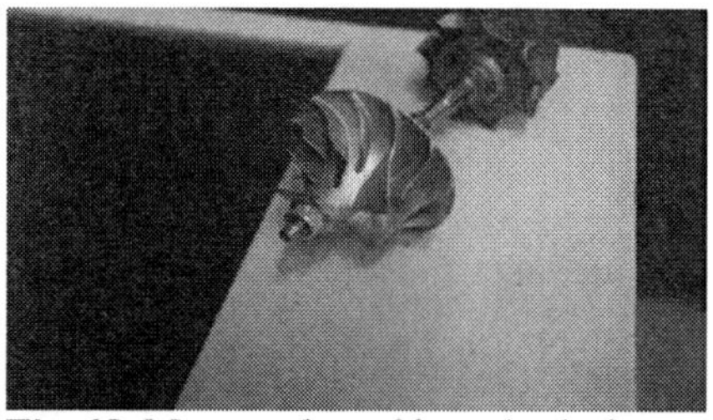

Fig. 13. Mass on the turbine wheel after compression test.

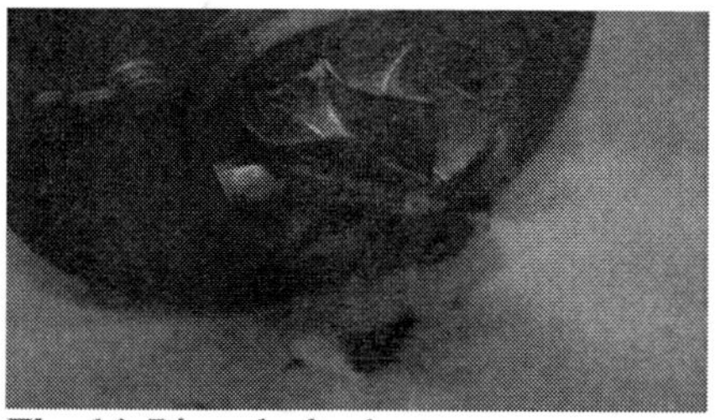

Fig. 14. Ripped wheel massss.

Turbochargers of properly operated car engines do not require frequent repairs and reach mileages from 100 000 to 300 000 km. They run reliably on heavy commercial vehicles, achieving mileages up to 500,000 kilometers and more. The reason of their possible earlier failures are usually mistakes in the exploitation and service of engines or damage to other systems cooperating directly with the turbochargers.

4 Rational use of the turbocharger

Turbocharger is an exhaust-driven compressor used for turbocharging combustion engines, which is used to increase engine power by increasing the amount of oxygen supplied to the combustion chamber - Fig. 15. The construction of the turbocharger is designed in such a way that it can cooperate with it without any problems during the engine lifetime.

The turbocharger does not need any specialist maintenance and its examination should be limited to a few basic steps included in the instructions for use of the turbocharger provided by the manufacturer. Typical damages to turbochargers are:

- damages caused by foreign material,
- Insufficient oil pressure,
- temporary delays in oil delivery,
- contaminated oil,
- overheating,

- ➢ construction faults in the process of turbocharger manufacturing.

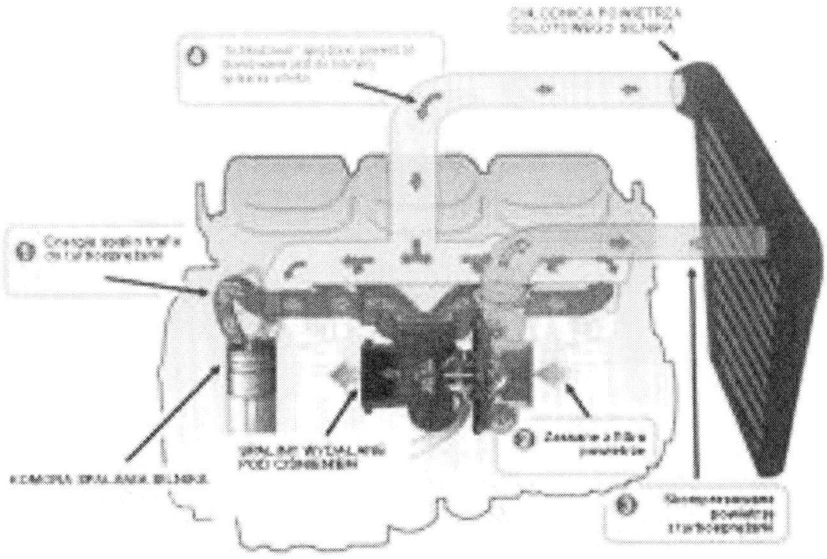

Fig.15. Diagram of turbo-diesel system [3].

There are obvious symptoms of turbocharger worn out:

- ➢ sudden loss of power;
- ➢ much louder operation (whistling);
- ➢ blue exhaust smoke combined with rapid oil loss in the lubrication system;
- ➢ oiling of the area around the turbocharger and in the air inlet ducts;
- ➢ („check engine") error and low engine rpm.

Approximately 90% of damages of turbochargers are caused by their incorrect operation (use and maintenance). The main causes of turbocharger damages are following:
overheating of the turbocharger (after a long period of use, it is recommended to leave the unit cooled down, lubrication decays, charring of the lubricating oil, wear of bearings, clogging of oil ducts);

- ➢ excessive load on the cold engine (cold oil is dense, the bearings are not lubricated);
- ➢ use of not recommended oil by the manufacturer (preferably synthetic, for high operating temperatures);
- ➢ exceeding the recommended oil change intervals (loss of lubricating properties);
- ➢ low oil level (bearings, tilting in the sloping);
- ➢ damaged air filter (sand, dust, high speed – wear);
- ➢ defective turbocharger control (worn actuator and valve);
- ➢ damage to the turbocharger seals (pumping oil into the engine, increase in rotation, uncontrolled damage).

Only the manufacturer of the turbocharger or authorized repair companies have the tools and personnel specialized in repairs of these devices.

The rational use of engines and turbochargers requires certain habits and simple maintenance to ensure the reliable operation of these units. The basic recommendations in this area are:

- ➢ avoiding sudden engine stop after prolonged use;
- ➢ do not use the full power of the engine immediately after it has been started up;
- ➢ use of recommended oil, resistant to high temperatures and carbonization;
- ➢ compliance with the timeliness of the oil change and the oil filter;
- ➢ regular inspection of the engine oil;

- on-time replacement of air filters;
- application of CERAMIZER (protective layer for shafts, rings, sleeves).

5 Summary

It is difficult to imagine a modern motor vehicle that is not equipped with an engine turbocharging system. The most common charging device is the turbocharger, which has gained its popularity thanks to its universality and flexibility. It is used both in internal combustion engines and in innovative hybrid drive units.

The most common cause of failure of turbochargers is unbalance of turbocharger's rotor. A characteristic feature of this problem is a linear increase in the amplitude of bearing vibrations, which depends on the rotational speed of the rotor. Incorrect balancing of the basic turbocharger-impeller element is caused by the following damages: cracked wheel of turbine or compression, damage to the turbocharger shaft, damage to thrust bearings and variable geometry system failure.

Analysing the proposed innovative method of balancing the rotor of turbocharger by adding heat-resistant mass we can conclude that the presented method is an interesting alternative to the current method of balancing turbochargers. Certainly, the probability of turbocharger failure during the reduction of unbalanced mass is minimized. Correction of the unbalanced mass of the rotor by the "mass addition" method does not affect the mechanical properties of the material and does not affect its structure. Thus, it does not cause cracks of rotating elements in the turbocharger.

Further research on this issue should identify the phenomenon of the loss of added mass from the compression wheel at a rotation of 200,000 min-1. In the initial analysis, this can be explained by the incorrect selection of glue for the material mass.

The proposed method aims to remove the problem of incorrect reduction of unbalanced weight. This problem appears in the form of cracks, i.e. damage to the basic components of the turbocharger's rotor. This involves expensive repairs to the main turbo-charging unit in the turbo-charging system.

The presented first results of research of the proposed solution confirm the fact that the right balance is the key aspect for the correct, failure-free use of the turbocharger.

Briefly indicated basic recommendations for rational use of car engines and turbochargers ensure safe and reliable operation of these systems.

References

1. J. Mysłowski, *Pojazdy samochodowe – doładowanie silników*. WKŁ Warsaw (2011).
2. H. Wojciechowski, *Analiza metod badania turbosprężarek pojazdów samochodowych*. Logistyka **No. 4** (2015).
3. P. Bojarczuk, *Doładowanie silników spalinowych - przegląd konstrukcji i diagnostyka*. Poradnik serwisowy, **No. 1** (2007).
4. J. Merkisz, M. Rychter, *System OBD II jako przyszłe sposoby diagnozowania pojazdów*. Eksploatacja i Niezawodność, **No. 1** (2002).
5. J. Jaskólski, G. Budzik, A. Marciniec, *Balancing of turbocharger rotors*. Journal of Kones (2007).
6. M. Idzior, T. Bieliński, T. Borowczyk, W. Karpiuk, *Analiza wpływu warunków eksploatacji na stan techniczny turbosprężarek doładowanych silników spalinowych*. Logistyka (2010).
7. J. Kaszkowiak, S. Borowski, E. Dulcet, M. Zastępowski, *Analiza uszkodzeń turbosprężarek*. Logistyka (2015).
8. B. Żółtowski, C. Cempel, *Inżynieria diagnostyki maszyn*. Polskie Towarzystwo Diagnostyki Technicznej. Instytut Technologii Eksploatacji. (2004).
9. B. Żółtowski, *Badania dynamiki maszyn*. ATR Bydgoszcz (2002).

10. B. Żółtowski, *Podstawy diagnostyki maszyn*. ATR Bydgoszcz (1996).
11. B. Żółtowski, M. Łukasiewicz, *Wibroakustyka maszyn w laboratorium*. ATR Bydgoszcz (2005).
12. M. Żółtowski, *Informatyczne systemy zarządzania w inżynierii produkcji*. ITE - PIB, Radom (2011).

Parametric evaluation of the technical object suitability based on analyze the quality of its maintenance and use taking into account the initial conditions

Rafał Grądzki [1,*]

[1] Bialystok University of Technology, Faculty of Mechanical Engineering, , Wiejska 45C, Bialystok 15-351, Poland

Abstract. In this paper, the comparison of three various technical objects (engines of public transport buses) exploitation research for different initial conditions are presented. Object researches were carried out in 2012 and then repeated in 2013. Gathered operational data is presented in three sets (1 – concerning object, 2 – concerning driving conditions, 3 – concerning driver, where set 1 is the collection of diagnostic information D_k, and sets 2 and 3 are the information about object environment U) in form of conventional points (experts numerical assessments). Relation between point information of object and point information of environment was described by coupled equations of state (describing relations between operation condition and technical condition including initial conditions for each analyzed exploitation period). That method allows to determine parameters of technical condition a_T and operation condition a_R and next, from the course of a_T parameter, set of parametrical damage $m_T(t)$ and from course of parameter a_R – set of momentary damage $a_R(t)$. Thus it is possible to evaluate exploitation, technical and operation conditions of each object (bus engine). Received reliability parameters allows to properly control exploitation and service of particular objects and set of objects (fleet of buses) and its elements.

1 Introduction

In the process of technical object exploitation occurs the gradual damage (decrease in its exploitation properties and mechanical wear). This damage consists on deregulation, wear and deterioration of its reliability. Hence the importance of constant supervision of the operation condition changes, technical condition changes and reliability changes during service and use of technical objects. [1÷16]

In this paper, current work on new method of exploitation analysis of object aptitude is explained. This is based on assumption that every information (diagnostic signals – set 1, use signals – set 2, use condition signals – set 3) is described in the form of point ratings (numerical assessment by expert in the detailed questionnaire).

Diagnostic signals in form of points were used for the first time in diagnostics of ship-based helicopter SH-2G (USA) [16]. In this method aptitude was analyzed only by summing all the points. Next, obtained value is compared to threshold set by manufacturer (USA). Based on relation between sum of the points and allowed number of points, the machine is qualified for further exploitation or repair.

During researches conducted for this article every diagnostic signal, symptom or expert evaluation, was registered in the questionnaire in point form (which is innovatory). Next, in this paper the experts assessment of the bus operation was used to determine required

* Corresponding author: r.gradzki@pb.edu.pl

regulation and technical condition parameters from model in form of influencing each other equations of state (including initial conditions) [6, 7, 9] and subsequently from its courses - set of damages that are base for determining reliability characteristics. [2, 8, 11, 12, 14, 15]

Presented practical method of exploitation analysis of object aptitude (operation condition, wear condition, reliability) might be, and partially is, applied to process of servicing technical objects. Such data allows to decide on object further destination (about continuation of its use).

2 Diagnostic research of technical object and its environment

The analyzed objects were three various MAN buses exploited in the years 2012 and 2013 in Municipal Transport Company of the City of Bialystok. The aim of the research was to obtain diagnostic signals D_k concerning condition of objects (data from diagnostician) and data on their environment (expert information from driver).

The examinations were conducted in the following months of the year. Thus, a database for 12 months in the year 2012 and 12 months in the year 2013.

Obtained determined (measurable), probabilistic and heuristic data was divided into two subgroups in line with the method:

- diagnosis signals concerning technical condition D_k
- signals concerning environmental impacts on the object U

All measured diagnosis signals (D_k) were obtained from diagnostician: D_{M1} - average fuel consumption without heating [l/100km]; D_{M2} - average fuel consumption with heating [l/100km]; D_{M3} - fogging value during engine blowout [1/m]; D_{M4} - fogging value during engine operation [1/m]; D_{M5} - peak value of noise level[dB]; D_{M6} - average value of noise level[dB]; D_{M7} - value of the braking force on the front axle - left wheel [kN]; D_{M8} – value of the braking force on the front axle - right wheel [kN]; D_{M9} - percentage difference between D_{M7} and D_{M8} [%]; D_{M10} - value of braking force on the rear axle - left wheel [kN]; D_{M11} - value of braking force on the rear axle - right wheel [kN]; D_{M12} - percentage difference between D_{M10} and D_{M11} [%]; D_{E1} – clatters; D_{E2} – spontaneous engine stalls (on neutral gear); D_{E3} – engine stalls (during operation); D_{E4}– bus mileage

In order to obtain knowledge about environment from specialists (experts) was developed a detailed questionnaire to communicate with users – bus drivers. This gave information about the quality and conditions of use of the object. [6]

Research was undertaken on group of 20 experts, whose tenure as drivers was at least 5 years. [6, 17, 18, 19]. The signals collected were:

U_{K1} – driver's tenure [years]; U_{K2} – number of work hours in month [h]; U_{K3} – driving fluency (braking, accelerating); U_{D1} – number of stops; U_{D2} – route length; U_{D3} – surface and formation; U_{P1} – environment temperature; U_{P2} – wind; U_{P3} – atmospheric pressure; U_{P4}– rainfall.

These signals are of varying physical nature and thus were in each case should be transformed to point value.

The conversion procedure for point values and summarizing tables with importance of information D_{ki} and U_i expressed in expert point weights for assessment of engine and its environment condition in 2012 is presented in the previous article [6].

Table 1 shows the recorded signals D_k and U of bus 301 in 2013.

Table 1. Combination of diagnostic and environment signals in physical form.

Operation signals – 2013 year																
Month	D_{M1}	D_{M2}	D_{M3}	D_{M4}	D_{M5}	D_{M6}	D_{M7}	D_{M8}	D_{M9}	D_{M10}	D_{M11}	D_{M12}	D_{E1}	D_{E2}	D_{E3}	D_{E4}
1	37,2	37,2	0,42	0,03	96	93	14	14,8	5,41	16,2	15,3	5,56	1	0	0	243747
2	37,2	37,2	0,42	0,03	96	93	14	14,8	5,41	16,2	15,3	5,56	0	0	0	246125
3	37,2	37,2	0,42	0,03	96	93	14	14,8	5,41	16,2	15,3	5,56	2	0	0	248577

4	37,2	37,2	0,42	0,03	96	93	14	14,8	5,41	16,2	15,3	5,56	0	3	0	250992
5	37,2	37,2	0,42	0,03	96	93	14	14,8	5,41	16,2	15,3	5,56	1	0	0	253490
6	37,2	37,2	0,42	0,03	96	93	14	14,8	5,41	16,2	15,3	5,56	0	0	1	255855
7	34,1	34,1	0,47	0,05	93	91	12,6	13,3	5,26	14,8	14,5	2,03	0	0	0	258256
8	34,1	34,1	0,47	0,05	93	91	12,6	13,3	5,26	14,8	14,5	2,03	1	1	0	260695
9	34,1	34,1	0,47	0,05	93	91	12,6	13,3	5,26	14,8	14,5	2,03	0	0	0	263183
10	36,2	36,2	0,45	0,03	94,4	92	11,9	12,3	3,25	13,3	15,7	15,29	0	0	0	265524
11	36,2	36,2	0,45	0,03	94,4	92	11,9	12,3	3,25	13,3	15,7	15,29	0	3	2	267835
12	36,2	36,2	0,45	0,03	94,4	92	11,9	12,3	3,25	13,3	15,7	15,29	0	0	0	270190

Environment signals – 2013 year										
Month	U_{K1}	U_{K2}	U_{K3}	U_{D1}	U_{D2}	U_{D3}	U_{P1}	U_{P2}	U_{P3}	U_{P4}
1	4	141	3	4	14,6	2	-2,67	15,91	992,30	little rainfall
2	4	148	3	4	14,6	2	-0,03	14,38	994,71	little rainfall
3	4	151	4	4	14,6	2	0,19	13,37	992,98	little rainfall
4	4	165	4	4	14,6	2	11,4	8,49	995,84	no significant rainfall
5	24	165	8	21	9,4	1	21,61	8,19	990,74	no significant rainfall
6	24	168	8	21	9,4	1	25,06	6,14	993,67	little rainfall
7	24	153	4	21	9,4	1	26,41	7,79	1009,49	no significant rainfall
8	24	153	4	21	9,4	1	26,35	6,95	1017,16	no significant rainfall
9	24	174	5	21	9,4	1	17,5	8,95	1012,36	little rainfall
10	24	182	5	21	9,4	1	14,1	8,64	999,67	no significant rainfall
11	24	157	5	21	9,4	1	7,23	8,96	993,58	no significant rainfall
12	24	153	6	21	9,4	1	2,70	11,97	998,91	no significant rainfall

Values of signals UP1÷ UP3 in Table 1 were taken as average values from whole month, whereas values of signal UP4 were most frequently occurring. In Table 2 total number of points were obtained by daily change of observed signals to point values which were subsequently, at the end of the month, summed and divided by number of days in the month.

Based on data from Table 1. and including experts weight [6] point forms of signals were obtained – Table 2 for 2013 year. Data for other two buses was filled analogically.

Table 2. Combination of diagnostic and environment signals in point form.

Diagnostic signals in point form – 2013 year													
Month	D_{M1P}	D_{M2P}	D_{M3P}	D_{M4P}	D_{M5P}	D_{M6P}	D_{M7P}	D_{M8P}	D_{M12P}	D_{E1P}	D_{E2P}	D_{E3P}	D_{E4P}
1	15	15	10	10	5	5	30	30	0	8	0	0	4
2	15	15	10	10	5	5	30	30	0	0	0	0	4
3	15	15	10	10	5	5	30	30	0	8	0	0	4
4	15	15	10	10	5	5	30	30	0	0	8	0	4
5	15	15	10	10	5	5	30	30	0	8	0	0	4
6	15	15	10	10	5	5	30	30	0	0	0	12	4
7	15	15	10	10	5	5	30	30	0	0	0	0	4
8	15	15	10	10	5	5	30	30	0	8	4	0	4
9	15	15	10	10	5	5	30	30	0	0	0	0	4
10	15	15	10	10	5	5	30	45	0	0	0	0	4
11	15	15	10	10	5	5	30	45	0	0	8	12	4
12	15	15	10	10	5	5	30	45	0	0	0	0	4

Environment signals in point form – 2013 year										
Month	U_{K1P}	U_{K2P}	U_{K3P}	U_{D1P}	U_{D2P}	U_{D3P}	U_{P1P}	U_{P2P}	U_{P3P}	U_{P4P}
1	3	12	9	4	4	12	2,00	1,00	1,32	2,84
2	3	12	9	4	4	12	2,00	1,00	1,54	2,75

3	3	12	9	4	4	12	2,00	1,03	1,39	3,03
4	3	12	9	4	4	12	1,60	1,00	1,57	2,27
5	9	12	9	12	2	6	1,71	1,00	1,23	2,32
6	9	12	9	12	2	6	2,07	1,00	1,43	2,53
7	9	12	9	12	2	6	2,16	1,00	2,00	2,06
8	9	12	9	12	2	6	2,26	1,00	2,13	1,84
9	9	12	9	12	2	6	1,33	1,00	2,13	2,67
10	9	18	9	12	2	6	1,13	1,00	1,77	1,57
11	9	12	9	12	2	6	1,77	1,00	1,37	2,23
12	9	12	9	12	2	6	2,00	1,00	1,68	1,77

3 Algorithm of determining parameters of technical and operation condition of technical object

Object (engine) in its environment is described by coupled state equations including initial conditions describing relations between technical condition and operation condition [6, 7, 9]. It should be noticed that, in these equations diagnostic signals D_k are environment for operation signals U and operation signals U are environment for diagnostic signals D_k.

$$\frac{dU}{dt} = a_R U + b_{R1} D_k + b_{R0} \tag{1}$$

$$\frac{dD_k}{dt} = a_T D_k + b_{T1} U + b_{T0} \tag{2}$$

where: U – operation signal; D_k - diagnostic signal of the bus; a_R - parameter of the usage condition depends primarily on object work and of the object diagnostic signals D_k impact on it;
b_{R1}, b_{R0} - the impact of diagnostic signals parameters and initial conditions on its operating status; a_T – parameter of the technical condition of the object, which is dependent on diagnostic signals and signals resulting from acting environment; b_{T1}, b_{T0} - the parameters of interaction between controlling signals (action) and initial conditions on the technical condition of the bus.
Equation (1) and (2) might be written as follows:

$$\frac{\Delta U}{\Delta \theta} = a_R U + a_R \hat{a}_{R1} D_k + a_R \hat{a}_{R0} \tag{3}$$

$$\frac{\Delta U}{\Delta \theta} = a_R (U + \hat{a}_{R1} D_k + \hat{a}_{R0}) \tag{4}$$

$$\frac{\Delta D_k}{\Delta \theta} = a_T D_k + a_T \hat{a}_{T1} U + a_T \hat{a}_{T0} \tag{5}$$

$$\frac{\Delta D_k}{\Delta \theta} = a_T (D_k + \hat{a}_{T1} U + \hat{a}_{T0}) \tag{6}$$

According to rules of statistical identification $\dot{D}_k = 0$ from equations (2, 5, 6) we obtain:

$$\hat{a}_{T1} = \frac{n \sum_{i=1}^{n} D_{k_i} U_i - \sum_{i=1}^{n} D_{k_i} \sum_{i=1}^{n} U_i}{n \sum_{i=1}^{n} U_i^2 - \sum_{i=1}^{n} U_i \sum_{i=1}^{n} U_i} \tag{7}$$

$$\hat{a}_{T0} = \frac{\sum_{i=1}^{n} D_{k_i} \sum_{i=1}^{n} U_i^2 - \sum_{i=1}^{n} D_{k_i} U_i \sum_{i=1}^{n} U_i}{n\sum_{i=1}^{n} U_i^2 - \sum_{i=1}^{n} U_i \sum_{i=1}^{n} U_i} \tag{8}$$

and ultimately sought parameter of technical condition a_T:

$$a_T = \frac{\Delta D_k}{\frac{\Delta\Theta}{resurs}(D_k + \hat{a}_{T1}U + \hat{a}_{T0})} \tag{9}$$

Parameter a_T characterizes object technical condition and depends on diagnostic signals (D_k) and signals coming from driver's and environment influence as well as initial conditions $\hat{a}_{T0}$.

According to rules of statistical identification $\dot{U} = 0$ from equation (1, 3, 4) we obtain:

$$\hat{a}_{R1} = \frac{n\sum_{i=1}^{n} D_{k_i} U_i - \sum_{i=1}^{n} D_{k_i} \sum_{i=1}^{n} U_i}{n\sum_{i=1}^{n} D_{k_i}{}^2 - \sum_{i=1}^{n} D_{k_i} \sum_{i=1}^{n} D_{k_i}} \tag{10}$$

$$\hat{a}_{R0} = \frac{\sum_{i=1}^{n} U_i \sum_{i=1}^{n} D_{k_i}{}^2 - \sum_{i=1}^{n} D_{k_i} U_i \sum_{i=1}^{n} D_{k_i}}{n\sum_{i=1}^{n} D_{k_i}{}^2 - \sum_{i=1}^{n} D_{k_i} \sum_{i=1}^{n} D_{k_i}} \tag{11}$$

and ultimately sought parameter of the object operation condition a_R:

$$a_R = \frac{\Delta U}{\frac{\Delta\Theta}{resurs}(U + \hat{a}_{R1}D_k + \hat{a}_{R0})} \tag{12}$$

Parameter a_R characterizes object operation condition and depends on operation (utility) signals coming from driver action and environment U, diagnostic signals D_k and initial conditions $\hat{a}_{R0}$.

Signals D_k and U in homogenous point form [6] were transformed to resultant diagnostic signal (D_k) and environment (U).

$$D_k = \sqrt{D_{M1P}^2 + D_{M2P}^2 + D_{M3P}^2 + D_{M4P}^2 + D_{M5P}^2 + +D_{M6P}^2 + D_{M7P}^2 + D_{1EP}^2 + D_{2EP}^2 + D_{3EP}^2 + D_{4EP}^2} \tag{13}$$

$$U = \sqrt{U_{K1P}^2 + U_{K2P}^2 + U_{K3P}^2 + U_{D1P}^2 + U_{D2P}^2 + U_{D3P}^2 + U_{P1P}^2 + U_{P2P}^2 + U_{P3P}^2 + U_{P4P}^2} \tag{14}$$

Using data from Table 2 and equations 1÷12, increases in technical condition described by parameter a_T and operation (regulation) condition described by parameter a_R are calculated.

Subsequently $\sum a_T$ describing level of object wear and $\sum a_R$ describing resource of proper operation possibility are determined.

In Table 3, row 1 data was implemented from previous year thus allowing determination of parameters a_T and a_R starting from February (position no. 3).

Results of courses of normalized parameters a_T and a_R changes for bus no. 301 were presented on Figure 1, for bus no. 303 on Figure 4 and for bus no. 304 on Figure 7.

Figure 2 (from 2012 year) and Figure 3 (from 2013) shows monthly changes (increases) values of these parameters values for the bus 301.

Figure 5 (from 2012 year) and Figure 6 (from 2013) shows monthly changes (increases) values of these parameters values for the bus 303.

Figure 8 (from 2012 year) and Figure 9 (from 2013) shows monthly changes (increases) values of these parameters values for the bus 304.

Table 3. Procedure of calculating technical condition and regulation parameters for bus no. 301 in 2012 year [6].

Month	Operation time θ	θ	D_K	U	$\Delta\theta$	ΔD_K	ΔU	a_T	a_R	Δa_T	Δa_R
1	213429	2200	54,28	22,98	2200	54,28	22,98				
2	215582	2153	50,16	21,05	4353	104,44	44,03	239,16	240,28		
3	217982	2400	50,32	22,52	6753	154,76	66,55	223,67	222,38	-15,49	-17,91
4	220102	2120	50,16	22,39	8873	204,92	88,93	227,26	225,99	3,59	3,61
5	222669	2567	51,58	21,30	11440	256,49	110,24	219,59	221,95	-7,67	-4,04
6	225125	2456	50,79	24,43	13896	307,29	134,67	215,74	207,02	-3,85	-14,93
7	227358	2233	50,79	28,70	16129	358,08	163,36	217,88	194,61	2,14	-12,41
8	229758	2400	50,16	23,12	18529	408,24	186,48	217,73	216,70	-0,15	22,09
9	232273	2515	50,16	25,33	21044	458,40	211,81	215,63	205,59	-2,11	-11,11
10	234571	2298	50,16	25,26	23342	508,56	237,07	215,89	206,84	0,26	1,25
11	236699	2128	93,20	23,35	25470	601,76	260,42	159,62	219,24	-56,27	12,40
	239047	2348	91,47	23,37	27818	693,22	283,79	166,81	218,42	7,19	-0,82
12	241392	2345	91,47	27,01	30163	784,69	310,81	166,77	199,93	-0,04	-18,48

Table 4. Procedure of calculating technical condition and regulation parameters for bus no. 301 in 2013 year.

Month	Operation time θ	θ	D_K	U	$\Delta\theta$	ΔD_K	ΔU	a_T	a_R	Δa_T	Δa_R
1	241392	2345	91,47	27,01	30163	784,69	310,81	166,77	199,93	-0,04	-18,48
2	243747	2355	50,79	20,61	32518	835,48	331,42	250,53	261,92	83,77	61,99
3	246102	2378	50,16	20,61	34896	885,64	352,03	249,19	260,24	-1,34	-1,68
4	248480	2452	50,79	20,64	37348	936,44	372,68	244,59	254,94	-4,60	-5,31
5	250932	2415	50,79	20,52	39763	987,23	393,20	242,99	252,92	-1,60	-2,01
6	253347	2498	50,79	22,38	42261	1038,03	415,58	230,47	239,28	-12,52	-13,64
7	255845	2365	51,58	22,44	44626	1089,60	438,02	227,55	235,88	-2,93	-3,40
8	258210	2401	50,16	22,45	47027	1139,76	460,47	229,43	237,92	1,89	2,04
9	260611	2439	50,95	22,45	49466	1190,71	482,92	226,60	234,65	-2,83	-3,27
10	263050	2488	50,16	22,46	51954	1240,87	505,38	226,95	235,00	0,35	0,35
11	265538	2341	60,34	26,04	54295	1301,21	531,42	193,24	197,93	-33,71	-37,07
12	267879	2311	62,04	22,38	56606	1363,25	553,80	205,76	210,65	12,52	12,72
13	270190	2355	60,34	22,38	58961	1423,59	576,18	208,97	213,95	3,21	3,30

First row from Table 4 was taken from year 2012 (last row in Table 4). It gives the opportunity to determine parameters a_T and a_R from January 2013 (position no. 2).

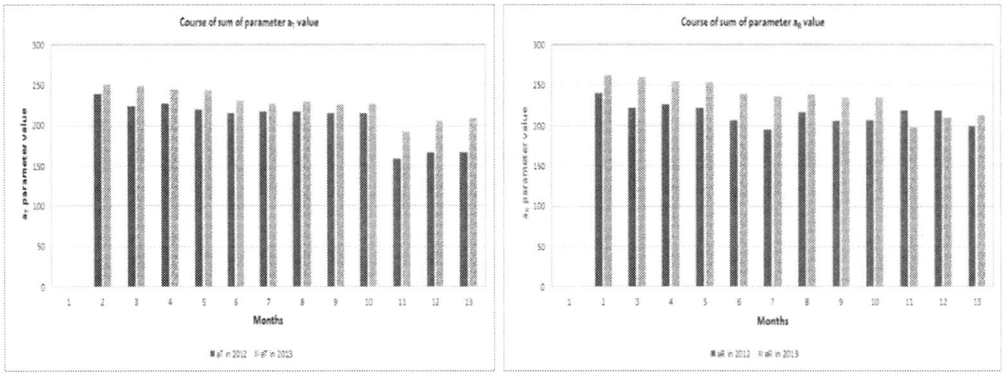

Fig. 1. Changes of normalized parameter of technical condition and regulation for bus no. 301 in year 2012 and 2013.

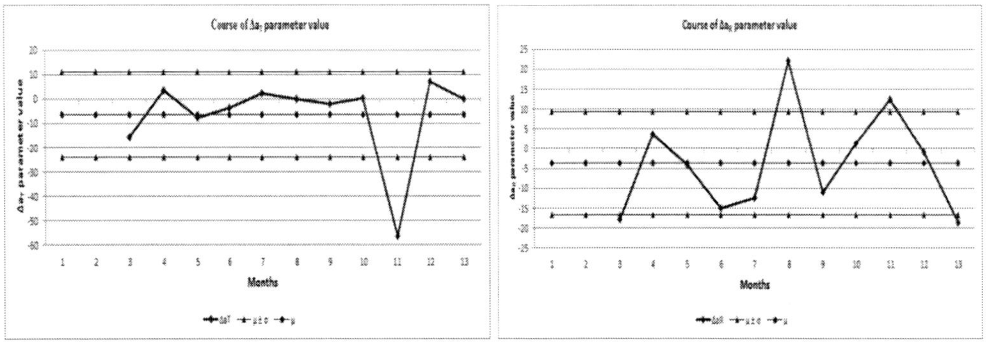

Fig. 2. Course of value of technical condition and regulation parameter increase for bus no. 301 in year 2012.

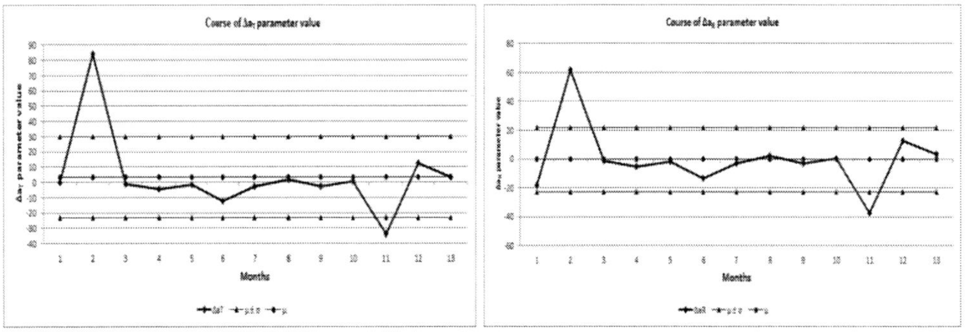

Fig. 3. Course of value of technical condition and regulation parameter increase for bus no. 301 in year 2013.

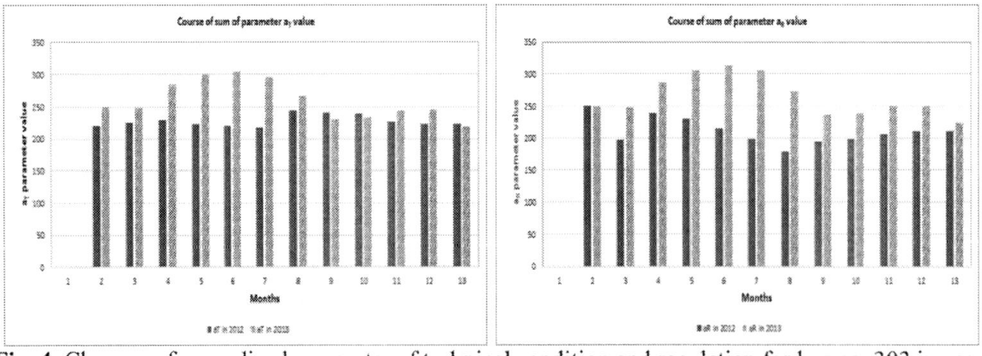

Fig. 4. Changes of normalized parameter of technical condition and regulation for bus no. 303 in year 2012 and 2013.

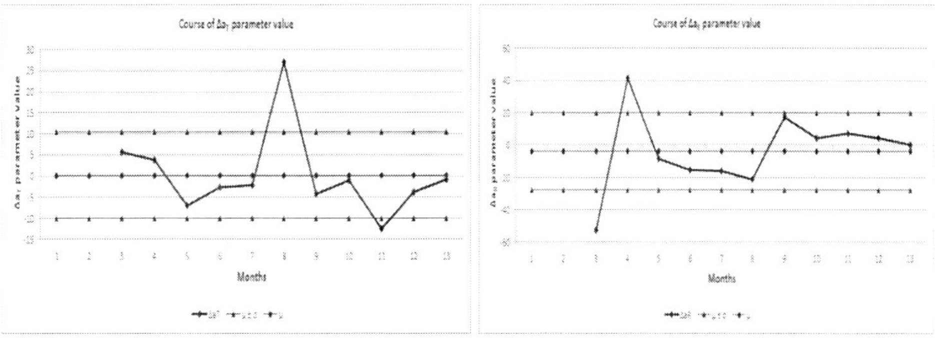

Fig. 5. Course of value of technical condition and regulation parameter increase for bus no. 303 in year 2012.

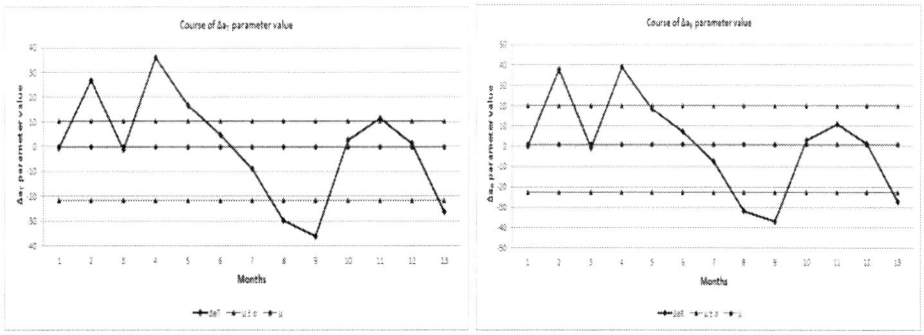

Fig. 6. Course of value of technical condition and regulation parameter increase for bus no. 303 in year 2013.

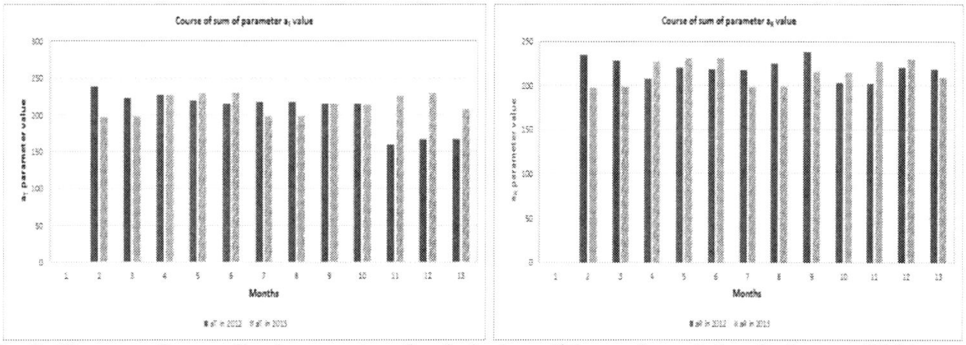

Fig. 7. Changes of normalized parameter of technical condition and regulation for bus no. 304 in year 2012 and 2013.

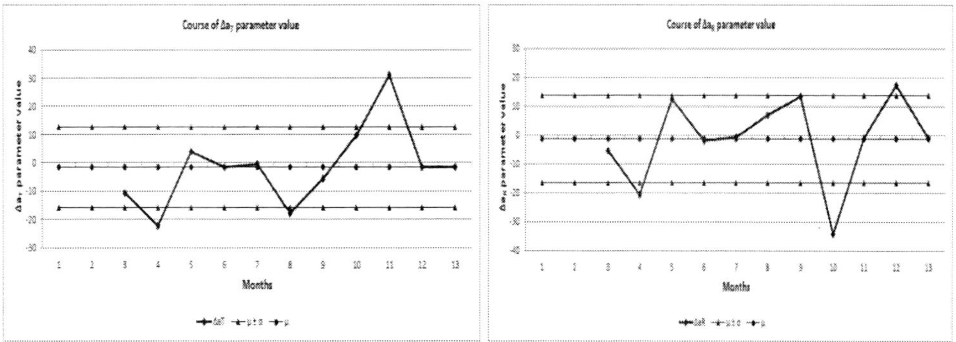

Fig. 8. Course of value of technical condition and regulation parameter increase for bus no. 304 in year 2012.

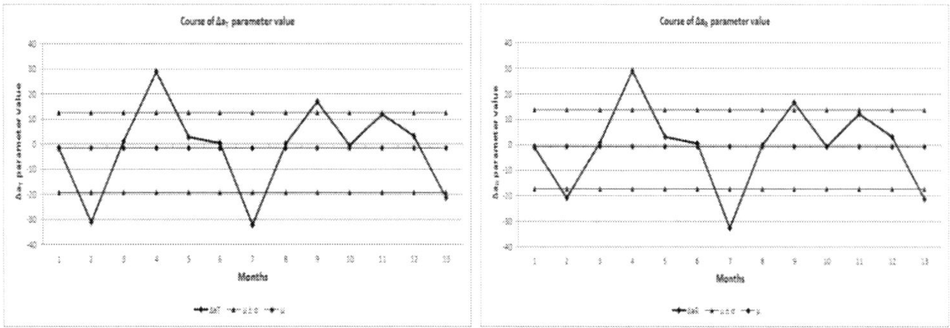

Fig. 9. Course of value of technical condition and regulation parameter increase for bus no. 304 in year 2013.

Table 5. Summary table.

Year 2012	bus no. 301	bus no. 303	bus no. 304
sum of a_T	166,77	222,40	228,44
parametrical damage $m_T(t)$	1	2	3
sum of a_R	199,93	211,03	219,00
momentary damage $m_R(t)$	4	2	3

Year 2013	bus no. 301	bus no. 303	bus no. 304
sum of a_T	208,97	219,47	224,23
parametrical damage $m_T(t)$	2	4	5
sum of a_R	213,95	224,04	237,12
momentary damage $m_R(t)$	2	6	5

On the base of data from Table 3, 4 and charts 1, 4, 7 the following might be noted: the best technical and regulation condition is that of bus no. 301 (2012 year). Values of sum of a_R and a_T are the least, thus statement can be made that wear of bus no. 301 is the least and regulation potential was used to the least extent. Its changes in the years 2012 and 2013 are resulting from their use. Buses 303 and 304 (still worse than 301) have improved their performance in 2013 compared to 2012 due to the improvement in operating conditions (more frequent diagnostic tests, better service)

Obtained results were confirmed by discerning analysis of analytical assessments of aptitude of aforementioned buses by direct method realized by experts (diagnosticians – drivers in years 2012 and 2013).

Excess of increases values of technical condition and operation condition from the value of standard deviation might be observed in Figures 2, 3, 5, 6, 8, 9. It means that in case when increase of technical condition is greater than standard deviation threshold value, parametrical damage occurred, whereas in case of operation parameter exceeding standard deviation threshold value, momentary damage occurred.

Parametrical and momentary damage do not disqualify object from further exploitation, but may be the base for determining reliability characteristics before the occurrence of dangerous catastrophic damage. The sum the parameters a_T, a_R and the number of parametric damages and failures and momentary damages in 2012 and 2013 are shown in Table 5.

The designated number of buses 301, 303 and 304 defects is slightly larger in 2013 than in 2012 year. That means that their condition deteriorates, but these changes are not large. Finally, it can be concluded that all tested buses (301, 303, 304) may still be safely used.

4 Conclusions

This paper is a continuation of the works contained in [6]. The basis of innovatory method of using varying diagnostic data (in form of points) are presented.
Each change of signal (depending on its value and moment of occurrence) corresponds with number of points predetermined by the experts. So far in such method of aptitude analysis, all the points were summed and subsequently a range was determined in which is the object, its condition and level of exploitation. Proposed method is based on using signals in form of points to determine parameters of regulation and technical condition (equations 9, 12) from coupled state equations [3, 5] with initial conditions [3,4] and subsequently reliability and safety of the object during exploitation (Figures 2, 3, 5, 6, 8, 9). The described method may prove to be very useful due to unequivocal presentation of object global exploitation condition using parameters a_T, a_R and the number of damages $m_T(t)$ and $m_R(t)$. Thus allows to predict how the object should be used and when the object should be serviced (repaired, overhauled). This method can be especially useful during the operation of the numerous set of objects.

This paper was supported by the research project No. S/WM/1/2016 and financed by the Ministry of Science and Higher Education.

References

1. L. Bukowski, *Prognozowanie niezawodnoś ci i bezpieczeń stwa systemów zautomatyzowanych.* Proceedings of the 21st School of Reliability (Szczyrk 2003)
2. C. Cempel, HG. Natke, *Damage Evolution and Diagnosis in Operating Systems.* Safety Evaluation Based on Identification Approaches Related to Time-Variant and Nonlinear Structures. Springer, pp. 44-61 (1993)
3. J. Filipczyk, *Faults of duty vehicles in the aspects of securing safety.* Transport Problems, **6**, Issue 1, pp. 105-110 (2011)
4. K. Golak, P. Lindstedt, R. Gradzki, *Studies of the jet engine control quality based on its response to the disturbance inflicted on the object, designated from its response to the set point inflicted to the controller.* In: Safety and Reliability: Methodology and Applications - Proceedings of the European Safety and Reliability Conference, pp: 137-140 (2015)
5. I. Gołda, M. Izdebski, A. Podviezko, *Assessment of efficiency of assignment of vehicles to tasks in supply chains: a case study of a municipal company.* Transport, **32** Issue 3, pp. 243–251 (2017) https://doi.org/10.3846/16484142.2016.1275040
6. R. Grą dzki, P. Lindstedt, *Method of assessment of technical object aptitude in environment of exploitation and service conditions.* Eksploatacja i Niezawodnosc - Maintenance and Reliability, **1**; 54-63 (2015), http://dx.doi.org/10.17531/ein.2015.1.8
7. R. Grą dzki, P. Lindstedt, K. Golak, *Premises of evaluation of the technical object suitability with including the quality of its maintenance and operation, and their initial conditions.* In: Safety and Reliability: Methodology and Applications - Proceedings of the European Safety and Reliability Conference, chapter 44; pp. 319 – 326 (2015)
8. AKS Jardine, D. Lin, D. Banjevic, *A review on machinery diagnostics and prognostics implementing condition-based maintenance.* Mechanical Systems and Signal Processing, **20**, Issue 7, pp. 1483-1510 (2006)
9. P. Lindstedt, T. Sudakowski, *The Method of Assessment of Suitability of the Bearing System Based on Parameters of Technical and Adjustment State.* Solid State Phenomena Mechatronic systems and materials V, **199** , pp. 73-78 (2013)
10. H. Madej, J. Filipczyk, *The methods of assessment of car technical condition regarding environmental protection.* Journal of KONES, **16**, Issue 2, pp. 103-108 (2009)

11. T. Nowakowski, *Reliability Model of Combined Transportation System.* Probabilistic Safety Assessment and Management. Springer, pp. 2012-2017 (2004) http://doi.org/10.1007/978-0-85729-410-4_323
12. H. Sarangaa, J. Knezevicb, *Reliability prediction for condition-based maintained systems.* Reliability Engineering & System Safety. Elsevier, **71**, Issue 2, pp. 219–224 (2001)
13. J. Rymarz, A. Niewczas, A. Krzyzak: *Comparison of operational availability of public city buses by analysis of variance.* Eksploatacja i Niezawodnosc – Maintenance and Reliability; **18**, Issue 3, pp. 373–378 (2016) http://dx.doi.org/10.17531/ein.2016.3.8.
14. J. Rymarz, A. Niewczas, J. Stoklosa, *Reliability evaluation of the city transport buses under actual conditions.* Transport and Telecommunication, **16**, no. 4, pp. 259–266 (2015), DOI 10.1515/ttj-2015-0023
15. E. Zio, *Reliability engineering: Old problems and new challenges.* Reliability Engineering and System Safety, **94**, Issue 2, pp. 125–141 (2009), http://dx.doi.org/10.1016/j.ress.2008.06.002
16. S. Szawłowski, *Przeglą d kontrolny ASPA w systemie obsługiwania ś migłowca pokładowego SH-2G.* In: AIRDIAG conference Warsaw, Poland, 27–28 October 2005.
17. J. Jaź wiń ski, J. Szpytko, *Zasady wyznaczania zespołu ekspertów w badaniach niezawodnoś ci i bezpieczeń stwa urzą dzeń technicznych.* XXXIV Zimowa Szkoła Niezawodnoś ci PAN "Niekonwencjonalne metody oceny trwałoś ci i niezawodnoś ci". Szczyrk: pp. 157-167 (2006)
18. M. Woropay, Ł. Muś lewski, M. Ś lę zak, M. Szubartowski, *Assessment of the impact of human on safety of transportation system operation.* Journal of KONBiN, **1** No. 25, pp. 97-106 (2013)
19. M. Zają c, *Wykorzystanie badań ankietowych do oszacowania niezawodnoś ci systemu transportu intermodalnego.* XXXIV Zimowa Szkoła Niezawodnoś ci PAN "Niekonwencjonalne metody oceny trwałoś ci i niezawodnoś ci". Szczyrk, 2006.

Methodology of machine destruction testing

Vladimir Boykov[1], *Aleksandr* Povarecho[1,*]

[1] BNTU University of Minsk, Faculty of Mechanical Engineering, Belarus

Abstract. This paper presents selected problems connected with automation of procedures involved in assessment of machine degradation degree using vibration method with special emphasis on the machine state prognosis. The current knowledge of these problems is not sufficient and needs further research on data processing, analysis of efficiency of diagnostic and prognostic procedures, collection and selection of diagnostic parameters and development of automatic procedures for recognition and prognosis of a machine state. New solutions and different aspects of diagnostic prognosis based on the proposed partial procedures focus on factors determining automation of procedures for identification of technical systems states. New automated procedures for acquisition and processing of symptoms indicating the machine state provide better possibilities of control and supervision of technical systems operation and maintenance through identification of their current states, and its good prognosis.

1 Introduction

Increasing demand for methods and means of technical diagnostics, that is tools for assessment of technical systems degradation states, meets the needs of contemporary entrepreneurs who use a variety of technical appliances to run their businesses. Destruction processes affecting technical systems have a direct influence on their operation safety and therefore it is necessary to monitor their technical state [1,10,23]. Methods and means of modern technical diagnostics are tools for diagnosing states of technical systems which provides the basis for making operational decisions [4].

Constantly changing demands and new possibilities have generated new categories of research problems, highlighted others and disqualified some issues which have lost their importance in terms of application due to [15,22]:

- access to advanced world technologies;
- availability of the newest generation research equipment;
- possibilities of the latest computer applications in the field of hardware and software;
- access to databases, capital market and vast possibilities of cooperation.

All this have diametrically changed the views and achievements in automation of detection and control of technical object state changes by methods of technical diagnostics, especially in the field of shaping the current and predicted safety of technical objects. This provides the possibility of computerization of the processes connected with an object operation, supervision of the state changes, location of faults and minimization of damage effects and safety threats [12,21,24].

Technical diagnostics of mechanical engineering is based on utilization of information about the machine changes which enables identification of safety threats and progressing destruction of a machine throughout its life cycle. Changes of the state, reflected by methods of technical diagnostics, prevent from occurrence of causes and effects of damage [2,3,16]. A factor which boosts development of machine diagnostics is the responsibility of the function it performs reduction of the following threats: threat to human life and health,

* Corresponding author: POVAREKHO@bntu.by

threat to natural and technical environment, threats to operation quality and economic risks [8,20,25]. The need to assess the technical condition of a machine results from the necessity of making decisions on the machine quality and its further operation. It can be a decision concerning its further use, taking preventive actions or altering its structure, use of technologies or maintenance methods.

This work deals with selected problems connected with the machine state control and monitoring in time, focusing the machine operation process and the final product quality shaping, diagnostic modeling and control of a machine usability in its operation and maintenance by means of prognosis methods – in terms of developing faults. Computer aided modeling of an object lifetime, degradation processes, operation states, operation control and decision support systems – are the main issues to be addressed to provide technical systems with efficient operation.

2 Description of the machine wear process

Occurrence of faults is one of events that accompany the process of machine operation and maintenance and determines its reliability, efficiency and the way it is operated. After being damaged a machine (assembly, component) becomes unusable. A usable machine can perform tasks it is designed with simultaneous maintenance of parameters specified in its technical documentation. Due to the impact of the environment and performance of the assigned tasks the object initial properties can be changed in the following way [26]:

- in result of irreversible aging and wear processes that take place in the machine;
- in result of occurrence of reversible processes with different intensity, caused by exceeding permissible values of one or more forcing factors;
- in a step - like manner which is manifested by a non-constant transition of one or two features beyond the limits accepted for a given machine.

The major causes of damage occurrence are classified as follows:

a) structural - faults due to design and construction errors, most often when extreme loads are neglected;

b) production related (technological) - faults which occur due to errors and imperfections of technological processes (no dimensions tolerance, insufficient surface smoothness, thermal treatment, etc.) or faults of materials used in the object or its structural components;

c) operational - faults which occur in effect of failing to observe respective rules of operation or due to the impact of external factors unexpected for the object service conditions;

d) due to aging and wear - always accompany operation of objects and are caused by irreversible changes that impair strength and cooperation of components.

In the energy-based approach, the object damage occurs under the influence of energy transmitted by it. Depending on the kind of energy prevailing in given conditions can be divided into:

a) mechanical — static stresses, creeping, fatigue, pitting, friction wear;
b) chemical — metal corrosion, rubber, paint and insulation aging, wood rotting;
c) electrical — electro-corrosion;
d) thermal — hyper melting, intensification of occurring phenomena.

The process of wear and degradation of functional qualities is not homogenous and can be divided into four physically different degradation processes. In the first place, it is fatigue wear that should be mentioned as it leads to loss of cohesion of the structure components and is caused by too high dynamic stresses – oscillatory, that is vibrations. The second is friction wear in all forms which occurs in kinematic pairs and leads to losses in mass – backlashes. Rheological properties of construction materials cause the third degradation process involving creeping (floating) leading to changes in the size and shape of the structure stressed elements. The fourth way of erosive wear of machine components involves physical (cavitation) or chemical (corrosion) degradation resulting in the material mass loss.

This work takes advantage of the basic assumption, already proven in many studies, that knowledge of relations between the machine model described by parameters: m, k, c (m – mass, k – stiffness, c – damping), and quantities a, v, x (a – acceleration, v – velocity, x – displacement) describing the vibration process provides big possibilities and vast applications of vibration-based diagnostics.

Tribology deals with all the phenomena connected with a machine wear. It assumes on proportionality of vibrations D to the wear processes advancement [4,5]. Creation of a mathematical model of a machine about all the acoustic and wear phenomena is a tribology approach. Advanced wear of a machine – tribological E_z is best measured by energy lost for irreversible wear processes - tribological. Having in mind that a new machine (with no wear) is already characterized by some level of rated vibration D_n, the constitutive relations of diagnostics can be expressed as follows:

$$\mathbf{D} = \mathbf{D_n} + \gamma\, \mathbf{E_z} \tag{1}$$

where: D - amplitude measurement of vibration – e.g. acceleration, velocity etc.; D_n - level of rated vibration of a new machine, E_z – energy lost for irreversible processes of wear, y – coefficient of tribovibroacoustic lossiness (TWA) depicting correlation between tribological processes and vibrations [4].

Knowing the machine mechanical efficiency η it is possible to estimate lost energy E_z, as:

$$\eta = \frac{N_{uż}}{N_{zas}} = \frac{N_{zas} - N_z}{N_{zas}} = \frac{E_{zas} - E_z}{E_{zas}} \tag{2}$$

where: $N_{UŻ}$ – is useful power , N_{zas} – feed power, E- respective energies.

Tribological processes which most frequently accompany machine motion include fatigue and friction, whereas their energy is generated proportionally to the fatigue crack length $\boldsymbol{l}$ and the volume of the worn material U. Focusing on the two phenomena it can be accepted:

$$E_z = E_{tarcia} + E_{zmęczeń} + \ldots \cong al + bU \tag{3}$$

whereas: a and b – certain coefficients of proportionality, l - computed from Paris formula [12], U - from Archard formula [17].

Computing fatigue crack length from Kocańda formula [8] is:

$$l = C_3(\sigma_0 + \sigma_a)^n N = C_3(\sigma_0 + \sigma_a)^n \cdot f \cdot \theta = C_3(\sigma_0 + eD)^n \cdot f \cdot \theta \tag{4}$$

where: e – coefficient of vibration proportionality, D - to variable stresses σa, Θ – time of their operation, f – vibration frequency.

When dynamic component of thrust $p = p_0 + g\,D$ with coefficient of proportionality to vibration amplitude g is introduced into Archard [17] equation we receive:

$$U = \frac{k_0 \cdot v \cdot \theta}{R_e}(p_0 + gD) \tag{5}$$

Substituting E_z to expression (3) describing energy of wear processes, and this in turn to constitutive relation (1) we receive:

$$D = D_n + \gamma\left[aC_3(\sigma_0 + eD)^n \cdot f + \frac{bk_0v}{R_e}(p_0 + gD)\right]\cdot\theta = D_n + \gamma N_z\theta$$

$$N_z = aC_3(\sigma_0 + eD)^n \cdot f + \frac{bk_0v}{R_e}(p_0 + gD) \quad (6)$$

into which the power of wear processes NZ = E_z/Θ depending on vibration amplitude D is introduced.

Using these dependencies for final determination of the machine vibration amplitude in a function of its operation time we can find:

$$D = \frac{\gamma N_z^0 \cdot \theta}{1-\gamma\theta \frac{dN_z}{dD}\big|_{D=D_n}} + D_n \quad (7)$$

Analysis of the final formula of this tribovibroacoustic model involves identification of two characteristic points. For time $\Theta = 0$, the above formula yields $D = D_n$, that is the amplitude of rated vibrations for a new object, whereas for:

$$\theta = \theta_{aw} = (\gamma\frac{dN_N}{d}\Big|_{D=D_n})^{-1} \quad (8)$$

we have D = ∞ , which can be interpreted as a failure. Value of time to failure:

$$\theta_{Aw} = \left(\frac{dN_z}{dD}\Big|_{D=D_n}\right)^{-1} \quad (9)$$

is inversely proportional to tribo-vibro-acoustics loss γ, and sensitivity of wear to vibrations: dN_z/dD.

The value of vibration amplitude D and the rate of their changes during operation Θ depend on these characteristics as well as on the total power of machine tribological processes.

Both the model and the insight into vibration feedback of wear processes it provides are very valuable. This model shows that the time of failure ΘAw can be predicted for a new machine knowing only the vibration gradient of lost power:

$$\frac{dN_z}{dD}\Big|_{D=D_n} \quad (10)$$

and its lossine TWA γ. Sequential utilization of the model for prediction of a machine state and its vibrations is possible as well [6,9].

In practice however, the current state of a machine can be identified by observing the object functioning, that is, its main output of converted energy (or a product) and dissipation output where residue processes e.g. thermal, vibratory, acoustic, electromagnetic are observed.

Observation of these outputs provides a vast range of possibilities of machine diagnosis through:

***observation of operational processes** by continuous monitoring their parameters or by testing the machine efficiency on special test stations (power, speed, pressure etc.),

* **studying of its product quality**, consistence of measurements, links, etc, as generally speaking the better the machine technical state the better the production quality,

* **observation of residual processes** which provide the basis for many attractive diagnostic methods based mainly on symptom-based models.

The above presented problems should be solved with the use of the newest technologies. This is an area for wide application of computer aid and automation of procedures in the field of: holistic and symptom-based modeling, inference, planning and performance of tests, utilization of artificial intelligence encompassing counselling systems and neuron networks with fuzzy logic [7,10,19].

Knowledge and technology of diagnosis are always connected with the kind of diagnostic model that can be developed for a given object.

Reduction of destructive influence of an object physical aging and fatigue wear is necessary in all stages of its existence. Desirable effects of the number of reductions in the number of technical object failures can be achieved in:

* **the field of construction** – by appropriate selection of materials and their shapes to withstand the loads, shaping unit compressions, choice of adequate materials and substances for friction pairs, elimination of dry friction, wide application of appropriate tightening, providing appropriate temperature;

* **in the field of technology** - by choosing optimal treatment, shaping an optimal surface layer, choice of appropriate thermal and thermal-chemical treatment, proper installation and adjustments;

* **in the field of operation** – by complying with frequencies and scope of services (smearing, regulation, protection from corrosion), avoiding overloads and abrupt speed changes, monitoring of technical condition.

Inappropriate operation and maintenance causes intensive impact of wear processes leading to premature failures and threats to safety.

3 Machine diagnostics throughout its life time

Each machine undergoes four stages of its existence: **validation, design, manufacturing and operation.** Increasing requirements from machines provide a set of criteria which are studied in each of these stages [9,22,24]. Methods and means providing such possibilities of control of quality of machines – fulfilment of these criteria – are the included in the field of technical diagnostics.

Technical diagnostics covers the following activities: assessment of the state, genesis and prognosis of the state. These activities are performed by intelligent and often automated diagnostic systems (mobile software and hardware with self-learning loop and risk assessment). The set of diagnostic parameters is identified from a set of input parameters accompanying the machine operation. Determination of a set of damage sensitive diagnostic parameters should take into consideration:

- ability of reflect the machine state changes in time during its operation and maintenance.
- amount of information on technical state of the transmission,
- sensitivity to parameters value during operation.

An advantage of the above presented methods is that they provide the possibility to choose, from a set of output parameters, both the one-element and multi - element sets of diagnostic parameters.

Criteria for optimization of the set of parameters include:

- parameters should characterize the destruction process and be closely related to it;
- parameters should be sensitive to changes in the object degradation process;
- the number of diagnostic parameters cannot be too big as too many of them make it difficult to identify the process of devastation and sometimes it becomes even impossible;
- diagnostic parameters should be measurable;
- the must exists reliable statistical and analytical data of the distinguished parameters.

The diagnostic system becomes a subject of separate considerations and diagnostic features of such systems require elaboration and description (with mathematical formalization). A diagnostic system is a set of diagnostic agents, set of methods and means for obtainment, processing, presentation, generation and collecting information and a set of engineering objects, their models and algorithms for diagnostics, generation and prognosis of states as well as relations between these elements. It is intended to support making reliable decisions on an object's belonging to a respective class of states. The structure of a diagnostic system proposed in fig.6 shows the basic relations between the test object and its diagnostic model as well as the system of its state identification and decision making.

Diagnostic systems belong to a class of automated computer systems and are characterized by:

- the goal of their operation is identification of the state other objects, actually they do not affect this state,
- this goal involves generating a diagnose on the object properties into information on its state.

For these reasons, the attention must be focused on the informative structure of the diagnostic system and it needs to be designed, optimized and evaluated in terms of the information flow possibility. Different forms of components and their different functions provide the possibility of creation of diagnostic systems with different structures and with diversified automation degree.

The process of a machine handling is as follows:

a) for a usable machine: testing the state – performance of necessary servicing activities –prognosis of the machine state – scheduling the next survey;

b) for an unusable machine: assessment of the machine state – finding causes of its failures (maladjustment, damage, poor quality of repairs or liquidation) – removal of failure – assessment of a repair quality – performance of servicing activities – prognosis of the state – date of the next diagnosis.

Problems connected with diagnostics are usually addressed by a diagnostic team. They define the scope, methods and means necessary to supervise machine operation mad maintenance (critical, important, unimportant). The tasks are performed in cooperation with the departments of the chief mechanic, production supervisor and manager of marketing department and accountancy department. Particular departments of a company are interested in different information items on the condition of machines, depending on the specificity machine use and production.

4 Examples of practical diagnosis of the state

The machine technical state prognosis process can be carried out by different methods [5,7,14]: forecasted symptom value, times of a machine operation or other prognoses the states of machine (e.g. extrapolation trend methods and adaptive methods) [16,20].

The selected problems of diagnosing the state of machines using forecasting methods have been validated in several different studies on operation and maintenance, and some examples of the issues and their solutions are presented selectively below.

Schematic diagram of the predictive diagnosis consists of: selection of the optimal set of diagnostic parameters, choice of the optimal method of forecasting, and assessment of predictive quality of the diagnosis. The optimal set of diagnostic parameters guarantees:

1. the minimum number of points needed to run the time prediction;
2. changes in units of the machine during operation resulting from replacement or adjustment of teams and variable operating conditions (road conditions, climate, quality of handling) that cause abrupt changes in the value of diagnostic parameters;
3. the maximum value of the time step;
4. the cardinality of the set of relevant parameters to the application.

The choice of a forecasting method is conditioned by: the forecast horizon; the minimum number of points needed to run the time prediction; the number of measurements before time t_b to calculate the predictive diagnosis; the variable degradation bands, causing abrupt changes in symptoms. Evaluation of the quality of diagnosis prognostic prediction error is determined.

Studies of the procedures of machine prognosis state in the aspect of determining methods for forecasting damage, examining the influence of horizon prognoses in the recognition of damage, and studies of the influence of the number of diagnostic parameters on the damage prognosis, were carried out on the basis of: determination of a set of prognosis methods for values of diagnostic parameters, and prediction of the next time of diagnosis - according to the process presented in [25].

Summing up the research performed for the state prognosis method, it can be stated that:

- considering low values of the curvilinear correlation coefficient (< 0.8), high values of prognosis of damage, and negative values for the next operation dates of the objects examined in analytical models (linear model, exponential model, model Brown-Mayer, and Holt model), for potential applications it is necessary to use the Brown–Mayer model and the Holt model;
- the accepted optimization criteria, and the presented algorithm identify unambiguously the prognosis methods and the method of estimating the next diagnosis term. This confirms the appropriateness of the formulated procedure and will provide the basis for methodology to be used for the machine state recognition and in the field of state prognosis for other objects.

The analysis of results of the research on the machine state prognosis methodology allows of formulate conclusions in the areas of: diagnostic parameters optimization and state prognosis. After appropriate verification the presented conclusion can provide the basis for application of dedicated software for a machine state recognition system in an online (board system) and off–line (for a stationary system) modes.

Determining the next term of diagnostics

Evaluation of state changes based on change monitoring of diagnostic parameters (Θ_{b1} determination) was carried out for a vehicles STAR 200, for the results of tests of peripheral clearance of the main shaft - shown in Table 2.

This procedure uses the already approved method of determining the limit value of the symptom [8,22]:

$$S_{jgr} = \bar{s} + \sigma_s \sqrt{\frac{P_g}{2A}} \qquad (11)$$

where: P_g – repair policy of the works, A= k (1-P_g), k ={1-9}.

The presented relationship allows determine the term of the next diagnostics in relation (11) to the current value of a measured symptom, its relation to the value symptoms of limiting with considering the repair policy of the works [22]:

$$\Theta_{b1} = \frac{(1-P_r)(S_{gr} - S_j)}{S_j} \qquad (12)$$

Determination of this value (12) based on the value of data on a circumferential backlash of the main shaft of t STAR 200 vehicle in operation is shown in Table 1.

Table 1. Measuring data of peripheral clearance of examined main shaft.

Clearance-S_j [mm]	6,6	7,5	8,1	9,5	10,5	14,2	15,7	16,5	16,9	17,1
Mileage Θ_j [x10³ km]	2	24	48	72	96	120	144	168	192	216

In accordance with relations (11) [7,10,15] it was evaluated: S_{jgr} = **18,99**, accepted in (17) k = 4, P_g = 0,95, A = k (1 – P_g) = 0,2. Evaluating further values Θ_{b1} according to (12) for accessible measuring data (Table 2) there was fixed a trend of transmission gear clearance changes (state symptom) in the following form:

$$S^*_j = 6{,}0091\, e^{0{,}1189\Theta_j} \qquad (13)$$

Thus, in table 2 there are presented values S^*_j (calculated according to (13) for determined mileages Θ_j, which allowed to determine forecast error e^*_p, defined as [24]:

$$e^*_p = (S^*_j - S_j) / S^*_j \qquad (14)$$

Table 2. Calculations of results.

Mileage Θ_j [10³ km]	2	24	48	72	96	120	144
S_j	6,6	7,5	8,1	9,5	10,5	14,2	15,7
Θ_{b1} [km]	488	4560	7776	7776	7644	720	Reversal of sign
S^*_j	7,27	9,75	10,4	13,19	11,16	11,36	X
e^*_p [%]	10	30	29	37	6	20	X

The average forecast error of a symptom value (S_j) for mileages Q_j accepted in the measurements is 20%. A relatively big forecast error of the proposed method may be decreased by optimising activities undertaken in measuring accessible data from servicing. The performed analytical studies, testing conformity of the measured values of symptoms for the determined mileages (table 2) by means of the proposed methods [22], allow of propose the following actions:

- S_{jgr} should be maximised in the procedure of step-by-step determination by periodical changes of P_g and k [22];
- good approximations are obtained for an increase of P_g, and k in a half of the vehicle operational time (according to data from table 2).

For S_{jgr} accepted from standard (or from operator's manual), results of theoretical calculations are well represented by operational data:

- approximation of Θ_j evaluated from operational data is satisfactory for relation (12);
- presentation of frequency the examination with registrations of exponential trend of the examined symptoms.

The results of the experiments allow to show a way of reasoning and inference in terms of the future diagnostic agent, an intelligent system monitoring critical technical systems. On fig.7 shows the general idea of an agent system to be used for diagnosis and observation of critical systems as well as for information flow handling in the future of diagnostics. We can also see a need to provide some other information concerning the system operation including: load, current time of system life, and also some previous records on the system history contained in the maintenance data base.

The last step of SAD is to develop a measure of further operation risk and make other decision of service type / repair resulting from the assessment of the status and information about the object and the environment. Thus, it can be implemented, as shown graphically, and as you can see, for innovative objects which lack experience in operation and have no operating history there is only one chance to fulfill this mission by using a self-learning mode.

5 Conlusions

Apart from tribology, reliability, security theory and the theory of operation technical diagnostics is one of the fundamental issues connected with rational operation and maintenance of objects. Understanding the physical phenomena occurring during the operation of the machine allows determine of the qualitative relationships taking place between destructive processes and the machine state. The growing demand for diagnostics goes hand-in-hand with applications of new inventions in microelectronics, computer technology, fractal theory, neural networks and artificial intelligence - effectively support diagnostics capabilities.

The basic condition for these considerations to achieve success is to establish a uniform wear of machinery or provide knowledge of the measured symptoms trend (knowledge of intensity of damage). The phenomena that cause wear of assemblies are very complex and their formation is usually affected by several factors, so including them all in the forecasting process is impossible.

The issue of machine tests for the process of prognosis to be used for estimation of the technical state, as well as legal acts related to safety of the user and the environmental protection, are a stimulus to search for new prognostic methods, determining new measures and tools that can describe the current state in the machines specific work condition. This paper attempts to summarize basis mathematical rules to be used for machine condition forecasting based on multi-symptom prognostic methods.

The presentation of the procedures for machine state prognosis allows of determine the parameters, according to the accepted criteria: diagnostic parameters set, diagnostic parameters values prognosis and machine operation date estimation.

References

1. R. Ahmad, S. Kamarudin, *An overview of time-based and condition - based maintenance in industrial application*. Computers & Industrial Engineering (2012)..
2. I. Birger, *Technical diagnostics*. Nauka, Moscov (1978).
3. C. Cempel, *Basics of vibroacoustic diagnostics of machines*. WNT, Warsaw (1982).
4. G. Box, G. Jenkins, *Time series analysis, forecasting and control*, London (1970).
5. L. Bowerman, R.T. O'Connel, *Forecasting and Time Series*. Doxbury Press, USA (1979).
6. C. Cempel, H.G. Natke, *An introduction to the holistic dynamics of operating systems*. Progress Report, **No.2** (1996).
7. C. Cempel, M. Tabaszewski, *Multidimensional condition monitoring of machines in nonstationary operation*. Mechanical Systems and Signal Processing **(21)**, (2007).
8. N.R. Draper, H. Smith, *Regression analysis of used*. BNInż., Warsaw (1973).
9. P. Eykhoff, *Identification in dynamical systems*, BNInż., Warsaw(1980).
10. W. Findeisen, *System analysis - base and methodology*. PWN Warsaw (1985).
11. K. Mańczak, *Methods for identifying the multidimensional of control objects*. WNT Warsaw (1971).
12. S. Niziński, R. Michalski, *Diagnosis of technical objects*. ITE Radom (2002).
13. T. Uhl, J. Giergiel, *Identification of mechanical systems*. PWN Warsaw (1990).
14. A.K.S. Jardine, D. Lin, D. Banjevic, *A review on machinery diagnostics and prognostics implementing condition-based maintenance*, Mechanical Systems and Signal Processing, **20** (2006).

15. J. Luo, M. Namburu, K. Pattipati, L. Qiao, M. Kawamoto, *Model-based prognostic techniques, in: Proceedings of the IEEE*. Systems Readiness Technology Conference (2003).
16. Ł. Muślewski, *Study and assessment of transport system operation efficiency*. Journal of KONES Powertrain and Transport, Vol. **17**, No. **4**, Warsaw (2010).
17. V.T. Tran, B.S. Yang, M.S. Oh, *Machine condition prognosis based on regression trees and one-step-ahead prediction.* Mechanical Systems and Signal Processing, **22**, (2008).
18. J. Qu, M.J. Zuo, *An LSSVR - based algorithm for online system condition prognostics*. Expert Systems with Applications, **39 (2)** (2012).
19. H. Tylicki, B. Żółtowski, *Determination methods of the next diagnosis term of transport vehicle*. Archives of Transport, Warsaw (2001).
20. A. Zeliaś, *Prognosis theory*. PWE Warsaw (1984).
21. B. Żółtowski, *The methods of virtual engineering in the research risks status, safety and environmental operated machines*. UTP Bydgoszcz (2012).
22. B. Żółtowski, *The study of environmental risks losing the suitability of technical systems.* UTP Bydgoszcz (2013).
23. M. Żółtowski, *Management information systems in engineering and manufacturing*. ITE - PIB, Radom (2011).
24. M. Żółtowski, *Investigations of harbour brick structures by using operational modal analysis*. Polish Maritime Research, No. **1/ (81),** vol.**21** (2014).
25. M. Żółtowski, *Assessment State of Masonry Components Degradation*. Applied Mechanics and Materials **Vol. 617** (2014).
26. B. Żółtowski, M. Żółtowski, *Vibrations in the Assessment of Construction State.* Applied Mechanics and Materials **Vol. 617** (2014).

The process of diagnosing an electrostatic precipitator's state during operation in the energy sector

Aleksandra Brzezińska[1,*], and *Bogdan* Żółtowski[1]

[1]University of Science and Technology, Mechanical Engineering Department,
Al. Prof. S. Kaliskiego 7, 85-796 Bydgoszcz, Poland

Abstract. The paper proposes procedures for diagnosing the electrostatic precipitator's technical state in terms of its key function as a facility in a complex waste gas dedusting system. The presented state diagnosing procedures concern the assessment, genesis and forecast of the electrostatic precipitator's state during its rational operation.

1 Introduction

The Polish energy sector started developing rapidly in the 1960s when forced industrialisation triggered a sudden demand for electricity. When power plants were constructed, no one realised that power units would have to be temporarily shut down for reasons other than emergencies or repairs [1]. To facilitate the process of planning service activities, modernisations and repairs of any failures of the facility, an appropriate methodology for controlling the electrostatic precipitator's operation and diagnosing its individual components' degradation should be applied. The actions taken will make it possible to improve the electrostatic precipitator's operation and will significantly reduce costs of maintaining its operability during the facility's operation [18].

2 Environmental protection in the energy sector – legislation

Long-term draft amendments to energy policy were issued by the European Commission as a Green Paper including issues related to a European strategy for sustainable, competitive and secure energy [2]. Directive 2010/75/EU of the European Parliament and of the Council of 24 November 2010 on industrial emissions (replacing Directive 96/61/EC (IPPC) and a BAT reference document) brought about key transformations in Polish law, more specifically environmental protection regulations [3]. The most important amendments relate to making regulations on emission standards for industrial lobbies more restrictive. In particular, the IED Directive provides for reducing sulphur dioxide, nitrogen oxides and dust generated by combustion plants. The Polish energy sector is also subject to two Regulations of the Minister of the Environment providing for intensifying pro-ecological activities:

- Regulation of the Minister of the Environment of 30 October 2014 on the requirements for the measurement of emissions and the measurement of the amount of water taken [4].
- Regulation of the Minister of the Environment of 4 November 2014 on the emission standards for certain types of the installations, fuel combustion sources and incineration and co-incineration equipment [5].

Activities to modernise energy facilities enable the energy sector to meet the requirements of EU directives and to achieve a high level of efficiency of individual components of a power unit. However, supervision to ensure the correct operation of power

* Corresponding author: aleksandra.brzezinska91@gmail.com

facilities, including their installations, and appropriate controls to comprehensively diagnose complex facilities, including electrostatic precipitators, are most important.

3 Structure of a thermal power plant

The thermal power plant has a complex structure which, apart from its main equipment (boiler, turbine, generator, transformer), consists of multiple circuits and pieces of auxiliary, control, signalling, security and measurement equipment. Its structure should ensure the lowest possible cost of electricity generation and the greatest possible efficiency of individual transformations [6]. Some part of energy generated in the power plant serves own needs. It is drawn directly from generator terminals and sent via a transformer to specific power plant equipment. The most important pieces of power plant equipment serving own needs include: water and steam circulation equipment (feed pumps, condenser equipment), combustion equipment (fans, dust collectors, including electrostatic precipitators), fuel transportation equipment and machines, coal pulverisers, turbine operating equipment, a generator cooling system, measurement, security, control and signalling equipment, lighting [7].

As certain equipment needs to be in constant operation, it has a back-up power supply system (two or three pieces of equipment are installed to serve a similar purpose, but normally only one of them is active) and some part of electricity in power plants is used to constantly power the most important equipment in the energy conversion process [15, 16].

3.1 Electrostatic precipitator as a power unit component

Apart from fuel combustion installations, power units include fuel dedusting equipment. Development trends in this area focus on making their structure and operation more efficient [8]. The electrostatic precipitator is among complex machines making up the power unit whose structure can be divided into five characteristic groups [9, 10]:

- Structural components: foundations for the ESP's heavy structure to be laid on them later, casing of dumping hoppers for ash and the equipment's frame, i.e. its support structure with pivot bearings and a chamber divided into an upper and lower part.
- Mechanical part: bearings mounted on pillars, rapper shafts with hammers, a diffuser and a confusor.
- Electrical equipment: transformer units and their installations, automation and control systems, auxiliary equipment.
- Thermal insulation group: mineral wool resistant to high temperatures.
- A system of stairs, landings and platforms to access inspection manholes located on the roof, chamber side walls and dumping hopper side walls.

Any damage to one or more of the ESP's indicated components does not have to lead to the facility's complete shutdown, but any undesirable event related to the occurrence of damage makes the operation of equipment less efficient in a certain way. Complex machines can be considered as the so-called "renewable facilities" only if such damage to the facility's components can be repaired and if performed activities lead to making such equipment operable again [17]. The repair of the facility by a qualified team should not normally interfere with key tasks performed on site.

4 Conditions for diagnosing electrostatic precipitators' state

Measurements in the energy industry are made regularly during a warranty period after which they are only made upon order by an investor in accordance with PN-Z-04030-7:1994 "Tests for dust content. Measurement of concentration and mass flow rate of particulate matter in waste gases by gravimetric method" [11]. There is neither a predetermined time interval, which would indicate deadlines for tests to diagnose the electrostatic precipitator's current state, nor a list of parameters to be checked during inspections. The checks to date have covered the values, which are set out in the

Regulations of the Minister of the Environment, relating to combustion equipment, i.e. NO_x, SO_x and dust. Laboratories, which make measurements, test processes, systems and energy equipment, have their own list of offered parameters which can be tested during a check, but it is a person ordering the test who ultimately chooses values to be measured.

If a new emission standard for dust cannot be met, the first idea is usually to modernise the existing electrostatic precipitator power supply system or to invest in a new facility. Enhancements or repairs of electrostatic precipitators are not the cheapest solutions and often make it necessary to shut down a boiler for a few months. Therefore, it is so important to use an appropriate system to diagnose electrostatic precipitators' technical state in order to set deadlines for the machine's repairs, modernisations or inspections, in order to systematically plan works to improve the facility's operation and ultimately to help reduce the electrostatic precipitator's operational costs.

5 Process of diagnosing the electrostatic precipitator's technical state

The process of diagnosing the technical state of the electrostatic precipitator as a complex facility, which is among critical equipment in the power unit, should clearly make it possible to identify certain components, parts, material in the context of recognising their degradation. The procedures for diagnosing the electrostatic precipitator's state and controlling the facility's operability, which are proposed in this paper, list individual steps dedicated to the Maintenance Services which develop maintenance, modernisation or service plans for the facility. Additionally, the specific procedures were supplemented with algorithms to ultimately localise damage and to control the progressive degradation of the existing ESPs' state [12].

5.1 Procedure for assessing the ESP's technical state

To correctly assess the electrostatic precipitator's technical state, it is important to use tools providing a full insight into its current operation. Given its important function (purification of waste gases from harmful dust particles resulting from the fuel combustion process), the electrostatic precipitator must be guaranteed failure-free operation. The proposed scope of the procedure includes an assessment of the machine's (electrostatic precipitator's) technical state performed during operation and control activities to analyse the facility's operation on an ongoing basis, and activities to detect damage: fatigue, aging, random damage, and the electrostatic precipitator's deteriorating structural and mechanical components. As part of the procedure for assessing the electrostatic precipitator's technical state, the following procedure for individual teams supervising the facility's operation should be distinguished:

1. ***Assessment of the technical state based on the visual inspection and detection of failures, damage, defects, changes in the waste gas purification process during the electrostatic precipitator's operation:***
   - diagnosis of failures by telephone based on mA and kV values, and information provision (company performing tests for the energy area);
   - internal and external inspection of electrostatic precipitators to detect the cause of failure (mechanic at the Maintenance Department together with the Head of the Maintenance Department) if damage prevents the facility from further operation;
   - technical supervision of repair works on site (moderniser/repairer responsible for the area of electrostatic precipitator repairs);
   - operational optimisation of mechanical and electrical equipment (moderniser/repairer and the company performing tests for the energy area);
   - implementation of a procedure for preventing electrostatic precipitators' inoperability (NZU) depending on changes in the combustion process (Head of the Maintenance Department);

- operational test – testing and monitoring electrostatic precipitators' operation (moderniser/repairer, the Head of the Maintenance Department and the company performing tests for the energy area);
- technical report together with documentation on the removal of the cause of failure or the repair of malfunction (moderniser/repairer, the Head of the Maintenance Department and the company performing tests for the energy area).

2. ***Assessment of the electrostatic precipitator's technical state based on an algorithm for selecting a test to control the technical state of and localise damage to the facility:***
   - preparation of data for their acquisition;
   - optimisation of a set of values of diagnostic parameters (only if set Y is large, e.g. $m>10$);
   - organisation of the data set by determining set $\{s_i(\Theta_k), i=1,\dots, 1; k=1, \dots, K\}$;
   - testing the significance of changes in values of diagnostic parameters $\{y_j(\Theta_k)\}$ depending on the state of set $\{s_i = f(\Theta_k); \Theta_k \in (\Theta_1, \Theta_b)\}$, i.e. which diagnostic parameter "best" describes the state of the equipment's inoperability – s_i;
   - specification of a Boolean matrix:
     - 1 – when a change in the state of the equipment's inoperability – s_i causes significant changes in values of diagnostic parameter y_j;
     - 0 – when a change in the state of the equipment's inoperability – s_i causes no significant changes in values of diagnostic parameter y_j.
   - determination of an optimal method for selecting a test to control the state and to localise damage depending on the size of the set of diagnostic parameters and the machine's operation time:
     - development of an optimal method for selecting a test to control the machine's state. The test to control the state (TCS) to be selected based on the Boolean matrix – the TCS as a vector of logical values (0, 1) of diagnostic parameters and its analogical vector of states: $\{(y_1, \dots, y_n, \dots, y_N)\} \Rightarrow S_0 \vee S_1 = \{(s_1, \dots, s_n, \dots, s_N)$ for the state of the equipment's operability S_0 and the state of the equipment's inoperability S_1;
     - testing the impact of the machine's operation time on an optimal diagnostic test;
     - development of an optimal method for selecting a test to localise damage to the machine (test to localise damage (TLD) to be selected based on the Boolean matrix).
   - interpretation of results of tests to control the electrostatic precipitator's (ESP's) state.

5.2 Procedure for the genesis of the ESP's technical state

The procedure covers the genesis of the ESP's technical state performed after assessing the state during the electrostatic precipitator's (ESP's) operation. Symptom genesis is an innovative approach which makes it possible to register values of the ESP's diagnostic parameters and states during operation. Based on collected data and the equipment's visual inspection, it allows for determining what caused and what conditions accompanied the state of the equipment's inoperability. As part of the procedure for the genesis of the electrostatic precipitator's technical state, the following procedure for individual teams supervising the facility's operation should be distinguished:

1. ***Genesis of the technical state based on registered values of the electrostatic precipitator's diagnostic parameters and states during its operation to date:***
   1. constant monitoring of the following: discharge electrode rappers, collection electrode rappers, hopper heating, stand-off or rotating insulator heating, hopper ash level sensors' state, high voltage systems in each zone, dust collection from under hoppers (separate installation);

2. continuous measurement of the ESP's diagnostic parameters: concentration of waste gas components, i.e. SO_2, NO_x, CO, H_2O, O_2 and dust, volumetric flow rate of waste gas, moisture content in waste gas, temperature of waste gas, temperature inside the ESP's chamber, static pressure;
3. when the state of the equipment's operability changes into the state of the equipment's inoperability, it is necessary to take dust samples and determine dust particle size and content of: Cd, Cr, Ni, Pb, V, Cu, Al, As, Zn, H_2SO_4, NH_3, TOC (total organic carbon), TC (total carbon), share of flammable and non-flammable parts;
4. collecting all measurement results, performing necessary calculations and determining the relationship between changes in the genesis value of the diagnostic parameter with a genesis error and the limit value of the diagnostic parameter;
5. determining the relationship between changes in the genesis value of the diagnostic parameter with a genesis error and the limit value of the diagnostic parameter (algorithm for the genesis of the ESP's technical state according to the scheme of estimating values of diagnostic parameters) [13, 14].

The set of parameters to be optimised: by calculating values of criterion functions and weights of diagnostic parameters based on input data for the selected facility (ESP) in the Dedicated Diagnostic System for the ESP (DDSESP). As a criterion for selecting the diagnostic parameter(s), one should assume the maximum weight values wj and select diagnostic parameters according to the criterion adopted. The procedure allows for selecting diagnostic parameters as prefered by a system operator (Head of the Maintenance Department) and updating weigth values for new sets of selected diagnostic parameters [12, 13, 14].

The genesis of the ESP's state to be performed by:

- determining the genesis value of the diagnostic parameter and the genesis error (only for a discrete event notation);
- determining the minimum distance of the genesis value of the diagnostic parameter to its limit value and testing the impact of operational conditions (number of parameters, number of time series) on the state's genesis;
- visualising and recording approximation or interpolation functions for the selected ESP's parameters;
- estimating the cause of damage to the machine based on the minimum distance of the genesis value of the diagnostic parameter to its limit value, values of the machine's process parameters, values of ambient parameters and additional events.

Collecting materials and information on causes of damage to the system of dust collection from under hoppers (taking into account hopper ash level sensors' state) together with taking a series of photographs to illustrate the actual state of individual mechanisms and parts of the dust collection system. It is also necessary to record the date and location of the damage. Information necessary for data generation based on the algorithm for the genesis of the ESP's technical state according to the scheme of estimating values of diagnostic parameters.

5.3 Procedure for forecasting the ESP's technical state

The basic purpose of the procedure for forecasting the ESP's technical state is to present a sequence of tasks necessary for presenting a track record of past events, which occurred during the ESP's operation, so that unfavourable events in the ESP's operation (ESP's damage putting the facility at risk of being shut down) can be predicted in the future based on the data collected. As part of the procedure for forecasting the electrostatic precipitator's technical state, the following procedure for individual teams supervising the facility's operation should be distinguished:

1. ***Forecast of the ESP's technical state based on values of parameters of the ESP's state during operation and the ESP's completed modernisations and repairs:***
   - developing a model process of diagnosing the ESP based on identified symptoms and collected information on: operation of discharge electrode rappers, collection electrode rappers, hopper heating, stand-off or rotating insulator heating, hopper ash level sensors' state, high voltage systems in each zone, dust collection from under hoppers (separate installation).

   Having analysed the process of diagnosing the ESP in detail, the following necessary diagnostic parameters for the ESP were determined: concentration of waste gas components, i.e. SO_2, NO_x, CO, H_2O, O_2 and dust, volumetric flow rate of waste gas, moisture content in waste gas, temperature of waste gas, temperature inside the ESP's chamber, static pressure. The model process of forecasting the ESP's state must be developed based on information on the course of modernisation, repair activities and documentation presenting the genesis and assessment of the ESP's technical state.
   - collecting local values of the measure of similarity between instantaneous states and model time series;
   - generating a model of the ESP's forecast state also based on an algorithm for forecasting the ESP's technical state;
   - calculating forecast values of the measure of probability based on collected data on the ESP's operation, modernisations and repairs;
   - coming to conclusions on future states of selected mechanisms and parts of the tested facility (ESP).
2. ***Algorithm for forecasting the ESP's technical state:***
   - forecasting the value of diagnostic parameter y_{j*}:
     - by Brown-Mayer adaptive method, grade 1 (B-M1), with coefficient α=(0.1-0.9) for forecast horizon τ=3ΔQ set for time interval (Q1,Qb) – the method used when a forecast variable is almost constant and minor random fluctuations occur;
     - by Holt adaptive method with coefficient α=(0.05-0.3) and β=(0.4-0.9) for forecast horizon τ=3ΔQ set for time interval (Q1,Qb) – the method used when persistent time trends in a time series are observed;
     - by analytic methods (linear, exponential and power methods of the first, second, and third grade) for forecast horizon τ=3ΔQ set for time interval (Q_1,Q_b) – the method allows for finding a trend function which will meet the set criteria and will optimally adapt to the forecast variable's time series (it is important to select an appropriate analytical form of the trend function);
     - setting a deadline for the next servicing and diagnosing of the machine Θ_d.

6 Summary

The coal power stations have the right number of starts machines (and exceeding permissible parameters that occur in such states), sooner or later there are problems and damages. Cracks appear, followed by failures of lower or higher severity. Paradoxically, older power units are technologically more adapted to load changes, because the steam parameters are not exorbitant. For some power units, further solutions are introduced enabling undervaluation lower of power. Unfortunately, most of the power units are currently probably already largely exploited. The number of work hours the power units goes in hundreds of thousands, slowly exceeding the thresholds allowed, and their current work is based on multiple modernizations. Unfortunately, the history of the power units exploitation is usually not collected, as a result, the actual state of the devices is not known.

Increases over the years, probability of occurrence of more failures and deterioration of the availability coefficients. Of course each of the consortium (manufacturing companies) reports internal reliability parameters and costs incurred for the repair of facilities, but the additional cost alone caused by large load changes and more boiler withdrawals from use, unfortunately it is a problem that is frequently off in the juxtaposition. Therefore, it is necessary to supervise the correct operation of the power facilities, including installations is entering in its composition, and conducting reliable controls aimed at comprehensive diagnosis of complex objects. For financial reasons, the introduction of a methodology to monitor the ESP degradation and control of the current operation of the facility is of key importance.

References

1. M. Kudełko, *Czynniki determinujące rozwój krajowego sektora energetycznego – analiza wrażliwości,* Journal of Management and Finance, **13**, 255-256 (2015).
2. Green Paper, *A European Strategy for Sustainable*, Competitive and Secure Energy, Brussels, 17-18 (2006).
3. EU, *Directive 2010/75EU of the European Parliament and of the Council of 24 November 2010 on industrial emissions (integrated pollution prevention and control),* (Recast), 1-2 (2010).
4. Ministerstwo Środowiska, *Rozporządzenie Ministra Środowiska z dnia 30 października 2014 r. w sprawie wymagań w zakresie prowadzenia pomiarów wielkości emisji oraz pomiarów ilości pobieranej wody,* **2014/1542**, 1-3 (2014).
5. Minister for the Environment in Poland, *Regulation of the Minister for the Environment of 4 November 2014 on the emission standards for certain types of installation, fuel combustion plants and waste incineration or co-incineration facilities,* Dziennik Ustaw, **2014/1546**, 1-4 (2014).
6. B. Zaporowski, *Technologie wytwarzania energii elektrycznej dla polskiej elektroenergetyki*, Energy Policy Journal, **18**, 29-44 (2015).
7. T. Chmielniak, A. Rusin, H. Łukowicz, *Modernizacja i rewitalizacja istniejących bloków węglowych ważnym warunkiem stabilności sektora wytwarzania elektryczności,* Energy Policy Journal, **19**, 49-59 (2016).
8. A. W. Yuen, *Collector Current Density and Dust Collection in Electrostatic Precipitators*, VDM Verlag Dr. Mueller, 1-7 (2008).
9. F. K. Fuliful, A. A. Hamood, *Collection Efficiency of Wire– Duct Electrostatic Precipitator,* LAP Lambert Academic Publishing, 11-20 (2017).
10. H. A. Ziedan, *Modeling of Corona Discharge in Wire-duct Electrostatic Precipitators,* LAP Lambert Academic Publishing, 35-44 (2016).
11. PN-Z-04030-7:1994, *Ochrona czystości powietrza. Badania zawartości pyłu. Pomiar stężenia i strumienia masy pyłu w gazach odlotowych metodą grawimetryczną,* PKN, 1-7 (1994).
12. B. Żółtowski, D. Landowski, B. Przybyliński, *Projektowanie Eksploatacji Maszyn,* Wydawnictwo Naukowe Instytutu Technologii Eksploatacji - Państwowego Instytutu Badawczego, 212-215 (2012).
13. H. Tylicki, B. Żółtowski, *Genezowanie stanu maszyn*, Wydawnictwo Naukowe Instytutu Technologii Eksploatacji - Państwowego Instytutu Badawczego, 29-45 (2012).
14. H. Tylicki, B. Żółtowski, *Rozpoznawanie stanu maszyn,* Instytutu Technologii Eksploatacji - Państwowego Instytutu Badawczego, 11-66 (2010).
15. M. Żółtowski, *Informatyczne systemy zarządzania w inżynierii produkcji.* Wydawnictwo ITE-PIB, Radom 2011.

16. B. Żółtowski, *Metody inżynierii wirtualnej w badaniach stanu, zagrożeń bezpieczeństwa i środowiska eksploatowanych maszyn*. Wyd. UTP, Bydgoszcz 2012.
17. B. Żółtowski, *Badania zagrożeń utraty zdatności środowiskowych systemów technicznych*. WU UTP, Bydgoszcz 2013, s.181.
18. A. Brzezińska, B. Żółtowski, *Dozorowanie procesu technologicznego pracy bloków energetycznych w diagnostycznej strategii eksploatacji*, Opracowanie wewnętrzne, 2017.

Diagnostics of supercapacitors

Valeriy Martynyuk [1*], *Oleksander* Eromenko [1], *Juliy* Boiko [1], *Tomasz* Kałaczyński [2]

[1]Programming and Computer, Telecommunication Systems Faculty in Khmelnytsky National University, Department of Telecommunication and Computer Integrated Technology, 29016, 11 Institutska street, Khmelnytsky, Ukraine
[2]Faculty of Mechanical Engineering UTP in Bydgoszcz, Department of Vehicle Engineering, Al. Prof. S.Kaliskiego 7, 85-796 Bydgoszcz, Poland

Abstract. The paper represents the mathematical model for diagnostics of supercapacitors. The research objectives are the problem of determining a supercapacitor technical condition during its operation. The general reliability of diagnostics is described as the methodological and instrumental reliabilities of diagnostics. The instrumental diagnostic reliability of supercapacitor includes the probabilities of errors of the first and second kind, α and β respectively. The methodological approach to increasing the reliability of supercapacitor diagnostic has been proposed, in terms of multi-parameter supercapacitor diagnostic by applying nonlinear, frequency dependent mathematical models of supercapacitors that take into account nonlinearity, frequency dispersion of parameters and the effect of transient processes in supercapacitors. The more frequencies, operating voltages and currents are applied in the supercapacitor diagnostics, the more methodological reliability of diagnostics will increase in relation to the methodological reliability of supercapacitor diagnostics when only one frequency, voltage and current are applied.

1 Introduction

Supercapacitors are polar electrochemical devices that are capable of storing and releasing electrical energy by internal redistribution of electrolyte ions. According to their electrical parameters, supercapacitors occupy an intermediate position between electrolytic capacitors of high capacitance and batteries, but according to the principle of operation differ from both of them [1, 2].

The quality and reliability of supercapacitors are largely determined by diagnostics support at all stages of their life cycle. At the production stages, where it is crucial to ensure zero-defect manufacturing technology and labour for producing supercapacitors, the volume of control and diagnostic operations reaches 50% of the total labour intensity of their production. A characteristic problem of diagnosing supercapacitor manufacturing defects under the conditions of their mass production is that the later the defects are detected, the more costs are necessary to localize the place of their emergence.

Hence, diagnostic support will be effective only when the high reliability values are ensured at minimum cost. At the operation stage, the purpose of diagnosing a superapacitor is to evaluate its technical condition, to determine the reasons of its operation failures, to predict changes in a supercapacitor technical condition and determine the reasons for such changes or time intervals after which processes causing undesirable changes in a supercapacitor technical condition might begin.

At present, the scope of supercapacitor application is very wide, namely electric vehicles, hybrid transport, devices for starting diesel and gasoline internal combustion engines,

* Corresponding author: martynyuk.valeriy@gmail.com

powerful electronics and telecommunications. Thus, the issues of developing methods and means of supercapacitor diagnostics have become urgent.

In most cases, supercapacitors are applied in parallel with chemical elements, batteries and accumulators for smoothing the peak values of load current, which greatly increases operation lifetime and improves the characteristics of electrochemical power sources.

Each year, supercapacitor mass production increases and we observe gradual improvement of the electrical parameters of supercapacitors that are produced by world leading companies such as Maxwell, AVX, Cooper, Elna (USA), Fuji Heavy Industries, NEC, Panasonic, Power Systems (Japan), Ness, LS Cable (South Korea), EPCOS (Germany), BatScap (France), Cap-XX (Australia), Tavrima (Canada), ESMA, EL Technology (Russia) and Yunasko (Ukraine).

The best samples of modern supercapacitors are characterized by capacities in thousands of Farads, high operating currents (hundreds and thousands of amperes), and low values of active resistance (several milliohms and hundreds of microhms).

2 Supercapacitor technical condition

Supercapacitor diagnostics is an information procedure the purpose of which is to reflect its technical condition in the form of a conclusion about the nature and significance of this condition. Such a conclusion regarding the diagnostic results is called a diagnosis [3].

Technical condition of a supercapacitor is a condition that is characterized at a certain time, under certain environmental conditions by the values of supercapacitor parameters set by its data sheet. According to the existing practice, it has been confirmed to define such mutually exclusive incompatible pairs of technical conditions for technical products: serviceable – unserviceable, operable – faulty.

For the serviceable – unserviceable conditions the basis of their comparison is compliance of all supercapacitor parameters with the requirements of normative data sheet. And if at least one of these parameters, regardless of its impact on the supercapacitor ability to perform its functions, will not fulfill these requirements, the technical condition is considered unserviceable. And when all the parameters meet all the requirements of proper documentation, the technical condition is serviceable.

A supercapacitor is in an operable technical condition if its relevant diagnostic parameters are within the limits where working functions can be performed. When the diagnostic parameters go beyond the specified limits, then the supercapacitor can no longer fully perform its functions, and it has to be considered faulty.

Evaluation of a supercapacitor technical condition is a complex task that allows solving the following problems:

- at the production stage – preventing disruption of technological processes;
- at the operation stage – detecting emerging defects at an early stage, thus preventing processes that would lead to undesirable change in a supercapacitor technical condition and cause emergency situations.

Principally, the problem of determining a supercapacitor technical condition can be solved in two ways: functional and test diagnostics. While performing functional diagnostics it is necessary to observe the behavior of an operating supercapacitor and compare it with the standard. For test diagnostics it is necessary to compile a description of a supercapacitor for various types of its malfunctions and to determine what influences they can be detected by.

Low reliability of supercapacitor diagnostics in the existing diagnostic systems is conditioned by a complex change in diagnostic parameters in case of supercapacitor malfunction. This circumstance does not permit to fully solve the problem of analyzing a supercapacitor technical condition neither in static nor in dynamic operating modes at the supercapacitor production and operation stages.

Reliability of diagnostics is the degree of objective conformity of the result of diagnostics to the actual technical condition of the supercapacitor, which is estimated by the probability of making the correct decision about the condition of the supercapacitor [4]. The components of the general reliability of diagnostics are methodological and instrumental reliabilities of diagnostics [5]:

$$D_G = D_M \cdot D_I , \tag{1}$$

where D_M is methodological reliability of diagnostics; D_I is instrumental reliability of diagnostics.

Methodological reliability of diagnostics is calculated by the following formula [5]:

$$D_M = \frac{N_1}{N} ; \tag{2}$$

where N_1 is the number of exponents which characterize the technical condition of the supercapacitor and are included into the mathematical model of diagnostics;

N is total number of exponents which characterize the diagnostics of the supercapacitor.

Instrumental reliability of diagnostics is defined by the probabilities of errors of the first and second kind respectively α i β [5]:

$$D_I = 1 - \alpha - \beta , \tag{3}$$

where α is probability of errors of the first kind; β is probability of errors of the second kind.

3 Mathematical model for diagnostics of supercapacitors

Let us consider a simplified mathematical model for diagnostics of supercapacitors in the stationary mode, which is based on a one-time measurement of capacitance C, active resistance R and self-discharge current Isd of supercapacitors.

Depending on the diagnostic method, capacitance C and active resistance R of supercapacitors are measured either at the alternating current at one probe effect frequency ω0 with one nominal constant voltage at the supercapacitor outputs Unom or at the direct charge (discharge) current of supercapacitors I_0. Self-discharge current Isd of supercapacitors is also measured at one nominal constant voltage at the supercapacitor outputs Unom. As a result of such measurements, we get three measured values: r, c and isd.

Next, two hypotheses are put forward.

1. The basic hypothesis H_0 (the supercapacitor is working):

$$H_0 : r \in [R_{\min}, R_{\max}] \cap c \in [C_{\min}, C_{\max}] \cap i_{sd} \in [Isd_{\min}, Isd_{\max}] , \tag{4}$$

where R_{min}, R_{max}; C_{min}, C_{max} and Isd_{min}, Isd_{max} are limits of permissible interval for r, c and i_{sd} respectively.

2. Alternative hypothesis H_1 (the supercapacitor is defective):

$$H_1 : r \notin [R_{\min}, R_{\max}] \cap c \notin [C_{\min}, C_{\max}] \cap i_{sd} \notin [Isd_{\min}, Isd_{\max}] . \tag{5}$$

The diagnostic reliability of the supercapacitor is defined as the product of the methodological and the instrumental diagnostic reliabilities of supercapacitor (1). The methodological diagnostic reliability of supercapacitor is defined by the ratio between the number of exponents characterizing the technical condition of the object of diagnostic and included to the mathematical model of diagnostics and the total number of exponents characterizing the technical condition of the supercapacitor.

From this we can conclude that the simplified mathematical model for diagnostics of supercapacitors, which is based on the one-time measurement of capacitance C, active resistance R and self-discharge current Isd of supercapacitors, is characterized by low methodological reliability of diagnostics.

It can be explained by the fact that for supercapacitors capacitance C≠const, active resistance R≠const and self-discharge current Isd≠const.

In case of a simplified diagnostic model of supercapacitor, the number of exponents characterizing the technical condition of supercapacitors $N_1=3$ (capacitance, active resistance and self-discharge current). As for the total number of exponents characterizing the technical condition of supercapacitors, frequency dispersion and non-linearity of supercapacitor parameters (R(ω,u), C(ω,u), Isd(t,u)) are equivalent to the increase in the total number of exponents characterizing the technical condition of supercapacitors. Hence, we can conclude that N≠N1 i N>>N1, and the methodological diagnostic reliability of supercapacitor for a simplified model of supercapacitor diagnostics is much smaller than one.

$$D_M = \frac{3}{N} << 1 . \tag{6}$$

The instrumental diagnostic reliability of supercapacitor is defined by the expression (3), which includes the probabilities of errors of the first and second kind, α and β respectively. The probabilities of errors of the first and second kind are defined separately for the capacitance C, active resistance R and self-discharge current Isd of supercapacitors according to the expressions:

$$\alpha_C = \int_{C_{\min}}^{C_{\max}} f(C) \left(\int_{-\infty}^{C_{\min}-C} f(\delta_C) d\delta_C \right) dC + \int_{C_{\min}}^{C_{\max}} f(C) \left(\int_{C_{\max}-C}^{\infty} f(\delta_C) d\delta_C \right) dC ; \tag{7}$$

$$\beta_C = \int_{-\infty}^{C_{\min}} f(C) \left(\int_{C_{\min}-C}^{C_{\max}-C} f(\delta_C) d\delta_C \right) dC + \int_{C_{\max}}^{\infty} f(C) \left(\int_{C_{\min}-C}^{C_{\max}-C} f(\delta_C) d\delta_C \right) dC , \tag{8}$$

where $C_{\max}$, $C_{\min}$ are permissible maximal and minimal capacitance values for supercapacitors; $f(C)$ is density of probability of supercapacitor capacitance; $f(\delta_C)$ is density of probability of supercapacitor capacitance measurement error.

$$\alpha_R = \int_{R_{\min}}^{R_{\max}} f(R) \left(\int_{-\infty}^{R_{\min}-R} f(\delta_R) d\delta_R \right) dR + \int_{R_{\min}}^{R_{\max}} f(C) \left(\int_{R_{\max}-R}^{\infty} f(\delta_R) d\delta_R \right) dR ; \tag{9}$$

$$\beta_R = \int_{-\infty}^{R_{\max}} f(R) \left(\int_{R_{\min}-R}^{R_{\max}-R} f(\delta_R) d\delta_R \right) dR + \int_{R_{\max}}^{\infty} f(R) \left(\int_{R_{\min}-R}^{R_{\max}-R} f(\delta_R) d\delta_R \right) dR , \tag{10}$$

where $R_{\max}$, $R_{\min}$ is permissible maximal and minimal active resistance values for supercapacitors; $f(R)$ is density of probability of supercapacitor active resistance; $f(\delta_R)$ is density of probability of supercapacitor active resistance measurement error.

$$\begin{aligned} \alpha_{Isd} = & \int_{Isd_{\min}}^{Isd_{\max}} f(Icp) \left(\int_{-\infty}^{Isd_{\min}-Isd} f(\delta_{Isd}) d\delta_{Isd} \right) dIsd + \\ & + \int_{Isd_{\min}}^{Isd_{\max}} f(Icp) \left(\int_{Isd_{\max}-Isd}^{\infty} f(\delta_{Isd}) d\delta_{Isd} \right) dIsd \end{aligned} ; \tag{11}$$

$$\begin{aligned} \beta_{Isd} = & \int_{-\infty}^{Isd_{\min}} f(R) \left(\int_{Isd_{\max}-Isd}^{Isd_{\max}-Isd} f(\delta_{Isd}) d\delta_{Isd} \right) dIsd + \\ & + \int_{Isd_{\max}}^{\infty} f(Isd) \left(\int_{Isd_{\min}-Isd}^{Isd_{\max}-Isd} f(\delta_{Isd}) d\delta_{Isd} \right) dIsd \end{aligned} , \tag{12}$$

where Isd_{max}, Isd_{min} are permissible maximal and minimal self-discharge current values for supercapacitors; $f(Isd)$ is density of probability of supercapacitor self-discharge current; $f(\delta_{Isd})$ is density of probability of supercapacitor self-discharge current measurement error.

The general probabilities of errors of the first and second kind, α and β respectively, are defined by the expressions:

$$\alpha = 1-(1-\alpha_C)(1-\alpha_R)(1-\alpha_{Isd}); \quad (13)$$

$$\beta = 1-(1-\beta_C)(1-\beta_R)(1-\beta_{Isd}). \quad (14)$$

Then, the general diagnostic reliability of supercapacitor for a simplified model is defined by the expression:

$$D_G = D_M \cdot D_I = \frac{3}{N}\cdot(1-\alpha-\beta) << 1. \quad (15)$$

Assessment of supercapacitor diagnostics for a simplified model can also be performed using the formula of average diagnostic risk:

$$\Psi \approx W_{1R}\alpha_R + W_{2R}\beta_R + W_{1C}\alpha_C + W_{2C}\beta_C + W_{1Isd}\alpha_{Isd} + W_{2Isd}\beta_{Isd}, \quad (16)$$

where W_{1R}, W_{1C} and W_{1Isd} are losses due to errors of the first kind for each parameter r, c and i_{sd} respectively;

W_{2R}, W_{2C} and W_{2Isd} are due to errors of the second kind for each parameter r, c and i_{sd} respectively;

α_R, α_C and α_{Isd} are probabilities of errors of the first kind for each parameter r, c and i_{sd} respectively;

β_R, β_C and β_{Isd} are probabilities of errors of the second kind for each parameter r, c and i_{sd} respectively.

Increase in the diagnostic reliability of supercapacitor is possible when using supercapacitor characteristics that contain information on the nominal values of technical condition of supercapacitors in the range of changes in nominal voltages in the operation frequency range or in the operation time range. Such characteristics include nonlinear complex resistance Z(jω, u) (or nonlinear resistance $Z(s,u)$ and nonlinear transient resistance $z(t,u)$ of supercapacitors, related to each other by inverse and direct Laplace transforms:

$$Z(s,u) = L^{-1}\left[z(t,u)\right]; \quad (17)$$

$$z(t,u) = L\left[Z(s,u)\right], \quad (18)$$

where $s = j\omega$ is complex frequency.

The supercapacitor's input is supplied with either the probe voltage $u_p(t)$ or probe current $i_p(t)$, and either the current $i_m(t)$ through the supercapacitor or the voltage $u_m(t)$ at the supercapacitor is measured. Operators Z or Y determine the complex of mathematical operations over $u_p(t)$ or $i_p(t)$ which allow obtaining $i_m(t)$ or $u_m(t)$ respectively:

$$i_m(t) = Z\left[u_p(t)\right]; \quad (19)$$

$$u_m(t) = Y\left[i_p(t)\right]. \quad (20)$$

Let's consider that the probe voltage $u_p(t)$ and the probe current $i_p(t)$ are observed without an error, and the measured voltage $u_m(t)$ and the measured current $i_m(t)$ are observed with errors:

$$i_m^*(t) = i_m(t) + \Delta i_m(t); \quad (21)$$

$$u_m^*(t) = u_m(t) + \Delta u_m(t). \quad (22)$$

where $\Delta i_m(t)$ and $\Delta u_m(t)$ are stationary random processes with known characteristics.

Hence, we can conclude that in the process of supercapacitor diagnostics it is necessary to obtain a rating of the defined operators Z_m and Y_m approaching the operators Z and Y which describe the properties of supercapacitors. Taking into account the fact that operators Z or Y are in most cases unknown, to describe supercapacitors their mathematical models characterized by operators Z_{mod} or Y_{mod} are used [6, 7].

The criterion for estimating the proximity of operators Z_m, Y_m, and Z_{mod}, Y_{mod} is approximation of the measured voltage $u_m(t)$ and the measured current $i_m(t)$ to the voltage $u_{\text{mod}}(t)$ and current $i_{\text{mod}}(t)$ of the model at the similar probe effects of $i_p(t)$ and $u_p(t)$ respectively.

Let's introduce positive functions of losses $\eta_u\left[u_m(t), u_{\text{mod}}(t)\right]$ or $\eta_i\left[i_m(t), i_{\text{mod}}(t)\right]$ which depend on the measured voltage and current of an supercapacitor and its model, and also determine the amount of losses connected with different combinations of $u_m(t)$, $u_{\text{mod}}(t)$ and $i_m(t)$, $i_{\text{mod}}(t)$. Average loss or average risk is defined by the mathematical expectations of loss functions $\eta_u\left[u_m(t), u_{\text{mod}}(t)\right]$ or $\eta_i\left[i_m(t), i_{\text{mod}}(t)\right]$:

$$\bar{\Psi}_u = M\left\{\eta_u\left[u_m(t), u_{\text{mod}}(t)\right]\right\}; \quad (23)$$

$$\bar{\Psi}_i = M\left\{\eta_i\left[i_m(t), i_{\text{mod}}(t)\right]\right\}. \quad (24)$$

On the other hand, average risks are defined by expressions:

$$\bar{\Psi}_u = \iint \eta_u(u_m, u_{\text{mod}}) f_u(u_m, u_{\text{mod}}) du_m du_{\text{mod}}; \quad (25)$$

$$\bar{\Psi}_i = \iint \eta_i(i_m, i_{\text{mod}}) f_i(i_m, i_{\text{mod}}) di_m di_{\text{mod}}, \quad (26)$$

where $f_u(u_m, u_{\text{mod}})$ and $f_i(i_m, i_{\text{mod}})$ are voltage and current probability functionals.

Expressions (25) and (26) can be simplified, if the nonlinear transient resistance $z(t,u)$ and nonlinear transient conductance $y(t,i)$ of supercapacitors are monotone functions, and probe signals $u(t)$ and $i(t)$ are stationary:

$$\bar{\Psi}_u = \int_{u \in U} \eta_u\left[z_m(t,u), z_{\text{mod}}(t,u)\right] f_u(u) du; \quad (27)$$

$$\bar{\Psi}_i = \int_{i \in I} \eta_i\left[y_m(t,i), y_{\text{mod}}(t,i)\right] f_i(i) di, \quad (28)$$

where U is definition domain of u;

I is definition domain of i.

Deviation of the defined nonlinear transient resistance $z_m(t,u)$ from the nonlinear transient resistance of the model $z_{\text{mod}}(u)$, as well as deviation of the defined nonlinear transient conductance $y_m(t,i)$ from the nonlinear transient conductance of the model $y_{\text{mod}}(t,i)$ may be estimated by the criterion of regular approximation (maximum deviation of characteristics):

$$m_u = \max_{u \in U}\left|z_m(t,u) - z_{\text{mod}}(t,u)\right|; \quad (29)$$

$$m_i = \max_{i \in I}\left|y_m(t,i) - y_{\text{mod}}(t,i)\right|. \quad (30)$$

Let's choose such mathematical models of supercapacitors $z_{\text{mod}}(t,u)$ and $y_{\text{mod}}(t,i)$ which will satisfy the criteria of regular approximation in the ranges of change in the probe voltage $u_p(t)$ and current $i_p(t)$ of supercapacitors.

Let's define the average risk of supercapacitor control for the following loss functions $\eta_u = \left|z_m(t,u) - z_{\text{mod}}(t,u)\right|^k$ and $\eta_i = \left|y_m(t,i) - y_{\text{mod}}(t,i)\right|^k$:

$$\bar{\Psi}_u = \int\limits_{u \in U} \left|z_m(t,u) - z_{\text{mod}}(t,u)\right|^k f_u(u)du \leq \int\limits_{u \in U} m_u^k f_u(u)du = m_u^k; \quad (31)$$

$$\bar{\Psi}_i = \int\limits_{i \in I} \left|y_m(t,i) - y_{\text{mod}}(t,i)\right|^k f_i(i)di \leq \int\limits_{i \in I} m_i^k f_i(i)di = m_i^k. \quad (32)$$

The advantage of the obtained expressions of the average supercapacitor diagnostic risks (31) and (32) in comparison with the expressions (25) and (26) is the possibility of calculating the maximum values of the average supercapacitor diagnostic risks that do not depend on the probability densities $f_u(u)$ and $f_i(i)$ of supercapacitor voltage and current probe effects.

4 Defining reliability of supercapacitor diagnostics

Let's consider the errors in supercapacitor diagnostics applying the developed supercapacitor diagnostics means. The law of changing supercapacitor capacitance, active resistance and self-discharge current depends on many factors, such as technological inaccuracy in manufacturing mechanisms involved in the production of electrodes, instability in time of the force applied to press the shafts, with the help of which the layer of activated carbon is applied to the aluminum foil of the electrodes.

It happens due to the vibration of equipment during the operation process, as well as changeable linear velocity of aluminum foil when it passes through the application area. It is difficult to distinguish a dominant one among all these external factors.

Relying on the analysis of the external factors affecting supercapacitor production, it has been found out that the laws of distribution of supercapacitor diagnostic parameters are normal and may be expressed as follows:

$$f(C) = \frac{1}{\sigma_C \sqrt{2\pi}} e^{\left(\frac{C - M_C}{2\sigma_C^2}\right)}, \quad (33)$$

$$f(R) = \frac{1}{\sigma_R \sqrt{2\pi}} e^{\left(\frac{R - M_R}{2\sigma_R^2}\right)}, \quad (34)$$

$$f(Isd) = \frac{1}{\sigma_{Isd} \sqrt{2\pi}} e^{\left(\frac{Isd - M_{Isd}}{2\sigma_{Isd}^2}\right)}, \quad (35)$$

where σ_C, σ_R and σ_{Icp} are mean squared deviations of capacitance value C, active resistance value R and self-discharge current value Isd of supercapacitor; M_C, M_R and M_{Isd} are nominal values of capacitance C, active resistance R and self-discharge value Icp of supercapacitor.

Using the Mathcad symbolic mathematics package and expressions (33) – (35), dependencies of the first-order errors and the second-order errors on such parameters as $A_C = \sigma_{\tau C} / \sigma_C$, $A_R = \sigma_{\tau R} / \sigma_R$ and $A_{Isd} = \sigma_{\tau Isd} / \sigma_{Isd}$. have been calculated separately for capacitance C, active resistance R and self-discharge current Isd of supercapacitor.

These parameters determine the ratio between the mean squared deviation of the total measurement error of capacity C, active resistance R and self-discharge current Isd of supercapacitor and mean squared deviation of these parameters.

In the calculations, it was assumed that the control increments of the tolerance zones for the upper and lower limits are zero and tolerance zones for capacitance C, active resistance R and self-discharge current Isd of supercapacitor were considered equal to 5% of their nominal values. The nominal values of capacitance C, active resistance R and self-discharge current Isd of supercapacitor were considered equal to $C = 100$ F, $R = 7$ mOhm and $Isd = 0{,}1$ mA correspondingly.

The general probabilities of supercapacitor diagnostic errors of the first and the second order – α and β correspondingly – are defined by the expressions (7) and (12). The graphs of dependencies between the diagnostic errors of the first and the second order – α and β correspondingly – are shown in the fig. 1 (a and b, respectively).

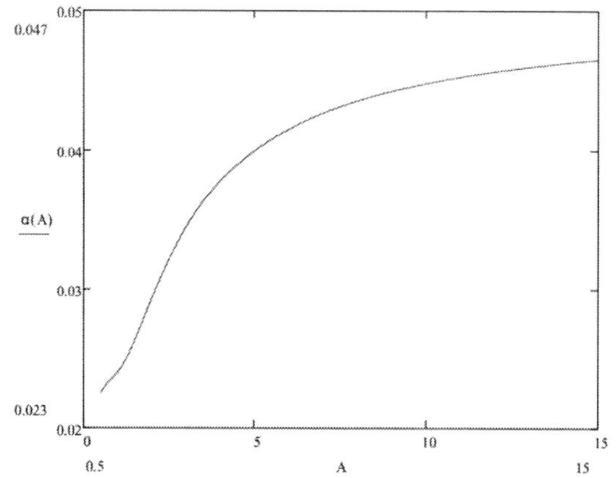

a)

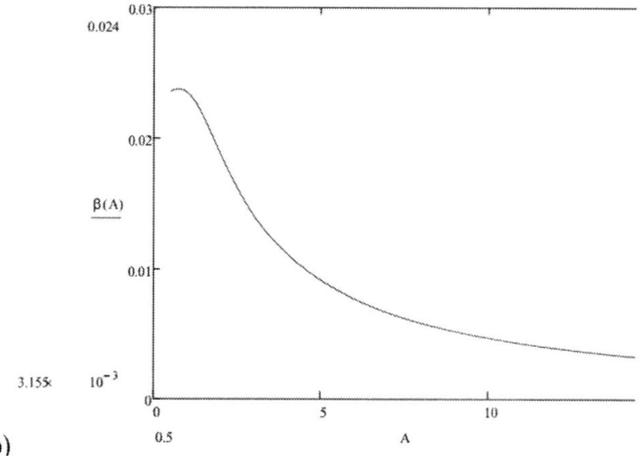

b)

a) graph of dependency between the first-order diagnostic error α and parameter A;
b) graph of dependency between the second-order diagnostic error β and parameter A

Fig. 1. Graphs of dependency between the first-order α the second-order β diagnostic errors and parameter A.

The instrumental reliability of supercapacitor diagnostics is determined by the errors of the first and second orders α and β according. The probability of making an erroneous decision in supercapacitor diagnostics will be equal to the sum of errors of the first and the second orders. The graphs of dependencies between the instrumental reliability of supercapacitor diagnostics and the probability of making an erroneous decision in the process of supercapacitor diagnostics and parameter A, are shown in fig. 2 (a and b, respectively).

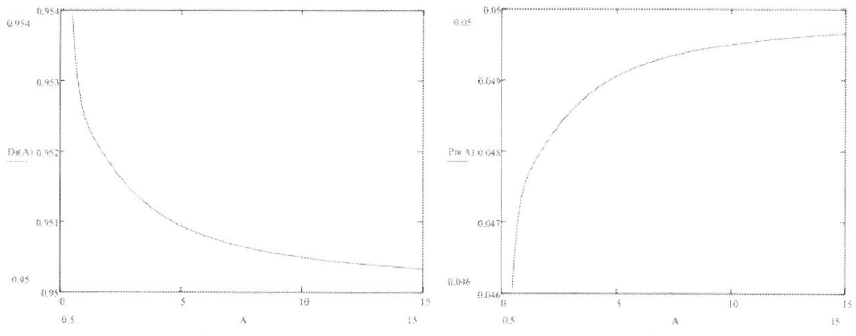

a) b)

a) graph of dependency between the instrumental reliability of diagnostics D_i and parameter A; b) graph of dependency between the probability of making an erroneous decision P_{II} and parameter A

Fig. 2. Graphs of dependencies between the instrumental reliability of diagnostic, the probability of making an erroneous decision and parameter A.

Analysis of the dependency between the instrumental reliability of diagnostics, which represents the probability of a correct decision for the above-set parameters, varies from 0,954 to 0,95.

Methodological reliability of diagnostics is determined by the ratio between the number of parameters that characterize technical condition of the object of control being included into the mathematical model of diagnostics and the total number of parameters that characterize technical condition of the object of diagnostics.

If the number of parameters that characterize technical condition of the object of diagnostics being included into the mathematical model of diagnostics is equal to the total number of parameters that characterize technical condition of the object of diagnostics, then methodological reliability of diagnostics is equal to one, and the total reliability of diagnostics will be equal to the instrumental reliability of diagnostics:

$$D_G = D_M \cdot D_I = 0,95\,. \tag{36}$$

In the case of using only one frequency, voltage and current, at which supercapacitor diagnostics is carried out, the methodological reliability of diagnostics sharply decreases and is defined by the expression

$$D_{MO} = \frac{3}{N}\,. \tag{37}$$

The more frequencies, operating voltages and currents are applied in the supercapacitor diagnostics, the more methodological reliability of diagnostics will increase in relation to the methodological reliability of supercapacitor diagnostics when only one frequency, voltage and current are applied. Increase in the methodological reliability of supercapacitor diagnostics while applying various frequencies, operating voltages and control currents can be defined by the expression

$$\mathrm{K} = \frac{D_{MF}}{D_{MO}} = \frac{L}{3}\,, \tag{38}$$

where D_{MF} is methodological reliability of supercapacitor diagnostics while applying L frequencies, operating voltages and currents.

7 Conclusions

Analysis of the expressions of the average supercapacitor diagnostic risks (31) and (32) shows that development of mathematical models of supercapacitors that most accurately describe their parameters and characteristics is extremely important for reducing the average supercapacitor diagnostic risks.

The methodological approach to increasing the reliability of supercapacitor diagnostic has been proposed, in terms of multi-parameter supercapacitor diagnostic by applying nonlinear, frequency dependent mathematical models of supercapacitors that take into account nonlinearity, frequency dispersion of parameters and the effect of transient processes in supercapacitors.

The procedure for increasing the reliability of supercapacitor diagnostic has been suggested on the basis of minimizing the average supercapacitor diagnostic risks that do not depend on the probability densities of probe effects of supercapacitor voltage and current.

References

1. Conway B. E. Electrochemical Supercapacitors: Scientific Principles and Technological Application / B. E. Conway. – New York : Plenum, 1999. – P. 435
2. Martynyuk V. V. Fractional Model of an Electrochemical Capacitor / V. V. Martynyuk, M. D. Ortigueira // Journal of Signal Processing. – V. 107, 2015, pp. 355-360
3. IEC Publication 50 (191) June 1988: Chapter 191 of the IEV: Reliability, maintainability and quality of service
4. Radford, George S. The Control of Quality in Manufacturing, New York: Ronald Press Co
5. D. C. Montgomery and G.C. Runger, Applied Statistics and Probability for Engineers. 2nd Edition, John Wiley & Sons, New York, 1999
6. Martynyuk V., Vdovin O., Boyko J. and. Vlasenko N. "Super-high capacitor analyzer with compensation of common-mode error," in Proc. IMECO TC-4, 2001, pp. 340–343
7. Martynyuk V., Makaryshkin D., Boyko J. "Frequency Domain Analysis for Electrochemical Supercapacitors" in Proc. IMEKO TC-4, 2007, pp. 357–361

Method for initial assessment of unit costs of public city transport means operation

Bogdan Landowski[1,*], *Łukasz* Muślewski[1], *Michał* Pająk[2], and *Oleh* Polishchuk[3]

[1] University of Science and Technology in Bydgoszcz, Faculty of Mechanical Engineering, Al. prof. S. Kaliskiego 7, 85-796 Bydgoszcz, Poland
[2] University of Technology and Humanities in Radom, Faculty of Mechanical Engineering, ul. Stasieckiego 54, 26-600 Radom, Poland
[3] Khmelnitsky National University, Instytutsʹka Street, 11, Khmelʹnytsʹkyi, Khmelʹnytsʹka oblast, 29000, Ukraine

Abstract. The study deals with the problems connected with modeling and simulation of technical objects operation processes carried out within a certain class of complex technical systems. The analyzed system is a city bus transport system. The study presents a method for assessment of unit costs involved in a technical object's being in selected operational states. One of the subsets of data to be used for modeling the process of a technical object operational state changes, includes data concerning the costs borne by the system when the object is in the analyzed operational states. In order to determine this data, it is often necessary to modify the system of costs registration used in a given operation system. Since it is a very expensive and log process, a method of initial assessment of indispensable data has been proposed, on the basis of a costs registration method and simulation of the operation process. The study also includes basic assumptions of the operation process model, being an important element of the developed method. The assumptions of the operation model have been formulated on the basis of identification of a real operation system of city bus transport system and experimental tests carried out in system.

1 Identification

The technical objects analysed in the study are buses used in a public city transport system. The complexity of its operation is caused by both a big number of its subsystems and elements situated at different levels performing different functions as well as dynamically changing processes and task the technical objects have to accomplish in the system they are used. Some processes are carried out in a predetermined and planned way, whereas others are of random character.

The analysed technical system operates buses of different types and makes. Services are provided on a several traffic routes. One of the most important decision issues is the choice of transport means to be used for particular transport routes. In work [1], an example is given of a method used for the choice of technical objects to provide transport services, on the basis of a homogenous Markov process as the operation process model, whereas in work [2], economic aspects connected with a purchase of a transport means to be used in a public transport system, are discussed.

The primary goal of a public bus transport system operation is to provide passengers with safe transport services within an assigned quantitative and territorial scope. However, due to limited financial means, to be provided to achieve the goal, economic efficiency of

* Corresponding author: bl-sluzbowy@wp.pl, bogdan.landowski@utp.edu.pl

the offered transport services needs to be constantly increased, thus reducing the outlays on public transport functioning.

In order to determine the values of selected decision variables to be used for rational control of the operation process carried out in a given system, an operation model and a software for the process simulation, have been developed [3, 4]. One of the subsets of data necessary to conduct simulation experiments, is the one which contains the values of unit costs borne by the system while a bus (of a given type and form the identified category of objects [3]) is in the distinguished operational states. For the purpose of accurate determination of this data, it is necessary to modify the system of costs registration used in this system. Since it is an expensive and time consuming process, a simple method for initial assessment of necessary data, developed on the basis of the costs registration method used in the analyzed system, has been proposed.

It needs to be emphasized that the obtained in such a way data, enables only rough estimation of the above mentioned costs and cannot replace necessary modifications of the costs registration system. The results of an experimental simulation are based on input data obtained in this manner, however, they do not enable its qualitative analysis.

2 Model of the operation process

The model of the operation process has been developed on the basis of an analysis of the operational states space and events from the operation of vehicles used in the public bus transport system.

The model of the operation process has been described in terms of its states and possible transitions between these states. The states of the process are characterized by distributions of times of being in particular states and profits (losses) connected with technical objects' being in the identified states [3].

Random process $X(t)$ with finite state space S and a set of parameters R+ (subset of real numbers >= 0) is a natural mathematical model of the operation process of many categories of technical objects [5, 6]. Homogeneous stochastic processes, including Markov and semi Markov processes, and decision Markov and semi Markov processes, are commonly used for modelling of operational state changes. [3, 4, 6-11].

Sets of source data, necessary to formulate the model assumptions and its initial verification, were obtained from experimental tests performed in a real research object by the method of passive experiment. Due to a limited volume of this study, only the major assumptions of the proposed model have been presented.

Let $S_1, S_2, ..., S_k$ $(k \in N)$ denote the analysed, significant operational states, which technical objects can go through.

Each of the operated technical objects can, in a given time t, $t \in R_+$, be only in one of the identified states. Being in a particular state generates incomes and/or costs connected with the system operation.

The operation process consists of a sequence of particular states. The order of their occurrence, time of being in particular states and frequency of their occurrence depend on individual features of particular technical objects, characteristics of the processes they undergo and the structure of subsystems cooperating during cooperation.

In result of a city bus transport system identification and the process of its functioning, on the basis of its states significance criteria, (criterion of profit), the following bus operational states have been distinguished [3]:

S_1 – state in which a technical object is functioning (while providing transport services),

S_2 – state in which a technical object is waiting for a technical emergency service,

S_3 – state in which a technical object is being repaired and the repair is performed within the transport system environment,

S_4 – state in which a technical object is waiting for a repair to be performed within the transport system environment,
S_5 – state in which an object's being repaired within the transport system environment,
S_6 – state in which a technical object is waiting to be diagnosed,
S_7 – state in which a technical object's being diagnosed,
S_8 – state in which a technical object's daily servicing,
S_9 – state in which a technical object waiting to undertake the transport task,
S_{10} – state in which a technical object stands idle due to inoperability of the environment,
S_{11} – state in which a technical object stands idle because of organizational reasons.

Stochastic process $\{X(t), t \geq 0\}$ with a finite set of S states is a mathematical model of the process of technical objects operation. If $X(t) = S_i$ ($i = 1, 2, ..., k$), then the considered technical object, in time t, is in state S_i. For the purpose of this study, a semi Markov process has been assumed to be $X(t)$ process.

Using semi-Markov processes for mathematical modelling, the following major assumptions have been accepted:
- semi-Markov process reflects the modelled real operation process well enough from the point of view of the research purpose,
- the modelled operation process has a finite number of states,
- random variables denoting duration times of the operation process have finite and positive mean values and variances,
- the distinguished states of the operation process are positively returning states,
- random process $X(t)$, being a mathematical model of the operation process, is of homogenous character.

Analysed process $X(t)$ (homogenous semi-Markov process) is fully defined when the following data is known [6]: initial distribution of process $\mathbf{P_0}$, matrix $\mathbf{P}$ of probabilities of transitions between the process states (probability of transitions of the inserted Markov chain) and distribution of random variables T_{ij} (T_{ij} - random variable denoting the time when the object is in state S_i under the condition that the next state is state S_j).

A computer program for a numerical simulation of the operation process simulation has been developed for the proposed model. The problems connected with simulation of operational state changes is presented, among others, in works [3, 4, 8-10, 12-13]. Data necessary for simulation of the stochastic process which is mathematical model of the operation process, has been assessed on the basis of initial experimental tests.

3 The research method

Values of synthetic indexes of costs involved in the system operation in respect to the accepted structure, in monthly periods, have been determined on the basis of the existing costs registration system.

The measure of transport tasks accomplishment degree by buses of a given type is the distance they have covered while providing their daily transport services. The so called 'vehicle kilometres' are used for financial settlements and economic analyses. One 'vehicle kilometre" corresponds to the distance of 1 km. covered by the vehicle of a given type.

The idea of the method involves simulation of the analysed operation process and determination of selected characteristics of the analysed process and next estimation of incomes and unit costs connected with technical objects' being in a given state, on the basis of economic data obtained from the analysed company and simulation results.

The operation process simulation of a single technical object, according to the accepted assumptions, involves numerical simulation of the analysed stochastic process, being a mathematical model of the process of a technical object operational state changes. Algorithms enabling numerical simulation of the stochastic process have been developed. Data necessary for determination of the analysed stochastic process is the data necessary for simulation. Indexes enabling analysis of the modelled state change process are determined during the simulation.

After determination of time interval, for which, appropriate components of the unit costs vector **N**, unit incomes **P**, need to be determined, on the basis of the existing costs registration system, particular components of costs (KSMZK). The next step involves carrying out a simulation of the operation process. Particular components of vectors **N** and **P** are determined on the basis of particular components and simulation results costs, according to the below dependency.

Applied denotations:

a) resultant data:

P=[p_i] – unit income vector,

N=[n_i] – unit costs vector,

p_i – income generated by a random bus for a time unit (given type and category), when it is in the i-th operational state,

n_i – cost per time unit, borne by the system, when a random bus (of a given time and a given category) of the system is in the i-th operational state,

$i = 1, 2, ..., k$ – index of the operational state,

k – number of identified operational states (for the analysed system $k = 11$).

b) data from the costs registration system:

T – time of experiments, days

La – real number of buses (of a given type) used in a given system,

KSMZK = {*MPA, MPO, P, PZU, OG, KPC, KRB, KW, KO, WK, CWK*} – data obtained from the costs registration system for a given type of buses in period of investigation T,

MPA – costs of fuel,

MPO – costs of the remaining materials,

P – costs of drivers' salaries with derivatives,

PZU – costs of insurance,

OG – costs of tires,

KPC – costs of daily services,

KRB – costs of current repairs,

KW – department costs,

KO – general costs,

WK – number of the so called 'vehicle kilometres',

CWK – income gained by the system for one 'vehicle kilometres'.

c) data necessary for simulation of the operation process (data for the analysed type and category of buses obtained from the experimental tests):

T = [T_i] – vector of mean times of a bus being in the distinguished operational states,

T_i – mean time of a bus being in the i-th operational state time,

$\mathbf{D^2T}$=[D^2T_i] – vector of standard deviation of times of being in a distinguished operational state,

D^2T_i – standard deviation of the mean time of being in the i-th operational state,

R=[R_i] – vector of codes of time distributions of being in the i-th operational state,

R_i – code of the type of distribution of random variable defining the time of a bus being in the i-th operational state, was accepted to be: R_i={1,2,3}, 1 - gamma distribution, 2 – Weibul distribution, 3 – logonormal code,

P=[p_{ij}] – matrix of probabilities of transitions between the distinguished operational states,

p_{ij} – probability of transition from state i to state j,

La – number of buses.

d) simulation results:

Ts – total time of simulation,
ST_i – summary time of being in the i-th state, $i = 1, 2, ..., k$,
LW_i – summary number of entrances to the i-th state.
Calculations:
- income p_i while being in states from S_2 to S_{11} (it is assumed that the system generates income only when the bus is in state S_1):

$$p_i = 0 \qquad \text{for } i = 2, 3, ..., 11 \tag{1}$$

- income p_1 in state S_1:

$$p_1 = WK \cdot CWK / ST_1 \tag{2}$$

- regular costs KS_i:

$$KS_i = (KW + KO) / Ts \text{ for } i = 1, 2, ..., k, \tag{3}$$

$$KS_1 = KS_2 = KS_k = KS \tag{4}$$

- unit cost n_1 born in state S_1:

$$n_1 = KS + KZ_1 \tag{5}$$

$$KZ_1 = (MPA + MPO + P + PZU + OG) / ST_1 \tag{6}$$

- unit cost n_i born in state S_i, for $i = 2, 4, 6, 9, 10, 11$:

$$n_i = KS \tag{7}$$

- unit cost n_3 borne in state S_3:

$$n_3 = KS_3 + KZ_3 \tag{8}$$

$$KZ_3 = (x \cdot KRB) / ST_3 \tag{9}$$

where:
x – share of current connected with technical emergency unit repairs in the total costs of repairs,
- unit cost n_5 born in state S_5:

$$n_5 = KS_5 + KZ_5 \tag{10}$$

$$KZ_5 = (\mathrm{z} \cdot KRB) / ST_5 \tag{11}$$

where:
z – share of current costs connected with renovation of the technical object, performed within the operation system, in the total repair costs,
- unit cost n_7 born in state S_7:

$$n_7 = KS_7 + KZ_7 \tag{12}$$

$$KZ_7 = [(1\text{-x-z}) \cdot KRB] / ST_7 \tag{13}$$

- unit cost n_8 borne in state S_8:

$$n_8 = KS_8 + KZ_8 \tag{14}$$

$$KZ_8 = KPC \,/\, ST_8 \tag{15}$$

The presented calculation procedure does not include an event when a bus needs to be replaced by another bus due to its failure, and there is a need to use a substitute bus instead of the damaged one. If the data obtained from the research object makes it possible to estimate probability values of occurrence of a necessity to provide a substitute bus for the one which is damaged, there is a possibility of the calculation procedure modification. For instance, the following dependencies can be used for states S_2 and S_3:

$$KZ_2 = (Pw \cdot LW_2 \cdot KW) \,/\, ST_2 + Pw \cdot n_1 \tag{16}$$

$$KZ_3 = (x \cdot KRB) \,/\, ST_3 + (Pw \cdot LW_3 \cdot KW) \,/\, ST_3 + Pw \cdot n_1 \tag{17}$$

where:
Pw – probability of occurrence of an event involving the necessity to provide a substitute bus to replace a bus which has been damaged and is in state S_2 (estimated on the basis of data obtained from the analysed system),
KW – regular cost connected with providing a substitute bus.

It is possible to say, on the basis of the research object identification, that assuming that $n_{10} = KS_{10}$ is not a significant error because there are no additional costs borne by the studied system due to delays caused by inoperability of the environment.

4 Conclusions

The procedures presented in this study refers to a system of a city bus transport. However, due to assumed generality of the provided description, they can be used for analysis of the operation processes applied in other transport systems.

It needs to be emphasized, that the data on unit costs provided by the proposed method, can only be used for estimation of the above listed costs. The proposed method cannot replace necessary modifications of the costs registration system used in similar transport companies. However, according to the authors of this study, the results of simulation experiments with the use of input data obtained in this way, enable their qualitative analysis.

References

1. B. Landowski, D. Perczyński, P. Kolber, Ł. Muślewski, An example of Markov model of technical objects maintenance process. Engineering Mechanics 2016, 22nd International Conference, may 9 – 12, 2016, Svratka, Czech Republic, Book of full texts, Institute of Thermomechanics Academy of Sciences of the Czech Republic, pp. 346-349 (2016)
2. B. Landowski, D. Perczyński, Ł. Muślewski, P. Kolber, Economic aspects of a city transport means purchase. Proceedings of 58th International Conference of Machine Design Departments - ICMD 2017, Publisher: Czech University of Life Sciences Prague, Czech Republic, pp. 194-199 (2017)
3. M. Woropay, L. Knopik, B. Landowski, *Modeling processes of operation in transport system (Modelowanie procesów eksploatacji w systemie transportowym)*, Library of Maintenance Problems (Maintenance Technology Institute, Bydgoszcz-Radom 2001)
4. B. Landowski, M. Woropay, A. Neubauer, *Controlling reliability in the transport systems (Sterowanie niezawodnością w systemach transportowych)*, Library of Maintenance Problems (Maintenance Technology Institute, Bydgoszcz-Radom 2004)
5. N. Buslenko, W. Kalasznikov, I. Kovalenko, *Theory of complex systems* (PWN, Warszawa, 1979)

6. F. Grabski, Theory of semi-Markov processes of technical objects operation and maintenance, Scientific Bulletins of AMW (Polish Naval Academy), no. **75 A**, Gdynia (1982)
7. M. Baykal-Gürsoy, Semi-Markov Decision Processes, Wiley Encyclopedia of Operations Research and Management Sciences (2010)
8. B. Landowski, Ł. Muślewski, Decision model of an operation and maintenance process of city buses, Proceedings of 58th International Conference of Machine Design Departments - ICMD 2017, Publisher: Czech University of Life Sciences Prague, Czech Republic, pp. 188-193 (2017).
9. B. Landowski, M. Pająk, B. Żółtowski, Ł. Muślewski, Method of building a model of operational changes for the marine combustion engine describing the impact of the damages of this engine on the characteristics of its operation process. Polish Maritime Research, no **4** (96), vol. 24, pp. 67-76 (2017)
10. B. Landowski, Ł. Muślewski, Numerical simulation of stochastic process as a model of technical object state changes. Engineering Mechanics 2018, 24nd International Conference, may 14 – 17, 2018, Svratka, Czech Republic, Book of full texts, Institute of Theoretical and Appiled Mechanics of the Czech Academy of Sciences, Prague, pp. 485-488 (2018)
11. L. Knopik, K. Migawa, Optimal age-replacement policy for non-repairable technical objects with warranty. Maintenance and Reliability, Vol. **19**, No. 2, pp. 172 - 178 (2017)
12. Ł. Muślewski, Evaluation Method of Transport Systems Operation Quality. Polish Journal of Environmental Studies 2A **18** (2009)
13. A. Bustilo, D. Yu. Pimenov, M. Matuszewski, T. Mikołajczyk, Using artificial intelligence models for the prediction of surface wear based on surface isotropy levels. Robotics and Computer Integrated Manufacturing, Vol. **53**, Iss. 5, pp. 215-227 (2018)

Diagnostics of the effect of prior cold deformation and aging time on intergranular corrosion resistance in case of austenitic stainless steel

Zdzisław Ławrynowicz[1,1*]

[1]University of Science and Technology UTP, Mechanical Engineering Faculty, Department of Materials Science and Engineering, Av. Kaliskiego 7, 85-796 Bydgoszcz, Poland

Abstract. Intergranular corrosion of commercial stainless steel type X6CrNiTi18-10 (1.4541, AISI 321) is frequently observed in several process environments. These localized attacks are normally attributed to the carbide precipitation and concomitant depletion of chromium near grain boundary due to steel exposure to sensitization temperature. Such undesirable microchemistry is expected to be changed further if the material undergoes deformation prior to sensitization. Present paper deals with the influence of a wide range of cold rolling (5, 10, 15 and maximum 40% cold deformation) and the sensitization time (aging at 700°C for 0.12, 0.5, 1, 4, 16 and 32 hours) on intergranular corrosion (IGC). The assessment of the degree of intergranular corrosion was based on the measurement of mass loss and observation of corroded surfaces on optical microscope. The consequences of deformation on IGC have been investigated by using EN ISO 3651-1methods (Huey test). Cold deformation before sensitization reduces the intensity of intergranular corrosion of this steel. The deformed structure created during the cold work process, numerous slip planes and the twins boundaries are just like the grain boundaries and the places where the chromium carbides preferentially precipitates.

1 Introduction

The stainless alloy X6CrNiTi18-10 (1.4541, AISI 321) is a general purpose austenitic stainless steel with a face centred cubic structure. It is essentially non-magnetic in the annealed condition, and can only be hardened by cold working. Titanium is added to suppress grain boundary chromium carbide precipitation, and to reduce susceptibility to intergranular corrosion, through the formation of very stable titanium carbide[1-3].

It knows that the most important corrosion form in case of austenitic steels is the intergranular corrosion. The well-known susceptibility of austenitic stainless steels to intergranular corrosion after heat-treatment in the temperature range of 500°–800°C (sensitization) has long been attributed to depletion of Cr from regions of the alloy matrix adjacent to grain boundaries in which $Cr_{23}C_6$ had precipitated. Those regions of the steel in which the local Cr composition falls below about 12% have a diminished ability to form a passive film and hence corrode preferentially [4] (Fig. 1).

Stainless steels are common material of construction as these offer a wide range of corrosion resistance along with good fabrication and mechanical properties to many industrial environments. At the same time, these alloys are prone to microstructural changes along with changes in chemistry when exposed to sensitization temperatures possibly due to faulty heat treatment or welding operation. This causes frequent catastrophic failure of engineering components.

* Corresponding author: lawry@utp.edu.pl

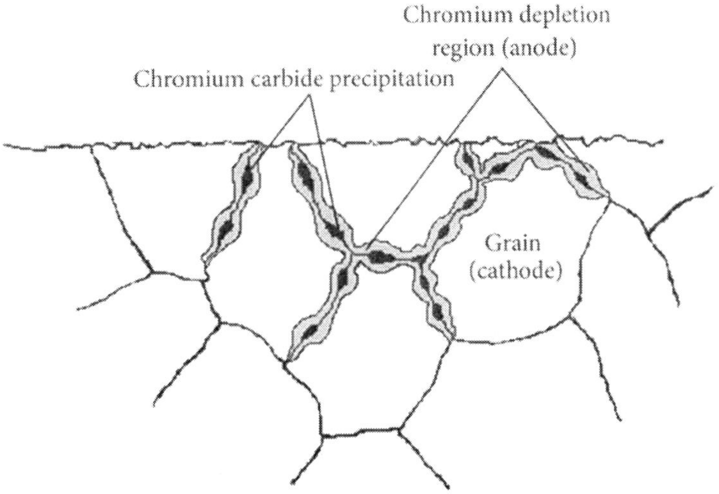

Fig. 1. Schematic representation of the carbide precipitation at grain boundaries in austenitic stainless steel [4].

Before shaping the component into a desired geometry, steels often undergoes a thermo-mechanical processing during which materials change certain properties. Cold deformations are assumed to alter the properties of steels through increasing dislocation density which apart from its effect on mechanical properties, may help in accelerating the diffusion mechanism, phase transformation, lowering the activation energy for faster nucleation [2]. There have been contradictory remarks in the literature about the net influence of deformation on the sensitization kinetics of stainless steels. It was first studied by Bain and Tedmon [1, 2] who reported the lower corrosion of strained material compared to that of annealed. On the other hand Briant [3] observed higher resistance of the unstrained material. One reason of such conflicting results could be the consideration of low level of deformations only. Consequently, the need was felt to investigate the sensitization and intergranular corrosion behavior of austenitic stainless steel with wide range of deformation in order to understand the mechanism.

The present paper reports the results of various deformation levels on IGC and the attempt has been made to establish the correlation of corrosion rates with sensitization time and microstructure.

2 Material

The material that was investigated was the commercially available X6CrNiTi18-10 (1.4541, AISI 321) stainless steel. The chemical composition of the investigated material, which was determined by optical emission spectroscopy, is shown in Table 1. The samples provided were in a supersaturated state.

Table 1. Chemical composition of the X6CrNiTi18-10 steel used in this study. All concentrations are given in wt.%.

%C	%Cr	%Ni	%Mn	%P	%Si	%S	%Ti
0.12	17.9	9.6	1.50	0.014	0.016	0.016	0.58

The starting material was in the form of steel strips, thickness g = 3mm and width b = 100mm. These strips were then cold rolled with various degrees: 5, 10, 15 and maximum 40%. Samples of size: 100 × 20 × 3mm were cut from the deformed strips and 6 batches of samples were prepared, each batch containing: a sample without deformation, and cold worked samples of 5, 10, 15 and 40%, which were subjected to a sensitizing anealing at 700°C during : 5 minutes, 30 minutes, 1h, 4h, 16h and 32h. Sensitization is a terminology used to describe the precipitation of carbides at grain boundaries of alloys during high-temperature exposure. The essence of sensitization is that the rate of attack of austenitic stainless steels under intergranular attack conditions is extremely sensitive to the carbides precipitated in the 500 to 800°C temperature range.

After the heat treatment sensitizing to intergranular corrosion, the samples were cooled in air. Two dimensions of samples were prepared to assess the intensity of intergranular corrosion:

a) 20 × 20 × 3 mm samples were used for the weight loss test by boiling in HNO_3 (Huey test),

b) metallographic samples were made on 10 × 20 × 3 mm samples and after boiling in the Huey reagent, observations were made on optical and scanning microscopes,

Huey test was aimed at determining the intensity of intergranular corrosion by measuring the loss of mass during boiling in HNO_3. Boiling consisted of 3 periods of 40 hours each with each exchange of acid after boiling. Before weighing, the samples were thoroughly rinsed in water and the corrosion products were removed with a soft rubber. After drying, the samples were weighed with an accuracy of 0.0001 g.

3 Experimental procedures

Optical microscopy was used to examine etched structures. Microscopic images were recorded using the inverted metallographic microscope Nikon MA100 equipped with the ERC5s digital camera and the ZEN 2011 archiving program.

The Huey test (EN-ISO 3651-1. Determination of resistance to intergranular corrosion of stainless steels-Part 1: Austenitic and ferritic-austenitic (duplex) stainless steels-Corrosiontest in nitric acid medium by measurement of loss in mass) [5] was used, consisting of boiling in concentrated HNO_3 (nitric acid) at 95°C for three periods of time, each 40 hours. This test can generate intergranular corrosion in the austenitic steel. Before immersing the test specimens in nitric acid, they were well degreased with acetone solution and weighed to the fourth point of decimal (grams). After each test period the specimens were cleaned and weighed. The corrosion rates were obtained by using the formula (1) and are reported in ($g \times m^{-2} \times h^{-1}$).

4 Results of Huey test

Corrosion losses of sample mass after each boiling cycle are presented in Table 2.

Table 2. Mass losses of samples after boiling in HNO_3 during 40, 80 and 120 hours

Lp.	Time of sensitization	After solution heat treatment, State supersaturated			Cold work 5%		
		Weight loss of the sample in grams after boiling in HNO_3 at the time:			Weight loss of the sample in grams after boiling in HNO_3 at the time:		
		40h	80h	120h	40h	80h	120h
		G_1	G_2	G_3	G_1	G_2	G_3
1.	5 min.	0.0101	0.0212	0.0473	0.0108	0.0255	0.0533
2.	30 min.	0.0185	0.1069	0.2671	0.0109	0.0273	0.0783
3.	1 h	0.0216	0.1506	0.3005	0.0138	0.0303	0.0770
4.	4 h	0.0219	0.0795	0.2053	0.0128	0.0307	0.0510
5.	16 h	0.0152	0.0305	0.0406	0.0123	0.0239	0.0443
6.	32 h	0.0170	0.0433	0.0473	0.0129	0.0248	0.0520

Continued Table 2.

No.	Time of sensiti-zation	Cold work 10%			Cold work 15%			Cold work 40%		
		Weight loss of the sample in grams after boiling in HNO_3 at the time:			Weight loss of the sample in grams after boiling in HNO_3 at the time:			Weight loss of the sample in grams after boiling in HNO_3 at the time:		
		40h	80h	120h	40h	80h	120h	40h	80h	120h
		G_1	G_2	G_3	G_1	G_2	G_3	G_1	G_2	G_3
1.	5 min.	0.0103	0.0205	0.0395	0.0103	0.0167	0.0388	0.0103	0.0198	0.0306
2.	30 min.	0.0115	0.0194	0.0310	0.0111	0.0152	0.0372	0.0020	0.0102	0.0341
3.	1 h	0.0101	0.0193	0.0332	0.0078	0.0158	0.0364	0.0099	0.0175	0.0239
4.	4 h	0.0114	0.0183	0.0378	0.0096	0.0147	0.0240	0.0068	0.0153	0.0222
5.	16 h	0.0125	0.0226	0.0359	0.0116	0.0210	0.0458	0.0123	0.0228	0.0338
6.	32 h	0.0132	0.0203	0.0427	0.0171	0.0249	0.0403	0.0126	0.0246	0.0257

Assessment of the intensity of intergranular corrosion after boiling test in HNO_3 was based on determining the average corrosion rate V_{cor} according to PN-EN ISO 3651-1 [5]:

$$V_{cor} = \frac{10000m}{S\,t} \quad (1)$$

where:

t - time of cooking the solution in hours,

S - surface area of the test sample, in square centimeters

m - the average loss in mass of the sample during the test in grams

$m = G_1 - G_2$

where:

G_1 - weight of the sample before boiling (g),

G_2 - sample weight after boiling (g).

The corrosion rate after each cooking cycle and the average corrosion speed are shown in Table 3.

Table 3. Corrosion rate V_1, V_2 and V_3 after the boiling cycle during 40, 80 and 120 hours and the average arithmetic corrosion rate V_{av}, ($g \times m^{-2} \times h^{-1}$)

No.	Time of sensiti-zation	After solution heat treatment, State supersaturated				Cold work 5%			
		Corrosion rate, ×10 V_{av}, ($g\times m^{-2} \times h^{-1}$)				Corrosion rate, ×10 V_{av}, ($g\times m^{-2} \times h^{-1}$)			
		V_1	V_2	V_3	V_{av}	V_1	V_2	V_3	V_{av}
1.	5 min.	2.4	5.3	11.6	6.4	2.6	6.3	13.0	7.3
2.	30 min.	4.4	25.7	64.3	31.5	2.7	6.7	19.2	9.5
3.	1 h	5.3	36.8	73.6	38.6	3.4	7.2	19.1	9.9
4.	4 h	5.1	18.7	48.3	24.0	3.0	7.2	12.0	7.3
5.	16 h	3.6	7.3	9.7	6.5	2.8	5.8	10.8	6.5
6.	32 h	4.1	10.4	11.4	8.6	3.1	5.9	12.5	7.2

continued Table 3.

No.	Time of sensiti-zation	Cold work 10%				Cold work 15%				Cold work 40%			
		Corrosion rate, ×10 V_{av}, (g× m^{-2} ×h^{-1})				Corrosion rate, ×10 V_{av}, (g× m^{-2} ×h^{-1})				Corrosion rate, ×10 V_{av}, (g× m^{-2} ×h^{-1})			
		V_1	V_2	V_3	$V_{av.}$	V_1	V_2	V_3	$V_{av.}$	V_1	V_2	V_3	$V_{av.}$
1.	5 min.	2.4	4.9	9.4	5.6	2.4	4.0	9.3	5.2	2.6	5.0	7.8	5.1
2.	30 min.	2.7	4.7	7.4	4.9					2.0	3.0	8.7	4.6
3.	1 h	2.4	4.8	8.1	5.1	1.9	3.8	8.9	4.9	2.5	4.4	6.1	4.3
4.	4 h	2.8	4.4	9.2	5.5	2.3	3.6	5.9	3.9	2.2	3.9	5.7	3.9
5.	16 h	3.0	5.4	8.6	5.7	2.8	5.0	10.1	5.4	3.1	5.8	8.6	5.0
6.	32 h	3.2	4.9	10.4	6.2	4.1	5.9	9.7	6.6	3.2	6.2	6.5	5.3

The dependence of changes in the intensity of intergranular corrosion from the time of sensitization and degree of cold work are shown in Figure 2.

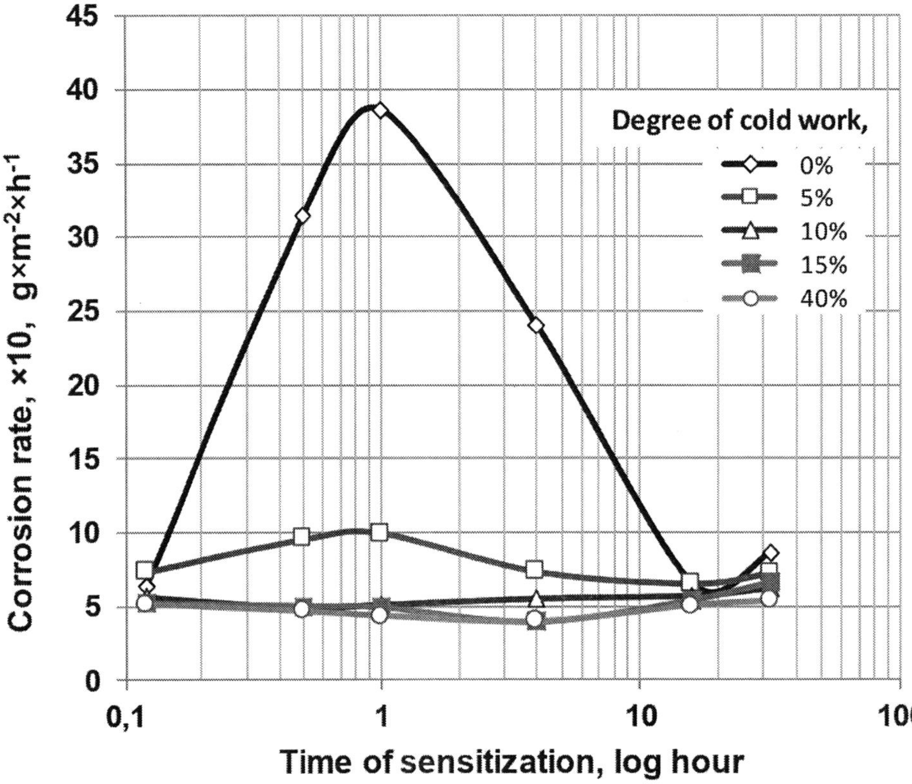

Fig. 2. Dependence of the intergranular corrosion rate in X6CrNiTi18-10 steel on time of sensitization for different degrees of cold work.

5 Microstructure and analysis

The microstructure of the investigated stainless steel sample (Fig. 3a), observed by an optical microscope, was austenitic. Metallographic investigation revealed austenitic polygonal grain features intersected with annealing twins and slips, produced by cold working (Fig. 3b).

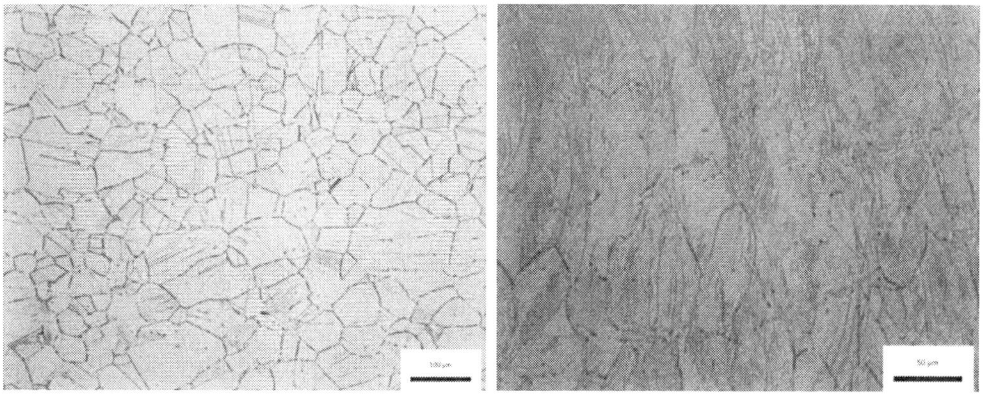

a) b)

Fig. 3. Microstructure of X6CrNiTi18-10 steel, a) supersaturated state, no deformation, b) cold work 40%.

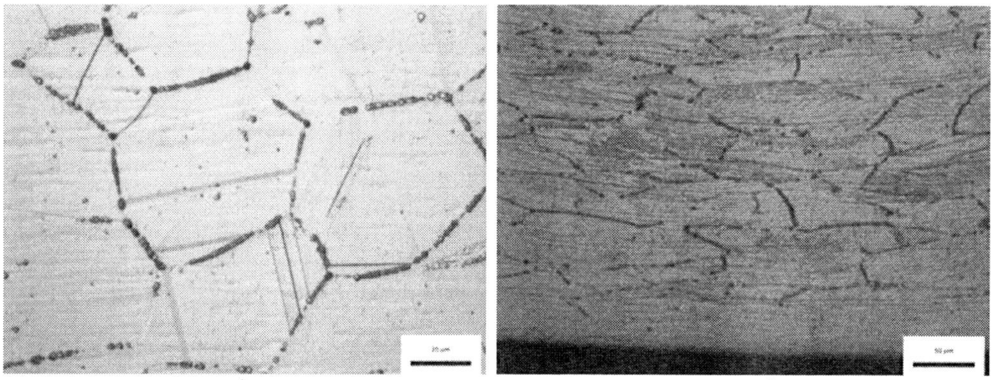

a) b)

Fig. 4. Microstructure of X6CrNiTi18-10 steel after sentizitation at 700°C/1h, a) cold work 0%, b) cold work 40%

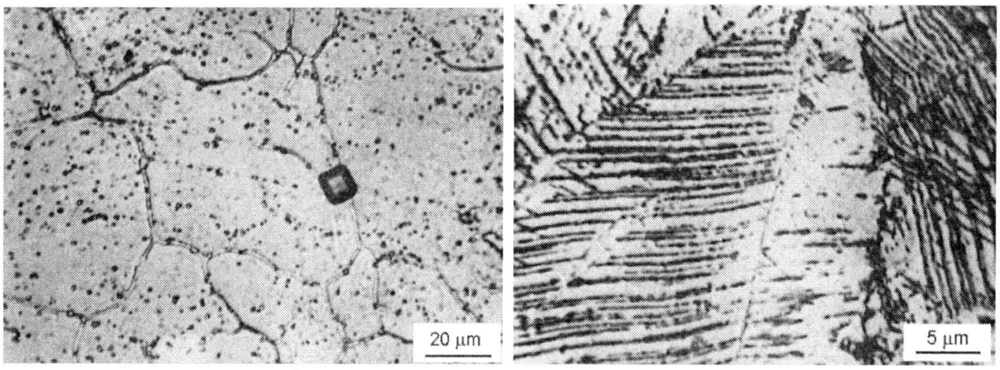

a) b)

Fig. 5. Microstructure of X6CrNiTi18-10 steel after sentizitation at 700°C, a) 32 hours, cold work 10%, b) 16 hours, cold work 40%.

An attempt to boil in HNO_3 as well as microscopic observations showed that the deformation (starting from just 5% small crushing forces) very clearly reduces the susceptibility of X6CrNiTi18-10 steel to intergranular corrosion (Fig. 2). This steel, in the supersaturated state, intensely corrodes intercrystalline, while the deformation of 5%, especially 10%, 15% and maximum 40%, significantly reduces the tendency of this steel to intercrystalline corrosion (Fig. 2).

For X6CrNiTi18-10 steel in the supersaturated state, the most intense corrosion occurs after heating at 700°C from 30 minutes to 4 hours. After prolonged heating periods at a temperature of 700°C, the intensity of corrosion decreases. This phenomenon may be a confirmation of the validity of the theory of depletion of chromium of grain boundaries, which explains the sensitization of austenitic steels to intergranular corrosion [6-8].
The Fig. 2 shows that as the heating time increases about 1 hour for steel in the supersaturated state, the intensity of corrosion decreases.
In this first stage of annealing (up to about 1 hour), due to the precipitation of the carbide phase at the boundaries of austenite grains, the process of depleting in chromium the border regions of the grain takes place, thus reducing their stability.
In the second stage, for longer boiling times, there is diffusion of chromium atoms from the internal grain regions towards the border zones, which were previously depleted in chromium. These zones, after the time required for the diffusion of chromium atoms to the grain boundaries, are again enriched with chromium, and when its content reaches 12.5%, they regain stability and thus become resistant to the corrosive effects of the environment.
Cold work of X6CrNiTi18-10 steel very clearly improves its resistance to intergranular corrosion. With increasing deformationy, corrosion resistance of this steel is also gradually increasing. The obtained results are consistent with the general view that cold work before sensitization, reduces the intensity of intergranular corrosion [7-9]. This is due to the deformation of the structure (twins, slip bands, stacking errors, grain refinement) (Fig. 4, 5). All these places are similar to the grain boundaries, in places of the privileged precipitation of carbide phases, as a result of which the concentration of the separated phases on the grain boundaries is smaller, and thus the corrosion rate decreases. Also the composition of steel has a large influence on the tendency of steel to intercrystalline corrosion in the case of the addition of titanium [10-11]. Titanium reduces the formation rate of $M_{23}C_6$ carbides to a very large extent, because a significant part of the carbon is bound in titanium carbonitride Ti(CN).

6 Conclusions

The effect of prior cold deformation and sensitization time on intergranular corrosion resistance in X6CrNiTi18-10 austenitic stainless steel were investigated. The following conclusions were drawn:

1. Intergranular corrosion of undeformed specimens increases with sensitization time of exposure up tp 1 hour. Sensitization of the X6CrNiTi18-10 steels accelerated the precipitation of the carbides which were also detected on the grain boundaries and in the austenitic matrix in the case of cold working samples.
2. High deformations reduces rate of intergranular corrosion. After~40% cold working, even the sites inside the grain matrix have high energy, and carbides can nucleate there easily.

References

1. E. C. Bain, R.H. Aborn, and J.J.B. Rutherford, Trans Amer. Soc. Steel Treating, **21**, 481-509 (1993)
2. C.S. Tedmon, Jr. D.A. Vermilye and D.E. Broecker, Corrosion, **27**, 104-106 (1971)
3. C.L. Briant and A.M. Ritter, Metall Trans A, **11A**, 2009-2017 (1980)
4. F. Haraszti and T. Kovacs, IOP Conf. Series: Materials Science and Engineering, **175** (2017) 012048 ((http://iopscience.iop.org/1757-899X/175/1/012048))
5. PN–EN ISO 3651-1:2004. Determination of resistance to intergranular corrosion of stainless steels-Part 1: Austenitic and ferritic-austenitic (duplex) stainless steels-Corrosion test in nitric acid medium by measurement of loss in mass (Huey test)
6. A.H. Advani, L.E. Murr, D.G. Atteridge, and R. Chelakara, Metall Trans A, **22A**, p.2917 (1991)

7. G.E. Deiter, Mechanical Metallurgy, Mc-Graw-Hill, 1988, p.139
8. H. Zhao, Z. Zhang, H. Zhang, J. Hu, J. Li, Journal of Alloys and Compounds, **672,** 147-154 (2016)
9. H. Luo, H. Sub, G. Yinga, C. Dongc, X. Li, Applied Surface Science **425,** 628–638 (2017)
10. L. Zhang, J. A. Szpunar, R. Basu, J. Dong, M. Zhang, Journal of Alloys and Compounds, **616**, 235–242 (2014)
11. M. Terada, M. Saiki, I. Costa, A. F. Padilha, Journal of Nuclear Materials, **358,** 40–46 (2006)

Effect of functional solutions applied in machinery on fuel consumption and emissions

Jerzy Napiórkowski [1,*], *Jarosław* Gonera [1] , *Piotr* Szczyglak [1] , and *Robert* Czarnecki [1]

[1]Department of Design and Operation of Vehicles and Machines, University of Warmia and Mazury in Olsztyn

Abstract. Because stricter norms of emission of hazardous substances are being introduced every few years, machinery manufacturers are also forced to introduce various modifications to reduce emissions. To this end, systems aimed at optimising the process of fuel combustion and reducing the amount of hazardous compounds are being applied. Three tractors used in agriculture, with various functional solutions, were examined in this study. The effect of various systems on reduction of hazardous emissions and fuel consumption was determined. The tests were conducted during the ploughing with a 4-bottom swivel plough. The effect of the selective reduction (SCR) system with AdBlue liquid, the exhaust gas recirculation (EGR) system, the particulate filter, the number of valves per cylinder and the type of a turbocharger in the engine's inlet system were determined. It was observed during the tests that there were differences in fuel consumption and emissions depending on the constructional solutions applied in the tractors.

1 Introduction

As increasingly strict norms are introduced regarding exhaust emissions from combustion engines, it is necessary to apply various systems reducing hazardous emissions. Successive TIER/EURO Stage norms (Fig. 1) which apply to machinery, have put a particular emphasis on the reduction of emission of NO_x, i.e. nitrogen oxides, and reducing the emission of particulate matter (PM) [4]. Emission of hydrocarbons is also reduced in successive norms (HC). In order to reduce emissions, manufacturers of vehicles apply modern systems to reduce hazardous emissions, both inside and outside of engines.

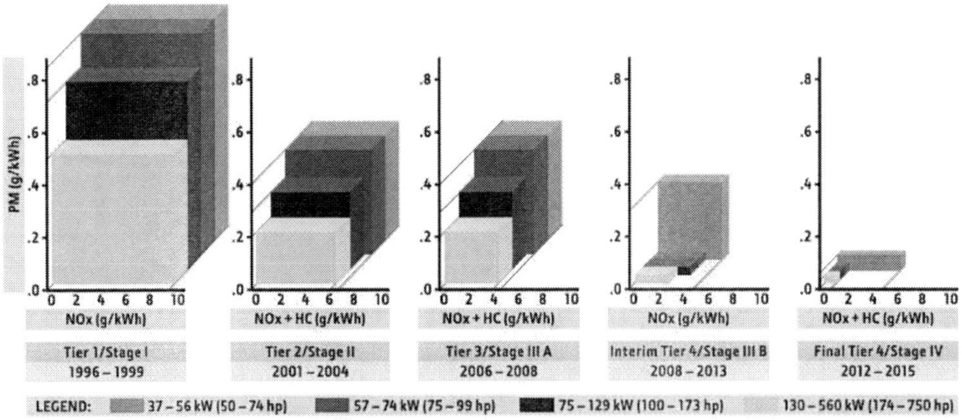

Fig. 1. Valid norms of emissions for agricultural tractors with the power ranging from 37 kW to 560 kW. [13]

* Corresponding author: jerzy.napiorkowski@uwm.edu.pl

The method of emission reduction in an engine consists in optimisation of fuel combustion, in particular in the environmental aspect. Such reduction also means lower fuel consumption, which also brings an economic benefit [1]. Norms of fuel emissions depend on the power of the engine installed in a tractor. Such tests are conducted according to the Stage norms in Europe and the Tiere norm in North America [5].

For example, a comparison of the Tiere 2/Stage II norm with the Final Tiere 4/Stage IV norm shows that the sum of NO_x and HC was reduced several dozen times in each of the four engine categories. Reduction of particulate matter was achieved mainly by applying non-engine methods, such as particulate filters. For NO_x emission, the main focus was on its reduction by engine-related methods, i.e. by optimising the combustion process. To this end, Common Rail systems were applied, in which a fuel dose is divided into several portions, as well as modern constructions of EGR systems together with advanced algorithms of controlling operation of combustion engines [5]. Exhaust emissions from agricultural tractors are usually reduced by the use of EGR systems, catalysts (catalytic converters), which are used for after-burning of hydrocarbons and carbon oxides, diesel particulate filters (DPF) and systems of selective catalytic reduction of NO_x using AdBlue reducing liquid (high purity, 32.5% aqueous solution of urea) [5, 9].

A high content of NO_x in exhaust gas has an adverse effect on human health because – when absorbed by the body – it reacts with haemoglobin and can cause weakness, numbness of limbs, pulmonary oedema and heart damage. Large amounts of CO also bind to haemoglobin, resulting in hypoxia and consequent brain damage and mental changes. Hydrocarbons have carcinogenic properties, they accumulate in the body and are lethal in high concentrations [1, 5, 10].

2 Functional solutions in agricultural tractors

Considerable changes have been made in the construction of machinery, particularly in agricultural tractors, since 2004. The machine fleet has been modernised owing to the bridge programme. Despite this, there are several machines with obsolete functional solutions operating in the Polish economy [7].

Large amounts of hazardous substances are particularly emitted by older machines, with traditional systems of fuel supply, based on injection pumps. In such machines, exhaust gas emission is exceeded many times compared to modern tractors. A reduction in the number of operated tractors of this kind can bring a considerable improvement in air quality [10]. For example, up to 55% of tractors registered in 2012-2016 were older than 20 years.

Complex processes take place during the operation of machinery (including agricultural tractors) in difficult and variable environmental and atmospheric conditions. These machines are operated both on public roads and – usually – under difficult field conditions. Depending on the conditions, various wear and destructive processes take place. Wear and destruction frequently results in an increase in emissions and fuel consumption by tractors [3].

Considering the fact that certification tests are conducted under laboratory conditions, their outcome may differ from the results obtained in real-life conditions, during normal operation of machinery. This is of particular importance for tractors which are operated under specific field conditions [8].

Proper setting up of the machine aggregate, for example, for ploughing, is one of aspects which may bring about a reduction of emissions and fuel consumption [5]. It is the most energy-consuming operation in agriculture, with a tractor working at variable engine rotational speed, load and torque demand [10].

A high level of purity of exhaust gases in agricultural tractors is also important because some of them may also work indoors: in poultry houses, pigsties, cowsheds. In such cases, exhaust gases have an effect both on the health of the operators and of farm animals. NO_x

and HC have a particularly adverse effect on animals because it results in a lower quality of milk and meat [6]. Therefore, only tests conducted in real-life conditions provide results illustrating the actual harmfulness of combustion engines.

An SCR exhaust gas treatment system (Fig. 2) decomposes harmful nitrogen oxides (NO_X) to water and nitrogen. AdBlue liquid – an aqueous solution of urea – is used in the treatment process [7]. AdBlue liquid is injected to the system by an injector. SCR systems are relatively failure-prone because AdBlue crystallises at approx. –11oC, clogging the injector and tubing, and subsequently damaging the liquid pump and the NO_X sensor.

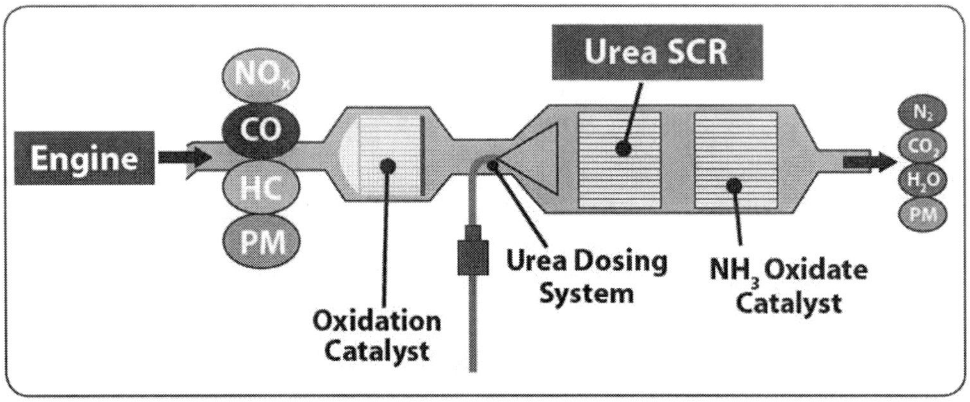

Fig. 2. Diagram of the SCR system. [14]

A particulate filter (Fig. 3) retains soot particles. A filter insert has porous walls, and canals are closed – alternately – on their inlet and outlet. The pore diameter is smaller than that of soot particles, which are then stopped inside the filter and burned. Cleaning a particulate filter is done by burning out [7].

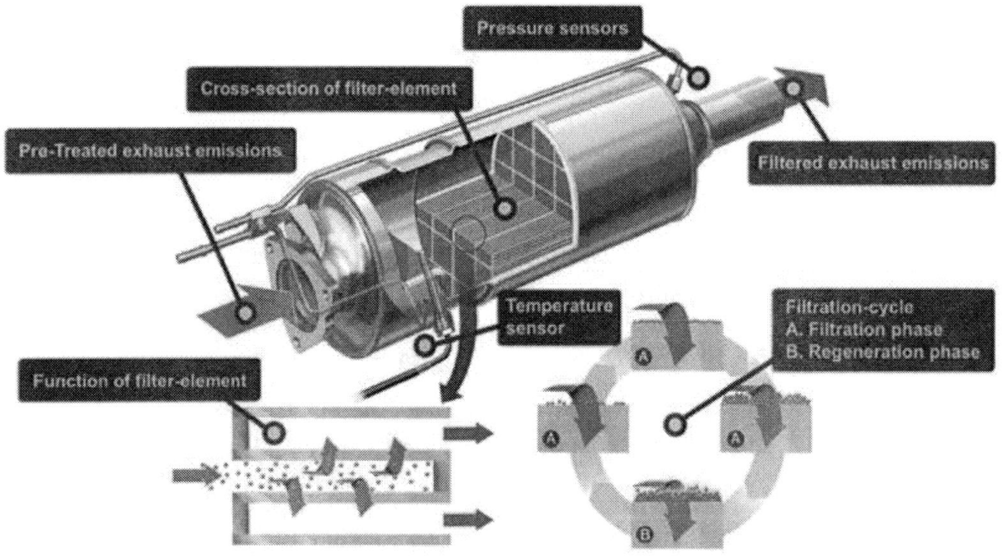

Fig. 3. A diagram of a particulate filter. [11]

Exhaust gas recirculation (EGR) – after-burning – involves reintroducing part of the exhaust gas into the combustion chamber [4]. The EGR system (Fig. 4.) accelerates fuel evaporation, decreases the combustion temperature and oxidises unburned hydrocarbons left in the exhaust gas. The principal aim of the EGR system is a reduction of nitrogen oxide (NO_x) emission by decreasing the combustion temperature of the oxygen-poor mixture and reducing HC emission by its oxidation.

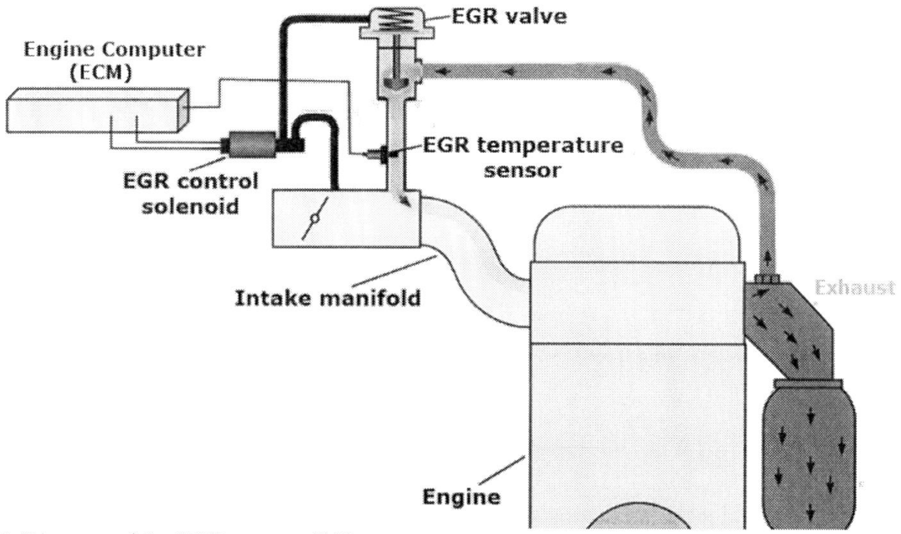

Fig. 4. Diagram of the EGR system. [12]

3 Study methodology

Three tractors used in agriculture, with various functional solutions regarding reduction of hazardous emissions, were examined in this study. The tractors were produced in the years 2012-2013. Their nominal towing force was 30 kN, and the nominal engine power ranged from 104 to 114 kW. The tractors were of the following types: (Table 1) Case Maxxum 140 Multicontroler, Claas Arion 620 and John Deere 6930.

Tab. 1. Construction parameters of the tractors.

Tractor model	**Case Maxxum 140 Multicontroler**	**Claas Arion 620**	**John Deere 6930**
Year of manufacture	2013	2012	2012
Engine capacity [dm^3]	6,7	6,7	6,7
Number of cylinders	6	6	6
Number of valves per cylinder	4	4	2
Nominal engine rotational speed [rpm]	2200	2200	2200
Engine power [kW]	104	114	114
Type of gear box	Powershift	Hexashift	Auto PowerQuad
Number of gears forwards/backwards	16/16	24/24	20/20
Power take-off speed [rpm]	540/540ECO/ 1000/1000ECO	540/540ECO/ 1000/1000ECO	540/540ECO/ 1000/1000ECO
Hydraulic pump rate of delivery [l/min]	119	110	138
Length [mm]	4532	4754	4454

Width [mm]	2000	2000	2000
Tyre size front/back	14.9R28/18.4R38	14.9R28/20.8R38	14.9R28/20.8R38

The parameters of all the tractors were comparable. They were powered by self-ignition engines and fitted out with a CommonRail fuel injection system. Engines with four valves per cylinder make it possible to achieve a higher air flow and a more rapid response to changes in load. This is a result of the quicker air flow into the combustion chamber and exhaust gas flowing out of it. Moreover, the injector in engines with four valves per cylinder is situated at the centre of a cylinder, which has a beneficial effect on the combustion process.

The Case Maxxum 140 Multicontroler is fitted out with a Powershift gearbox and a SCR (Selective Catalytic Reduction) exhaust gas treatment system and it complies with the exhaust gas Stage IV norm. The Claas Arion 620 has a Hexashift gearbox with 4 groups of gears (A, B, C and D), with 6 half-gears in each. A particulate filter which complied with the Stage IV exhaust gas norm was applied in the tractor as an exhaust gas treatment measure. Unlike the other two, John Deere 6930 had only two valves per cylinder and a variable geometry turbocharger was not applied. The Auto PowerQuad gear had five operation modes, each with four half-gears. The tractor had an EGR system for exhaust gas after-burning.

All of the tractors were tested during the ploughing operation with the same Vogel&Noot 4-bottom swivel plough. It was fitted out with height and depth regulation capability. The plough weighed 2,230 kg and it had a working width set at 1.6 m. The plough had divided shares and openwork mouldboards.

The tests were conducted during the autumn ploughing, in a 720m-long field. After wheat was harvested, it was disked to the depth of 8 cm. The ploughing was done at a depth of 25 cm. The constant ploughing speed of 8 km/h at a crankshaft rotational speed of 1,700 rpm ensured the selection of the right gear ratio.

The fuel consumption was measured by the full tank method. The fuel tank in each tractor was filled up at the beginning of work. After an hour's work, the tanks were filled up again with a 1 l measuring vessel with a measuring error of 5 ml.

The analysis of exhaust gas was carried out with Kane Auto Plus 5-2 – a portable, compact exhaust gas analyser (tab. 2). It is a device analysing the content of five gases: CO, CO_2, O_2, NO_x and HC. The analyser was connected to a dedicated printer which enabled the results to be printed directly. Measurements could be done 90 seconds after the device was switched on. Kane Auto Plus 5-2 has a Bluetooth module, which enabled the wireless transfer of measurement data to the computer. The Kane Live software was installed on a portable computer, on which the data were archived and visualised. Exhaust gas emission was measured in each tractor during a stop at an idle gear and during work under load.

Tab. 2. Measurement range and measurement resolution of individual components of exhaust gas.

Component	Measurement range	Resolution
CO	0-10 %	0,01 %
CO_2	0-16 %	0,1 %
O_2	0-21 %	0,01 %
HC	0-5000 ppm	1 ppm
NO_x	0-5000 ppm	1 ppm

4 Results

4.1 Fuel consumption

The data (Fig. 5) show that the fuel consumption was the lowest in the Claas Arion 620 tractor. It was lower by approx. 15% than in the other tractors, in which it was at a similar level.

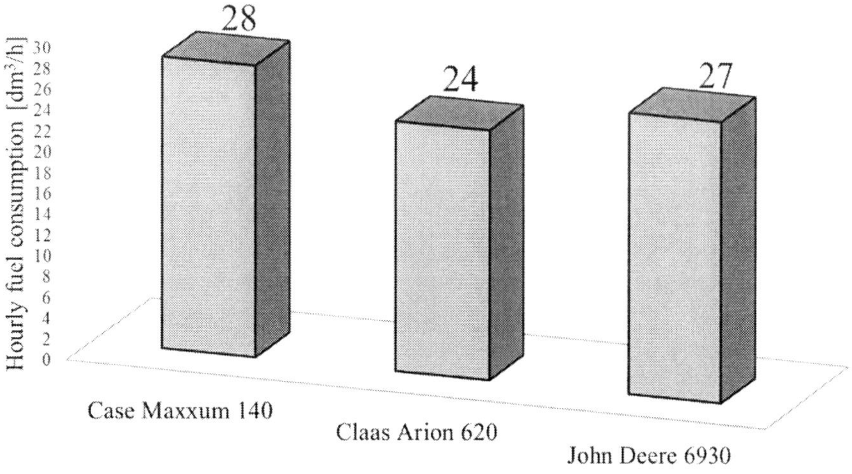

Fig 5. Hourlyfuel consumption.

The same relationship was observed when the fuel consumption was divided by the ploughed area. The theoretical productivity was the same for all the tractors: 1.3 ha/h. The fuel consumption was 18.46 l/ha in Claas Arion 620, 20.77 l/ha in John Deere 6930 and 21.54 l/ha in Case Maxxum 140.

The largest fuel consumption was found for the Case Maxxum 140, in which the SCR system with AdBlue liquid was applied. Fuel consumption was also high for the John Deere 6930. This was caused by the lack of a variable-geometry turbocharger and the presence of only two valves per cylinder. Therefore, the engine characteristics and the maximum torque were different, available within a more narrow range of the rotational speed.

4.2 Exhaust gas emission

Measurements of exhaust gas emission included determination of CO_2, CO, O_2 and NO_x contents (Fig. 6 – 9).

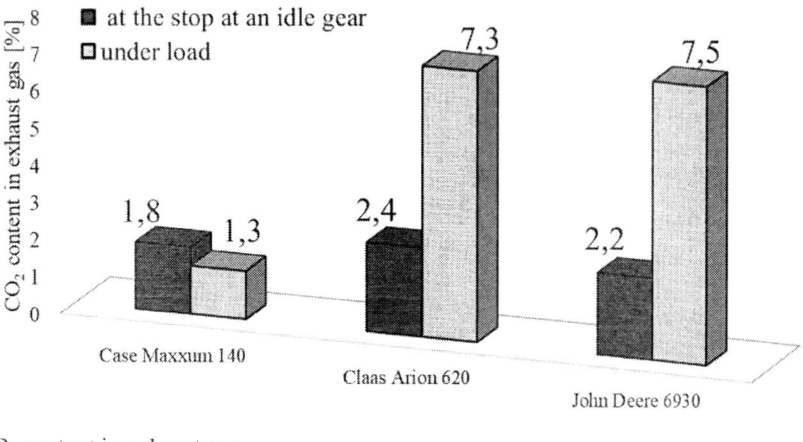

Fig. 6. CO_2 content in exhaust gas.

The SCR system with AdBlue liquid in the Case Maxxum 140 tractor proved to be the best option (Fig. 6) for CO_2 reduction in exhaust gas. The content of CO_2 was lower (by as much as six times) compared to other tractors – both at the stop at an idle gear and under load. Particularly large differences in exhaust gas emission were visible under load. The CO_2 content in Case Maxxum 140 was several times lower compared to the other two tractors. The content of CO_2 in exhaust gas from Claas Arion 620 and John Deere 6930 was similar.

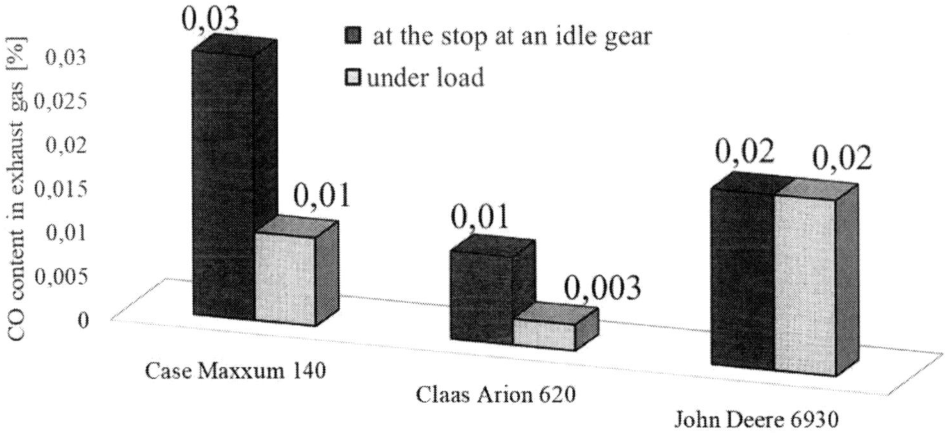

Fig. 7. CO content in exhaust gas.

Data presented in Fig. 7 show that exhaust gas from the Claas Arion 620 tractor contained the lowest amounts of CO. In this case, much lower amounts of CO were found both during the stop, at an idle gear, and under load. This was a consequence of applying a particulate filter; moreover, the engine had four valves per cylinder. Owing to the latter, the fuel combustion process could be optimised. Moreover, installation of a varied-geometry turbocharger allowed for the maximum torque within a broader range of rotational speeds. A particularly high content of CO in exhaust gas from Case Maxxum 140 was observed during the engine work at a stop at the idle gear. The CO content was associated with the fuel consumption. A tractor which consumes less fuel emits less CO.

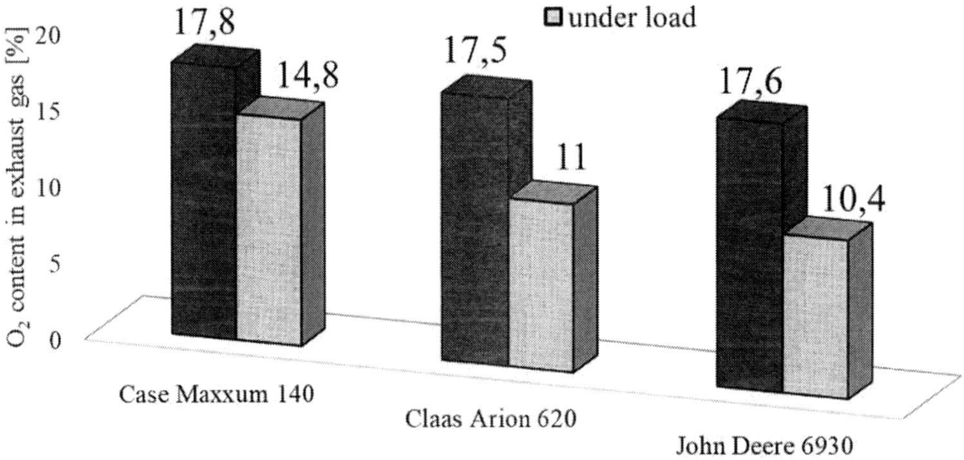

Fig. 8. O_2 content in exhaust gas.

The data presented in Fig. 8 show that the content of O_2 in exhaust gas was similar during a stop at an idle gear. On the other hand, the lowest amounts of O_2 during work under load were found in exhaust gas from Claas Arion 620. Oxygen initiates reactions that take place during combustion. It is also used in exhaust gas treatment systems for oxidation of the CO and HC in exhaust gas. The smaller the amount of O_2 in exhaust gas, the more efficient the combustion and oxidation process.

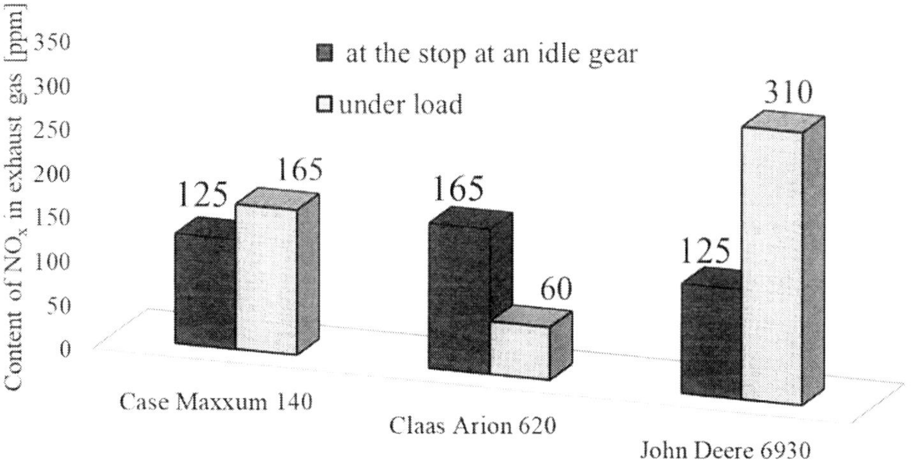

Fig. 9. Content of NO_x in exhaust gas.

Fig. 9 shows the content of NO_x in exhaust gas. It reveals a much higher content of these substances in exhaust gas from the John Deere 6930 tractor. This was a consequence of the fact that the engine did not have a particulate filter or an SCR system with AdBlue liquid. The most effective reduction of NO_x was achieved in the Claas Arion 620 tractor, which was fitted out with a particulate filter.

5 Conclusion

The results of these experiments have shown that functional solutions in agricultural tractors have a great effect on fuel consumption and exhaust gas emission. The tractors complied with the same emission norms and did the same work, although the differences in exhaust gas emission between them were considerable.

The Claas Arion 620 tractor had the lowest fuel consumption. Savings of 15% contribute to the economic benefits in agricultural activities. Fuel consumption by the Case Maxxum 140 Multicontroler and John Deere 6930 was similar.

An analysis of all the components of exhaust gas shows that the reduction of hazardous substances content was the least effective in the John Deere 6930 tractor. The exhaust gas treatment system was the least complex, with no particulate filter or the SCR system with AdBlue liquid, and the engine supply system was the least advanced. The exhaust gas in the Case Maxxum 140 tractor, which was fitted out with an SCR system with AdBlue liquid, was found to contain several times less CO_2 compared to the other two tractors. Exhaust gas from Claas Arion 620, which was fitted out with a particulate filter, contained the smallest amounts of CO and NO_x.

This experiment has not given a conclusive answer about the system with the best capability for reduction of hazardous substance content. This is because a particulate filter is more effective in reducing the content of CO and NO_x, whereas the SCR system with AdBlue liquid is better at reducing the content of CO_2.

Bibliography

1. K. Baumgard, S. Cooke, *John Deere Power Systems Symposium* (2005)
2. A. Bieniek, Agricultural Engineering, **143**, 31 (2013)
3. G. Chomik, Z. Chomik, *Nowoczesna Obsługa Techniczna Ciągników* (2016)
4. J. Gromadko, V. Hong, P. Miler, Maintenance and Rel. **4**, 63 (2008)

5. S. Jung, M. Ishida, S. Yamamoto, H. Ueki, D. Sakaguchi, International J. of Autom. T. **11**, 611 (2010)
6. A. Karbowy, A. Koniuszy, P. Sędłak, Teka Com. Mot. Ener. in Agric. **12**, 69 (2012)
7. A. Mazanek, J. Jakóbiec, W. Piekarski, Logistyka **5**, 381 (2015)
8. J. Merkisz, P. Lijewski, P. Fuć, S. Weymann, Maintenance and Rel. **15**, 364 (2013)
9. J. Merkisz, J. Pielecha, S. Radzimirski, *Emisja Zanieczyszczeń Motoryz.* (2012)
10. A. A. Naji, *A Comparison of Measured Diesel Emis. in Agr.* (2013)
11. J. Skudlarski, Roczniki naukowe Stow. Ekon. Rol. i Agrob. **19**, 65 (2017)
12. www.maasradiator.com/dpf-services
13. www.samarins.com/diagnose/diagrams
14. www.silniki.info.pl/ochrona_srodowiska.html
15. www.tomorrowstechnician.com/how-long-should-a-brake-job-last

The diagnostic information valuation method in servicing tractors

Żaneta Staszak[1,*], *Jarosław* Selech[1], *Jacek* Marcinkiewicz[1], *Dawid* Romek[1], *Konrad* Włodarczyk[1], *Łukasz* Gierz[1], and *Dawid* Wojcieszak[2]

[1]Poznań University of Technology, Faculty of Machines and Transport, ul. Piotrowo 3, 60-965 Poznan, Poland
[2]Poznań University of Life Sciences, Institute of Biosystems Engineering, ul. Wojska Polskiego 28, 60-637 Poznan, Poland

Abstract. The objective of the article was to develop a method of evaluating diagnostic information basing on the existing systems of agricultural tractors diagnostic used in authorized service centers, which will improve of the quality of the evaluation process. Initially, available tractor diagnostic systems were identified in order to gather information about individual diagnostic signals. Following that, the information was verified by empirical evidence concerning a given failure. Thus, the work presents an original evaluation method of the information obtained from the agricultural tractor diagnostic. Obtaining diagnostic information generates certain costs, hence, it is necessary to rationalize and make the processes efficient. Further part of the thesis contains a description and logical verification of the developed method using empirical database. Finally, we presented the method's application for evaluating diagnostic information obtained from the tractor computer diagnostic, in order to examine its effectiveness and for a final verification.The described method of validating diagnostic information creates new cognitive and pragmatic possibilities.

1 Review of the literature

Tractors are complex technical objects which structure consists not only of mechanical elements but also electronic, IT and mechatronic ones. A complex structure of modern tractors raises efficiency of their exploitation measured by cost reduction connected with performing some field works and the ones performed in farms. Electronic and IT modules installed in tractors make it possible to optimise work of mechanical units under given environmental conditions. Such activities increase also tractor reliability decreasing simultaneously effects of a negative influence of machine work on environment.

Tractor work conditions are also special. They are characterised by seasonality which is determined by a specific agrotechnical period during which machine usability necessary to perform certain works is required. The time of the beginning of those works, their finishing time and applied technologies depend mainly on atmospheric conditions. Therefore it is crucial for technical service to be prepared in a wider range and assortment than it is in fact required [1].

Contemporary tractors are equipped with modern drive units fitted with microprocessor control systems, multirange, stepless gears and hydraulic systems as well as board computers for continual machine diagnostics [2-7]. In agriculture, it is crucial for individual treatments (ploughing. fertilisation, sowing etc.) to be performed in a strictly determined period, called an agrotechnical period [8]. It is connected with continuous sustaining condition of machine efficiency by using authorised technical service centres having spare

* Corresponding author: zaneta.staszak@put.poznan.pl

parts, sufficiently staffed stands and equipped with computer software for repairing tractors [9-14]. Actions planned in such a manner related to tractor servicing and their state-of-the-art structure made it possible to increase machine efficiency and simultaneously to reduce exploitation costs [15].

The presented above tendencies in automation and diagnostics of working modules and elements of tractors are not free from flaws. Broader application of automation in tractors leads definitely to an increase in controlling their working parameters and enables a prompt reaction in emergencies. However, flawed information gained from one or a few sensors installed in certain tractor units may be the reason for making incorrect decisions by technical servic or unnecessary tractor downtime e.g. during higher intensity of field works.

A crucial element of an efficient tractor automatic control system and diagnostic functioning is to provide a proper manner of transmission and prioritisation of sent information between particular drivers and sensors as well as between a central driver of diagnostic management and an external receiver (at a vehicle diagnostic station, in a producer's data base etc.).

In such a context it is advisable to develop a data valuation method obtained during tractor diagnostics regarding assumed criteria e.g. minimum repair time of certain tractor units or a minimum cost connected with a vehicle downtime in a service station. A developed method may be applied to support and optimise decision making processes in diagnosing, servicing and exploiting tractors.

2 The aim of ther article

The aim of the article is to develop the method enabling valuation of diagnostic information and its verification on the example of tractors.

3 Method

The article is of a methodical character and as a result of it a universal method of valuation diagnostic information in servicing processes of tractors has been developed.

The information was obtained from tractor construction documentation and technological documentation of their servicing. Following its analysis, enriched with own practical experiments, there was created the knowledge base of a process structure, its objects, relations with environment and problems with functioning which are going to be a base for the developed method.

An integral part of the developed method was to create a data base of actual damage and diagnostic information.

In order to obtain that, observations and surveys were conducted. They involved identification of actual damage and diagnostic information about it. The criterion of tractor selection for the research was their share in a general number of tractors exploiting in agriculture. Five tractor makes were selected to the research: Valtra, John Deere, New Holland, Zetor and Massey Ferguson which according to AgriTrac data 2005-2015 account for more than 50% of the market share of the new machine market.

The tractors which were investigated had either experienced a failure or their malfunction had been discovered during scheduled servicing. In case of a failure, a tractor was delivered to a service station in which a malfunction was found during servicing by a diagnostic system. Information about damage is visible on a board computer as a failure code or only after connecting an external computer with software compatible with a serviced tractor. A failure code or oral information is a signal which generates a string of variants of possible failures. Unfortunately, a few variants of failures are often assigned to one signal. In such a case, a person performing diagnostics needs to make a decision which variant is to be checked as first. Such machines after removing a malfunction will have work parameters of particular units the same or very similar to values before a failure has occurred. The diagnostic process of a tractor is presented in Fig. 1.

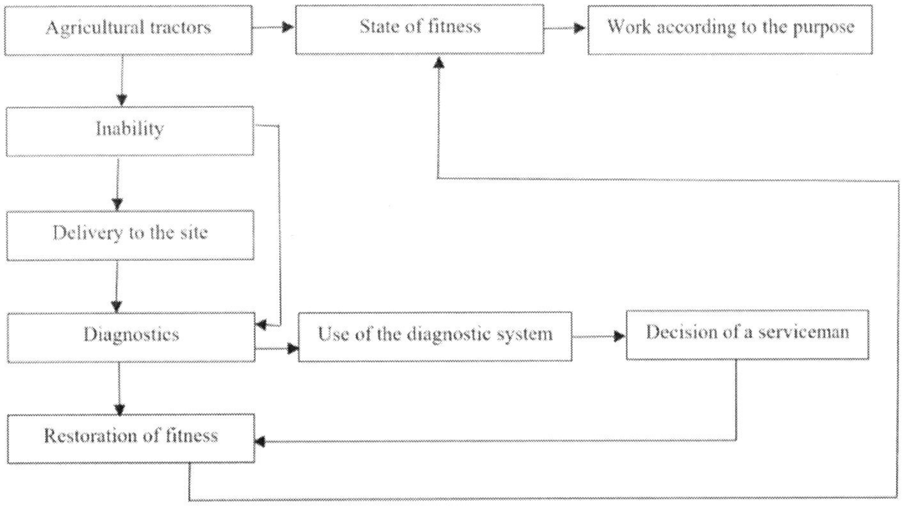

Fig. 1. A scheme of tractor diagnostics (own work).

4 Mathematical model of the method

In order to create the base of possible damage variants gained from one signal it is essential to collect proper data. The collected information was recorded using a relation in which one signal ***S***, is a set of information variants $\boldsymbol{n_w}$ of a proper quantity (1):

$$S = \{n_{wj}\} \qquad (1)$$

where: S – a single diagnostic signal,
n_w – a number of possible information variants.

From a gained diagnostic signal it is possible to generate a certain set of information $\boldsymbol{n_w}$ with a determined probability of occurring each of it.

Such an empirical system with a set of probable variants of its realisation being information in technical diagnostics can be described using information entropy.

The examples of information entropy application are included in the following works (Rzeźnik and Molińska, 2003; Ludwikowski, 2008; Bennington and Cummame, 1998; Dowdy et al., 2004; Hellwing, 1998), among the others.

The statistical information theory describes information entropy taking into account probability of event occurrence (information variants). The theory is applied to machine diagnostics as during diagnostic tests a machine generates a certain amount of information which should be used during recognition of its technical condition. The source of diagnostic information possible to be obtained is the knowledge about a machine structure, experience stemming from exploitation of that type of machines as well as a properly conducted interview with an operator. The information collected in such a manner may be used to diversify probability of damage occurrence of machine parts. In order to calculate an amount of lacking diagnostic information (information entropy) one uses the statistical information theory. It is mainly used during valuating logical experiment efficiency.

Applying the described scheme of valuating information, at the beginning a set of possible damage variants obtained from a certain signal on the basis of data from repairing documentation and data from diagnostic computer systems was created. At that stage of the method formulation it was assumed that the occurrence of each information variant generated by a certain signal was equally probable.

On the basis of the set of information variants developed in such a manner, the amount of lacking information – thus information entropy of each signal occurrence – was calculated. The equation of Shannon structural information theory was used to assess the amount of lacking information – information entropy at the same probability.

Servicing documentation and also computer diagnostic systems do not include all possible variants of a malfunction as in the case of a printed version it would be too unclear to include all information and in the case of computer diagnostic systems waiting time for complete diagnostics would be very long. Producers included the most common damage variants in the documents. However, as practice shows, there happens to be tractor models in which failures are different than the ones described by a producer. It may result from negligence at design or production stages. Therefore in order to create more precise sets, there was conducted empirical research among workers of authorised tractor service stations selected to the analysis of makes in the doctoral dissertation and of a possible malfunction to occur. Damage variants which had not been investigated before and were added by workers enabled creation of a more precise and more numerous information set. The amount of information entropy for a given signal in which damage variants of parts or units pointed by service workers were added varies. Their values were determined on the basis of service cards of damage being the reason for an investigated malfunction.

Following an analysis of service cards and included data, information variant probability was differentiated and information entropy was calculated using a statistical information theory equation.

The conducted analysis shows that it is possible to determine numerically the amount of lacking diagnostic information (information entropy) using the information theory. Additionally, due to providing a machine operator with a proper training or protecting properly wire connectors it is possible to reduce the amount of lacking diagnostic information necessary to be gained by instrumental diagnostic methods. It will increase efficiency of a tractor servicing process not only in terms of diagnostics.

Such enriched instrumental diagnostic information will help to eliminate the most unreliable spots and to check them as first and only when verification is positive, a service worker will proceed to control particular variants arranged according to probability of their occurrence. The least probable variants and the ones which occurrence depends on faulty machine operation by an operator will be eliminated from the variant list and their probability will be assigned in an even proportion to remaining variants. Final results are obtained calculating information entropy for variants which have been left after eliminating the most fallible ones.

While developing the mathematical model of valuating diagnostic information in servicing tractors, information theory and practical realisation of servicing processes were used.

As a result, the logical model describing that exploitation area was described which its logical correctness and practical utility required to be checked.

5 Logical verification of the method

The logical verification of the developed method requires checking whether a diagnostic information value obtained from it is pertinent to actual damage.

The research was performed for 100 tractors in which a signal was a malfunction of a rear power take-off (PTO) during diagnostics using a diagnostic computer system. Following selection and division of data three possible signals were obtained informing about damage of a rear power take-off where each of them contained information about a few variants of damage.

Table 1 presents the verification of the above method. The results of all stages are presented in one table.

Table 1. Comparison of lacking diagnostic information (information entropy) for a rear power take-off.

l.p.	Signal type / The manner of obtaining information	Lack of possibility to start a rear power take-off	Lack of possibility to stop a rear power take-off	Lack of possibility to change the number of rotations of a rear power take-off
1.	on the basis of repairing documentation and information from diagnostic computer systems	2.32	2.00	0.00
2.	from item 1 extended by information from employees of service stations	2.81	2.32	1.00
3.	on the basis of actual repairing cards	2.55	2.01	0.81
4.	after eliminating damage variants possible to be averted	1.96	1.00	0.00

On the basis of the data presented in the table it can be observed that the biggest information entropy occurs when all variants are equally probable and they include possible damage given by mechanics of authorised service stations whereas after eliminating variants likely to be averted we need the least information necessary to diagnose damage of a rear power take-off.

On the basis of the obtained data we can develop a list of recommendations for the order in which particular damage should be localized. In case of a failure to start a rear power take-off, the following diagnostic checks are recommended:

- damp or soiled wire connections,
- damage of an automation switch in a service console,
- damage of a switch in a rear wing,
- damage of a rotation sensor,
- damage of a rotation valve 1000,
- damage of a rotation valve 540,
- damage of clutch disks.

Other recommendations regard a sequence of diagnostic checks in identifying the reason for the lack of possibility to disconnect a rear power take-off. They include:

- damp or soiled wire connections,
- damage of a switch in a rear wing,
- damage of an automation switch in a service console,
- the number of journal rotations is flawed,
- damage of a rotation sensor.

Whereas for a signal about no possibility to change the number of rotations of a rear power take-off firstly we check whether connections of wires are not damp or soiled and then we check whether a switch of speed selection has not been damaged.

6 Summary

The logical verification of the developed method confirmed that numerical valuation of diagnostic information value results logically from technical condition of an assessed unit. The base for proper functioning of the developed method is to create a pertinent variant set for a particular signal and a proper verification of the most unreliable parts of a tested tractor unit. All of the above allow a hypothesiPs to be formulated that the method can be also applied to other units of a tractor.

After a positive verification of the diagnostic information valuation method, it has been applied to a few other signals obtained during tractor computer diagnostics in order to check its practical usefulness.

References

1. P. Rybacki, *Agricultural machines servicing quality assessment system,* (Wyd. Uniwersytetu Przyrodniczego w Poznaniu, 2012)
2. S. Juściński, W. Piekarski, Agricultural Engineering, **2(100)**, 57-65, (2008)
3. S. Juściński, M. Szczepaniak, Eksploatacja i Niezawodność, **2(38)**, 45-52, (2008)
4. T. Pokusa, *Logistyczna obsługa i lojalność klienta jako orientacje rynkowe,* (Wyd. Instytut Śląski, 2001)
5. A. Skrobacki, A. Ekielski, *Pojazdy i ciągniki rolnicze,* (Wyd. Wieś Jutra, 2006)
6. P. Rybacki, C. Rzeźnik, K. Durczak, TROL, **2**, (2011)
7. L. Dwiliński, IBMER, **7**, 7-36, (1973)
8. L. Dwiliński, Rocznik Nauk Rol., **69(C2)**, 25-28, (1972)
9. M. Ciesielski, *Logistyka w praktyce-studium przypadków,* (Wyd. Akademii Ekonomicznej, 2002)
10. D. Kempny, *Logistyczna obsługa klienta,* (Wyd. Ekonomiczne, 2001)
11. S. Juściński, W. Piekarski, Eksploatacja i Niezawodność, **2(38)**, 25-33, (2008)
12. S. Juściński, W. Piekarski, Agricultural Engineering, **8(117)**, 23-30, (2009)
13. J. Banasiak, *Agrotechnologia,* (PWN, 1999)
14. S. Juściński, M. Szczepaniak, Agricultural Engineering, **2(100)**, 67-74, (2008)
15. Z. Grześ, Agricultural Engineering, **2(100)**, 37-42, (2008)

The research was supported by statutory resources appropiated to Institute of Machines and Motor Vehicles, Poznan University of Technology as „The Grant for Young Researchers" No. 05/51/DSMK/3537.

Impact of the operating time and loading of an engine with spark ignition powered with biogas on selected parameters of engine oil

Jerzy Kaszkowiak[1,*], *Marietta* Markiewicz-Patalon[1] , and *Ewa* Kaszkowiak[1]

[1]UTP University of Science and Technology, 85-796 Bydgoszcz Al. S.Kaliskiego 7 Poland

Abstract. Engines powered with biogas obtained from wastes obtained from a municipal waste water treatment plant were subject to studies. The engines powered the electric current generators. It was found, that after 800 operating hours, differentiation in the contents of iron, zinc and copper depending on the mode of their loading occurred in the engine oil. In the group of engines loaded in about 60% of power, higher contents of iron, copper and zinc in the engine oil than in the engines loaded in 85% of power, was found. Occurrence of other statistically important differences in the contents in oil of manganese, chromium and aluminum was found.

1 Introduction

The use of biogas as the source of energy is the manner both for lowering of the level of consumption the fossil energy sources, as well as for the biological wastes' recycling. Biogas may be generated from plant products cultivated specially for the power purposes. It is a high-yield manner, however it requires designation for production of the cultivation areas on which food may be produced [1]. Also, seasonal availability of plant substrates requiring their preservation and necessity of their preparation already at the harvesting level, is its disadvantage. It requires more intensive shredding of the harvested plants, what results in the increase of power outlays [2, 3]. Biogas' generation is possible also from organic wastes [4]. Depending on substrate, the generated biogas may have different composition. Generation of biogas is possible also from municipal wastes, remains from animal production. More and more often there is also the use of seal plants (algae) for biogas generation. The composition of the obtained biogas depends on the used substrate and there occurs a relatively high variability of biogas' composition. The biggest differentiation occurs in the content of methane. The scope of the biogas' composition for selected substrates is presented in the table 1. For direct combustion and use as a source of heat, biogas is most often not purified, but in case of its use as a fuel for combustion engines, its purification is required [5]. The engines with spark ignition powered with biogas, belong to a special group of units. They are most often used as power transmissions in cogeneration systems. Such a manner of biogas' use enables both easy transfer of power with the use of power networks as well as for temporary storing in tanks with the possibility of their later usage. Supplementation of possible biogas' shortages for powering of cogeneration units with natural gas is often used.

Table 1. Composition of biogas depending on the substrate from which it has been generated.

Source \ Component	From power plants	Waste water treatment plants

* Corresponding author: jerzy.kaszkowiak@utp.edu.pl

CH_4	57-62%	64-75%
CO_2	33-38%	20-35%
H_2S	-8000 ppm	≤8000 ppm
N_2	3,4-8,1%	3-4%
O_2	0-0,5%	0,5%

The costs of work of the cogeneration systems powered with biogas depend on outlays incurred on biogas' generation (value of substrates, costs of their transport, costs of the biogas geneation process) [6] and the costs of the units' operation (amortization, repairs, servicing). In case of failure-free operation, the outlays connected with periodical oil's exchange are one of the biggest costs. Extension of the periods between the subsequent exchanges, lowers the outlays incurred on servicing, but at the same time it may result in the motor's durability decrease [7]. The change of the content of selected compounds and chemical elements in the oil, may be also used for the motor's condition assessment. In particular the contents of copper, mangane, aluminium and iron in the engine oil prove of that [8]. However, the impact of the oil filtering system, which may to a considerable degree change the contents of impurities in the oil, should be taken into consideration.
Determination of the biogas powered motor loading's impact and the impact of the time of operation on the contents of selected metals in the engine oil, was the purpose of the studies.

2 Materials and methods

The studies were conducted on the group of two motors powered with biogas propelling electric current generators. The tested units had an identical construction with 12 cylinders placed in the V system. The engines were equipped with turbo-charging of a constant geometry. The engine cubic capacity amounted to approximately $12000cm^3$, and their power quoted by the producer amounted to 380 kW. The engines were in a very good technical condition, till the tests' commencement they had been exploited for not longer than 2 years from the moment of their production. The engines were equipped with the by-pass oil filters. The highest quality dedicated synthetic oil designed for engines powered with biogas, was used for the engines' lubricating. At the time of the tests conducting, one of the engines was loaded in a constant mode, on the level of 85% nominal power at the variation not exceeding ±5%. The second tested engine was loaded in a variable mode at the average loading level of 65%, where the range of changes of the temporary loading changed within the range from 85 kW to 330 kW. The cycle of the tests each time lasted for about 800 operating hours (±10 hours). The testing cycles were repeated three times. After the end of each operating cycle, at the time of the engine oil's exchange its samples were taken for testing. There was examined the contents of iron, chromium, nickel, aluminum, copper, lead, manganese and zinc. Biogas with which the engines were powered, was generated from the wastes remaining after the treatment of municipal liquid wastes. After its generation, it was subject to purification. The impurities were removed mechanically, and the contents of sulphur and its compounds in biogas were removed. The composition of biogas was controlled on a continuous basis with a biogas analyser BIOTEX MultiPoint. The biogas' calorific value following its purification amounted from 17,8-18,1 MJ/m^3 . The composition of biogas following its treatment is presented in table 2.

Table 2. Main components of biogas powering engines following its treatment.

Component	Contents
CH_4	65,25%
CO_2	34%
H_2S	11 ppm

The engines operated in a common, not heated room. Combustion gases were carried off outside. The temperature in the room changed within the range 15-28^{0}C. The view of one of the engines is presented in figure 1.

Fig. 1. View of the engine during the studies.

The obtained results were subject to statistical analysis with the Tukey's test with the use of the STATISTICA programme. In case of studying the contents of lead in oil, its presence in the oil was not found and it was not included in the further analysis description.

3 Results

As a result of the conducted statistical analysis, occurrence of differentiation in the contents of iron, zinc and copper in the engine oil was not found.

For an engine loaded in a continuous mode in 85% of rated power, the average contents of iron amounted to 5,0 ppm. For an engine loaded in a variable mode, at an average load of 65%, the average contents of iron had a higher value and amounted to 9,3 ppm. The difference in the content of iron was statistically significant. The contents of iron depending on the mode of the engine's loading is presented in figure 2.

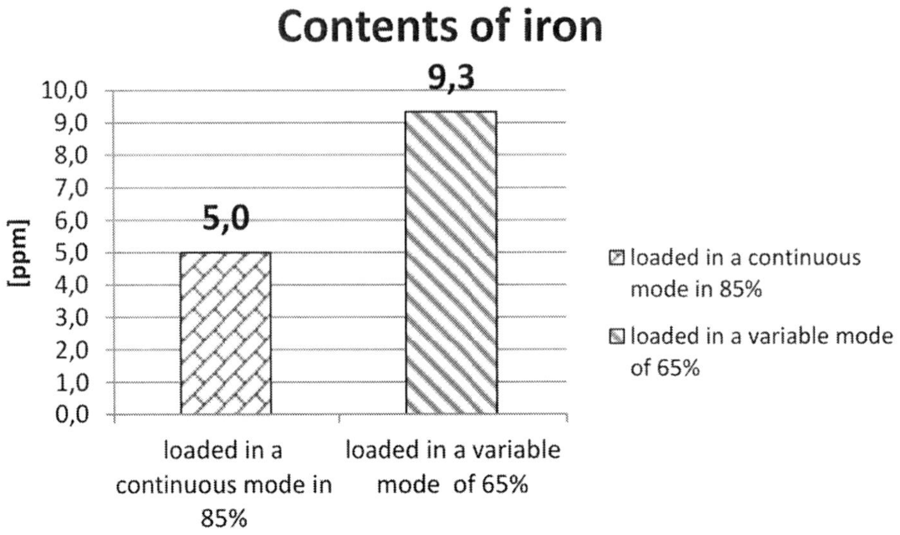

Fig. 2. The contents of iron in the engine oil depending on the mode of loading.

The average contents of aluminum in the oil for the engine loaded in a continuous mode, had a lower value and amounted to 2,0ppm. The average contents of aluminum for an engine loaded in a variable mode, had a lower value and amounted to 2,33 ppm. The differences in the contents of aluminum in the engine oil were however statistically insignificant. Their values are presented in figure 3.

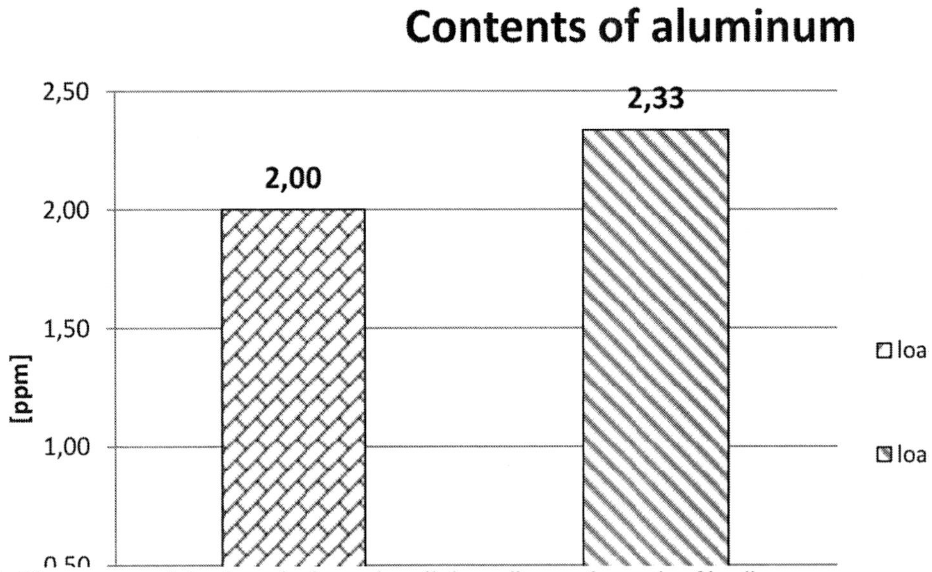

Fig. 3. The contents of aluminum in the engine oil depending on the mode of loading.

The mean contents of copper in the engine oil for an unit loaded in a continuous manner on the level of 85% of rated power, had the value of 2,33 ppm. For an engine loaded in a variable manner, at average loading of 65%, the contents of copper in the oil had a higher average value and amounted to 5,33 ppm. The average contents of copper in the engine oil for the examined cases are presented in figure 4. The differences were statistically significant.

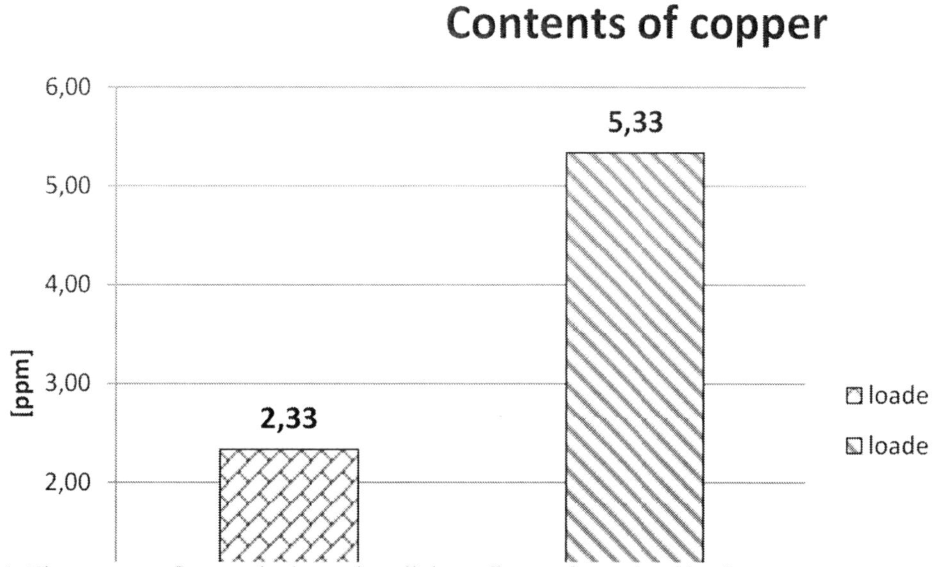

Fig. 4. The contents of copper in the engine oil depending on the mode of loading.

The contents of manganese and chromium in the engine oil were not found in both the examined cases. However, statistically significant differences occurred for the contents of zinc. The average contents of zinc in the oil for the engine loaded in a constant mode on the level of 85% of power, amounted to 356,3 ppm. For an engine loaded in a variable mode, the contents of zinc in the oil had a higher value and amounted to 389,3 ppm. The average contents of zinc in the engine oil is presented in figure 5.

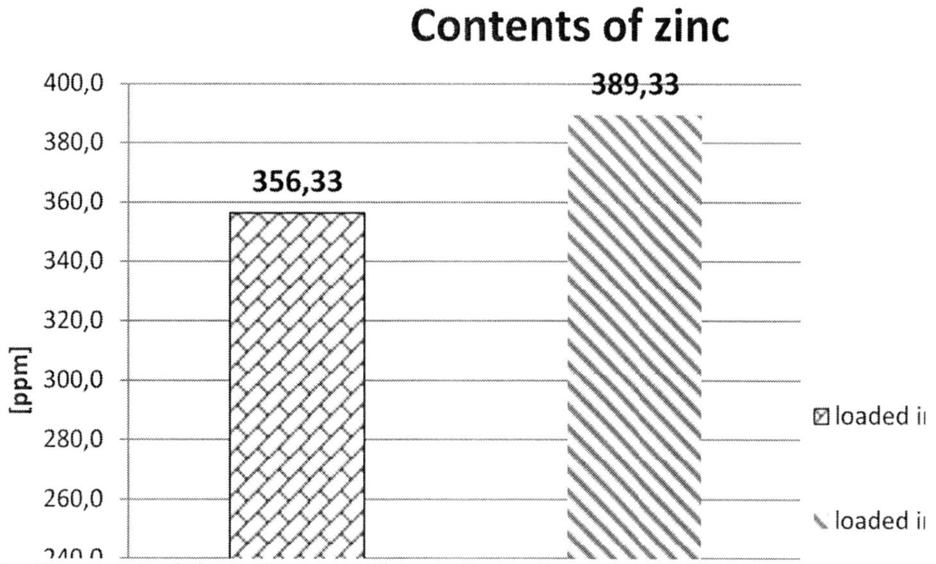

Fig. 5. The contents of zinc in the engine oil depending on the mode of loading.

For engines loaded in a constant mode and at simultaneous higher average value (85% rated power), the contents of iron, copper and zinc in the engine oil after the period of 800 operating hours was determined and was significantly statistically lower than for engines loaded in a smaller degree but in a variable mode (65% rated power).

4 Conclusions

Presence of the examined metals in the engine oil signifies occurrence of the engine elements and its accessories' (turbo-compressor) wear and tear processes. Copper found in oil is most often the effect of abrasive wear of slide bearings' bushings of the crankshaft, distribution shaft and the turbo-compressor's bushings. The presence of iron in the engine oil signifies the piston rings, cylinder sleeves and the crankshaft's wear and tear processes. The presence of zinc in the engine oil results from its use as an anti-corrosive additive, but the increase of its contents is the effect of the wear and tear of the slide bearings' bushings in which it is the component of the bearing alloy. Zinc may also come from refining additives, but its content should in such a case decrease [9, 10]. For a biogas-fueled engine, higher content of zinc, copper and iron engine oil was found in the case of a variable-load engine. The presence of copper and iron zinc in a 85% power-loaded engine was lower. The differences found are statistically significant. The tests carried out will be supplemented in the further research cycle by measuring the wear of selected engine components

References

1. A. Bilcan, O. Le Corre, A. Delebarre, Environ. Technol. Vol. **24** , No. 9, (2003)
2. M. Zastempowski, S. Borowski, J. Kaszkowiak, *Trends in Agricultural Engineering Prague*, 673-676 (2013)
3. E. Dulcet, J. Kaszkowiak, S. Borowski, J. Mikołajczak. Biosyst. Eng. Vol **3** 379-384 (2006)
4. W. Czekała, S. Bartnikowska, A. Lewicka, A. Bugała, Z. Zbytek, (2016)
5. F. Annika, T. Andersson, A. Karlsson, H. Svensson, J. Ejlertsson, J. Air Waste Manage. Assoc Vol. **54** , No. 7, (2004)
6. S. Borowski, L. Knopik, M. Markiewicz-Patalon, A. Brzostek, *Trends in Agricultural Engineering Prague* (2016)
7. M. Janosova, A. Petrovic, V. Vozarova, et al. *Conference: 23rd International PhD Students Conference (MendelNet)* (2016)
8. L. Hujo, Z. Tkac, J. Tulik et al. Adv. Mater. Res Volume: 801 137-142 (2013)
9. Z. Yang, W. Xie Fuel Process. Technol. Vol. **88**, Issue 6 (2007)
10. J.XDong, Z.Shu, Tribol. Int. Vol. **31**, Issue 5, (1998)

Wind turbine condition monitoring system as a source of diagnostic information

Sylwester Borowski[1,*], *Mirosław* Szubartowski[2], *Leszek* Knopik[2] and *Klaudiusz* Migawa[1]

[1]Faculty of Mechanical Engineering, UTP University of Science and Technology, Kaliskiego 7, 85-796, Bydgoszcz, Poland
[2]Faculty of Management, UTP University of Science and Technology, Kaliskiego 7, 85-796, Bydgoszcz, Poland

Abstract. The issues relating to the necessity of monitoring the wind turbines' condition and operation are presented in the study. The wind turbines of high powers, are very expensive and complicated machines requiring appropriate control and high quality technical service. The idea of appropriate technical turbine's maintaining, combines its high availability and productivity, as well as minimization of the costs related to failures and unexpected damages. Remote diagnostic systems allow obtaining the basic information, making it possible to maintain and appropriate control, use and high quality technical service.

1 Nuisance of wind power plant

The renewable energy sources are an important element of current life. Their spreading in the environment is consistent with the idea of the scattered energy sources. However, such scattering causes, that the local population coming into contact with them, may have negative feelings. The scientists have been conducting many studies to decrease the negative impact of the renewable energy's sources on the human being's environment. Apart from direct interactions, the manners of the interaction's decreasing by their energy-consumption's decreasing are also being tested [1].

A wind power plant is a complicated system composed of a wind motor connected with an electric generator, together with all the types of systems needed for its functioning. A wind power plant, at the time of its operation, is exposed to variable loadings resulting from wind gusts having an effect both on the rotor's planes but also on the tower and nacelle. These interactions are moved on the foundation's plate, and then to the ground. Considerable speeds of the blades' ends and a rotor's operation in the airstream lapping it, are also a source of noise. Spinning of huge masses consisting of the mass of a rotor, mass of wheels and shafts and bearing, as well as the electric generator's rotor, are the sources of oscillations and noise. Flickering of sunlight and shadow caused by considerable rotor's dimensions, are an additional source of dangerous interaction on living organisms. Even in the domestic environment, a human being and animals are exposed to wind interactions. These are most of all the infrasounds, which may penetrate the concrete barriers, as well as flickering of the electric light powered from the network that the power station is connected to.

Noise is the result of different sounds' overlapping, which we receive as having no order. As it is commonly understood, it is every sound which disturbs us. The impact of noise on a human body is different, however most of all it attacks the nervous system. It is a serious risk as its results very rarely become apparent at once – they more often cumulate in time. Noise affects live organisms in two ways. The first and the most often observed one is its impact on mental state, and consequently the physical health of a human being. The

* Corresponding author: sylwester.borowski@utp.edu.pl

second mode is the physical noise's affecting on the hearing organ and the central nervous system. Sound waves, which we receive as sounds or noise, are the changes of the air's pressure surrounding us having a direct contact with the delicate elements of the hearing organ. Exposure to excessive noise resulting in occurrence of an acute or chronic acoustic injury, accompanied by may defensive responses just like for example changes in the heart's action, of the breathing rhythm, of blood pressure, of body temperature, etc. As a result of staying in the environment burdened with excessive noise, there may occur the so-called post-noise syndrome covering: headaches and vertigoes, weakness, increased nervousness, sleep disorders, increased sweating, hearing impairment. Noise received by the ear through nerve connections with the cerebral cortex may interact on other centres located in the brain. It concerns most of all the central nervous system and the endocrine glands system. 75 dB is the boundary value of the physiological functions' disturbance, after exceeding of which there may occur their clear disturbances. For the value of the acoustic pressure within the limits of 55 ÷ 75 dB, there may occur the lack of concentration. Such a bodily's reaction is its defense against the long-term exposure to noise, which may contribute to different types of diseases (for ex. pressure disease, ulcer disease, neuroses).

Infrasounds are the sounds, in the spectrum of which there appear constituents of frequencies from 0 to 20 Hz. The sounds within such a band of frequencies, are theoretically not heard by a human being. However, the level of their reception depends on individual personal features and the level of acoustic pressure. The acoustic pressure thresholds for infrasounds' audibility is high (6 ÷ 8 Hz about 100 dB), and they lower with the frequency's increase. Infrasounds are also received by the vibration sensory receptors. The feeling thresholds are for 20 ÷ 30 dB higher than the hearing thresholds [2].

For the acoustic pressure on the level of 100 dB, the infrasound may be perceptible as an unpleasant feeing of internal vibration caused by the internal organs' resonance phenomenon. Another occurring symptom is the feeling of compression in ears and the occurring state of excessive tiredness, discomfort, drowsiness, balance disorders, psychomotor physical fitness and physiological functions' disorders. Changes caused by infrasound in the central nervous system are similar to lowering of the standby.

For the acoustic pressure above 140 - 150 dB, the infrasound may result in permanent, harmful changes in the organism [3].

Infrasound come into existence in a different manner. In the nature it is waving of oceans, leaves' rustle. The anthropogenic sources are most of all the vehicle transport, industry, ventilation in buildings. Huge flow machines just like wind turbines, have also become a noticeable source of infrasound.

The issues related to the noise are the object of a special interest of scientists. Particularly important, as far as the noise is concerned, there becomes the correct assessment of a wind farm's condition. It's used elements may cause the noise's emission increase. As presented by [4,5] wind farm plants' noise may influence the animal's productivity, their stress, fodder's consumption. Noise is also important in case of another green energy's processor, just as agricultural biogas plants [6].

Mechanical vibrations generated by a wind farm plant are transmitted through foundation on the surrounding land. A human subject to that factor is exposed to damage of tissues and blood vessels. Long-term exposure to vibration may result in disease changes in the vascular system, musculo-articular system and nervous system.

Changes in the musculo-articular system of a hand arise mainly as a result of local vibration of frequencies lower than 30 Hz. There are observed among the others, deformations of joint spaces, calcification of joint capsules, changes in bone texture.

Disorders in the internal organs' functioning are mainly the result of individual organs' stimulation for resonant vibrations (frequencies of own vibrations of most of the organs are within the range 2 ÷ 18 Hz).

The following ones are the most important constructional elements deciding on a typical wind turbine's efficiency:

- Main transmission with a generator (gear wheels, bearings, shafts and intermediary couplings),
- Turbine's blades with their angle's adjustment mechanism,
- Turbine's tower (foundation on the base, the tower's connections).

Incorrect operation or these elements' damages should be disclosed earlier, in the initial phase of their damage. When disclosed too late, it may result in damaging elements or whole installations by destructive feedback.

The main transmission is exposed to variable, cyclical loadings [7]. Their impulse nature results in occurring of fatigue damages and revealing of material defects. Exposed to damages in the main transmission are most of all the main bearings and the planetary transmission.

The wind turbine's blades are the elements converting the wind energy into the rotor's rotational motion. Due to big dimensions and moments having an impact on a blade, they should be under special supervision. Incidental influences such as strokes in birds, icing, are also disadvantageous.

Due to that, slight damages in a blade's critical places may result in very rapid increase of a damage's scope till a blade's complete destruction.

A tower is the next constructional element of a wind turbine, supporting the whole construction. Damages or loosening of the fixing elements in the tower result in the increase of the risk of the complete turbine's damaging [8]. Figure 1 shows the course of ground vibration at a distance of 50 m from the tower of the power plant.

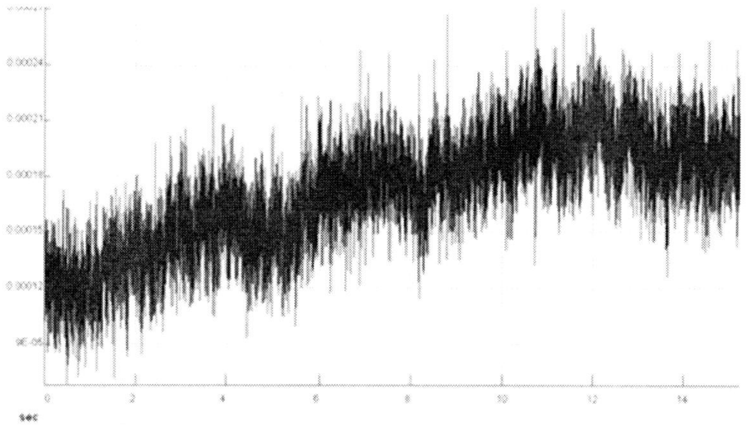

Fig. 1. The time course of vibrations generated by the wind turbine at a distance of 50 m from the tower of power plant.

2 Detection of damages in wind power plants

A properly designed machine in modern CAD/CAM systems seems to be free of defects. However, each operating process must be monitored and monitored on-line in order to avoid major failures [9].

The purpose of monitoring is early detection of damages with the use of the below listed methods:

- measurements of vibration and vibrodiagnostics,
- monitoring the condition of the blades based on the elastic waves' propagation and deformations' measurements,
- monitoring of the tower with the use of the own vibration frequencies' measurement techniques and monitoring of the screw connections' condition.

The systems of the transmission vibrations' monitoring enable recording and analyzing of vibrations from the transmission. Such a system consists of the set of sensors (vibration, process values such as power, speed and direction of wind, etc.) and the module of data acquisition and processing. The data is collected and then processed by a data server. In order to increase the performance, the data registering system has been divided with respect to the type of data and their priority, there are also recorded the events connected with crossing of the alarm thresholds. From the events like these, there is drawn up a protocol which, in an electronic form is automatically sent to the diagnostic centre.

The systems of the blades and the tower's monitoring make it possible to:

- detect the blade structure's damages (stratification of the composite material, cracking, comings off) – systems based on generation and measurement of elastic waves,
- detect the problems with unequal blades' burdening (icing, loosening of connections with a hub) – systems based on measurements of deformations with the use of optical fibre sensors,
- easy integration with the transmission's monitoring system – data are sent to the same database and are processed by the expert algorithms.

The complement of the whole monitoring system is most often the system of the measurement type SCADA data acquisition and visualization. It is the system supervising the course of the process. The following may be considered to be the basic tasks of such a system:

- remote controlling of the measuring devices,
- control of the measurement process,
- collecting and recording of the current measurement data in the database,
- analysis of the signal data,
- steering the technological process,
- alarms and notifications' generating,
- visualization of the parameters measured on the synoptic screens,
- visualization of the current and historical measurement data on the diagrams,
- export of data to other systems.

The following ones are also the indispensable features of the software:

- modeling of an object's kinematics,
- automatic setting of the controlled machines' characteristic frequencies,
- automatic setting of the alarm thresholds' values,
- creating reports.

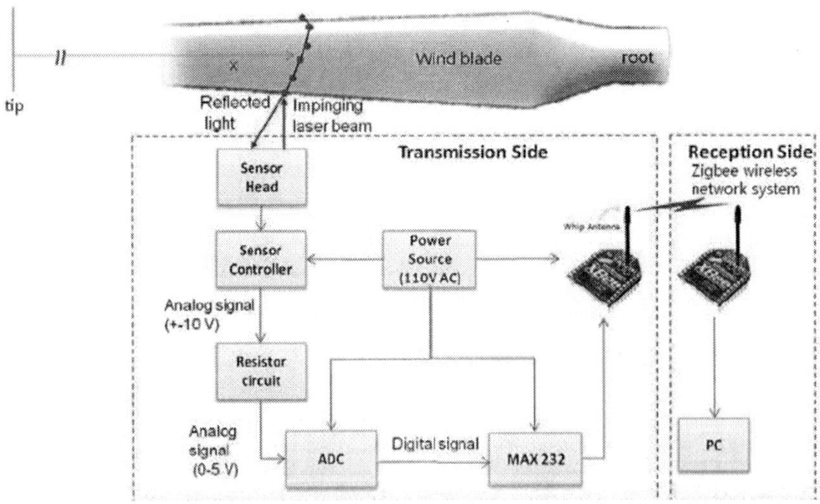

Fig. 2. Block diagram for wireless deflection monitoring system of wind turbine blades [10].

To protect the turbine blades, various types of control systems are used. The experimental wireless system is shown in Figure 2 [10].

3 SCADA systems in wind power plants

The SCADA systems consist of several crucial components. The basic one is the controlling computer located at the area of a wind farm, permanently connected to the operating turbine. In case of installation on one power plant, such a computer most often is installed in its interior. With the use of the internet links, the collected data from monitoring are transferred to a final recipient – a customer – operation of a wind farm, who reads them out with the use of a special computer application.

SCADA has the input data entered to the system, which with the use of the sensors located in the power plant, monitor its condition. Data is used not only for monitoring purposes, but also to different types of analyses often showing dependencies between given parameters. Thanks to algorithms, it is possible to detect the causes of failures of the elements' wear and tear. The following ones are the other important functions of the SCADA systems: data archiving, alarming of diagnostic centers, transfer of information to a wind farm's owner, diagnostics, cooperation with power generation plants, analysis of data already mentioned.

The service, thanks to SCADA software, is able to perform remotely diagnostics of the conditio, what reduces the costs and speeds up the work. The technical division, thanks to performing of the cause-effect relations' analyses, may avoid underestimation of a small defect, knowing on its future negative impact of a turbine's operation.

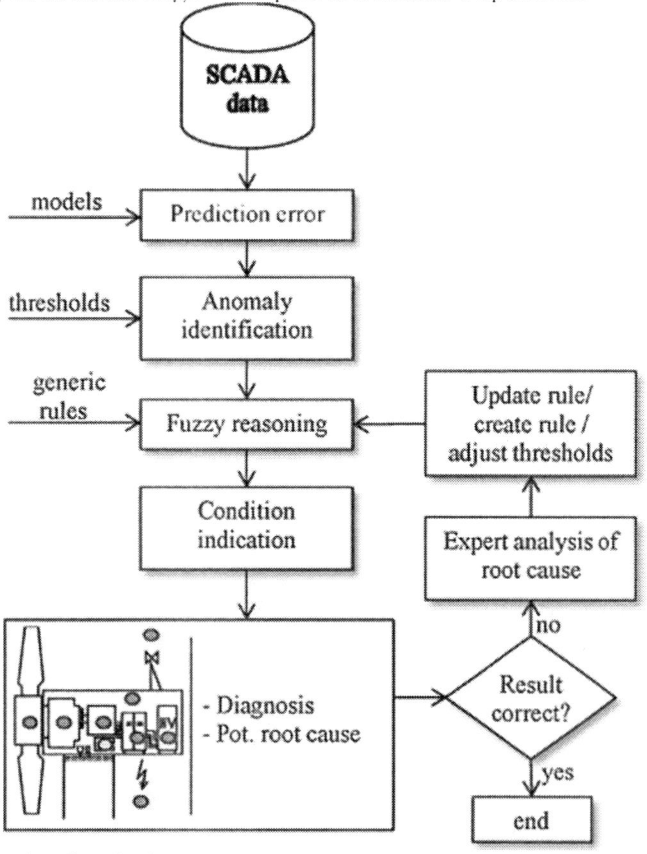

Fig. 3. The Information flow [11].

Thanks to the management systems, it is possible to choose the settings aiming at reduction of the noise emitted by wind turbines. One of the basic methods of noise reduction is to manipulate the speed of rotation of the wind turbine rotor's propellers.

Slowing down of the rotor's rotation is connected with the acoustic emission's reduction. The angle of the turbine blade's inclination also plays an important role here.

Its appropriate increase shall result in the decrease of the air masses' angle of incidence. So, selection of such settings of the inclination angle may considerably reduce the noise generated by the influence of air masses on the wind turbine rotor's surface. Such a method is very often used to reduce the acoustic interaction.

The diagnostic systems of that type may minimize considerably the number of the service's field inspections, and consequently, reduce the time of a turbine's shutdown from use. On the market there are available different types of systems – the author's SCADA of the manufacturers of wind turbines and independent systems [11].

The general flow of information using SCADA systems and the optimization of diagnostics has been highlighted in Figure 3.

3 Summary

A modern monitoring system enables not only measurements and signaling of a current machine's condition. They are most often used to analyze and forecast its condition in the subsequent moments. Such a task requires most often the simultaneous analysis of many parameters and using of different, often computationally advance methods.

References

1. M. Zastempowski, S. Borowski, J. Kaszkowiak, *Trends in agricultural engineering*, 673-676 (2013)
2. A.A. Nuno, C. Branco, *Ageing Mechanisms and Control: Part A – Developments in Computational Aero- and Hydro-Acoustics,* 3-12 (2001)
3. M. Karwowska, J. Mikolajczak, Z.J. Dolatowski, S. Borowski, Ann. Anim. Sci. **15,** 1043-1054 (2015)
4. M. Karwowska, J. Mikolajczak, S. Borowski, Z.J. Dolatowski, J. Marc-Pienkowska, W. Budzinski, Ann. Anim. Sci. **14,** 441-451 (2014)
5. J. Mikolajczak, S. Borowski, J. Marć-Pienkowska, G. Odrowąż-Sypniewska, Z. Bernacki, J. Siodmiak, P. Szterk, Pol. J. Vet. Sci. **16,** 679-686 (2013)
6. J. Kaszkowiak, M. Markiewicz-Patalon, S. Borowski, P. Dorszewski, *Engineering Mechanics,* 438-441 (2017)
7. J. Tulik, J. Kosiba, L. Hujo, J. Jablonicky, V. Sinsky, Trends in agricultural engineering, 617-621 (2013)
8. G. Szala, B. Ligaj, *Fatigue failure and fracture mechanics,* 69-76 (2012)
9. M. Macko, J. Flizikowski, Z. Szczepanski, K. Tyszczuk, G. Smigielski, A. Mrozinski, J. Czerniak, A. Tomporowski, *Proceedings of the 13th International Scientific Conference: Computer Aided Engineering,* 343-351 (2017)
10. G. Paritosh, L. Jung-Ryul, *The 1st International Conference on Advances in Structural Health Management and Composite Structures,* 29-31 (2012)
11. S. Meik, S. Ilmar, Appl. Soft. Comput. **14,** 447–460 (2014)

Estimation of the stress-strain state of ultra-low pressure tires

Vladimir Boykov[1,*] , *Cheslav* Zhdanovich[1], and *Andrey* Dzioma[1] and Joanna Wilczarska[2]

[1]Belarusian National Technical University, Department "Tractors", 220013 Minsk, Nezavisimosti, 65, Belarus
[2]Faculty of Mechanical Engineering UTP in Bydgoszcz, Department of Vehicle Engineering, Al. Prof. S.Kaliskiego 7, 85-796 Bydgoszcz, Poland

Abstract. To study the stress-strain state and evaluate the operability and durability of the ultra-low pressure tire F118A ns2, its three-dimensional model was developed on the basis of which a finite element model was developed, load regimes were established, and calculations were performed. We consider such modes as tire landing on the disk, pumping the tire to the nominal pressure, loading the tire with the weight of the chassis taking into account contact with the support surface to determine the static contact spot, rolling the tire at a constant speed, and analyzing the non-stationary rolling.

1 Introduction

Agricultural machinery with its own wheeled propellers and used on fields, meadows and pastures should meet both the requirements of agrotechnics (the pressure of propulsors on the soil and the normal mechanical stress in the soil should not exceed the permissible values), and the requirements for ensuring the specified tire life (tire deformation under action load on the wheel should not exceed the normative values) [1]. The pressure on the soil in the early spring period with its moisture content of 0.9 - 1.0 HB should not exceed 40 ... 50 kPa [2 - 4]. For this purpose, ultra-low pressure tires with an operating air pressure of 10-60 kPa are used on light transport-technological aggregates of agricultural purpose working on the field in the early spring period. When such tires with an air pressure of 10 - 15 kPa are fixed, creases are formed (Fig. 1) [5], which leads to their intensive wear. With insufficient internal pressure, the tire loses its carrying capacity, crushing the tire.

The performance and durability of ultra-low pressure tires depends significantly on their load-bearing capacity. In the complex design of pneumatic tires, there are stress concentration zones that can cause premature failure during operation. Fatigue failure leads to a decrease in the life of tires to 20-25% of the estimated value of the resource for wear. Therefore, analysis of stress-strain state and prediction of longevity in zones of stress concentration of pneumatic tires at the design stage is an actual and important problem [6]. To analyze the stress-strain state of tires, the finite element method [6 - 9] is widely used, which makes it possible to model a complex design of the tire.

* Corresponding author: boykov_vip@mail.ru

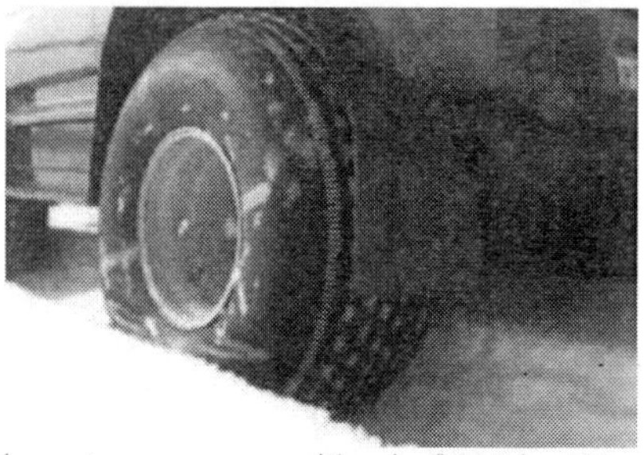

Fig. 1. Formation of folds on tires at a pressure and load of 350 kg [5].

2 Main part

Based on the finite element method using modern software tools, a methodology has been developed and the load capacity of ultra-low pressure tires has been calculated. The model of the model F118A ns2 (22,0 / 70-20) was chosen as the modeling object. The tire has a two-layer textile cord. Construction of the general geometry of the three-dimensional model of the bus was carried out in the software package Creo (Pro / ENGINEER) on the basis of drawings of the section of the tire and the profile of the tread. The main structural elements of the bus F118A ns2 are the frame, bead, belt, protector, sidewall. The construction of each constructive element was carried out in the "Detail" environment, the positioning of each element was carried out in the "Build" environment.To create a common geometry of the tire and tread, a sketch of a repeating tread element was constructed, then using the "Pull" command, each element was constructed. To create a certain number of elements of the tread pattern, the command "Array" was used. Using the "Toroidal bend" command, the flat geometry of the tread was bent according to the shape of the tire. Using the "Rotate" command, the form of the finished tire after vulcanization is given on the basis of the cross-section of the tire of the model.

For three-dimensional modeling of individual layers of the tire, we use the "Rotate" command and a cross-sectional drawing of the tire, which indicates the location of each layer in the body of the tire, creating each layer separately by rotating its section 360 degrees relative to the axis of the tire. The construction of the "Side Ring" part was carried out using the "Stretching" command, knowing the number of turns and the diameter of the wire. Ribbons that cover the joints of the bead ring are built using the "Rotate" command, while the angle to which the section sketch is rotated is determined from the condition that their length is 100 mm.When constructing the carcass plies and the carcass ply, we also took into account the arrangement of the layers in the zone of the bead ring. Figure 2 shows the 3D model of the tire.

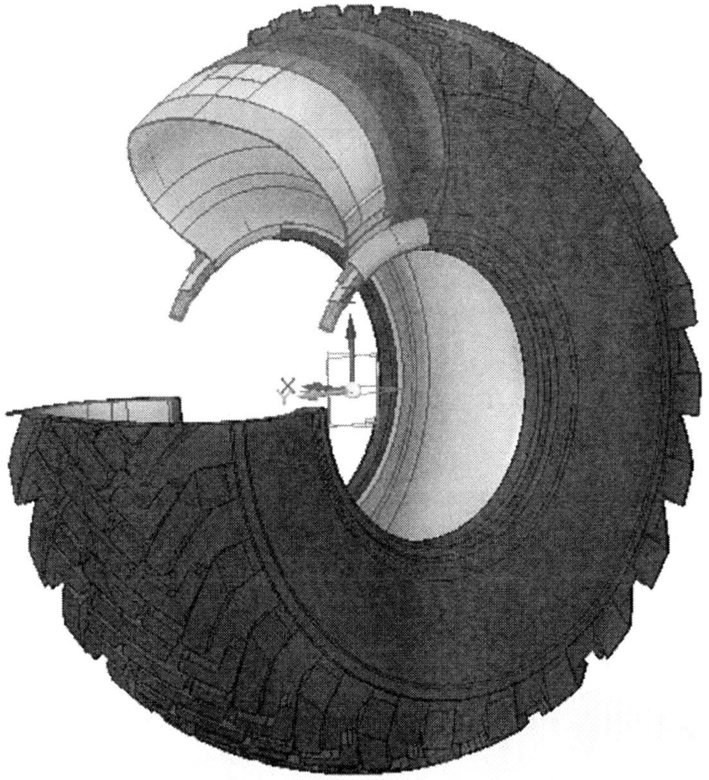

Fig. 2. Three-dimensional model of the bus, taking into account the layers.

In the study of stress-strain state (SSS) tires considered the following main tasks:

1) putting the tire on the disk;

2) pumping the tire to nominal pressure;

3) loading of the tire with the weight of the machine, taking into account contact with the support surface to determine the static contact spot;

4) rolling the tire at a constant speed, while you can get results for effort and torque on the spindle;

5) analysis of non-stationary rolling.

The methodology for solving the tasks is similar for the known universal software complexes used for the study of SSS models, such as ABAQUS, MSC.Software, ANSYS, etc.

Each subsequent step uses the results obtained at the end of the previous step. Whenever possible, it is recommended to use design symmetry to reduce the size of the design model and increase the computing efficiency.

Fig. 3. The scheme of application of loads for the study of SSS when the tire is placed on the disk.

Task of tire landing on the disk.

In the study of the first problem, a uniform load is applied to the tire surface in the area of the bead ring placement, while the axis of the tire itself remains stationary.Load is applied until the deformation of the tire exceeds the values determined by the parameters of the rim. Further, the value of the applied force becomes zero and the tire tends to return to its original position. Restriction of its movement will be the surface of the rim of the disc.

This task to reduce the calculation time can be solved in a flat section of the bus, and the results can be propagated to the entire circumference of the tire by means of transfer values.

The scheme of application of loads for the study of SSS when landing the tire on the disk is shown in Figure 3.

Pumping the tire to nominal pressure.

The scheme of application of loads and restrictions for the investigation of SSS when the tire is pumped up to the nominal pressure after landing it on the disk is shown in Figure 4. In this case, the movement restrictions are superimposed on the tire surface in contact with the rim of the disk, and the internal surfaces are loaded with uniformly distributed pressure along the internal surfaces tires. The pressure value during the analysis varies from initial to nominal.

Fig. 4. The scheme of applying loads and imposing constraints in the study of the problem of pumping the tire to the nominal pressure.

Loading of the tire with the weight of the chassis, taking into account contact with the supporting surface for determining the static contact patch.

To solve the task, unlike the previous one, an additional plane is used, loaded so that the magnitude of the applied force at each point of the plane corresponds to the load from the weight of the chassis. The scheme of application of loads and restrictions for the investigation of SSS when the tire is pumped up to the nominal pressure after landing it on the disk is shown in Figure 5.

To study the tire's SSS under the influence of lateral forces, the results of the study of the static contact patch are used and the restrictions on the movement of the tire in the transverse direction are additionally removed and the load applied to the center of the wheel is equal to the lateral load.

The task of rolling a tire is simulated by using a coordinate system associated with the tire axis. The coordinate system moves together with the tire axis, but does not rotate, while the tire rotates relative to the given coordinate system.

Fig. 5. The scheme of applying loads and imposing constraints in the study of the problem of determining the static contact spot.

For the calculation of the stress-strain state of a tire based on a three-dimensional model, it is necessary to develop an adequate finite element model (FEM). When designing a finite element model (Fig. 6), the type of the problem being solved is taken into account to study the stress-strain state of the tire. For example, to solve the problem of determining the static contact patch, a more detailed grid was formed in the lower part of the tire, and for the stationary rotation problem a detailed grid is only needed near the zone of the contact spot.

Before the construction of the finite element model, materials for each element were assigned. The main material in the tire are rubber compounds. Rubber refers to hyperelastic materials with the inverse convexity of the tension-compression curve. To describe the properties of the real material, the density of the material was set and the data obtained on the basis of the experiment were introduced. They can be specified as a table in the preprocessor by default or through a separate file. In the table, points (nominal stresses and deformations corresponding to them) are introduced, along which the stretching diagram for rubber is constructed (Fig. 7).

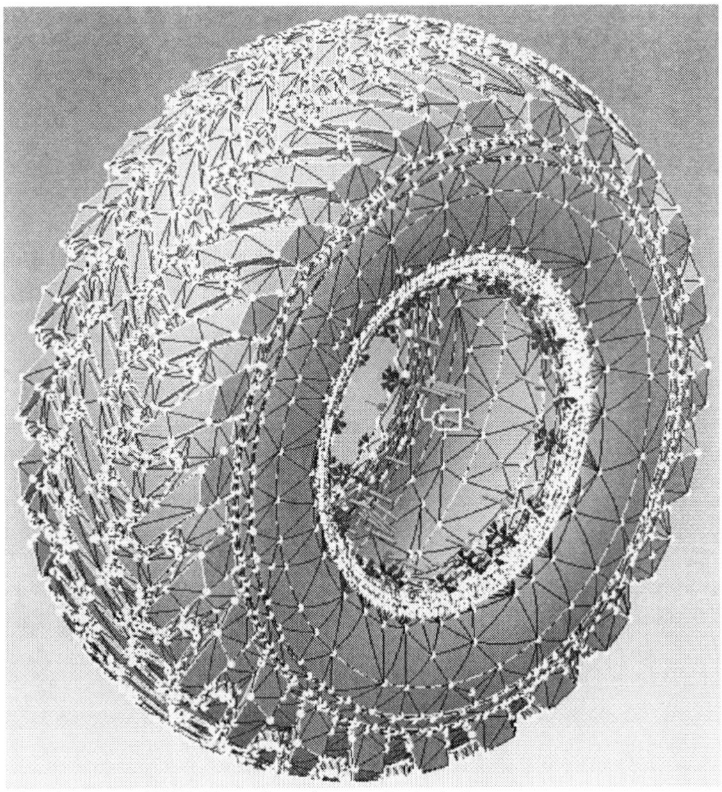

Fig. 6. Automatically created FEM based on the tetrahedral elements.

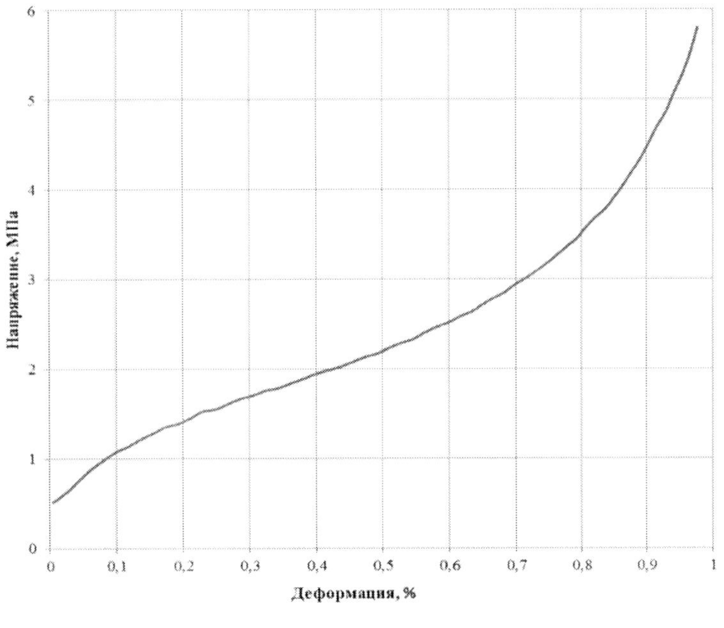

Fig. 7. General view of the rubber extension diagram.

Thus, to study the stress-strain state and evaluate the operability and durability of the ultra-low pressure tire F118A ns2, its three-dimensional model was developed on the basis of which a finite-element model was developed, loading regimes and calculations were performed. The basic operational parameters of the tire are established: the permissible load is 6500 N; the maximum load is 8000 N; the tire pressure corresponding to the maximum permissible load is 15 ... 20 kPa; the maximum speed of movement on unpaved roads with air pressure in the tire is 25 kPa - 50 km / h.

References

1. V.A.Rusanov, V.N. Belkovsky. Rubber and Rubber, **9**, 6 (1985)
2. V.A.Rusanov, *The problem of soil re-consolidation by propulsors and effective ways of its solution* (Moscow: VIM, 1998)
3. A.G. Bondarev, V.A. Rusanov, P.M. Sapozhnikov, *Overconsolidation of arable soils. Causes, consequences, ways of decreasing,* 166 (Moscow: Nauka, 1987)
4. V.V. Medvedev, V.G. Tsybulko, *Overconsolidation of arable soils. Causes, consequences, ways of decreasing,* 173 (Moscow: Nauka, 1987)
5. L.V. Barakhtanov, *Creation of an experimental sample of special vehicles of the northern version on ultra-low pressure tires for operation on low-bearing bearing surfaces* (Nizhny Novgorod: NSTU, 2012)
6. 8. S.L. Sokolov, *Computational and experimental methods for studying the stress-strain state and cyclic durability of pneumatic tires* (Moscow, 2011)
7. 9. LB Barakhtanov, VI Kotlyarenko, SE Manianin, IA Sokolov, Proceedings of the Nizhny Novgorod State Technical University. R.E. Alekseeva, **2 (95)**, 172 (2012)
8. 10. VO Kondrashov, M.Yu. Sokolova, Izvestiya of the Tula State University Natural Sciences, **4**, 144 (2015)
9. 11. Ch.I. Zhdanovich, A.A. Dzioma, Science - education, production, economics. Materials of the 15th International Scientific and Technical Conference, **2**, 29 (Minsk, BNTU, 2017)

Review of car operation safety system elements

Tadeusz Dąbrowski[1,*], *Adam* Rosiński[2]

[1] Military University of Technology, Faculty of Electronics, 00-908 Warsaw, gen. W. Urbanowicza 2, Poland
[2] Warsaw University of Technology, Faculty of Transport, 00-662 Warsaw, Koszykowa 75, Poland

Abstract. The paper includes a review of the following terms: anthropotechnical system, a layered model of an object operation process, safe car operation system structure. The current and predicted solutions in terms of car equipment impact on the safety of a driver-vehicle-road system were presented in detail.

1 An anthropotechnical system model

The operating process in an anthropotechnical system (Fig. 1) can be understood as the flows and transformations of substances, energy and information. The quantitative and qualitative values of these flows and transformations in a distinguished moment, determine the instantaneous state of a system. It is not hard to notice that an operating system of technical objects is formed by anthropotechnical systems with specific operating tasks (use or maintenance) [1].

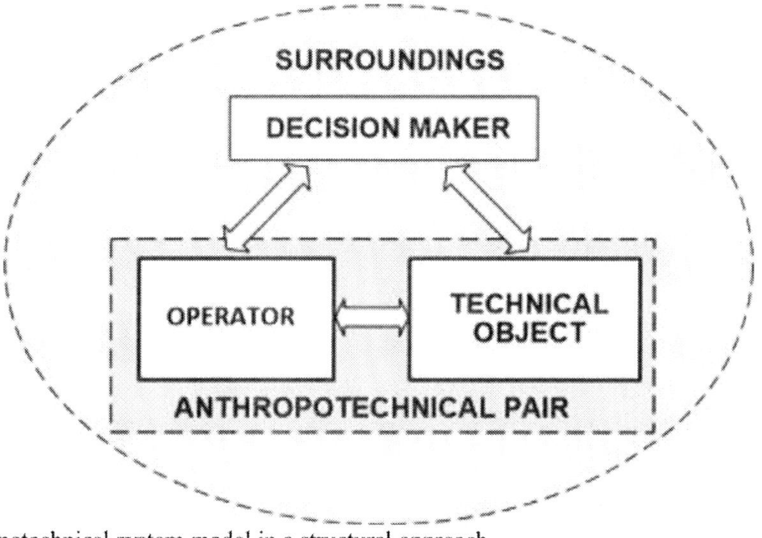

Fig. 1. Anthropotechnical system model in a structural approach.

2 A three-layer operating process model

The purpose of the utility phase of operation is – of course – generating the required operating effect, namely, execution of a operating task.
Three main processes may be distinguished in the operating phase (Fig. 2) [1, 2]:

1) the operating process, leading to the execution of an operating task;
2) the destructive process, counteracting task execution, which can be divided into three component sub-processes: initiating process, damage process, failure process;
3) anti-destructive process, which can be divided into three component sub-processes: shielding process, intervention process, rescue process [3,4].

* Corresponding author: tadeusz.dabrowski@wat.edu.pl

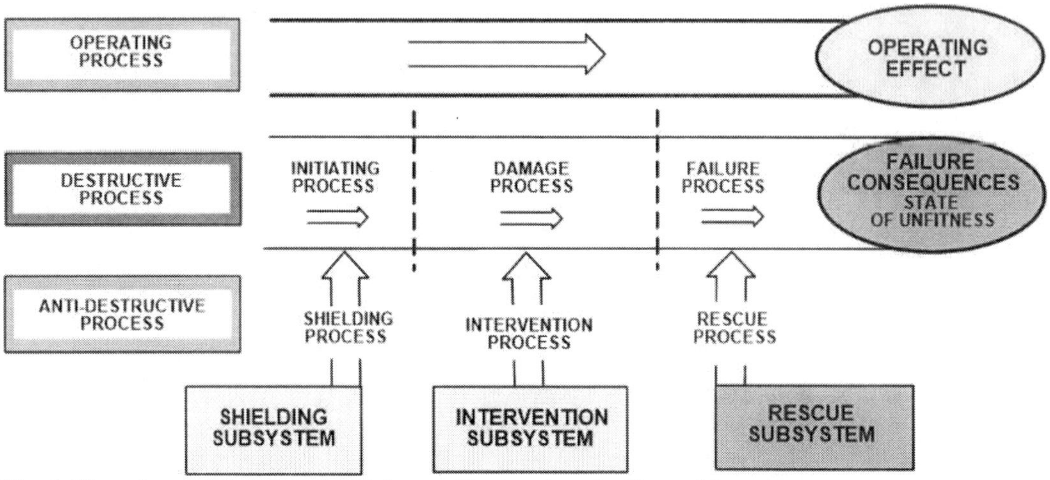

Fig. 2. Three-layered operating model for an object equipped with a safety system.

Destructive process component characteristics

Initiating process involves the activation of trigger factors, which may initiate a destructive process.

The triggering factors may be of "neutral" nature, associated with the operating process or "aggressive" nature, in the form of overriding, interference, etc.

Damage process precedes and leads to the state of task unworthiness, i.e., a state, which makes it impossible to execute (complete) a task.

Reaching the state of task unworthiness may result in the occurrence of two situation types:

– operating – as well as a destructive process – are interrupted in a "natural" manner;

– apparent operation (although the task is not being executed) is ongoing and the destructive process is developing; in this case, the destructive process becomes a failure process.

Failure process involves the destruction of consecutive elements of a technical object and its surrounding, and the destruction of previously generated effect.

In order to maintain task worthiness of an anthropotechnical system (SAT) one can – and often should – equip an object's operating system with a safety system (security system), designed to generate anti-destructive activities.

In an operating system containing a safety system, a destructive process may be "halted" (or at least slowed down) at the level of the trigger factors or the damage process level, and the failure consequences may also be limited.

The characteristics of anti-destructive process components

A shielding process involves supervision, namely, observation and detection (if present) of trigger factors and their deactivation.

An intervention process involves supervising a damage process (if present) and activating intervention, halting or slowing down factors.

A rescue process involves supervising a failure process (if present) and the activation of activities protecting a selected object and environment zone, and a previously generated effect against destruction.

These actions are shown in figure 3.

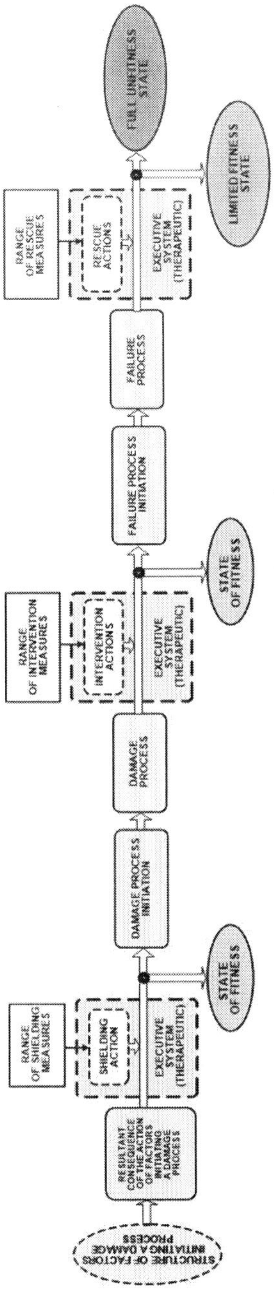

Fig. 3. Safety system action model

The operating effect is achieved as a consequence of executing an operating task, which is accompanied by integral destructive processes, triggered by a hazard system. The lack of preventive actions may result in a failure or even a disaster [5, 6]. In order to effectively prevent the effects of destructive processes, the selected objects and systems are fitted with **safety systems**. Their task is to counteract the occurrence of trigger factors, and to stop, or at least limit, destructive processes [7].

3 Structure of a car operation safety system

A system, which implements the car operating process is an anthropotechnical system. The general structure of the system is shown in figure 4.

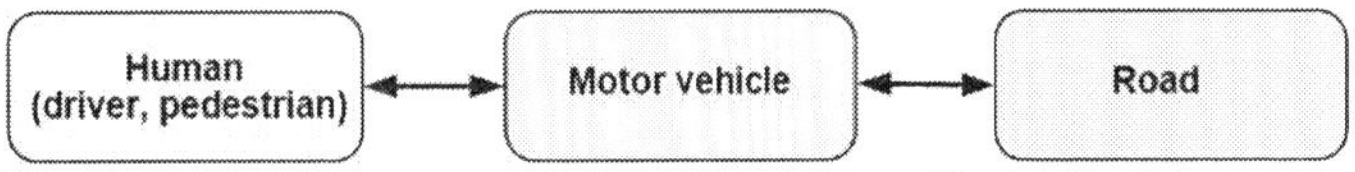

Fig. 4. Car operating process execution system illustration.

Proper functioning of the three mentioned system components determines comprehensive safety of a vehicle operation process [8].

The following elements impacting road traffic safety are, among others, **human-related** [9, 10, 11]:

- the skill of correct driving motor vehicles (ability of correct behaviour in traffic and to correctly assess a situation), also in the case of occurring hazards;
- familiarity with legal regulations applicable to road traffic;
- response time to occurring events.

The following elements **road-related** within the analysed system [12, 13]:

- road form (descent and ascent slope, sharpness of turns, profile);
- condition of the surface, its cohesiveness and bumpiness;
- pavement type (concrete, asphalt, boulder, gravel and other);
- road signage (lighting, horizontal, vertical) warning drivers against potential dangers;
- atmospheric conditions (strong insolation, dawn, water, snow or ice covering the road pavement, sidewind gusts, fog);
- traffic intensity (vehicle travel density);
- intelligent transport systems [14, 15].

The safety of road traffic participant is, however, mainly dependant of a **motor vehicle** and includes the proper functioning of three subsystems [16]:

- vehicle active safety subsystem;
- vehicle and passenger passive safety subsystem;
- preventive safety subsystem.

As per figure 2, the terms "*safety system*" shall be understood as a task-oriented structure in the form shown in figure 5.

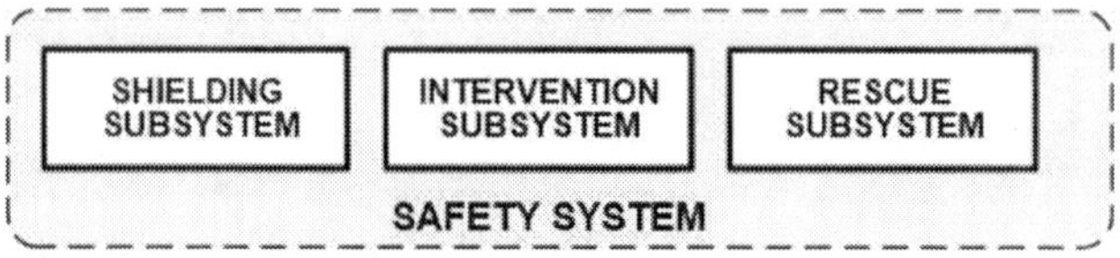

Fig. 5. Task-oriented subsystems of the safety system.

A shielding subsystem is a set of measures and actions generating shielding factors preventing the appearance of factors exciting destructive processes within an operating system.
Example: Road traffic regulations, limiting the permissible speed over a given road segment.
An intervention subsystem is a set of measures and actions stimulating intervention factors, which completely stop, or at least slow down, destructive processes leading to permanent unfitness (failure/disaster).
Example: ABS (Anti-Lock Braking System), preventing wheel locking when braking.
A rescue subsystem is a set of measures and actions activating rescue factors, which are aimed at limiting the consequences of unfitness (failure/disaster).
Example: AirBag system, minimizing the consequences of a road incident.
At this point, it is worth noting that the structure of each of the aforementioned subsystems must include not only specified executive modules (sub-assemblies, procedures), but also diagnostic systems (supervising), which provide information about process ongoing in the system: operation and destructive.

A vehicle and passenger passive safety subsystem

A vehicle and passenger passive safety subsystem **executes rescue actions** and includes vehicle subsystems with automatic functioning, **regardless of the driver participation.** This means, that these systems are aimed at mitigating the effects of an already occurred collision.
Their main objective is to limit the probability of injuries or death among a vehicle driver, its passengers or other road traffic participants (e.g. pedestrians, cyclists), as well as increase the safety of transported cargo. The operation of these systems takes place when a driver, regardless of the performed actions, has no impact on the road situation any more. A passive safety system includes:

- appropriate design of the bodywork, chassis, seats, head rests, windows, pedals, steering wheel and more.
- a basic passenger safety system:
  - safety belts (equipped with three mounting points, a self-rolling mechanism and a brake).
- additional passenger restrain and protection system:
  - SRS – Supplementary Restrain System [additional restriction for passenger movement];
  - AIR BAG – a system of air bags.
- additional safety systems:
  - Mobile communication and GPS (Global Positioning System) systems,
  - Automatic accident notification system,
  - On-board extinguishing systems,
  - Systems isolating engine fuel supply,
  - Systems decoupling the battery from the vehicle's electrical system.

Motor vehicle active safety system

Vehicle active safety system **performs intervention functions** and covers all systems, which are **impacted by the driver**. They enable decreasing or avoiding a collision risk, which means, they are systems that allow the driver to take action prior to a collision, aimed at avoiding it.

Such systems (subsystems) include [17]:

- braking system and the cooperating:
  - system counteracting brake locking (ABS),
  - electronic braking force distribution (EBD, EBV),
  - sudden braking aiding system (BAS, EBA, AFU),
  - sudden braking preparation function (EBP),
  - brake disc drying function (BDW),
  - automatic parking brake (HHC),
  - electro-hydraulic braking system (EHB);
- power transmission system and cooperating systems:
  - system preventing drive wheel slip (ASR, ATC, ASC, TCS, TRC, ETC, DSTC),
  - automatic lock of differential mechanisms (ABD, EDS, EDL),
  - bi-axial drive (4WD, AWD),
  - system improving engine braking quality in various road situations (EDC MSR, HDC),
  - driving speed controlling system (cruise control, CCS),
  - system for maintaining distance from a preceding vehicle (ACC, ICC, AICC, AHS, ADR);
- steering system and cooperating systems:
  - driving path stabilization system (ESP, DSC, DSR, ASMS, FDR, FSR, VSC, VCA, PSM),
  - speed-dependent wheel steering corrector (AFS),
  - four-wheel steering system (4WS),
  - electronic steering wheel;
- wheelset and cooperating systems:
  - hydro-pneumatic suspension with active stabilization (ASF),
  - adaptive damping system (ADS, ADC),
  - active bodywork control system (ABC),
  - electronically controlled suspension (ECS);
- vehicle lighting equipped with, i.a.:
  - road illuminating (driving, passing, daytime, anti-fog lights) and signalling lights (braking, turn signals, navigation, clearance lights), often activated automatically and with automatic illumination direction change (DLWR, AFL, AFS ALC, DLA);
- window and mirror wipers and heaters, often automatically activated;
- air conditioning system and ergonomic seats;
- vehicle bodywork design (enabling visibility in any direction).

Preventive safety system

The term preventive safety system (i.e. a system executing **shielding functions**) is used to call all systems warning the driver and initiating his/her actions aimed at removing the possibility of a collision. These are – usually – systems functioning independently from the driver, and their task is to mitigate or remove a risk of potential collision probability [16, 18].

The preventive system impacts:

- driver behaviour and prevents:
  - the driver falling asleep when driving,
  - starting a car in the event of the driver's intoxication;

- other cars present in a traffic lane:
  - visible directly (mentioned ACC system),
  - based on information exchange between vehicles (IVCS, V2V);
- other objects present in a traffic lane:
  - the presence and behaviour of other road users,
  - the presence of stationary obstacles;
- crossing the lines separating traffic lanes (AFIL, LDW, ALKA);
- presence and mixing of other vehicles or objects near the planned parking spot (PA, PAV).

The preventive actions and procedures also include, of course, such elements of an anthropotechnical transport system as: highway code, road signs, police patrols, etc.

It is worth noting that the aforementioned classification of systems associated with car operation safety is sort of arbitrary, since more and more often these systems are structurally and technologically integrated, which makes their operation significantly inter-dependent. An example of such a situation may be the ABS, which is often an element of ASR, which in turn, is an ESP element, often being a component of ABC.

4 Forecasting the development tendencies for systems increasing the operating safety of motor means of transport

EyeSight[1]

The leading safety technology is introduced to Subaru vehicles in the form of a stereoscopic camera, aimed at equipping a driver with a digital "second pair of eyes". The basic assumption of this system is constant observation of the road based on a 3D image, which is processed in terms of identifying other object present in the vicinity of the vehicle (cars, motorcyclists, cyclists, pedestrians and other), also taking into account their speed and distance. In the event of detecting a potential hazard, the numerous functions the EyeSight is equipped with, may be used in two areas:

- driver assistance:
  - traffic lane assistant,
  - adaptive cruise control,
  - information about the previous vehicle moving;
- accident prevention:
  - pre-collision emergency braking,
  - anti-collision choke valve management,
  - anti-collision obstacle by-passing assistant.

Pedestrian Airbag Technology[2]

A high number of traffic accident deaths and injuries induce to designing systems aimed at improving the safety of pedestrians.

A pedestrian airbag can often be found in modern cars. Its task is to mitigate the consequences of some head-on collisions with pedestrians.

An airbag is installed by the windshield of the vehicle, under the engine compartment cover. Sensors placed in the front bumped constantly transmit a signal to the control unit. Should it change, and the hit object is verified as similar to a human leg, the system is activated, and its operation can be divided into four stages:

---

[1] Subaru website: https://www.subaru.pl/2015outback_eyesight.html [1/1/2018].

[2] Volvo website: https://support.volvocars.com/pl/ [1/1/2018].

- raising (by 10 cm) and locking the rear part of the engine compartment cover,
- filling of the external airbag,
- turning on emergency lights,
- braking system switching into the emergency braking standby state.

The full sequence of actions from the moment of activation until complete filling takes several hundredths of a second. A filled airbag takes the lower part of the front poles, one third of the windscreen and the entire wiper cavity.

eCall

Manufacturers of cars previously leaving the dealerships, in order to obtain a homologation, were obliged to include the following systems as standard equipment: tyre pressure control, ABS, ASR. Since March 2018, the group of obligatory safety-increasing systems was expanded with the *eCall* system (*Emergency Call*, a European road accident notification system) [19].

The fundamental task of the system, also called "the black box", is to limit the number of road accident deaths, and also improve the functioning of emergency services in sudden situations. The assumption of the initiative is to provide a Pan-European service, available for all EU Member States, as well as for other countries, wanting to join the project. All new vehicles, regardless of the country, brand and location are equipped with the system.

The fit-out of vehicles is expanded for the needs of the project with the following elements:

- SSB measuring module (sensor set),
- GPS vehicle positioning device,
- digital camera (or a set of cameras),
- additional sensors,
- a processor with fixed memory and an operating interface,
- manual notification device,
- GSM transceiver.

After recording a serious road incident, the sensors in the vehicle automatically activate the eCall system, establishing a phone connection with the area-relevant Publish Safety Answering Point (PSAP), and at the same time, sends an MSD (minimum set of data) alarm message, which includes: car description (type of vehicle, type of fuel), number of passengers, time of incident, detailed location and driving direction[3]. There is also a possibility to manually activate the system through the SOS panel installed inside the passenger compartment, which enables contact with the 112 emergency number. Such application enables calling the services also by the people in other vehicles, who are the victims of an accident or any other random events (e.g. fire).

Automatic event recognition takes place on the basis of sensor indications, which send a message in the event of any of the vehicle axes exceeding its limit load (acceleration).

Despite the protests, vehicle drivers will not have the possibility to deactivate eCall, while being ensured that during normal operation, it does not collect any data.

Due to the technical components of the eCall system, there is an opportunity to develop other additional services, already existing or planned (e.g. electronic road toll collection system or a digital tachograph). The integration and optimization of a broader range of solutions within an open and consistent system architecture may positively influence cost reduction, increasing of usefulness and efficiency, as well as future improvement and expansion of their functionality.

According to the experts, the use of eCall will significantly improve the response time of emergency services by 50% in the cities and by as much as 60% in undeveloped areas. This may translate into several thousand more people saved per year.

Autonomous cars

---

[3] Based on the data obtained from satellite navigation systems, such as: EGNOS, Gallileo.

Undoubtedly another phase in the development of the automotive industry – after the current stage of drive assist and facilitation – will be replacing the driver and introducing autonomous cars on public roads. Ultimately, the role of the "driver" shall be limited only to selecting the route or its destination. A car will be responsible for safe transport of passengers from point A to point B, without their intervention.

For many years, this idea had been deemed a fantasy, but it is now slowly turning into reality. Already in 2013, the American state of Nevada allowed prototype versions of such vehicles for road traffic. In September 2016, first Uber[4] cars without a driver entered the streets of Pittsburgh, and next, attempts were made in congested San Francisco. In the case of serially manufactured cars, Tesla and Volvo (V90 and S90) are in the lead, offering models, which could travel long distances on autopilot only. Tesla manufacturers took a step further introducing an option for a car to autonomously leave a garage and wait for the passengers in front of the house front door.

Taking into account the handling properties, autonomous cars do not differ from traditional cars. Although the intention of the manufacturers was to limit the role of a man to a minimum, currently available vehicles are equipped with a steering wheel, making it possible to take control over at any moment.

Autonomous vehicles are equipped with an advanced on-board computer, as well as a great deal of sensors. Necessary equipment of such cars includes: a built-in GPS module, laser rangefinders mapping the area, as well as cameras detecting heat and motion, and cameras monitoring the behaviours in the immediate vicinity. The most important used tool is advanced terrain mapping, and the GPS maps themselves are rather a supporting element.

The main assumption in regard to autonomous cars is safety, achieved through, among others, eliminating the human factor errors and a more efficient clash-prevention. However, a computer is not always able to appropriately react to a road situation, therefore, the currently available vehicles with an autopilot, constantly monitor whether a driver is concentrated on the road. The advantages increasing safety include a fact that cars without drivers strictly follow speed limits and other applicable regulations.

5 Conclusions

Based on the presented review, the following conclusions can be formulated:

- The task of a safety system is to maintain an anthropotechnical system in a state of safety. It is achieved through counteracting the appearance of trigger factors, as well as preventing or at least mitigating the effects of destructive processes.
- A motor vehicle safety system includes the following task-oriented modules: shielding (e.g. road traffic regulations), intervention (e.g. ABS), emergency (e.g. AirBag).
- The safety of road traffic participant depends on the interaction of individual traffic components: human – car – road. The safety depending on a motor vehicle covers the correct functioning of the following subsystems:
    - a preventive safety system (shielding function) – systems warning a driver and initiating his/her actions aimed at avoiding a collision, acting independently from the driver;
    - vehicle active safety subsystem (intervention function) – systems impacted by a driver, thus allowing to decrease or avoid a collision risk;
    - vehicle and passenger passive safety subsystem (rescue system) – systems aimed at mitigating the effects of an already occurred collision, with automatic operation, independent from the driver.

[4] Uber – a company providing a mobile app for matching drivers and passengers aimed at ordering transport services.

- The current technological trends are aimed at solutions, which aid and facilitate driving. Modern car systems can "see" in the dark, identify objects in the vicinity, and if necessary, stop autonomously, in order to avoid a road collision.
- Another stage in the development of the automotive industry will be, undoubtedly, replacing the driver himself and introducing autonomous cars on public roads. Their task will be to deliver passengers from point A to point B, without any intervention from the passengers.

Both for car companies, as well as legislative bodies, safety is currently the main priority, and the cars leaving the showrooms are characterized by extensive equipment in this field.

References

1. T. Dąbrowski, *Diagnosis of anthropotechnical systems in potential and effect terms* (in Polish) (Publishing House MUT, Warsaw, **2676** 2001)
2. L. Będkowski, T. Dąbrowski, *The basis of exploitation, part II: The basis of exploational reliability* (in Polish) (Publishing House MUT, Warsaw, 2006)
3. J. Paś, *Operation of electronic transportation systems* (in Polish) (Publishing House University of Technology and Humanities in Radom, Radom, 2015)
4. A. Rosiński, *Modelling the maintenance process of transport telematics systems* (in Polish) (Publishing House Warsaw University of Technology, Warsaw, 2015)
5. J. Paś, A. Rosiński, Selected issues regarding the reliability-operational assessment of electronic transport systems with regard to electromagnetic interference, Eksploatacja i Niezawodnosc – Maintenance and Reliability, **19(3)**, pp. 375–381 (2017). doi: 10.17531/ein.2017.3.8
6. M. Siergiejczyk, J. Paś, A. Rosiński, Issue of reliability–exploitation evaluation of electronic transport systems used in the railway environment with consideration of electromagnetic interference, IET Intelligent Transport Systems **10(9)**, pp. 587–593 (2016). doi: 10.1049/iet-its.2015.0183
7. M. Bednarek, T. Dąbrowski, L. Będkowski, *Diagnozowanie bezpiecznościowe systemu kierowca-samochód,* Logistyka, **4** (2010)
8. P. Pyłko, *Systemy bezpieczeństwa w pojazdach samochodowych*, Praca dyplomowa, WAT Wydział Elektroniki (2018)
9. M. Siergiejczyk, K. Krzykowska, A. Rosiński, L.A. Grieco, Reliability and viewpoints of selected ITS system, in: Proceedings 25th International Conference on Systems Engineering ICSEng 2017, IEEE, Conference Publishing Services (CPS), pp. 141–146 (2017). doi: 10.1109/ICSEng.2017.68
10. B. Kalupova, I. Hlavon, Intelligent transport systems in the management of road transportation, Open Engineering **6(1)**, pp. 492–497 (2016). doi: 10.1515/eng-2016-0062
11. L. Janusova, S. Cicmancova, Improving safety of transportation by using inteligent transportation systems, 9th International Scientific Conference Transbaltica 2015, Procedia Engineering 134, pp. 14–22 (2016). doi: 10.1016/j.proeng.2016.01.031
12. J. Młyńczak, T. Hejczyk, B. Wszołek, A. Gałuszka, D. Surma, R. Ogaza, R. Burdzik, Passenger safety and information module in intelligent integrated traffic management system, Vibroengineering PROCEDIA **6**, pp. 234–237 (2015)
13. U. Ehlers, E. Ryeng, E. McCormack, F. Khan, S. Ehlers, Assessing the safety effects of cooperative intelligent transport systems: A bowtie analysis approach, Accident Analysis & Prevention **99**, pp. 125 – 141 (2017). doi: 10.1016/j.aap.2016.11.014

14. M. Rychlicki, Z. Kasprzyk, Increasing performance of SMS based information systems, in „Proceedings of the Ninth International Conference Dependability and Complex Systems DepCoS-RELCOMEX", given as the monographic publishing series – „Advances in intelligent systems and computing", vol. 286, Springer, pp. 373–382 (2014). doi: 10.1007/978-3-319-07013-1_36
15. F. Losurdo, I. Dileo, M. Siergiejczyk, K. Krzykowska, M. Krzykowski, Innovation in the ICT infrastructure as a key factor in enhancing road safety: a multi-sectoral approach, Proceedings 25th International Conference on Systems Engineering ICSEng 2017, IEEE Computer Society Conference Publishing Services (CPS), Las Vegas, Nevada, USA, pp. 157–162 (2017). doi: 10.1109/ICSEng.2017.69
16. T. Dąbrowski, M. Wiśnios, *Koncepcja pokładowego systemu monitoringu kierowcy,* Logistyka, **4**, pp. 2902-2909 (2015)
17. G. Boruta, A. Piętak, *Mechatronika samochodu. Układy bezpieczeństwa czynnego i biernego* (Wyd. Uniwersytetu Warmińsko-Mazurskiego, Olsztyn 2012)
18. J. Paś, T. Dąbrowski, J. Dyduch, *Metoda oceny bezpieczeństwa eksploatacji systemów elektronicznych użytkowanych w transporcie.* Prace Naukowe Politechniki Warszawskiej - Transport, **77**, pp. 107-120 (2011)
19. T. Kamiński, R. Borowski, *eCall - europejski system powiadamiania o wypadkach drogowych,* Instytut Transportu Samochodowego (2016)

Use of biogas to power diesel engines with common rail fuel systems

Sławomir Wierzbicki[1,*], *Michał* Śmieja[1]

[1]University of Warmia and Mazury in Olsztyn, Faculty od Technical Science, 11 Oczapowskiego str, 10-957 Olsztyn, Poland

Abstract. The limited resources of fossil fuels, as well as the search for a reduction in emissions of carbon dioxide and other toxic compounds to the atmosphere have prompted the search for new, alternative energy sources. One of the potential fuels which may be widely used in the future as a fuel is biogas which can be obtained from various types of raw materials. The article presents selected results as regards the effects of the proportion of biogas of various compositions on the course of combustion in a dual-fuel diesel engine with a Common Rail fuel system. The presented study results indicate the possibility for the use of fuels of this type in diesel engines; although changes are necessary in the manner of controlling liquid fuel injection.

1 Introduction

The ongoing climate change related to, inter alia, global warming and growing public awareness combined with a constant increase in demand for energy have prompted the search for new energy sources. In recent years there has been a strong emphasis on increasing the use of energy obtained from renewable sources. Particular attention is being paid to second generation renewable fuels that are manufactured from non-food products, which means that fuel production does not compete with food production.

One of the fuels of this type is biogas, which is generated spontaneously in nature as a result of natural processes and can also be produced in special biogas plants from various types of biomass [1, 2, 3, 4, 5]. The produced biogas contains considerable amounts of methane (40-75%) which is the main combustible component, and of non-combustible carbon dioxide (25-55%) which is a ballast. In addition, depending on the origin, biogas may contain small (at a level of a few per cent) amounts of other components, such as hydrogen, carbon monoxide, nitrogen and oxygen, as well as trace amounts of hydrogen sulphide and other chemical compounds. Given that biogas may be obtained using various production technologies, its chemical composition is not constant. The basic and most desirable biogas component is methane, i.e. the simplest hydrocarbon. The percentage of methane in biogas is determined by the technology of biogas production and ranges from 35% to approx. 75% [1, 2, 6]. Table 1 presents the compositions of biogas, including methods of obtaining it.

Table 1. Composition of biogas, depending on its origin [1, 2, 6].

Component	Content		
	agricultural biogas	**treatment plant biogas**	**landfill biogas**
Methane CH_4	45–75%	57–62%	37–67%
Carbon dioxide CO_2	25–55%	33–38%	24–40%
Oxygen O_2	0,01–2,0–2,1%	0–0,5%	1–5%
Nitrogen N_2	0,01–5,0%	3,4–8,1%	10–25%

* Corresponding author: slawekw@uwm.edu.pl

Hydrogen sulphide H_2S	10–30 000 ppm	24–8 000 ppm	15–427 ppm

The major advantage of fuels of this type, e.g. biogas, is primarily the relatively short time of their renewal in nature [1, 2, 3]. Carbon dioxide introduced to the atmosphere as a result of the combustion of these fuels is, most frequently, of plant or animal origin, and, assuming that it will be used by plants in the next vegetation cycle, the process of the use of such a fuel becomes a component of the natural carbon circulation and generally lasts no longer than one year. An example of carbon circulation when using all types of organic wastes for the production of biogas used to generate electricity and heat is presented in Figure 1 [7].

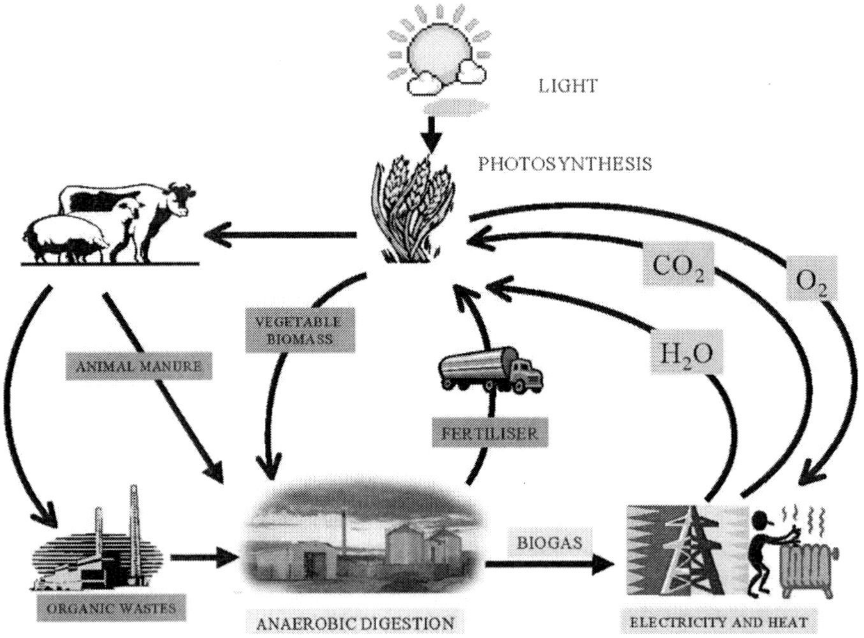

Fig. 1. Natural carbon circulation with biogas used as a fuel in cogeneration systems [7].

It is also worth stressing that the production of biogas is the most efficient form of the conversion of organic raw materials into a fuel. This is due to the fact that the production of liquid fuels such as biodiesel or ethanol use various types of grains, e.g. rape, while the production of biogas uses whole plants. What is most important is that there is a possibility of the production of biogas from wastes, even those generated in the production of liquid fuels. Figure 2 presents, in a diagrammatic form, the amount of energy that is possible to be produced from agricultural crops yielded from an area of 1 ha when converting them into various fuels, with reference to the range of a vehicle with an internal combustion engine consuming 7.4 l fuel, or the range of a vehicle with a diesel engine consuming 6.1 l fuel [7]. The presented comparison demonstrates that the energy obtained from biogas enables a 3-fold increase in the range of a vehicle compared to a vehicle powered by biodiesel or bioethanol obtained from the same area of crops. Therefore, biogas production generates much less waste than liquid fuel production does.

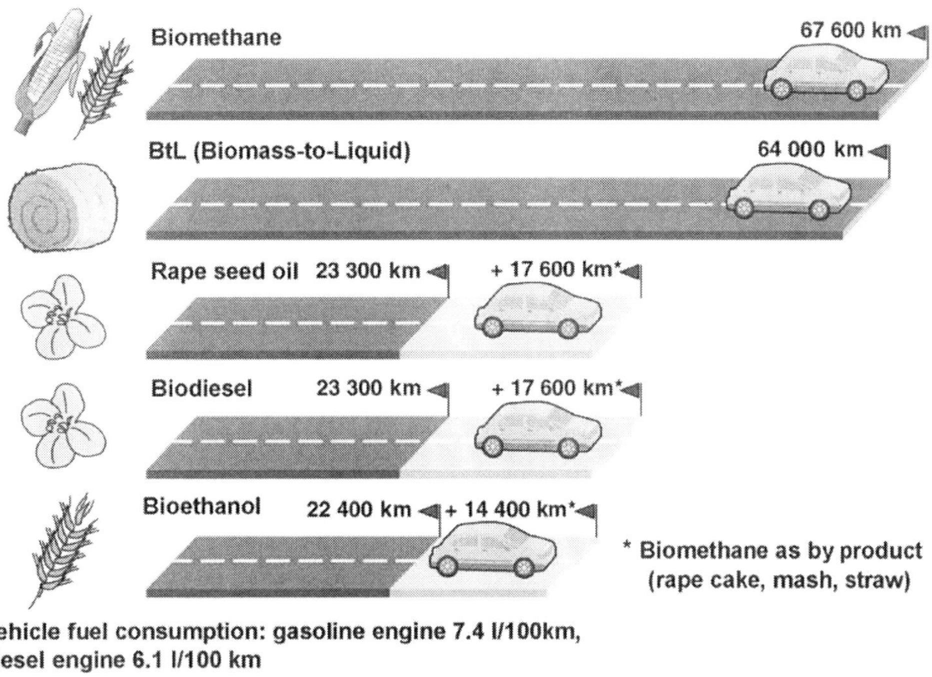

Fig. 2. A comparison of the potential range of a vehicle using different fuels produced from 1 ha of agricultural crops [7]

The properties of biogas as a fuel for internal combustion engines are strictly dependent on the content of particular compounds. They are affected the most by methane, i.e. the basic combustible component of biogas.

Biogas may be used as a fuel for both spark-ignition engines and diesel engines [3, 8, 10]. The use of biogas to power diesel engines is relatively unexplored and many research centres are currently conducting research into the development of efficient methods of using this fuel. A dual-fuel supply system is required for these engines. This is due to the fact that in order to ignite the air/biogas mixture in the combustion chamber, it is necessary to inject a dose of a liquid fuel whose spontaneous ignition initiates the combustion of a gaseous fuel.

This article presents the results of an analysis of the effect of the proportion of a biogas dose in the supply dose on the course of combustion in a diesel engine with the Common Rail fuel system.

2 Objective of the research and test bench description

This paper presents the preliminary test results of a compression ignition engine run in a dual-fuel system, fuelled with gaseous fuel with varying chemical composition.

The tests described were performed on an ADCR engine manufactured by Andoria-Mot. For the engine technical specifications, refer to table 2 [11].

The engine examined was installed in an AVL engine test bed ensuring proper engine load and measurement of basic operating parameters, including the rotational speed, torque and fuel consumption.

Furthermore, as the engine was running, the course of pressure changes in the first cylinder chamber were recorded using the 6056A type probe made by Kistler, placed in the glow plug seat with the aid of a special adapter. The probe was combined with the 5018A load amplifier, which was connected via DAQ card to a PC. The crankshaft rotation angle was measured by an optical sensor installed on the crankshaft. The angle marker resolution was 720 points/revolution in this case, which enabled pressure recording every 0.5° of crankshaft revolution, within the full range of the engine's work cycle [11].

Table 2. Technical data of the ADCR engine [11].

Engine	**ADCR**
Engine type	Diesel, 4-cycle, turbocharged with charge air cooler
Number and arrangement of cylinders	4 cylinders in-line, vertical
Bore diameter / piston stroke	94 / 95 mm
Swept volume	2636 cm3
Compression ratio	17,5 : 1
Rated power*/speed	85 kW / 3700 rpm
Peak torque*/speed	250 Nm / 1800-2200 rpm
Specific fuel consumption at peak torque*	210 g/kWh
Injection system /	Common Rail CR2.0 (Bosch)
Turbocharger	radial, with exhaust extraction valve

* in accordance with ISO 1585 / wg normy ISO 1585

Fig. 3. Test bench.

To complete the testing programme planned, a system was developed that enabled fuelling the engine with gaseous fuel with the assumed chemical composition, created artificially in a laboratory [11, 12]. Based on available literature data, it was assumed that the fuel fed to the engine during the tests will be a mixture of two main components of agricultural biogas, namely methane (CH_4) and carbon dioxide (CO_2). The percentage of other chemical compounds present in the biogas derived from agricultural biogas plants is negligible (below 1%), and thus, they were not considered in the fuel being tested.

Methane-rich natural gas type E from the municipal gas network, the composition of this class of natural gas is as follows:

- methane (CH_4) - ca. 97.8 %;

- ethane, propane, butane - ca. 1%;
- nitrogen (N_2) - ca. 1%;
- carbon dioxide (CO_2) and other components - 0.2 %.

The gas available in the gas network, at the time of the tests, had a caloric value of 36.335 - 36.374 MJ/m^3, as declared by the supplier – the Polish Gas Company (PSG).

As was assumed, in addition to methane as the main combustible component, the gaseous fuel should contain carbon dioxide. Compressed carbon dioxide with a purity class of 2.2, thus containing 99.2% CO_2, was used in the tests. This is a typical industrial gas commonly used for welding purposes.

To create fuel with the assumed chemical composition, the above gases were fed to the engine inlet manifold through a special system made for the purpose of the project, ensuring the designed flow rate of individual components of the gaseous fuel [13].

Figure 4 shows a general diagram of the system supplying gaseous fuel.

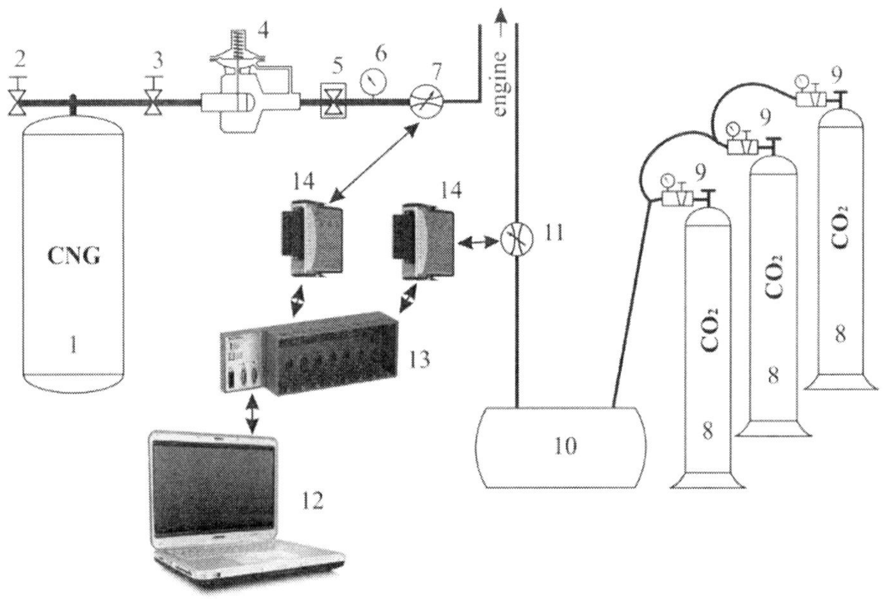

Fig. 4. Block diagram of the gaseous fuel supply system: 1 - CNG cylinder, 2 - filling valve, 3 - cut-off valve, 4 - two-stage reducer, 5 - solenoid valve, 6 - pressure gauge, 7 - MasStream gas mass flow regulator, 8 - CO_2 cylinders, 9 - reducers with pressure gauges, 10 - expansion tank, 11 - mass flow regulator, 12 - PC, 13 - CompactRio programmable controller, 14 - i/o cards to control the flow regulators.

During the study, an ADCR engine was controlled by an original Bosch EDC16C39 device with original engine-controlling software adjusted to single fuel operation. Depending on the current engine operating conditions, e.g. the engine speed or the load, the controlling device follows various control strategies. During the study, the engine controls maintained constant engine load by selecting a fuel dose depending on the engine operating conditions, taking into account the gaseous fuel dose. This resulted, with the increase in the proportion of a gaseous fuel, in a reduction in the liquid fuel dose injected into the combustion chamber.

3 Test results

At the first stage of the tests, the gaseous fuel dose was increased until the engine ran steadily without signs of knocking combustion. Table 3 shows selected results of this stage of the tests.

The energy share of diesel fuel U_{df} in the total engine fuel dose was determined from the relationship:

$$U_{on} = \frac{\dot{m}_{df} \cdot W_{df}}{\dot{m}_{df} \cdot W_{df} + \dot{m}_{g} \cdot W_{g}} \cdot 100\% \tag{1}$$

where:

$\dot{m}_{df}$ - diesel fuel mass stream;

$\dot{m}_{g}$ - stream of CNG supplied to the engine;

W_{df} - diesel fuel caloric value;

W_{g} - CNG caloric value.

Figure 5 presents diagrams of the pressure course in the engine combustion chamber at n=1500 rpm and T=100 N·m for various gaseous fuel compositions when supplying to the engine a gaseous fuel dose corresponding, in energy terms, to Ug=50% of a liquid fuel dose under conditions of a single fuel supply. Figure 5 presents the pressure courses for the same load conditions and the 75% energy share of a gaseous fuel in relation to the liquid fuel dose under conditions of a single fuel operation.

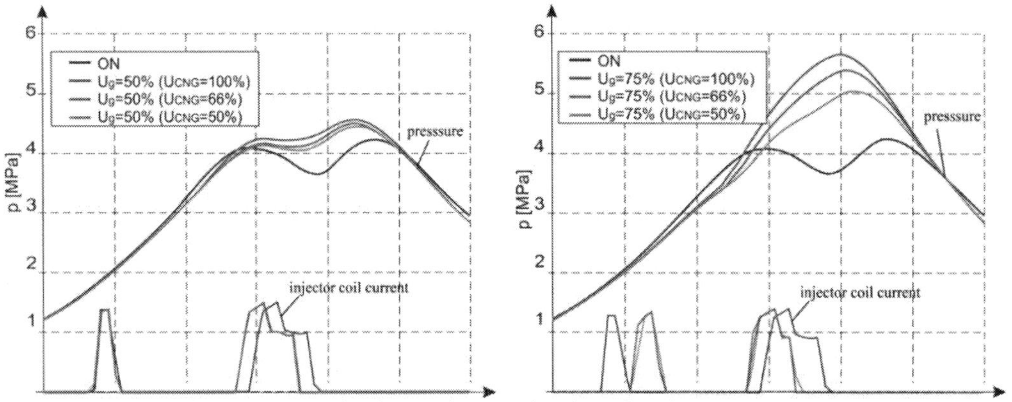

Fig. 5. Pressure change courses in the engine combustion chamber at n=1500 rpm, T=100 N·m, various compositions of gaseous fuels for gaseous fuel doses corresponding to: a) 50% dose of a liquid fuel under conditions of single fuel operation, and b) 75% dose of a liquid fuel under conditions of single fuel operation

As can be seen in fig. 5, supplying a gaseous fuel to the combustion chamber significantly changes the course of combustion, which is associated with the initiation of the combustion of gaseous fuel by the ignition of the pilot dose, which consequently leads to a greater increase in pressure in the initial combustion phase. At the same time, the effects of chemical composition of a gaseous fuel on the course of charge compression in the combustion chamber and on the course of its combustion can be observed. In addition, the courses presented in fig. 5a clearly indicate a significant effect of the gaseous fuel dose volume on the course of combustion. At lower proportions of this fuel, the mixture present in the combustion chamber has a relatively high λ value and, as soon as the combustion of a liquid fuel pilot dose stops, the flame goes out and the further combustion of the gaseous fuel only takes place along with the combustion of the liquid fuel main dose. On the other hand, with a richer mixture, the combustion of the gas and air mixture is intensified.

The courses presented in fig. 5 also enable the observation of the effects of the proportion of carbon dioxide in the supply dose on the course of pressure changes in the combustion chamber. As the proportion of CO_2 in the supply dose increases, the maximum pressure in the combustion chamber drops, which is associated with an increase in the specific heat of the charge.

Figure 6 presents the courses of pressure changes in the combustion chamber at n=3000 rpm and T=100 N·m for single fuel operation and for dual fuel operation, at various proportions and compositions of the gaseous fuel. At this engine speed, one dose of fuel is injected into the combustion chamber.

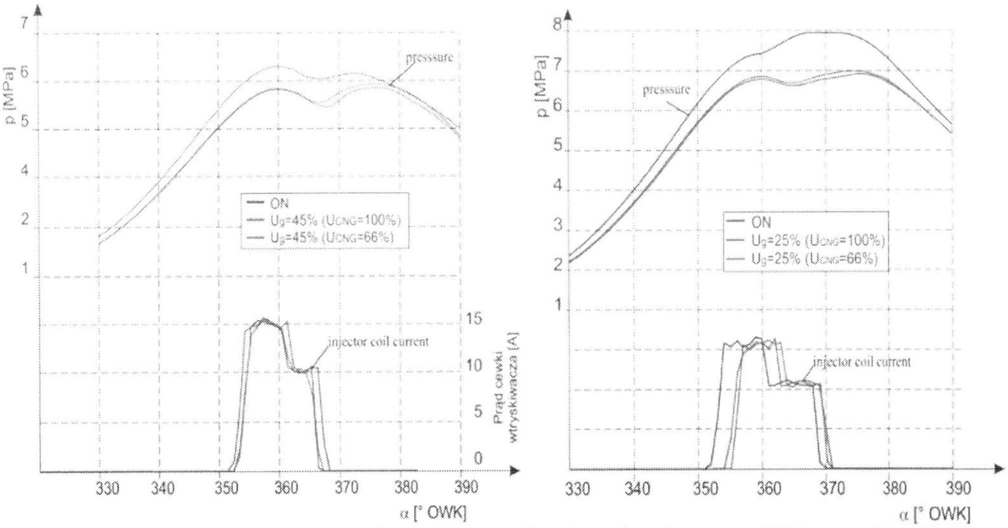

Fig. 5. Pressure change courses in the engine combustion chamber at n=3000 rpm, M=100 N·m, various compositions of gaseous fuels for gaseous fuel doses corresponding to: a) 45% dose of a liquid fuel under conditions of single fuel operation, T=100 N·m , and b) 25% dose of a liquid fuel under conditions of single fuel operation, T= 200 N·m

An analysis of the presented courses of pressure changes in the combustion chamber clearly shows the significant effect of the proportion of a gaseous fuel on both the course of combustion and the compression of the mixture. An increase in the proportion of a gaseous fuel in the supply dose results in a reduction in the pressure during the compression due to an increase in the specific heat of the gas and air mixture, which leads to a reduction in charge temperature and contributes to a delay in the spontaneous ignition of the liquid fuel.

Conclusions

The above-presented examples of the effect of the proportion of biogas in a dose supplied to a diesel engine, and the wider study results presented in [3] enable the formulation of the following conclusions:

- there is a possibility for the efficient use of biogas to power diesel engines;
- the course of gaseous fuel combustion process in dual-fuel diesel engines can be controlled by changing the parameters of liquid fuel dose injection;
- due to the high overall efficiency of the engine within a narrow load range, it is suitable for operation in stationary cogeneration systems in hybrid systems and in agriculture to power agricultural tractors;

On the other hand, major disadvantages of such diesel engine powering include:

- the supply of a gaseous fuel to the engine cylinder during the intake stroke prevents the control of the course of combustion in the engine. The combustion of a gaseous fuel starts at the moment of spontaneous ignition of the first liquid fuel dose, which prevents combustion control;
- at partial loads, the overall efficiency of the engine decreases;

Further research into the use of biogas and other gaseous fuels should be carried out in order to select the parameters of the diesel fuel pilot dose and to investigate the possibilities for the application of gaseous fuel direct injection to enable control of its combustion.

References

1. J. Cebula J, ACEE Journal, **2**, 95-103 (2009)
2. S. Wierzbicki, Journal of Kones **19**, 3, 477-482 (2012)
3. D. Barik, S. Murugan, Energy, **72**, 760-771 (2014).
4. B. J. Bora, U. K. Saha, Renewable Energy, **87**, 782-790 (2016).
5. S. Wierzbicki, *Studium możliwości wykorzystania biogazu do zasilania silników o zapłonie samoczynnym* (UWM Olsztyn, 2016)
6. http://lemvigbiogas.com/
7. J. Skorek, J. Kalina, *Gazowe układy kogeneracyjne*. (WNT Warszawa, 2005)
8. M. Mikulski, S. Wierzbicki, M. Śmieja, J. Matijošius, Transport, **30** 2, 162-171 (2015)
9. M. Mikulski, S. Wierzbicki, A. Piętak, Eksploatacja i Niezawodność – Maintenance and Reliability, **17**, 1, 42-48 (2015)
10. S. Wierzbicki, M. Śmieja, Solid State Phenomena, **210**, 136-141 (2014)
11. D. Szpica, Flow measurement and instrumentation, **59**, 147-156 (2018)
12. D. Szpica, Transport, **33**(1), 186-196 (2018)
13. S. Wierzbicki, M. Mikulski, M. Śmieja, Combustion Engines, **162**(2), 91-101 (2015)

Selected problems of motor vehicle maintenance after side impact collision

Jarosław Zalewski[1,*]

[1]Warsaw University of Technology, Faculty of Administration and Social Sciences, Pl. Politechniki 1, 00-661 Warsaw, Poland

Abstract. Over the years certain amount of works have been devoted to the selected aspects of motor vehicle collision modeling. Also, in some previous works by the author of this paper the selected problems related to collision modelling have been included. Those papers concerned mostly relations of the obtained crash simulation results to analytical results based on the Routh method, including restitution of the velocities normal and tangential to the plane of crash common for both vehicles involved in the accident. The aim of this paper is to present the selected problems related to the side impact motor vehicle collision modeling, which can be considered as a cause for certain diagnostic problems related to further maintenance of such vehicle after repair.

1 Introduction

Over the years certain amount of works have been devoted to the selected aspects of motor vehicle collision modeling. Also, in some previous works by the author of this paper the selected problems related to collision modelling have been included. This can be referred e.g. to [4], [5], [7] and [12]. Those papers concerned mostly relations of the obtained crash simulation results to analytical results based on the Routh method, including restitution of the velocities normal and tangential to the plane of crash common for both vehicles involved in the accident. The Routh method has been presented e.g. in [6].

Other more specific problems in road vehicle collision modeling were, for example, influence of the trailer towed by a vehicle [11] or assumption, that the road plane is not flat during collision [11], or even the problem of crashing into a concrete road barrier [9]. Almost all of these problems have been related to the frontal impact of the vehicle.

Not many works however have been devoted to the side impact collision. Some examples of this problem are presented e.g. in [1]. There are two reasons for the small amount of results for side impact. First, it is difficult to specify the direction of the impulse force vector for both vehicles involved. Second, there are not many works showing both experimental and theoretical results concerning the tangential coefficient of restitution, which is necessary to describe the phenomena occurring between the surfaces of the colliding vehicles. Such analyses have been presented e.g. in [3] and [6].

The aim of this paper is to present the selected problems related to the side impact motor vehicle collision modeling, which can be considered as a cause for certain diagnostic problems related to further maintenance of such vehicle after repair.

2 Selected aspects of road collision modeling

To properly describe the side impact crash, it seems necessary to make certain assumptions:
- quasi-stiff models of the vehicles involved in a collision;
- bodies of the vehicles represented by sets of cuboids of certain mass and stiffness, which do not change during the collision;

* Corresponding author: jzalewsk@ans.pw.edu.pl

- the motion of the vehicles is realised on a flat road surface. However, in further research both uneven and sloping road may be considered;
- in more complicated vehicle models nonlinear suspension may be included, but in the presented analysis it will be omitted.

Some additional assumptions may be related to the coordinate systems. For example, following [6], two basic, inertial coordinate systems may be included in the collision model, in which the planar motion is considered. First, marked in fig. 1 as *Oxyz*, the inertial global system may be assumed. Second, also shown in fig. 1 the inertial *O'ntb* system may be adopted as a local axis system, related only to both vehicles, located at the geometric center of the crash (the point of initial contact between the two vehicles).

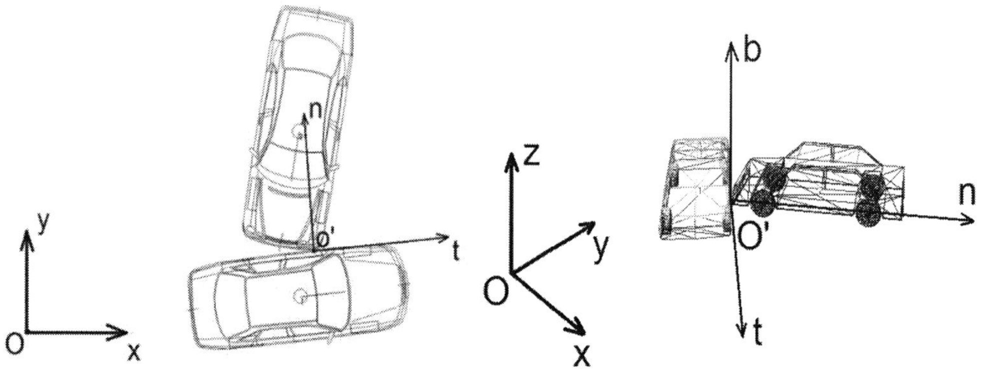

Fig. 1. Example of the possible coordinate systems assumed in the mathematical model of side impact crash between two vehicles [own research based on PC-Crash 8.0].

As for the resultant motion, the proposition of coordinate systems is also presented in fig. 1, where two systems are shown, with vertical axes (z and b) included. As for the axes symbols, t marks the tangential direction, n – normal, and b – binormal (normal to the n – t plane).

Additionally, some conditions related to the location of the force impulses in the colliding vehicles may be considered:
- in the planar motion both impulse components may be located along the tangential (t) and the normal (n) direction (fig. 1). The returns of these vectors should be opposite for each vehicle. i.e. if for one vehicle the vectors of normal and tangential impulse component point along the n and t axes respectively, then for the other vehicle they should point in the opposite direction;
- in the resultant motion the vertical component of crash impulse may be located along the binormal axis (b), with the return of its vector pointing upwards in accordance with the return of the $O'b$ axis, for both vehicles, or, like in case of normal and tangential, in the opposite directions.

Such parameters, as wheel camber angles or the vehicle body tilt in relation to the vertical axis can be omitted for simpler cases but may be included in the more complicated mathematical model of the crash.

It is important to remember, that the main goal of collision modeling is to enable calculation of the transverse and rotational velocities for both vehicles at the end of the crash, which lasts for a very short period of time. It means that in certain situations only the basic mathematical formulas are sufficient to do it. However, when there is a necessity to consider many parameters which would make it possible to fully and thoroughly reconstruct or simulate the specific road accident, then the obtained mathematical model may be more complicated and more difficult to solve, especially when the side impact crash is discussed.

3 Selected problems in side impact collision modeling

When analysing the side impact crash, two cases have been considered:
- both vehicles realise the planar motion during the impact;
- both vehicles realise the resultant motion during the impact, where the impulses of inertia, centrifugal and transportation forces have been considered.

It has also been assumed that for the description of motion for both vehicles a local coordinate system will be used, which in fig. 1 marked as *O'ntb* and its origin is placed in the point of initial contact of both vehicles. For the planar motion only normal (*n*) ant tangential (*t*) axes have been used, while for the resultant motion also so called binormal (*b*) axis has been considered.

3.1 Side impact model with vehicles realising planar motion

In fig. 2 the impulses of the impact force are presented with respect to both the planar (left part of the figure) and the resultant motion (right part). Let us accept that, regarding the difficulty in specifying the direction of the crash impulse vector, we can divide it into two components for planar, and three for the resultant motion, for each vehicle. Let the impulses be market with *S* and the lower index means tangential, normal and binormal as in the marking of the axes in the adopted coordinate system. Hence the normal impulse will be directed along the normal axis *n*, the tangential components will be located along the *t* axis and the binormal one – along the *b* axis. However, for the considered example the binormal component of the crash impulse will be returned in the same direction for both vehicles.

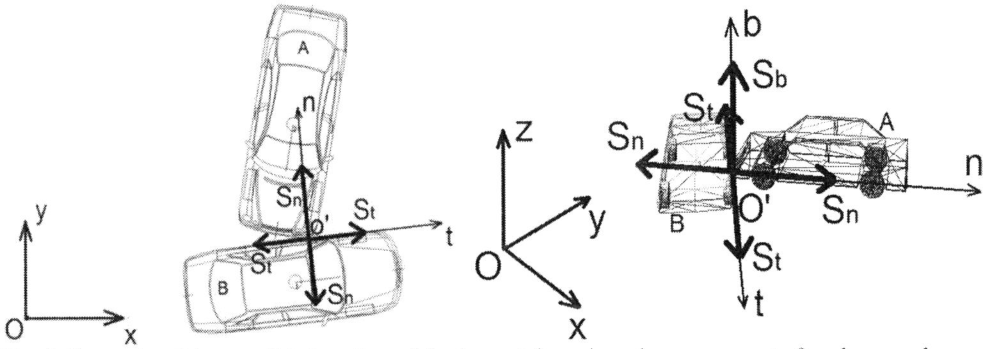

Fig. 2. Example of the possible location of the impact force impulse components for planar and resultant motion during the collision [own research based on PC-Crash 8.0].

3.2. Side impact model with vehicles realising resultant motion

For the resultant motion the adopted description will be the same as for the planar case (fig. 2). However, the consequences of such phenomena, which allows a momentary loss of contact between at least one wheel and the road surface for each vehicle, may be different than for the planar motion. It is not the intention of the author to examine the consequences so thoroughly, that they will be divided into these two cases (planar and resultant motion). The main goal of this part is to present the differences in the attempt towards collision modeling. Adopting a mathematical model that would describe the resultant motion during the impact seems to be more exact when the broader scope of phenomena is to be examined, because it offers more realistic attempt e.g. in accident reconstruction or collision analysis.

The scheme of the mutual location of both vehicles involved in a side impact crash with the impulses generated during resultant motion is presented in fig. 3.

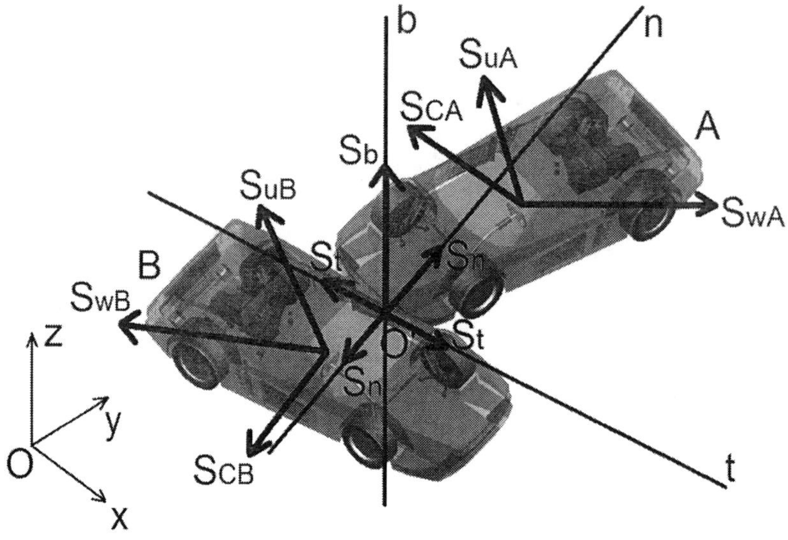

Fig. 3. The scheme of a mutual location of both vehicles with impulses occurring in resultant motion [own research based on MSC Adams/Car].

In fig. 3 certain parameters have been marked for vehicle A crashing into the side part of a vehicle B. Apart from the previously discussed impulses located in the geometric center of the impact, additional impulses characteristic for the resultant motion have been assumed. S_{wA} means the impulse of the inertial force of vehicle A relative to the point O', S_{uA} marks the impulse of the transportation force of a vehicle A, while S_{Ca} is the impulse of the Coriolis force of a vehicle A. The same vectors have been assumed for the vehicle B, only the markings in the lower index should refer to the vehicle B.

More thorough analysis of these impulses for the front impact has been presented, among others in [5] and [7], where attention has been paid to their location relative to the $O'ntb$ axis system (like in fig. 3). Like in the given example, the vectors of these impulses have been attached to the centers of mass of both vehicles.

Other attempt towards vehicle crash modeling allows assuming the additional factors, such as momentary vehicle oversteer, road conditions or mass distribution in a colliding vehicle, which can generate additional impulses attached not necessarily to the center of mass of a given vehicle. Hence the process of vehicle collision may be more complicated than previously imagined, especially from the point of view of mathematical modeling. Such attempt has been previously considered e.g. in [7] and [10].

In further part certain approach to the consequences of a side impact crash have been discussed.

4 General mathematical model for the side impact crash between two vehicles

In general, the vehicle side impact crash model can be presented as a set of equations (1) related to the scheme presented in fig. 3. The apostrophe in the upper index marks the value after the collision. Note, that by the collision duration should be understood as a period of contact between two vehicles, which corresponds to the sudden decrease or increase in velocity at a little time step.

$$
\begin{aligned}
& m_A(v'_{An} - v_{Ain}) = \sum_{i=1}^{k} S_{nA} + \sum_{i=1}^{k} S_{iwnA} + \sum_{i=1}^{k} S_{iunA} + \sum_{i=1}^{k} S_{iCnA} \\
& m_A(v'_{At} - v_{At}) = \sum_{i=1}^{k} S_{tA} + \sum_{i=1}^{k} S_{iwtA} + \sum_{i=1}^{k} S_{iutA} + \sum_{i=1}^{k} S_{iCtA} \\
& m_A(v'_{Ab} - v_{Ab}) = \sum_{i=1}^{k} S_{bA} + \sum_{i=1}^{k} S_{iwbA} + \sum_{i=1}^{k} S_{iubA} + \sum_{i=1}^{k} S_{iCbA} \\
& I_A(\omega'_A - \omega_A) = \sum_{i=1}^{k} S_{tA} n_{iA} + \sum_{i=1}^{k} S_{nA} t_{iA} + \sum_{i=1}^{k} S_{bA} b_{iA} \\
& m_B(v'_{Bn} - v_{Bin}) = \sum_{i=1}^{k} S_{nB} + \sum_{i=1}^{k} S_{iwnB} + \sum_{i=1}^{k} S_{iunB} + \sum_{i=1}^{k} S_{iCnB} \\
& m_B(v'_{Bt} - v_{Bt}) = \sum_{i=1}^{k} S_{tB} + \sum_{i=1}^{k} S_{iwtB} + \sum_{i=1}^{k} S_{iutB} + \sum_{i=1}^{k} S_{iCtB} \\
& m_B(v'_{Bb} - v_{Bb}) = \sum_{i=1}^{k} S_{bB} + \sum_{i=1}^{k} S_{iwbB} + \sum_{i=1}^{k} S_{iubB} + \sum_{i=1}^{k} S_{iCbB} \\
& I_B(\omega'_B - \omega_B) = \sum_{i=1}^{k} S_{tB} n_{iB} + \sum_{i=1}^{k} S_{nB} t_{iB} + \sum_{i=1}^{k} S_{bB} b_{iB} \\
& i = 1, \ldots, k
\end{aligned}
\tag{1}
$$

where:
- m_A, m_B – mass of the vehicle A and B respectively;
- v_{At}, v_{Bt} – a component of vehicle velocity (A and B respectively) tangential to the common plane of contact during collision;
- v_{An}, v_{Bn} – a component of vehicle velocity (A and B respectively) normal to the common plane of contact during collision;
- v_{Ab}, v_{Bb} – a component of vehicle velocity (A and B respectively) binormal to the plane of motion (road surface) during collision;
- ω_A, ω_B – angular velocity of a vehicle A and B respectively;
- I_A, I_B – moment of inertia relative to the vertical axes passing through the center of mass of each vehicle;
- S_{nA}, S_{tA}, S_{nB}, S_{tB}, S_{bA}, S_{bB} – normal, tangential and binormal components of the impulse of the impact force for the vehicle A and B respectively;
- S_{iwn}, S_{iwt}, S_{iwb} – normal, tangential and binormal components of a force resulting from the inertial force of vehicle A and B respectively, relatively to the point O' (geometric center of the impact);
- S_{iun}, S_{iut}, S_{iub} – normal, tangential and binormal components of a transportation force of vehicle A and B, resulting from the motion of O' (geometric center of the impact) versus the origin of a global axis system $Oxyz$ (fig. 3);
- S_{iCn}, S_{iCt}, S_{iCb} – normal, tangential and binormal components of a Coriolis force of a vehicle A and B, assumed in the resultant motion of those vehicles during the impact;
- n_{iA}, t_{iA}, b_{iA}, n_{iB}, t_{iB}, b_{iB}, – coordinates of the centers of mass of both vehicles (A and B) in relation to the point O' (geometric center of the impact) according to the direction of the axes $O'n$, $O't$ oraz $O'b$ (fig. 3).

The main goal of each vehicle collision model is to derive a kinematic state of before or after the impact. The kinematic state, allowing calculation of the velocities after the crash, has been presented among others in [4] and [6].

For both planar and resultant motion it is necessary to be able to define the values of impulses both for the impact force and the resultant motion. It is also important to know the values of the necessary coefficient of restitution used to define these impulses, which has been the subject of such works as for example [5] and [8].

Road vehicle collision modeling can be useful in understanding the process of vehicle collision and the possible influence of certain factors on the results. However, the other attempt may provide understanding the possible consequences of vehicle collision, especially side impact, on the diagnostic aspects in the process of post-accident vehicle repair and allowance for further maintenance.

5 Possible considerations on the phenomena occurring in contact area during the side crash

One of the most important parameters in vehicle crash modeling is the coefficient of restitution. As for the front impact collision it may seem that the phenomena occurring in the area of contact between the vehicles are of less importance due to the fact, that the vehicles involved in such collision penetrate each other and the deformations are mainly related to the damage of certain, crashworthy and energy absorbing structures of the front or the back part of a vehicle. Hence the only coefficient of restitution often taken into account is the coefficient related to the normal velocities, i.e. the components of velocity parallel to the direction of motion of a given vehicle.

Quite differently is for the side impact crash, where deformations can be immediately transferred to the parts of a vehicle body which do not absorb the kinetic energy of a collision. If those deformations reach the passenger compartment, then people being in such vehicle are posed to the risk of health or life loss. In side impact collisions modeling two coefficients of restitution should be taken into account: the one for normal (here called R_n) and the other for tangential (here called R_t) components of velocity. In resultant motion tangential coefficient of restitution should be considered, e.g for *O't* and *O'b* directions in fig. 3.

As for the direction of the impulse of impact force between two colliding vehicles, certain assumptions have been made in [6], where the first part of the following hypothesis, concerning the planar motion during a crash, has been adopted. However, the second part of this hypothesis has not been previously published. Note, that these simple calculations are only the example to simplify selected aspects of the more than usual complicated vehicle crash mathematical model.

If the angle between the colliding vehicles is close to 90 degrees, then the component of the impulse in direction *O't* can be omitted, because, given that the impulse can be decomposed into components along the three axes of the *O'ntb* system (fig. 3), the sine of this angle close to 90 degrees will be close to zero.

If, however, the angle between the colliding vehicle (*A* in fig. 3) and the plane of the mutual contact will not be close to 90 degrees, then its sine will not produce a value close to zero, and hence it may not be omitted in the equations describing the kinematic state during the crash.

In relation to the considerations, provided that the specific example of side impact crash is discussed, certain formula describing relations between the normal and tangential component of the impact force impulse has been presented, according to fig. 4:

$$S = \sqrt{S_n^2 + S_t^2},\ S_n = Scos\alpha,\ S_t = Ssin\alpha \quad (3)$$

where:
- S – impulse of the impact force;
- S_n – the normal component of the impact force impulse along the *O'n* axis (fig. 4);
- S_t – the tangential component of the impact force impulse along the *O't* axis (fig. 4);
- α – the angle between the colliding vehicle and the perpendicular to the plane of a common contact (fig 5).

It should be mentioned that for the small values of α:

$$S_n = S \cdot cos\alpha = S \text{ and } S_t = S \cdot sin\alpha = S \cdot \alpha \approx 0 \quad (4)$$

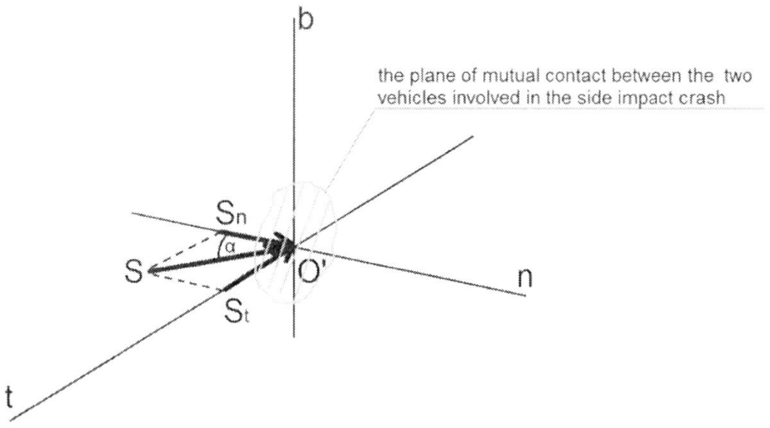

Fig. 4. Components of the impulse of impact force in planar motion side crash [own research]

As for the binormal component of the impact force impulse, its presence cannot be described in such simple manner as for the tangential direction, because the phenomena may change during the collision period, provided the resultant motion is taken into account and some wheels of at least one vehicle involved in the collision will lose contact with the ground.

However, adopting the same assumptions given above, in general the binormal component in case of resultant motion during the crash can be defined using another angle. Let β be the angle between the n-t plane and a vector of the impact force impulse. Then, according to fig. 5 all three components of the crash impulse will be:

$$S = \sqrt{S_n^2 + S_t^2 + S_b^2},\ S_n = Ssin\beta cos\alpha,\ S_t = Ssin\beta sin\alpha,\ S_b = Scos\beta \tag{5}$$

where:

- S_b – the binormal component of the impact force impulse along the $O't$ axis (fig. 5);
- β – the angle between the colliding vehicle and the perpendicular to the road (fig 5).

It also should be mentioned that for the small values of β:

$$S_n = S{\cdot}0{\cdot}cos\alpha \approx 0,\ S_t = S{\cdot}0{\cdot}sin\alpha \approx 0,\ S_b = S{\cdot}cos\beta \approx S, \tag{6}$$

while for the large values of β:

$$S_n = S{\cdot}sin\beta\ cos\alpha,\ S_t = S{\cdot}sin\beta\ sin\alpha,\ S_b = S{\cdot}cos\beta \neq 0. \tag{7}$$

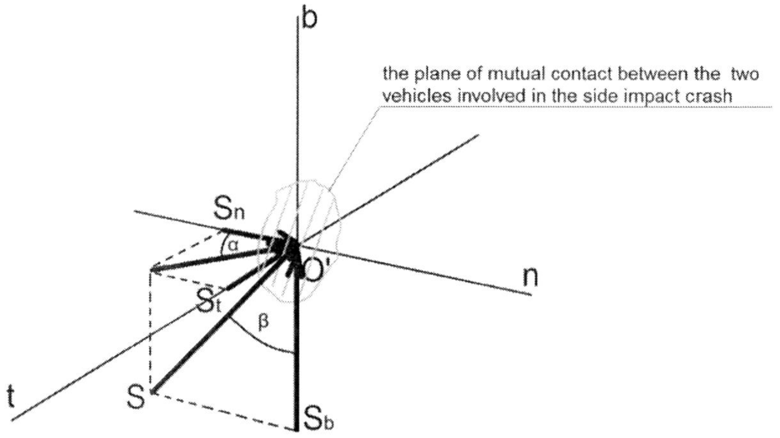

Fig. 5. Components of the impulse of impact force in planar motion side crash [own research]

If, however, the angle β was adopted between horizontal direction and the impulse S, then the components S_n, S_t and S_b would be calculated differently.

6 Possible consequences of the side impact collisions

As a side impact crash seems to be one of the most dangerous incidents in road traffic, it is necessary to consider the possible consequences in the aspect of a vehicle construction and maintenance, especially after post-accident repair.

The main result of a side impact crash, depending on the point of application of the impact force impulse, may be different disturbances in geometry of any given motor vehicle body. If the deformations would cover the lower parts of a body, i.e. the floor board, which is made of the closed, empty inside, omega-like profiles, then this qualitative assessment would be a possible factor compelling the given vehicle to be checked towards the location of the specific points in its body. However, there is a question whether these specific points, whose location informs only about a proper geometry, is sufficient enough to be used as a criterion allowing the given vehicle, repaired after accident, for further maintenance.

It seems obvious that the certain, specified criteria is necessary to be able to assess the vehicle involved in a collision to be either repaired, or utilisation.

References

1. Abe M., Morisawa M., Arai K. Sato T. B., Three-dimensional Behavior of Vehicle at Oblique Side Collision - Computer Simulation Using Dynamical Model, JSAE,1995.
2. Brach R. M., Restitution in Point Collisions, Contribution to Computational Aspect of Contact, Impact and Penetration, Ronald F. Kulak and Leonard E. Schwer, Editors, Elmepress International, Lausanne, Switzerland, 1991.
3. Ishikawa H., Impact model for accident reconstruction - normal and tangential restitution coefficients. SAE paper 930654, 1993.
4. Kisilowski J., Zalewski J., Analysis of Chosen Aspects of a Two-Car Crash Simulation, Telematics in the Transport Environment, Communications in Computer and Information Science 329, Springer, 2012.
5. Kisilowski J., Zalewski J., Analysis of the Selected Aspects of a Crash Between Two Vehicles, Advances in Intelligent Systems Research, Atlantis Press, 2013.
6. Kisilowski J., Zalewski J, Modelowanie zdarzeń w ruchu drogowym, Wydawnictwo Instytutu Technologii Eksploatacji, Radom – Warszawa 2016. (in Polish)
7. Kisilowski J., Zalewski J., Wybrane aspekty analizy i modelowania zderzenia samochodów, Technika Transportu Szynowego, 10/2013. (in Polish)
8. Wicher J., O współczynniku restytucji w zderzeniach samochodów, On the coefficient of restitution in the vehicle collisions, Zeszyty Naukowe Politechniki Świętokrzyskiej. Mechanika, 2006. (in Polish)
9. Wicher J., Rekonstrukcja zderzenia samochodu z drogową barierą betonową, Zeszyty Naukowe Instytutu Pojazdów, 2(57)/2005. (in Polish)
10. Wicher J., Stawicki R., Modelowanie zderzenia samochodów, Zeszyty Naukowe Instytutu Pojazdów, 1(31)/99. (in Polish)
11. Wicher J., Sztenke M., Oddziaływanie przyczepy i nachylenia drogi na parametry ruchu podczas zderzenia samochodów, Zeszyty Naukowe Instytutu Pojazdów, 3(29)/98. (in Polish)
12. Zalewski J., Analysis of a road accident in the aspect of mechanics of a frontal crash between two vehicles, Diagnostyka, vol 16 (4), 2015.

Algorithms for diagnostics of the hydraulic pressure modulators of ABS / ESP systems in stand conditions

Andrzej Gajek[1,*], *Piotr* Strzępek[1] and *Krzysztof* Dobaj[1]

[1]The Faculty of Mechanical Engineering, Cracow University of Technology

Abstract. The article presents possibilities of extending the scope of ABS / ESP systems control as part of periodic vehicle technical inspection. It was proposed to check the functioning of the ABS / ESP system based on the evaluation of its effects in the form of dynamic changes of braking forces measured on the roller diagnostic stand. The tested system is started by a diagnostic tester through the OBD socket in the car during the implementation of active tests. The effect of the system is checked on the basis of the analysis of the course of braking forces. This allows to detect those mechatronic system failures that are not detected by on-board diagnostic systems. The algorithms for control of ABS and ESP modulators using diagnostic testers are presented.

1 Introduction

In recent decades, the car's braking system has become the mechatronic system. The development of techniques and scope of diagnosis in the field of periodical technical tests should take into account this fact [6].

Mechatronic systems, whose main idea is to control the object through a programmable controller that uses signals from sensors, require a different approach to their diagnosis in relation to strictly mechanical systems. The self-diagnosis program built into the controller should be used in control tests. For this purpose, it is necessary to use testers through which any error codes registered by the self-diagnosis system can be read, as well as signals from the sensors and verify the correctness of their operation. The second purpose of using testers is the ability to force acting the circuits of the actuators through the electrical signals from the tester and obtaining answers. The use of testers allows to speed up the process of diagnosis mechatronic systems, as well as to more widely use existing stands for vehicle control tests.

2 Extending the scope of control tests of ABS / ESP systems

The current regulations regarding periodic technical inspections of vehicles include tests of electronically controlled safety systems in a very narrow range [1, 2]. The diagnostician only uses the signaling information on the car instrument panel. On the other hand, the use of diagnostic testers is required to control the operation of mechatronic systems. This allows to significantly extend the scope of diagnosis. Controlling actuators through the tester, with simultaneous recording of the response of the tested system, allows to detect failures that are not registered by the on-board diagnostic system [3]. For ABS and ESP braking systems, the integration of the ABS / ESP modulator valves control with the braking force measurements on a roller stand allows to check this system as a whole. This method allows to use the stands currently used in periodic technical tests.

* Corresponding author: gajeka@mech.pk.edu.pl

Currently, in the diagnosis of ABS / ESP systems in the scope of periodic technical tests, only signaling lights on the instrument panel of the vehicle are used, indicating the efficiency or damage of electronic and electrical components. The second level of diagnosis is the use of a tester that allows reading error codes, sensor indications, actuation of actuators - valves and ABS pump. This level is currently not used in the study of safety systems. In order to carry out such diagnostics, vehicle inspection stations should be equipped with a tester with a dedicated program for these stations [4,5].

The next level is checking the functioning of the ABS / ESP system by running it with a diagnostic tester. This level of testing allows you to evaluate the operation of the system as a whole. The main purpose of its application is to detect disabilities that are not detected by on-board diagnostic systems. It includes tests of the operation and indications of wheel speed sensors, measurement and registration of variation of braking forces at the roller stand during operation of the pump and ABS / ESP valves started-up with the tester. It is also possible to control the zeroing of the steering angle sensor, check the operation of the ω_z speed sensors and lateral acceleration a_y.

Checking the braking system with ABS / ESP at this level allows detection of the following failures which are not signaled by the OBD on-board diagnostic system: incorrect or unstable vehicle wheel speed sensor readings, possibility of connecting electric or hydraulic wires by mistake, leaks or mechanical immobilization of the modulator valves, obstruction or blockage of the brake fluid flow to the brake calipers, improper ABS pump operation, improper zeroing of the steering angle sensor, improper zeroing of the the angular velocity sensor and acceleration sensor, lack of reaction of these sensors on the dynamic forces.

3 Algorithms for diagnostics of the hydraulic pressure modulators of ABS / ESP systems on the roller stand

The ABS and ESP hydraulic modulator includes a high-pressure hydraulic pump, pressure reservoirs and a set of solenoid valves and mechanical valves, Fig. 1. The efficiency of the system depends both on electrical and mechanical systems. The electrical components are covered by the on-board diagnostic system. However, the mechanical part: tightness, flows, speed of operation, is not controlled by the diagnostic tester.

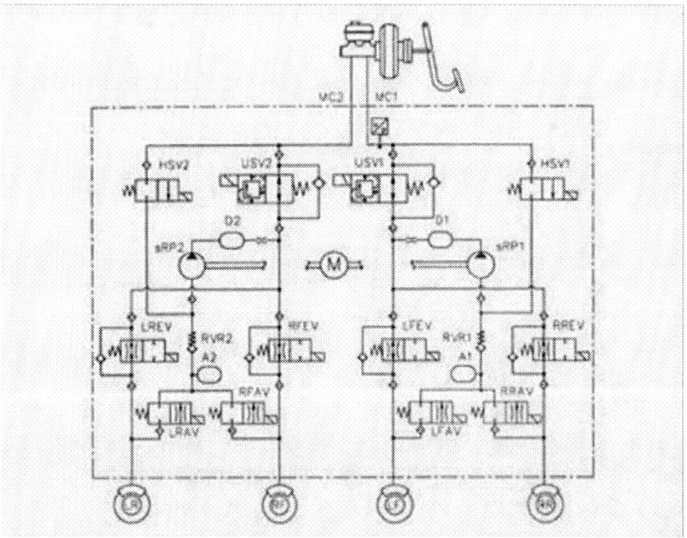

Fig. 1. Diagram of the ESP electrohydraulic modulator (ABS Bosch 5.7). RREV, RRAV, LFEV, LFAV, RFEV, RFAV, LREV, LRAV - valves of ABS, HSV1, USV1, HSV2, USV2 - ESP valves, sRP1, sRP2 - ABS pump.

The purpose of the research presented in the article was to assess the possibility of controlling the operation of valves and ABS / ESP pumps by measuring the variation of braking forces at the roller stand.
The test objects were hydraulic brakes of various passenger car brands equipped with ABS or ESP systems.

The valves and ABS / ESP pump were started by signals from the diagnostic tester. Testers from different manufacturers were used (KTS, TEXA, CDIF). Measurements of braking forces were carried out on a roller stand enabling registration of dynamic changes of braking forces.

The possible algorithms for controlling these systems using a roller brake stand and the results of these tests are presented.

3.1 Subsequent actuation of the ABS/ESP valves

3.1.1 ABS system

The diagnostic procedure allows to evaluate the operation of valves and ABS pump for subsequent car wheels based on the course of braking forces. In the graph (Figure 2), the braking forces are shown during the next steps of actuating the pressure control valves in the right front wheel. During the test the rate of rise and decrease of the braking force during operation of the ABS valves can be assessed. The next steps of the test concern testing of individual valves for subsequent vehicle wheels in accordance with the instructions given by the tester program and performed by the diagnostician. This way of communication between the program and the diagnostician causes a significant extension of the time of diagnosing the entire system.

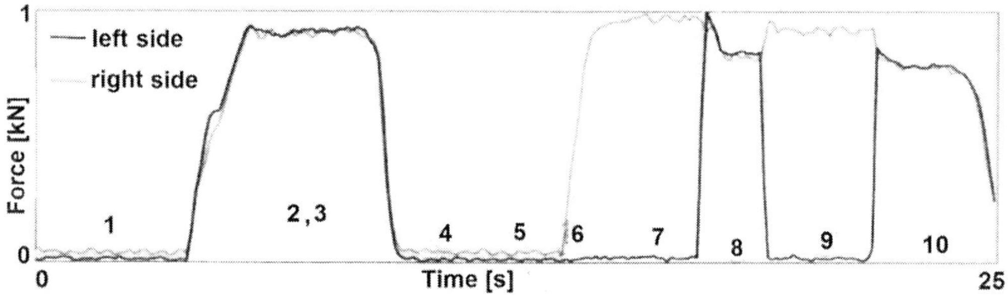

Fig. 2. The course of the braking forces of the front wheels when checking the operation of ABS valves. Next steps have been marked - commands of the diagnostic program: 1 – start the station, 2, 3 - checking braking forces, 4, 5, 6, 7 - checking the intake valve tightness, 8, 9 - intake and exhaust valve control, 10 - end of the test. Opel Astra II.

Another example of the control procedure shown in Figure 3 is to check the operation the ABS valves without first checking the intake valve for leaks.

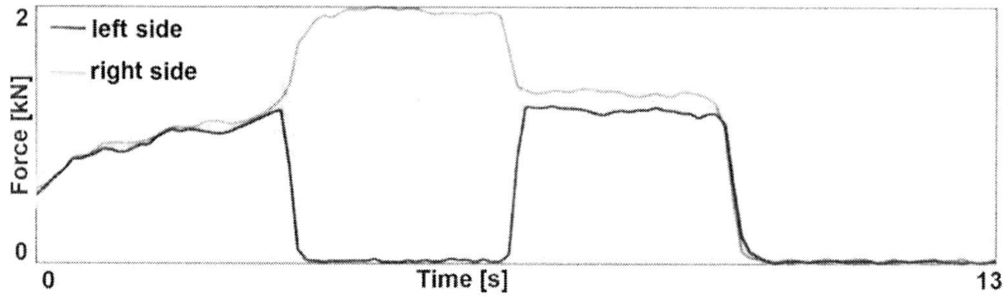

Fig. 3. Result of the ABS front left and right valve control. Bosch KTS tester, Ford Focus car.

Next procedure used in tests is simultaneous actuation of ABS valves in the brakes of the wheels of the tested axle. Its advantage is the acceleration of control by simultaneous actuation of valves on both wheels of the tested axle, Fig. 4. The test allows to evaluate the rate of decrease and the increase of the braking forces and tightness of ABS inlet and outlet valves. The difference in the braking forces of the left and right wheels in the tested car presented in the figure results from the mechanical failures of the brake.

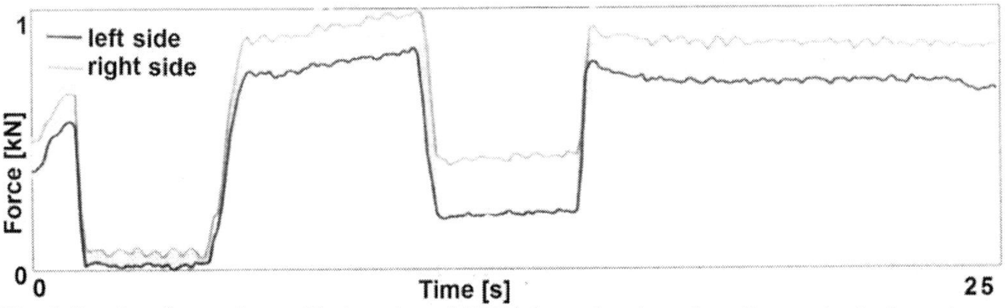

Fig. 4. Results of operation and leak testing of ABS inlet and outlet valves. Front wheels. Renault Scenic car.

3.1.1 ESP system

The procedure includes examining the operation of the ESP valves and pump without pressure on the brake pedal.

• The procedure of gradual pressure decrease

The next steps of the test and their effect are presented in the braking force diagram, Fig. 5:

1 - ABS pump activation, pressure increase in the brake caliper, brake force (without pressure on the brake pedal)

2 - Operation of ESP and ABS valves - pulsed pressure reduction, step changes in braking force

3 - Switching off the ABS pump - reducing the pressure and braking force to the minimum value

The next points, from 4 - 6, characterize the operation of the ABS system.

The course of the braking force in the figure 5 shows the control of the ESP and ABS pressure control valves in the brake caliper of the right front wheel. The same algorithm is used to control the valves of other wheels.

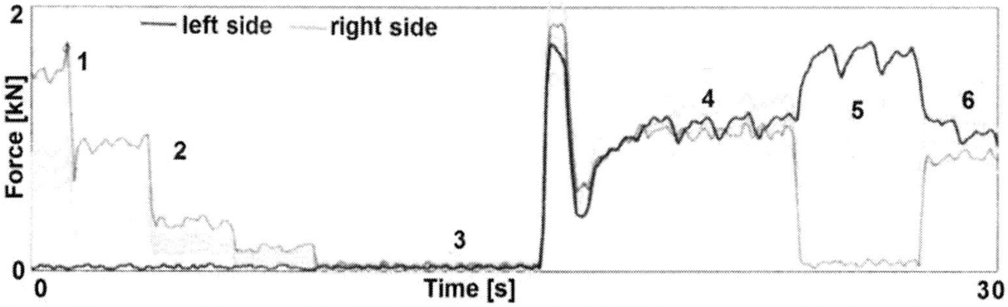

Fig. 5. The course of the test of the ABS/ESP valves and ABS / ESP pump test of the front right wheel. Audi A4. KTS tester. Description in the text.

• The procedure of gradual pressure decrease

This algorithm allows precise control of ESP valves without pressing the brake pedal. During the test, the tester program starts the ABS pump and controls the valves, gradually increasing the pressure in the tested brake caliper (fig. 6). The stability of the braking force and the speed of force changes are evaluated. This allows to check both the operation of the

ESP valves of the tested wheel and the tightness of valves servicing the adjacent wheel. The ABS valves are tested in the next steps, as in figure 5.

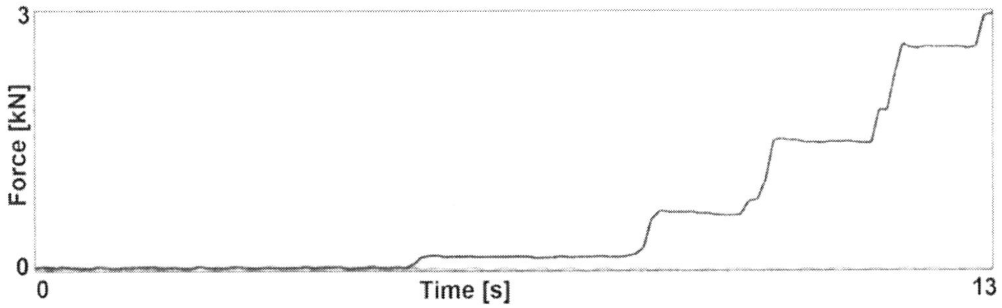

Fig. 6. The result of testing the operation of ESP valves and pump. Front left wheel. Ford Focus.

• Control of ESP valve operation - blocking the wheel

Fig. 7 shows the result of registration of braking forces during the test the valves of the left and right wheel of ABS system. The final part of the graph concerns the control of the dynamic activation of ESP valves, which led to the blocking of the right wheel on the stand.

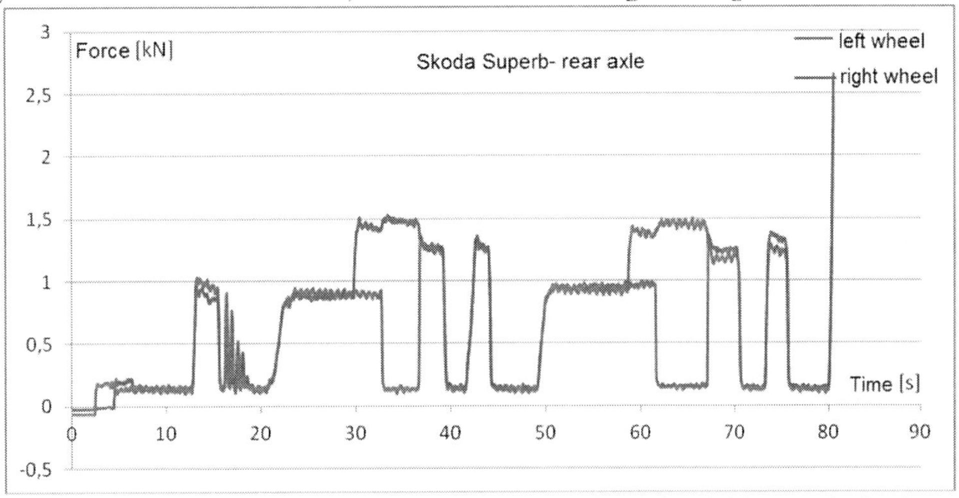

Fig. 7. Test of the inlet and outlet valves of the ABS system of the left and right rear wheel and control of the ESP valves. Skoda Superb. The results developed in Excel.

The above algorithms allow to check the operation of all ABS / ESP modulator solenoid valves. The disadvantage is the long test time, over 2 minutes for one axis.

During testing of ABS / ESP valves mechanical failures of the brakes, unrelated to the operation of ABS / ESP valves, are visible. Fig. 8 shows a clear pulsation of the braking force caused by the side beating of the brake disc.

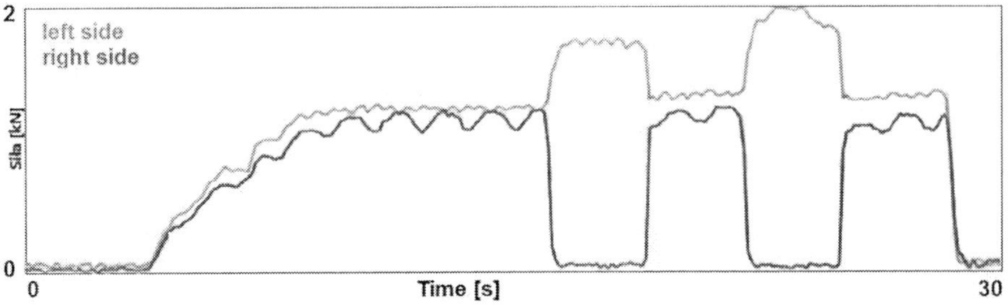

Fig. 8. The course of the braking force of the front left wheel after closing the intake valve and opening the outlet valve. Mitsubishi ASX.

3.2 Automatic actuation of the ABS/ESP valves

After starting the roller stand and opening the test program in the "Activation" mode, the diagnostician presses the brake pedal and starts the program. The program automatically performs the test by automatically executing the valve control sequence: closes the intake valve, opens the outlet and activates the ABS pump, fig. 9a). This cycle is repeated four times. Then the valves of the subsequent wheels are inspected, Fig. 9 b), c), d). The test allows to assess the rate of decline and the increase of braking force on the tested wheel and the repeatability of these actions.

a)

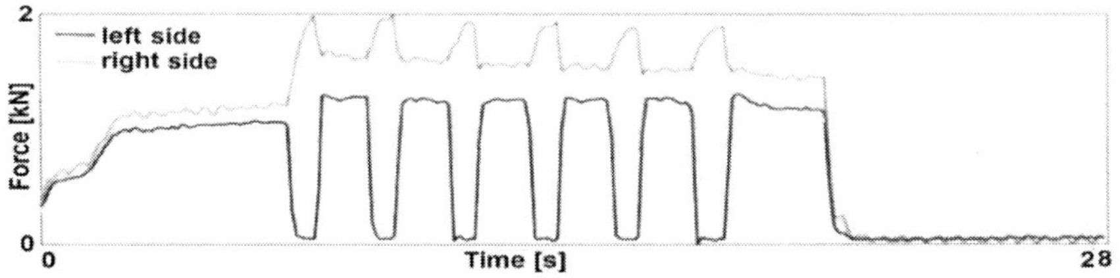

b)

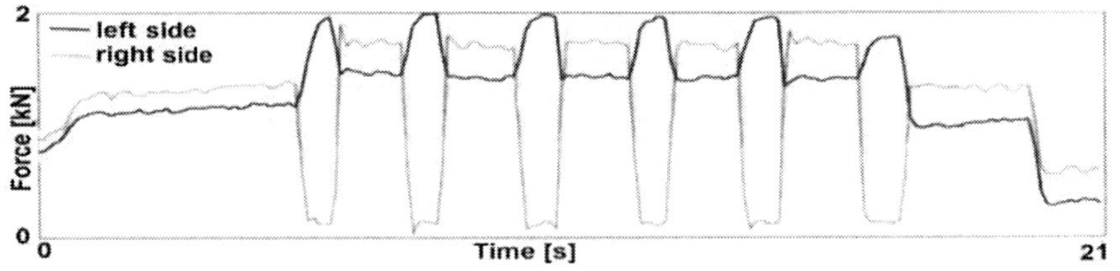

c)

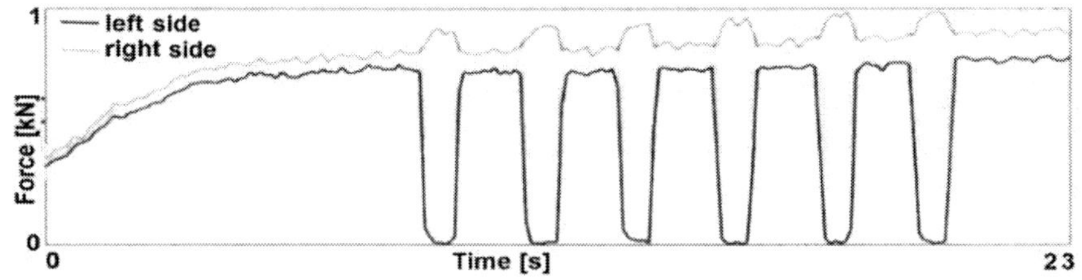

d)

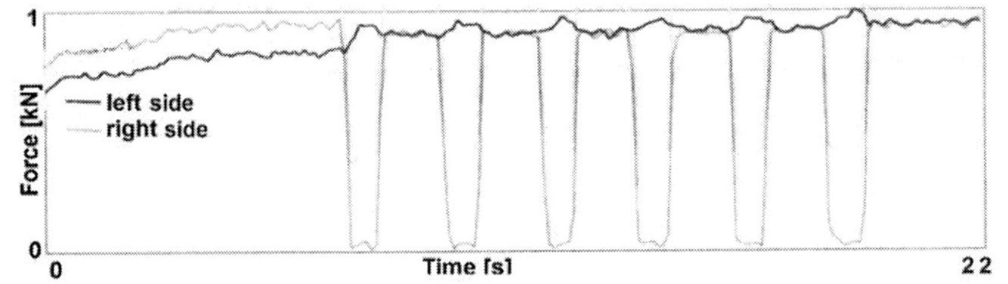

Fig. 9. Results of the ABS/ ESP pump and valves control. Automatic valve operation, a) front left wheel, b) front right wheel, c) rear left wheel, d) rear right wheel. Renault Scenic car.

The above algorithms allow to check the In the ABS / ESP diagnostic software of the Renault Talisman car there is an automatic control mode of ABS inlet and outlet valves for use on a roller stand, fig. 10. The ESP valves test works by activating the ABS pump and observing a dynamic increase in braking force to block the wheels, fig. 10.

The automatic valve actuation program allows to shorten the time of diagnosing the valves of ABS/ESP modulator on the roller stand to a few minutes. It is simple to implement for the diagnostician and can be used in periodic technical tests. The adoption of one diagnostic algorithm (automated) for different vehicle makes the diagnostic work easier during periodic technical tests.

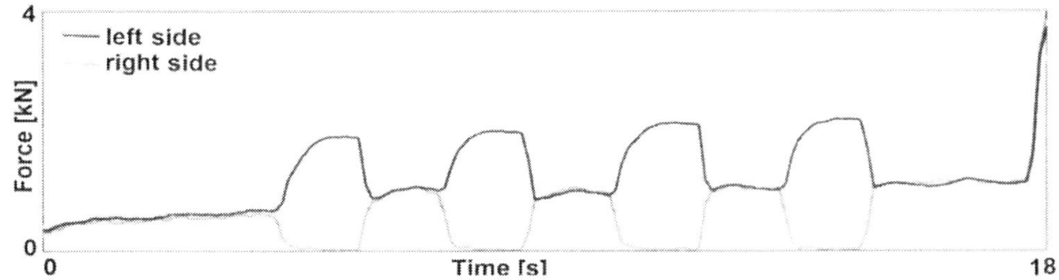

Fig. 10. The result of testing the inlet and outlet valves ABS of the right front wheel and the operation of the ESP valves. Renault Talisman.

4 Study of ABS / ESP valves actuation times and analysis of results

The method of testing the operation of the modulator of the ABS and ESP system on the roller stand allows to evaluate the response of the system in the form of changes of braking forces to forced activation of valves and ABS pump. Because on a low-speed roller stand with a circumferential speed of rollers up to about 5 km / h, the ABS system does not start itself, it is necessary to force its operation with an electric signal sent from the diagnostic tester to the valve controller.

Diagnosing the ABS / ESP system at the roller stand by measuring and recording the braking forces allows the evaluation of this system in the following scope:
- control of the operation of individual valves and ABS pump,
- checking the correct connection of hydraulic pipes to the ABS / ESP modulator
- assessment of the tightness of the valves based on the evaluation of the stability of the braking forces at a fixed pressure on the brake pedal.
- assessment of the rate of rise and decrease of braking forces when using ABS / ESP valves.

Investigations of the rate of change of braking forces, measured at the roller stand, under the action of ABS valves showed differentiation between braking systems of different manufacturers. The change of braking forces by about 1 kN took place in the time 0.1 - 0.2 s, Fig. 11. For the ESP system, the time of force increase by 1 kN was about 0.1 s, Fig. 12. Based on the performed tests of the ABS/ESP modulators and controllers, it can be concluded that for efficient ABS systems the rise time and deceleration of the braking force under the electromagnetic valve operation should be 5 - 10 kN / s, and for the ESP valves approximately 10 kN / s.

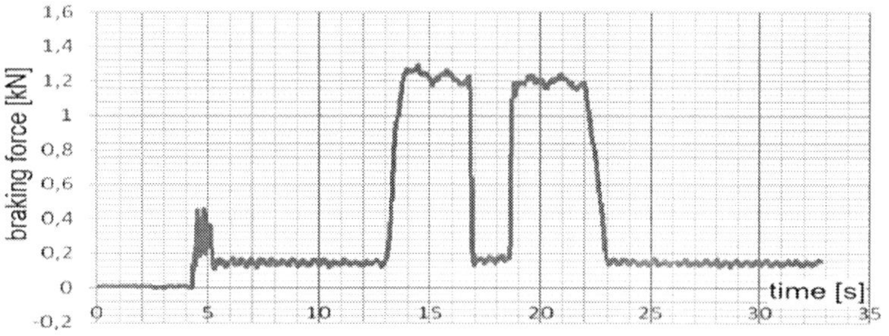

a)

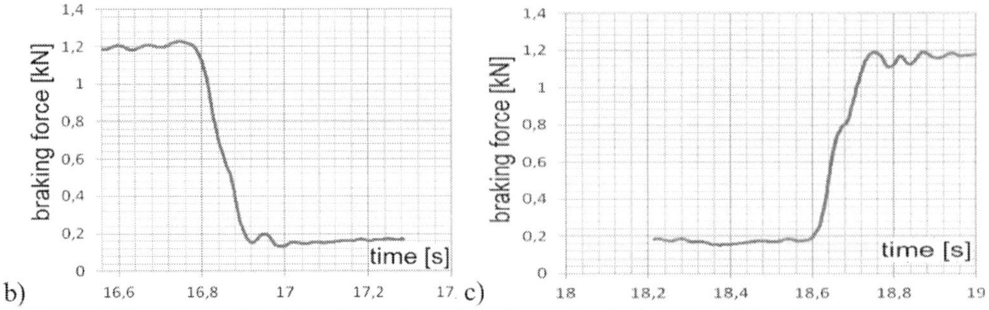

Fig. 11. a) The course of braking force of the right front wheel when the ABS valves are activated, decreasing b) decreasing c) increase of the braking force of the front right wheel when the ABS valves are activated. Honda

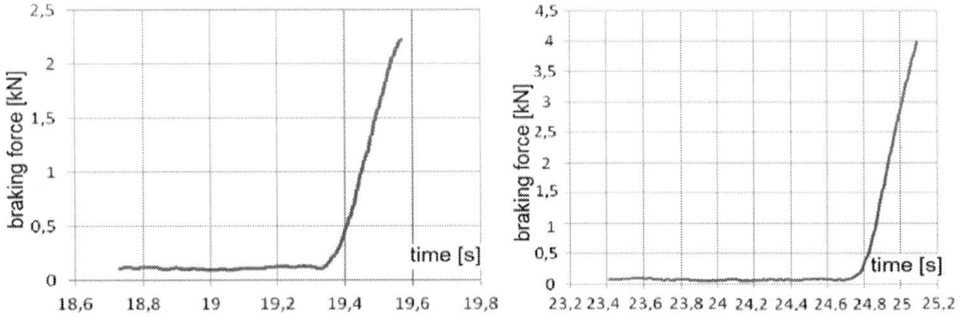

Fig. 12. a) The course of the braking force after the pump and ESP valves are activated

Studies of ABS / ESP systems from different manufacturers showed different diagnostic software. It can be listed the following types of this software:

- individual actuation of valves in turn for individual car wheels,
- simultaneous start of valves for wheels of the tested axis,
- automatic procedure for starting valves with repetitions,
- control of ESP valves operation with increasing pressure pulses
- control of ESP valves operation by a high pressure impulse that causes blocking wheels.

For diagnostic purposes, the diagnostic tester software that automatically starts the control procedure is suitable. This makes it possible to control the system on a roller stand by one person and significantly shortens the time of this control compared to the individual actuation of the valves.

It should be noted that testers from different manufacturers have software that allows to control the ABS / ESP valves of the same vehicle to a different extent.

Diagnostic software for some ABS / ESP systems contains information on the inability to control valves on a roller stand. There is also diagnostic software to check the valves and ABS pump only in terms of electrical efficiency.

Such diversity makes it difficult to introduce this scope of diagnosis into the practice of periodic vehicle technical inspection. For these purposes, it is necessary to harmonize the control procedure.

5 Conclusions

1. The conducted research revealed a variety of diagnostics software for testing the operation of the valves and pump of the ABS / ESP system. The scope of diagnosing the ABS / ESP system on a roller stand depends not only on the ABS / ESP driver software, but also on the diagnostic program implemented to the tester. The same ABS / ESP system in a particular car can be diagnosed in a slightly different range and in a different way with the participation of testers from different manufacturers.

2. There are two approaches to checking ABS valves in the diagnostic software:
- electrical inspection,
- control by evaluating the operation of the entire braking system.

The electrical tests are not related to the mechanical control of the system. Adding the measurement of the braking forces on the roller stand to this procedure extends the scope of the ABS / ESP system evaluation. This allows detection of not only an incorrect electrical connection, but also too low dynamics of changes in braking forces caused by mechanical malfunctions.

3. The condition for using the roler stand for the evaluation of the ABS / ESP system is its software allowing to record the dynamics of the braking forces during the test and to present the test results graphically.

4. Extending the scope of tests of ABS and ESP systems during periodic vehicle inspection requires the development of unified software for vehicle control stations, so that the diagnostician has easier and uniform IT access to the ABS / ESP controllers and actuators of individual systems. A reference database is also required to evaluate the operation of the system being tested.

5. New diagnostic programs for ABS / ESP systems (in new vehicle models) should enable control of ABS / ESP valves on a roller stand in an automatic cycle. This option facilitates and speeds up the diagnosis process.

References

1. Rozporzą dzenie Min. Transportu, Budownictwa i Gospodarki Morskiej w sprawie zakresu i sposobu przeprowadzania badań technicznych pojazdów oraz wzorów dokumentów stosowanych przy tych badaniach. Dz. U. z 7.09.2012 r., Poz. 996 (2012)
2. Directive 2015/45/EU of the European Parliament and of the Counsil of 3 April 2014 on periodic roadworthinees tests for motor vehicles and their trailers and repealing directive 2009/40/EC (2009)
3. Raport CITA: Study on a new performance test for electronic safety components at roadworthiness tests – Final Report, 2014, Summary 2006.
4. Chr. Theis, *Test methods and data for the PTI of ECSS and their international provision Federal ministry of transport and digital infrastructure, Germany.* (CITA Conference – Dubai , 2015)
5. Periodic Technical Inspection with the PTI Adapter 21 Plus. FSD Germany - 2015
6. A. Gajek, *Directions for the development of periodic technical inspection for motor vehicles safety systems.* (The Archives of Automotive Engineering, 2 , 2018)

Construction and tests of the research stand with built-in correctors for the detection of gas outflow from long pipelines

Karol Golak[1*], *Paweł* Lindstedt[2], *Błażej* Bartoszewicz[1]

[1]Faculty of Mechanical Engineering, Bialystok University of Technology, Wiejska 45C street, 15-351 Białystok, Poland
[2]Air Force Institute of Technology, Księcia Bolesława 6 street, 01-494 Warszawa, Poland

Abstract. The localization and identification of small gas leaks from a damaged gas pipeline is a very important but at the same time problematic and difficult process to carry out. Quick identification, estimation and precise localization of the leak leads to minimizing the financial and equipment losses resulting from damage. The most commonly used continuous internal methods of leak identification are based on pressure and mass intensity measurements flow at various places in the pipeline. The article presents a description of the construction of a test stand characterized by the correctors built in it and the examples of the signals obtained from its tests for different values of simulated leaks. The results obtained will be used to develop a new method for identifying and locating leakages from long pipelines. This method is based on the cross power spectral density of measured pipeline standard signals (pressure and mass flow) and the membrane displacement signals delivered from the test equipment (correctors) connected to the system.

1 Introduction

The gas pipelines are used for decades and during this period, due to progressive corrosion, mechanical leaks may occur. Gas leaks from long pipelines can cause serious accidents and may damage the equipment, hence it is important to detect, estimate and locate them as soon as possible.

There are a number of methods to detect the occurrence of a gas leak, generally they can be divided into non-continuous (e.g. smart pigging, helicopter inspection) and continuous monitoring. [1, 2]

Non-continuous methods do not allow for constant observation and assessment of the pipeline's technical condition. Individual inspections are carried out at certain period of time which often does not allow for instant leak detection.[4, 7]

Continuous methods are used without interruption during the entire operation of the pipeline. They could be divided into external and internal based systems. [2, 3, 6] Externally based systems detect leaking product outside the pipeline (acoustic systems, fiber optic, dielectric cables, laser sensors). Internal based systems (e.g. pressure analysis, mass balance method), known as computational pipeline monitoring (CPM) uses different types of sensors to monitor internal pipeline parameters which are a basis for different methods of gas leak detection, however, in order to use them, it is necessary to use them in order to use them. installation of leak detection systems before the pipeline is started. [9, 11, 12, 13,14]

The article presents the construction and test results of a test stand for simulating gas leaks from long pipelines. The stand is characterized by two correctors attached to the

* Corresponding author: k.golak@pb.edu.pl

pipeline which provide an additional diagnostic signal in the form of weak interaction signals. [5, 8, 10] The connection of correctors to the pipelines does not affect the quality of the pipeline and is easy to implement.

2 Construction of a test stand with built-in correctors to detect gas leaks from the pipeline

2.1 Construction of a test stand

Test stand built on the basis of the scheme in Fig.1. is shown in Fig. 2. The stand was built with about 27 meters of welded PPR pipes with an inner diameter of Ø 45mm mounted on a frame of aluminium structural profiles. At the beginning of the pipeline there is an air tank (V) with a capacity of 0,025m^3, to which atmospheric air is pumped using a compressor. Between the compressor and tank there is a compressed air preparation unit for adjusting pressure to the required level of working and for cleaning the. Along with the tank, the entire system has a capacity of about 0,075m^3.

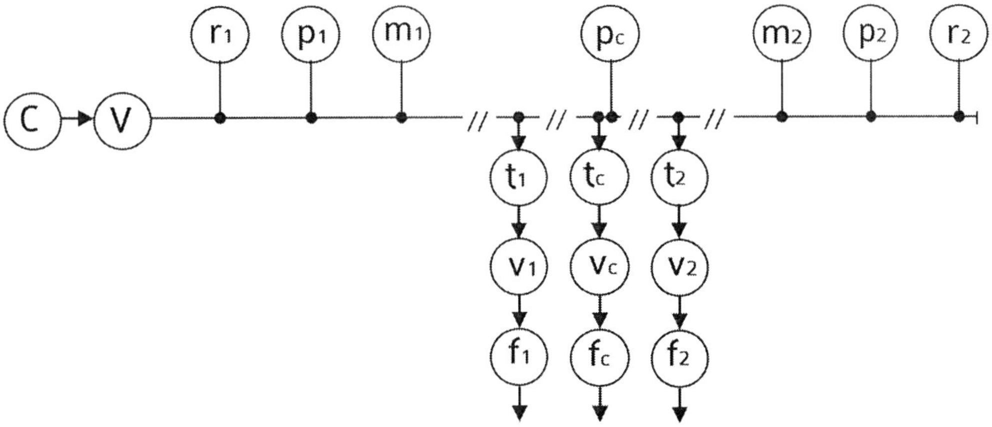

Fig. 1. Scheme of the test stand for leakage simulations from pipelines, where: *C* - compressor, *V*-pressure tank, *r* - correctors, *p* - pressure sensors, *m* - mass flow sensors, *t* - taps, v - adjustable valves, f-flow meters

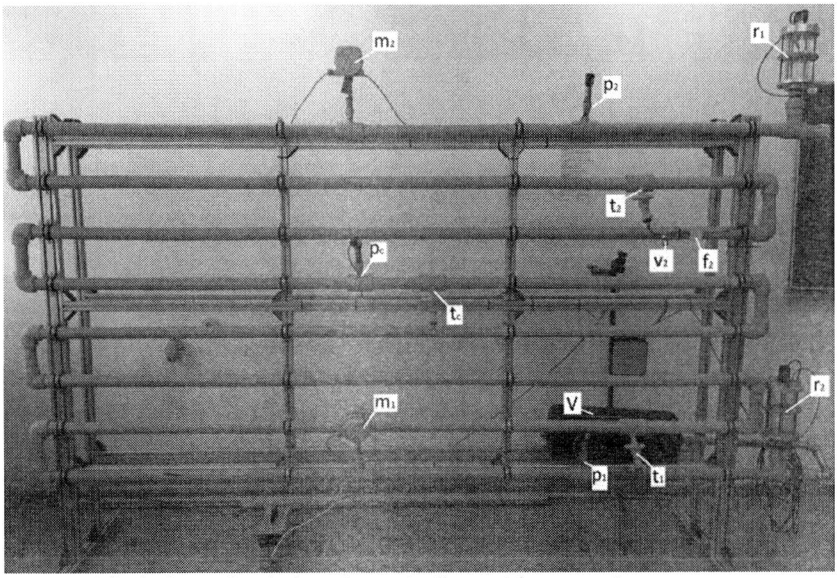

Fig. 2. Test stand for leakage simulations from pipelines with mounted correctors.

Behind the tank and at the end of the system are mounted correctors (r_1, r_2). Pressure sensors (p_1, p_c, p_2) and mass flow sensors (m_1, m_2) have been installed in the pipeline in the order shown in Fig. 1. The taps at the beginning (t_1), centre (t_c) and end of the pipeline (t_2) are used to simulate gas leaks. In order to enable precise control and measurement of the amount of leakage behind the tap from which the leak is simulated, a solenoid valve and a flow meter are installed.

Data instruments from National Instruments were used to collect data and control the simulated leaks volume.

2.1 Construction of the correctors

At the beginning and the end of the test stand, two additional devices - correctors (r_1 and r_2) were built using the tees. The corrector was built in the form of two chambers separated from each other by a rubber membrane with a mass installed in the middle (brass sheet). The membrane and its mass have been selected in such a way that there is visible continuous resonance vibrations. The chambers are connected with each other by a plastic pipe with an internal diameter of Ø 4mm, which allows pressure equalization between two correctors parts. On the membrane with the mass mounted on its centre, as a result of the gas leak its vibrations will be visible. A high-precision laser distance sensor was used to measure the vibrations of the membrane. The measurement is carried out by glass installed on the top of the corrector which transmitting the light with wavelength emitted by the sensor.

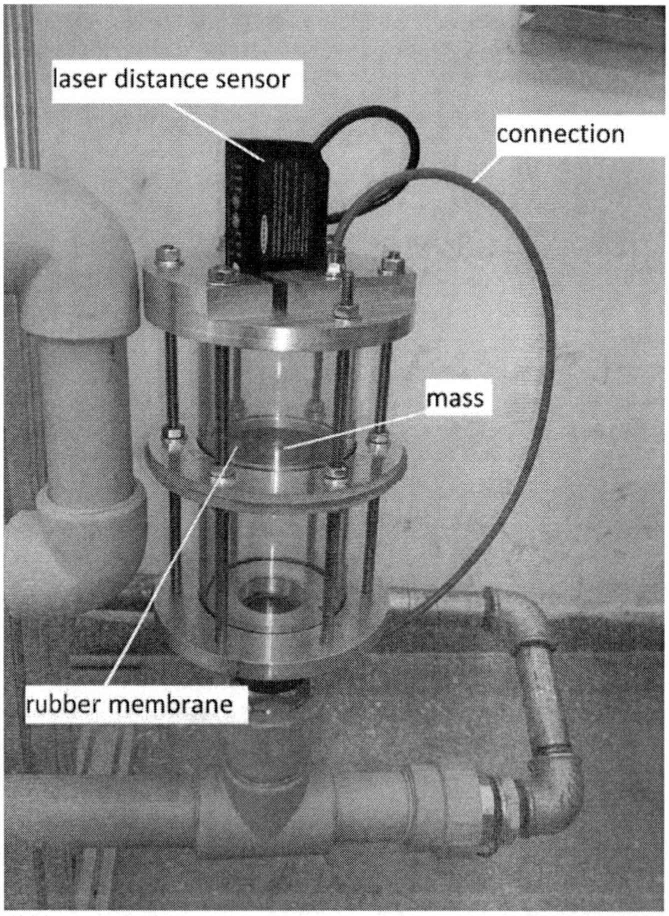

Fig. 3. Corrector build-in test stand.

The correctors built into the test stand are to provide an additional diagnostic signal - it is a device that transforms the pressure change that appears in the pipeline as a result of a leakage on the membrane displacement signal.

3 The test results of the test stand

The test stand tests were carried out at the pressure of about 0.6 MPa. The leakage control solenoid was set to obtain different leakages. The research was carried out by simulating leakages from three places in the pipeline (t_1, t_c and t_2). Each measurement lasted 6.5 seconds from the moment of the leak occurrence and during this time the following data was recorded with indicators describing their position in the pipeline *1* for the beginning, *c* - centre, *2* end of the pipeline:

- p_1, p_c, p_2 - pressure in the pipeline [Pa];
- r_1, r_2 - membrane displacement [mm].

Leak volumes:

- *f10* – 10 litres per minute (0.000166 m_3/s);
- *f20* - 20 litres per minute (0.000333 m_3/s);
- *f30* - 30 litres per minute (0.000499 m_3/s);
- *f40* - 40 litres per minute (0.000666 m^3/s);
- *f50* - 50 litres per minute (0.000833 m^3/s);

The sample results for the different gas leakage volume simulation from t_2 (valve on the end of stand) is presented from Fig. 4 to Fig. 8

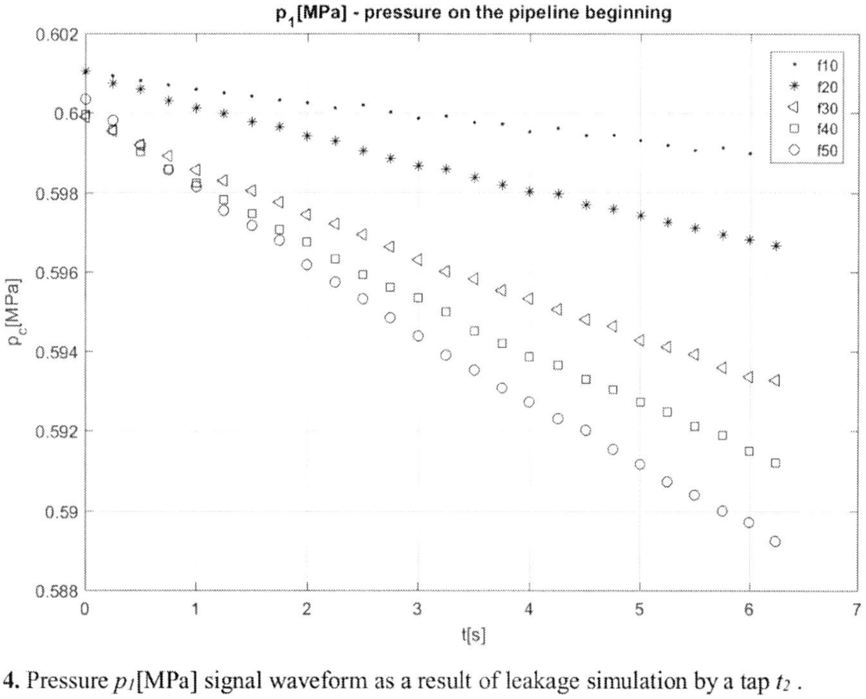

Fig. 4. Pressure p_1[MPa] signal waveform as a result of leakage simulation by a tap t_2 .

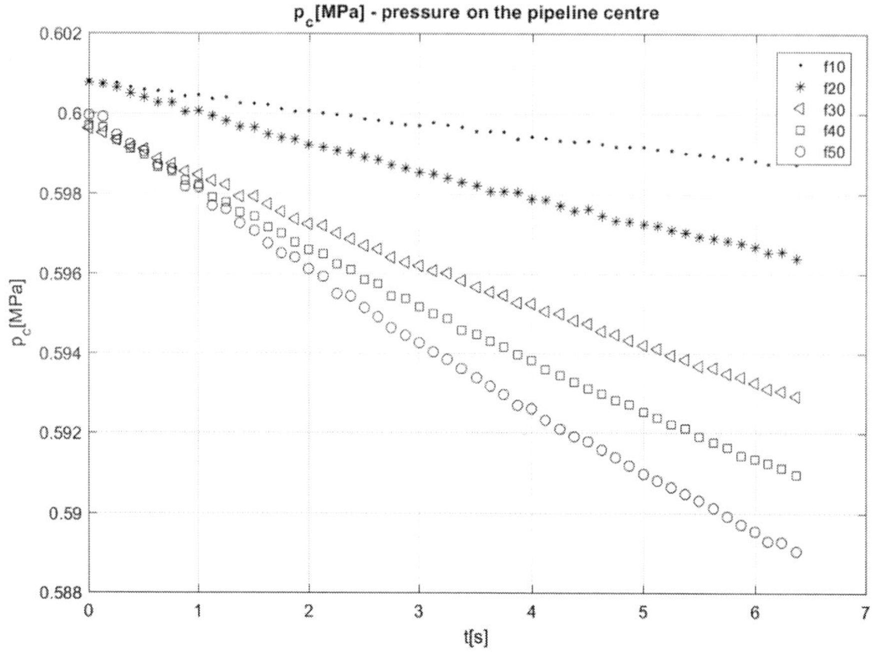

Fig. 5. Pressure p_c[MPa] signal waveform as a result of leakage simulation by a tap t_2 .

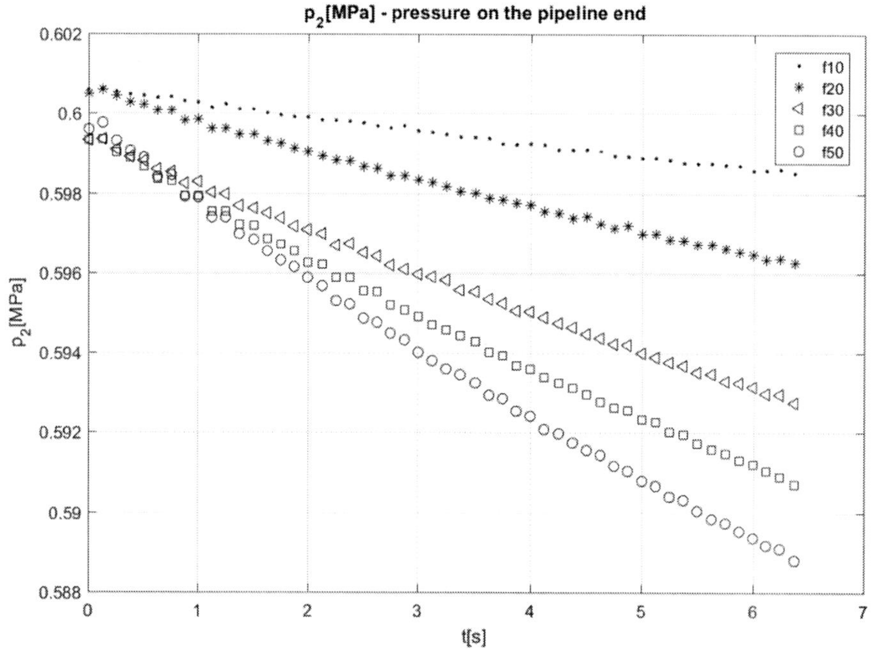

Fig. 6. Pressure p_2[MPa] signal waveform as a result of leakage simulation by a tap t_2 .

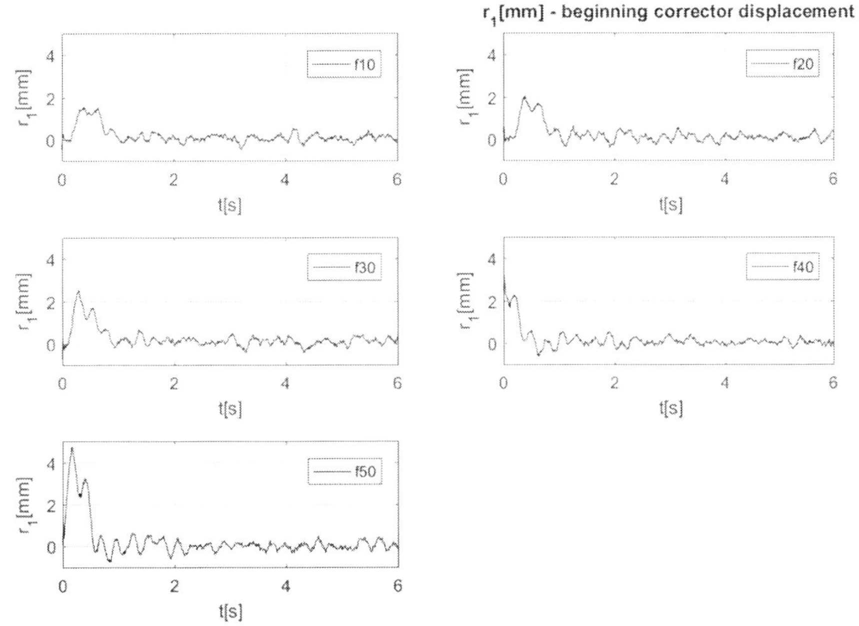

Fig. 7. Membrane displacement in corrector r_1 [mm] as a result of leakage simulation by a tap t_2 .

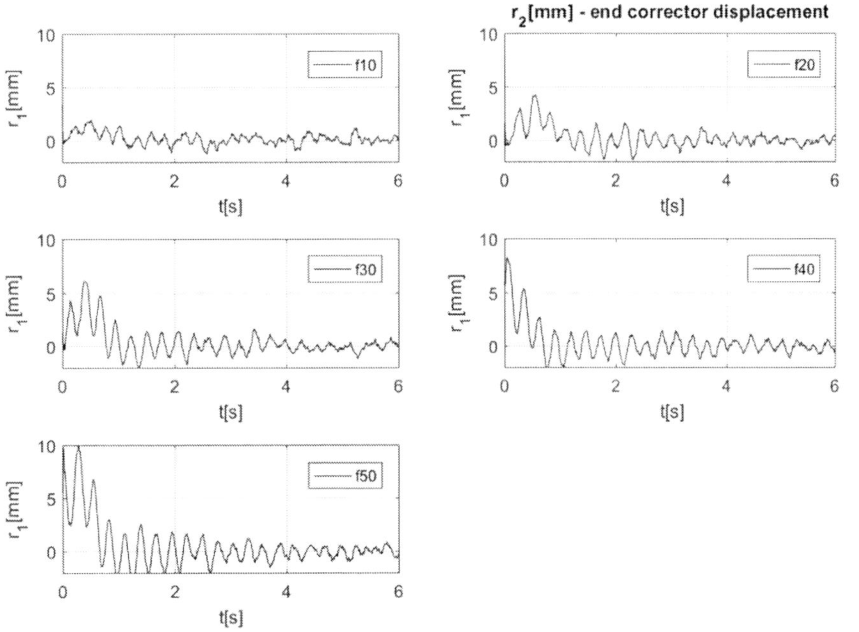

Fig. 8. Membrane displacement in corrector r_2 [mm] as a result of leakage simulation by a tap t_2 .

From the Fig. 4. to Fig. 8. the results of measurements with simulated small leakages are presented. This results confirm that correctors connected to the pipelines, can provide the new valuable diagnostic information that could increase the effectiveness of leaks identification, estimation and localization.

Summary

The article presents a research stand for simulating the leakages of gas from long gas pipelines characterized by additional two research equipment - correctors placed at the inlet and outlet of the pipeline and standard mounted pressure sensors and mass flow sensors.

Attaching correctors to the gas pipeline have a purpose to obtain additional valuable diagnostic information, which are weak interactions signals. Unlike standard pressure signals, the signals of weak interactions are more sensitive and more resistant to noise. The use of weak interaction signals can complement existing mass flow and pressure signals and can be used to identify the location and size of gas outflow.

Signals from the system and research devices (correctors) will be the basis for the development of a new method for testing the outflow from the gas pipeline based on the quotient of the spectral power density of signals generated by the corrector and measured signals of pressure and mass flow.

Karol Golak part of paper was supported by the research project No. S/WM/2/2017 and financed by the Ministry of Science and Higher Education.
Rafał Grądzki part of paper was supported by the research project No. S/WM/1/2016 and financed by the Ministry of Science and Higher Education.

References

1. L. Boaz, S. Kaijage, R. Sinde, *An overview of pipeline leak detection and location systems*. In: Science, Computing and Telecommunications (PACT), 2014 Pan African Conference on. IEEE;133–137 (2014).
2. API 1130, Computational pipeline monitoring (2nd ed.), American Petroleum Institute, Washington, DC, USA (2002)
3. W. Liang, L. Zhang, *A wave change analysis (WCA) method for pipeline leak detection using Gaussian mixture model*, Journal of Loss Prevention in the Process Industries, **25**, Issue 1, pp. 60-69, (2012)
4. L. Billman, R. Isermann, *Leak detection methods for pipelines*, Automatica, **23**, no. 3, 381-385, (1987).
5. P. Ostapkowicz, *Leakage detection from liquid transmission pipelines using improved pressure wave technique*, Maintenance and Reliability, **16**, no. 1, 9-16 (2014)
6. K. E. Abdulimen, A. A. Susu, *Liquid pipeline leak detection system: model development and numerical simulation*, Chemical Engineering Journal, **97**, Issue 1, (2004).
7. Z. Kowalczuk, K. Gunawickrama, *Detecting and locating leaks in transmission pipelines*, In J. Korbicz, J. Koś cielny, Z. Kowalczuk, W. Cholewa (Eds.): Fault diagnosis. Models, Artifical Intelligence, Applications, Chapter 21, 821-864 (2004)
8. P. Lindstedt, *Weak interactions between objects in the signal-based and parametric diagnostics of transport-dedicated complex engineering systems*, Aircraft Engineering and Aerospace Technology, **77**, no. 3, 222-227 (2005).
9. A. E. Liu,. Overview: *Pipeline Accounting and Leak Detection by Mass Balance*, Theory and Hardware Implementation (2008)
10. R. Grą dzki, K. Golak, P. Lindstedt, B. Bartoszewicz, *Reasons for the experimental research of gas outflows based on the signals of weak interactions between the tested model of the gas pipeline, and tested equalizer*, Journal of Konbin, 2, 39-48, (2015)
11. R. Grą dzki, P. Lindstedt, *Method of assessment of technical object aptitude in environment of exploitation and service conditions*. Eksploatacja i Niezawodnosc - Maintenance and Reliability, **1**; 54-63 (2015), http://dx.doi.org/10.17531/ein.2015.1.8
12. R. Grą dzki, P. Lindstedt, K. Golak, *Premises of evaluation of the technical object suitability with including the quality of its maintenance and operation, and their*

initial conditions. In: Safety and Reliability: Methodology and Applications - Proceedings of the European Safety and Reliability Conference, chapter 44; 319 - 326 (2015)
13. S. L. Scott, M. A. Barrufet, *Worldwide Assessment of Industry Leak Detection Capabilities for Single & Multiphase Pipelines*, Project Report Prepared for the Minerals Management Service, OTRC Library Number: 8/03A120, (2003).
14. Y. Sivathanu, *Natural Gas Leak Detection in Pipelines*. U.S. Department of Energy, National Energy Technology Laboratory, (2003).
15. Tomasik P. *Zastosowania modeli dokładnych sieci do detekcji i lokalizacji wycieków*. In: Diagnostics of Processes and Systems, 507-512, (2011).

An analysis of the active safety of a passenger car body during car use

Jarosław Gonera[1,*] and *Jerzy* Napiórkowski[1]

[1]Department of Design and Operation of Vehicles and Machines, University of Warmia and Mazury in Olsztyn

Abstract. Passenger car bodies made nowadays are most often self-bearing constructions. This is the reason why a change in the parameters of a car body during its use also affects other sets of component parts, which are attached to it. While a passenger car is driven, the car body is exposed to various factors which can cause sudden damage or gradual wear of the car's technical condition. This paper contains an analysis of the influence of changes in the geometry of a car body on changes in the geometry of wheels. The study was carried out on C-segment cars produced by the same manufacturer. While making an assessment of the car body's geometry, the location of base points on the floor pan and the upper parts of the car bodywork, within the motor chamber, etc. was assayed. All vehicles were constructed on the same floor pan. Measurements were performed with a Gysmeter made by the company Gys and supported by Autorobot Datasheet software, while the geometry of wheels was measured with a computerised laser device Autoboss A-860. Camber, castor and steering axis inclination were some of the wheel geometry parameters measured in our study.

1 Introduction

Mandatory periodical inspections of the technical condition of passenger cars do not necessitate any detailed assessment of the technical condition in terms of the car bodywork geometry and wheel alignment [13]. Assessment of the car body geometry is made only in cars that have been engaged in road collisions and crashes. During mandatory periodical inspection, a mechanic only briefly inspects the basic elements of the car body's bearing structure, to verify that they are not damaged and do not impair the road traffic safety [15, 16, 17, 18]. During the useful life of a passenger car, the automobile does not undergo periodical, multi-step investigations of the geometry of its bodywork [1]. The geometry of wheels is only measured after damaged or worn elements of the suspension system or the steering wheel system have been replaced. It is observed that quite many cars currently on our roads have been involved in various accidents or collisions, which invariably cause some damage to the car body [8, 11]. There are no detailed studies which would describe dependences between the scope of bodywork repairs and the technical condition of the systems responsible for the vehicle's safe use [5, 8, 14].

The use of elements connected with the active and passive safety in passenger cars contributes significantly to protecting the driver or passengers and other road users [9, 10]. Costs incurred by adding safety-enhancing elements are disproportionately low compared to the costs of removing consequences of an accident [3]. While constructing a vehicle, manufacturers follow two directions in the development of systems responsible for safety: reducing the risk of a traffic accident (active safety) and reducing injuries during an accident (passive safety) [4, 7].

* Corresponding author: jaroslaw.gonera@uwm.edu.pl

2 Problems associated with measurements of the car bodywork geometry and wheel geometry

There is a set of base points on a floor pan and in the upper parts of a body of any passenger car. Their location is determined by the car manufacturer. These points are typically construction holes, which help fix subunits of a vehicle, and auxiliary holes, made specially for measuring geometry [1]. The location of base points is illustrated in charts issued for mechanical measurement tools and in software databases develop to support electronic devices. When making measurements of the geometry of a car body, the required position of base points is compared with their actual location, which is determined through measurements [12]. Measurements of the geometry of a car body are made both in the area of the floor pan and in the upper parts of the body. For the sake of such investigations, it is presumed that the maximum difference between the required and actual location of base points must not exceed 3 mm [1, 11].

The geometry of the body is directly connected with the geometry of the suspension and steering systems. The reason is the fact that components of the suspension system and steering system are mounted onto the body. The moment the base points responsible for fixing these elements is shifted on the body, the parameters related to setting the wheels (their geometry) will change, too. The most important parameters associated with the geometry of wheels and affecting the way a car is driven are: front and rear axle toe, front and rear axle camber, values of the steering axis castor, steering axis inclination angles, deviation of the driving axis from the axis of symmetry, wheelbase difference, front and rear axle wheel setback [5, 11, 14].

Many of the parameters connected with the geometry of the suspension system and steering wheel system in most of the passenger cars manufactured nowadays cannot be regulated. Often it is impossible to adjust, for example, wheel camber, steering axis inclination or the angle of castor. Whenever certain irregularities are detected in this respect, it is necessary to repair the car bodywork, as the proper geometry of the suspension and steering systems cannot be restored unless the geometry of the car body is restored first [6, 13, 14].

3 Research methodology

The study included 10 passenger cars manufactured by the same motor vehicle maker. All the cars belonged to the C-segment according to the European classification systems. They were equipped with a MacPherson's strut and a semi-float rear axle suspension with a torsion beam. In addition, they had a rack and pinion steering system. The cars had been produced 12 to 14 years before the tests, and they had a mileage of 250 000 to 350 000 km. All had been driven on roads in Poland only. None had been in a car crash or collision.

A Gysmeter made by Gys (fig. 1) was used to measure the geometry of the car bodies. This is an electronic, telescope device used for measuring the geometry of a floor pan and upper elements of a car shell. It is not necessary to disassemble any of the elements of a car in order to make measurements with this tool. The accuracy of measurements taken with a Gysmeter was 1 mm, while the measurement range was within 400 to 2650 mm. During the measurements of the car body geometry, the position of base points located on the floor pan and upper parts of the bodywork was analysed. The information about the obligatory location of base points was derived from the database of the software named Autorobot Data System.

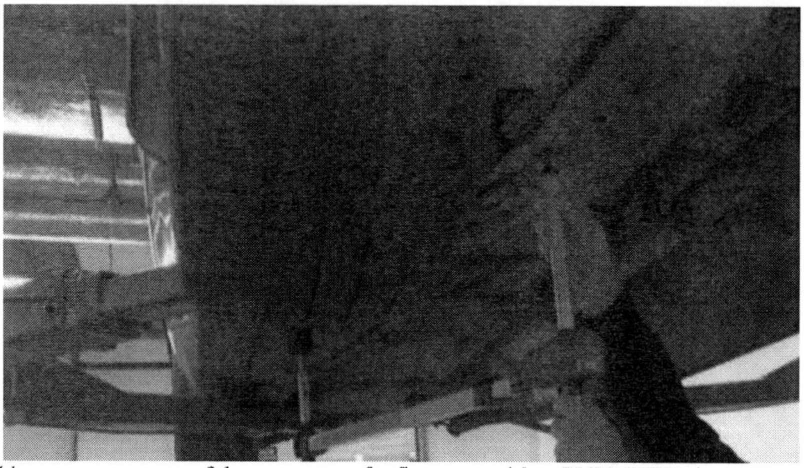

Fig. 1. Taking measurements of the geometry of a floor pan with a GYSMETER by GYS.

Measurements of the floor pan geometry and upper body parts were carried out mainly at the points connected with the mounting of the front suspension, the rear suspension and the front bench. These points were selected because they directly affect the position (geometry) of the suspension system and steering system. In addition, the displacement of the base points located on the front and rear frame rails and in the vicinity of the bulkhead was determined. During the measurements, the position of all analysed points relative to the initial ones, which were found in the rear part of the passenger compartment, were analysed. In none of the analysed cars, changes in the geometry of the floor pan were observed with respect to this aspect. The analysed sites in a car are illustrated in figs. 2 and 3.

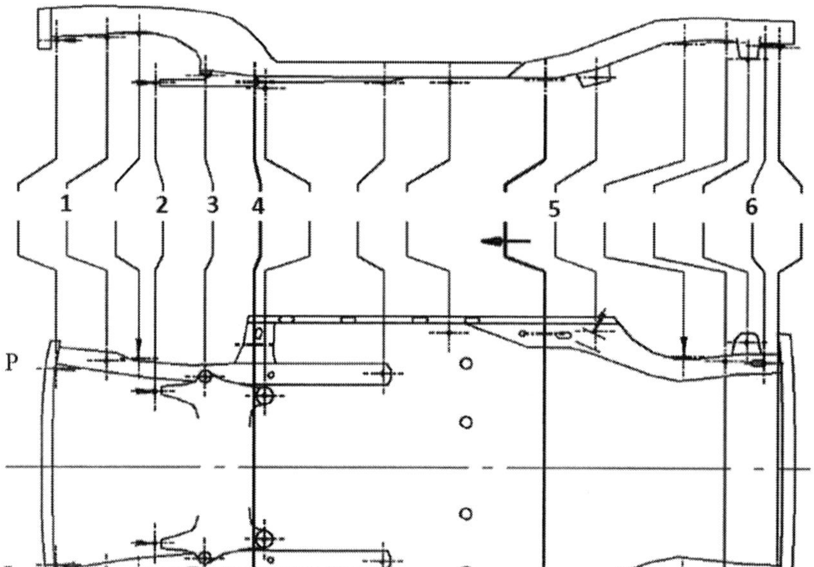

Fig. 2. A view of the floor pan originating from the Autorobot Data System programme, where 1 – points lying on the front railsm 2 – points fixing the front suspension, 3 – points fixing the front bench, 4 – points located in the vicinity of the bulkhead, 5 – points fixing the rear suspension, 6 – points located on the rear rails.

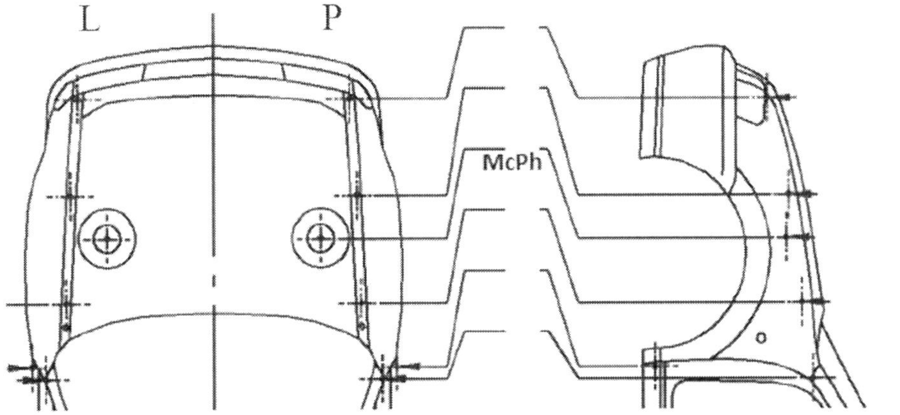

Fig. 3. A view of the upper bodywork parts (the motor chamber) plotted in an Autorobot Data System programme, where McPh – mounting of a MacPherson's strut.

The measurements of the geometry of the suspension and steering systems were carried out using an Autoboss A860 device. It is a computerised laser device for measuring practically all parameters connected with the geometry of wheels. It was equipped with a special programme and four heads, which were mounted on individual wheels of a car during the measurements. The communication between the heads and between each head and the steering unit was remote. The resolution for all measurements was 1'. The instrument enabled the user to measure the geometry of cars having wheels with a diameter from 11" to 21". As a result of the wheel geometry measurements, the parameters shown in Figure 4 were determined.

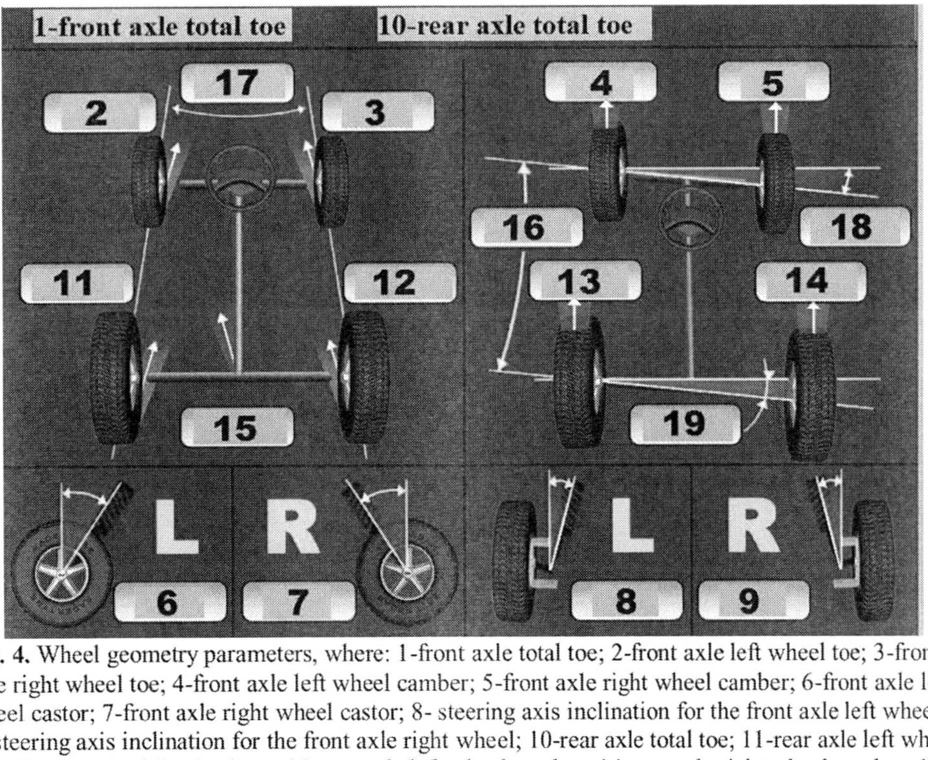

Fig. 4. Wheel geometry parameters, where: 1-front axle total toe; 2-front axle left wheel toe; 3-front axle right wheel toe; 4-front axle left wheel camber; 5-front axle right wheel camber; 6-front axle left wheel castor; 7-front axle right wheel castor; 8- steering axis inclination for the front axle left wheel; 9- steering axis inclination for the front axle right wheel; 10-rear axle total toe; 11-rear axle left wheel toe; 12-rear axle righ wheel toe; 13-rear axle left wheel camber; 14-rear axle right wheel camber; 15-driving axis from symmetry axis deviation; 16-wheelbase difference; 17-track width difference; 18-setback (front axle); 19-setback (rear axle).

4 Results of measurements of the car bodywork geometry

Based on the results (Tab. 1), it can be concluded that the smallest alterations in the geometry of a car body were noted in the vicinity of the bulkhead. In this area, the displacement of base points was no more than 1 mm. As for the points located on rear rails, the shift of the base points in most cars did not exceed 3 mm. However, this difference in 1 car was larger, around 11 mm. Changes in the geometry measured on front rails were usually within a range of 0 to 5 m, and in one case reached 7 mm. Larger differences were observed in the other areas of car bodies analysed, for example differences between the required and actual location of base points mounting the front suspension were within an interval of 2 to 31 mm. In turn, for points fixing the front bench the respective differences fell within the range of 2 to 9 mm. Points mounting the rear suspension were displaced from 1 to 6 mm. Considerable changes in the geometry were also determined in the motor chamber. With respect to the points representing the upper fixing of MacPherson's struts, the differences ranged from 3 to as much as 16 mm.

Tab. 1. Results of measurements of the geometry of the car body in the analysed vehicles.

No of car		S1	S2	S3	S4	S5	S6	S7	S8	S9	S10	Value of the shift of a given base point [mm]
Base point		Value of the shift of a given base point [mm]										
1P	Front rail	2	0	3	2	2	0	-4	2	3	-5	2,3
1L		0	1	5	2	2	-4	3	2	-5	-7	3,1
2P	Mounting of the front suspension	-7	-8	-4	-9	-7	-8	-11	-7	-11	-31	10,3
2L		-5	-7	2	-7	6	-6	-4	-7	-13	-20	7,7
3P	Mounting of the front bench	-6	-2	-3	-6	-5	-7	-7	-6	-7	-9	5,8
3L		-4	2	2	-5	4	-5	-3	-5	-7	-7	4,4
4P	Vicinity of the bulkhead	0	1	0	0	0	0	-1	1	1	-1	0,5
4L		1	0	0	0	-1	0	-1	0	1	-1	0,5
5P	Mounting of the rear suspension	1	5	3	-3	4	5	5	6	-3	2	3,7
5L		-4	-4	5	-4	-4	2	3	4	5	-6	4,1
6P	Rear rail	2	2	3	1	-2	0	3	2	1	-2	1,8
6L		1	2	2	0	1	-1	-2	11	3	3	2,6
McPh	Top mounting of a MacPherson's strut	-3	-3	-7	-3	-5	-12	-6	-13	-9	-16	7,7
Average change in the geometry in a car		2,77	2,85	3,00	3,23	3,31	3,85	4,08	5,08	5,31	8,46	

In order to determine the influence of the type and location of a base point on changes in the geometry of the floor pan and top mounting of MacPherson's struts, analysis of variance was applied (fig. 5).

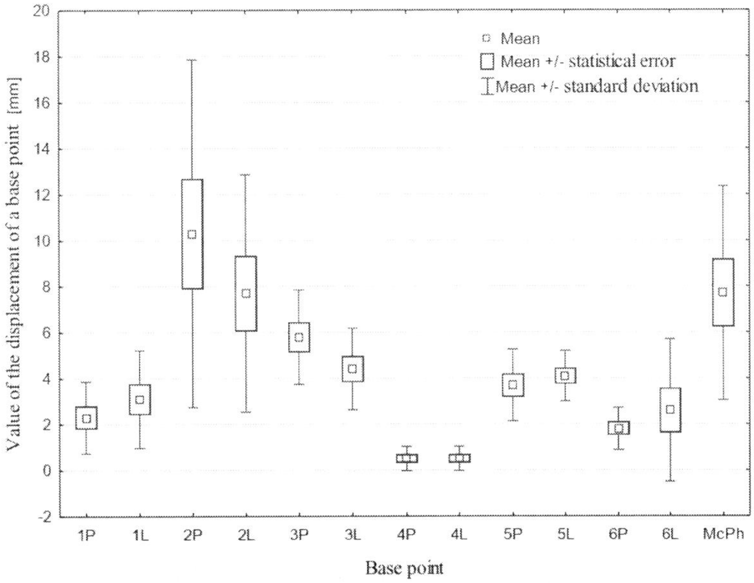

Fig. 5. Comparsion of changes in the geometry at particular base points on the floor pan and within the motor chamber.

A null hypothesis was formulated for each of the analysed points, suggesting the lack of differences between the values of changes in the geometry, as opposite to an alternative hypothesis stating significant differences in the geometry changes. Should the null hypothesis be rejected in favour of the alternative one, the Duncan test was used to distinguish homogeneous groups (Figure 6).

No of subclass	Duncan's test: alpha = 0,05000						
	No of base point	Means	1	2	3	4	5
8	4L	0,5	****				
7	4P	0,5	****				
11	6P	1,8	****	****			
1	1P	2,3	****	****			
12	6L	2,6	****	****	****		
2	1L	3,1	****	****	****		
9	5P	3,7	****	****	****		
10	5L	4,1		****	****		
6	3L	4,4		****	****		
5	3P	5,8			****	****	
13	McPh	7,7				****	****
4	2L	7,7				****	****
3	2P	10,3					****

where **** - illustrates statistically significant differences at $p<0.05$ between changes in

Fig. 6. The Duncan's test statistics regarding detailed differences between mean values of changes in the geometry of the car body.

The data presented above justify the conclusion that there are significant differences between individual points for which measurements were taken. Particularly large displacements were noted for the mounting points of the front suspension, which are the upper fixing of the MacPherson's struts. Big geometry changes were also visible for the fixing points of the front bench and the rear suspension.

5 Results of measurements of the geometry of wheels

Table 2 presents the values of changes in the wheel geometry parameters. These changes were related to all parameters describing the geometry of the suspension and the steering systems. For many of these parameters, the measured values significantly exceeded the values of an acceptable change. In the cars tested, the only wheel alignment parameter that can be adjusted is the front axle toe. The other parameters cannot be regulated. The biggest changes, far above the acceptable ones, appeared in the steering axis inclination angle (fig. 6), for both the right and the left wheel on the front axle. Comparably large changes were also found in the geometry of the rear axle wheels. In this case, big changes were detected in many cars concerning both the camber and the toe values for each of these wheels.

Tab. 2. Values of changes in the parameters of the geometry of wheels in the analysed cars.

No of car	Value of an acceptable change of a parameter +/-[°]	S1	S2	S3	S4	S5	S6	S7	S8	S9	S10	Mean value of the change of a given parameter [°]
Parametr		Value of the change of a given parameter in a given car [°]										
1 - Total toe	0.16	-0,51	-0,30	-0,04	+0,28	-0,09	-0,35	-0,13	-0,35	-0,03	-0,02	0.21
2 - Left wheel toe	0.08	-0,79	-0,51	+0,19	+0,15	-0,09	+0,19	+0,04	+0,11	+0,09	+0,16	0.232
3 - Right wheel toe	0.08	+0,28	+0,21	-0,23	+0,13	0	-0,54	-0,17	-0,46	-0,12	-0,18	0.232
4 - Left wheel camber	0.5	-0,29	-0,2	+0,22	-0,54	-0,56	-0,24	-0,01	+0,24	+0,48	+0,19	0.297
5 - Right wheel camber	0.5	+0,47	+1,54	+0,29	+0,16	-0,69	-0,22	-0,41	+0,32	-1,12	+0,06	0.528
6 - Castor left wheel	0.5	-0,08	-0,89	+0,68	+0,25	-0,38	+1,03	-0,08	-0,76	+0,31	-1,27	0.573
7 - Castor right wheel	0.5	+0,02	-0,41	+0,57	+0,42	+0,11	+0,84	-0,55	-0,41	-1,35	-1,53	0.621
8 - Steering axis inclination left wheel	0.2	-0,92	-0,06	-0,75	-0,52	-2,66	+1,33	-7,21	-0,29	-8,56	+8,12	3.042
9 - Steering axis inclination right wheel	0.2	-2,06	-0,40	-1,27	-0,74	-3,29	-7,04	-6,34	-7,49	-7,53	-7,21	4.337
10 - Total toe	0.16	-0,09	-0,10	-0,82	-0,87	-0,50	+0,50	+0,11	-1,69	-0,05	+0,08	0.481
11 - Left wheel toe	0.08	+0,15	+0,29	-0,29	-0,11	-0,02	+0,21	-0,18	-0,53	-0,20	+0,14	0.212
12 - Right wheel toe	0.08	-0,25	-0,40	-0,54	-0,77	-0,49	+0,28	+0,28	-1,17	+0,14	-0,07	0.439
13 - Left wheel camber	0.33	+0,05	-0,52	+0,89	-1,72	+0,11	+0,68	-0,65	-0,91	+0,56	+0,13	0.622
14 - Right wheel camber	0.33	+0,62	-0,51	+0,33	-2,00	+0,17	+0,26	-0,57	-0,45	-0,20	+0,19	0.53
15 - Driving axis from symmetry axis deviation	0	-0,20	-0,35	+0,33	-0,33	-0,24	+0,04	+0,21	-0,32	+0,17	-0,10	0.229
16 - Wheelbase difference	0	+0,04	-0,06	-0,14	0	-0,09	-0,26	-0,29	-0,37	-0,18	-0,12	0.155
17 - Track width difference	0	-0,73	-1,66	-1,12	-0,06	-0,33	-0,38	-0,74	+0,01	-1,07	-0,59	0.669
18 - Wheel setback (front axle)	0	-0,06	+0,05	+0,19	+0,04	+0,08	+0,24	+0,17	+0,29	+0,12	+0,25	0.149
19 - Wheel setback (rear axle)	0	-0,02	-0,01	-0,10	+0,04	0	-0,02	-0,11	-0,08	-0,05	+0,13	0.056
Average change in wheel geometry		0,40	0,45	0,47	0,48	0,52	0,77	0,96	0,86	1,18	1,08	–

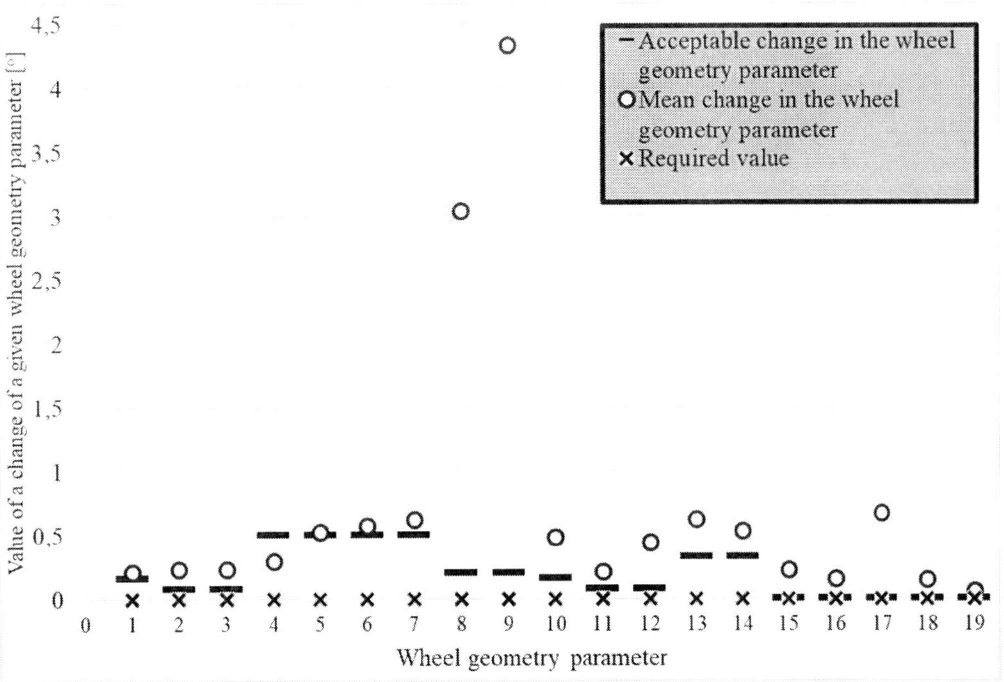

Fig. 6. Specification of changes in the wheel geometry parameters.

In order to determine the influence of the type of a wheel geometry parameter on changes in the geometry of the suspension system and steering system, analysis of variance was applied (fig. 7). A null hypothesis was adopted For each of the analysed points claiming that there was a lack of differences between the values of changes in geometry changes, and an alternative hypothesis suggesting the occurrence of significant differences in changes in geometry. It was found that the values of changes determined for the analysed parameters exceeded the thresholds. The biggest changes appeared for the parameters: the front axle left (8) and right (9) wheel camber.

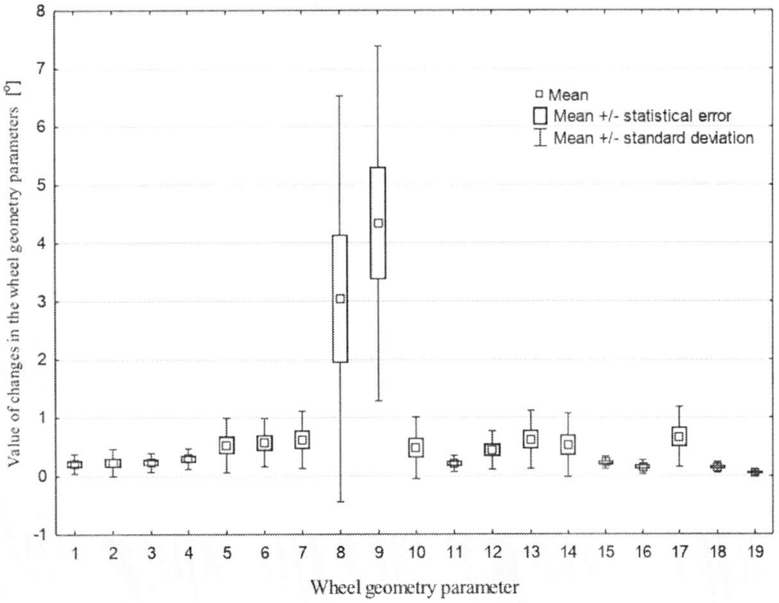

Fig. 7. Diagram of changes in the geometry for particular wheel geometry parameters.

Figure 8 illustrates the impact of changes in the car body geometry on changes in the wheel alignment. It appears that as changes in body geometry grow larger, changes in the wheel geometry increase as well, compared to standard values. In one case, a collapse of the geometry curve (fig. 10) was associated with the direction of the steering axis inclination.

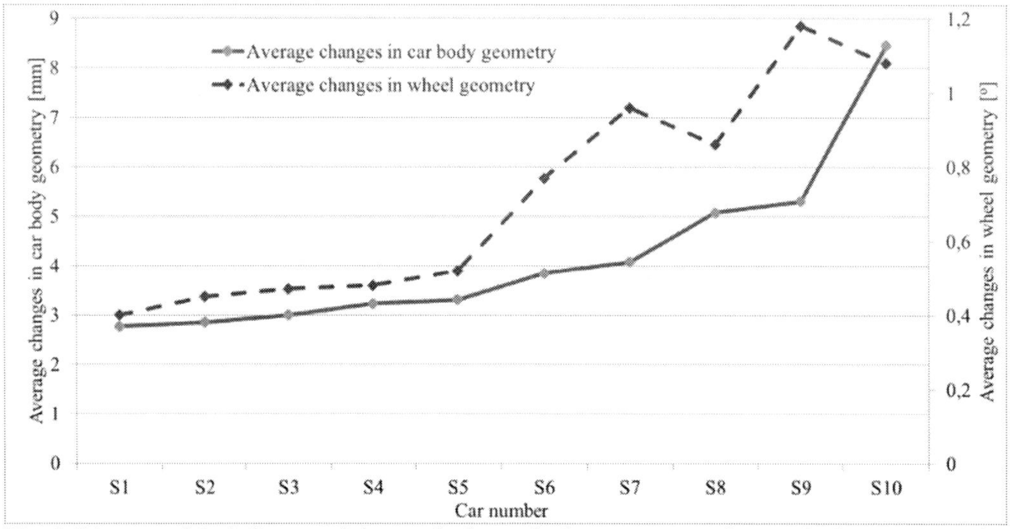

Fig. 8. Diagram of the influence of changes in the car body geometry on changes in the geometry of wheels.

6 Conclusions

The tested passenger cars, each with a mileage above 250,000 km, significant changes were detected in the geometry of both the body and the suspension and steering systems. In either case, the changes demonstrated in most of the automobiles significantly exceeded the threshold values. It was found that larger changes in the body geometry tended to coincide with ap greater the changes in wheel alignment.

The biggest changes were determined for points mounting the front suspension (over 10 mm) and representing the top mounting of the MacPherson's struts (around 8 mm). Smaller changes appeared for the points mounting the front bench (ca 6 mm) and rear suspension (over 4 mm).

In turn, the biggest changes in the geometry of wheels appeared with respect to the steering axis inclination angle. A large change within this parameter was a consequence of the fact that it was under the influence of a large number of base points, i.e. points mounting the front bench, points being the top fixing of the MacPherson's strut as well as points mounting the front suspension. A shift of these base points also caused changes in the front wheel camber values. In some cars, changes in this parameter were thrice above the acceptable value. On the other hand, changes in castor were induced mainly by a change in the position of points mounting the front bench and the front suspension. In the case of this parameter as well, acceptable values were exceeded by as much as threefold. A change in the geometry of rear axle wheels was mostly caused by a shift of points fixing the torsion beam of the rear suspension. These points were displaced by different values on the left and on the right side of cars (a 10% difference), which means that the beam was no longer set parallel to the transverse symmetry axis of the vehicle. In addition, displacement of individual base points fixing the front bench, front suspension and rear suspension by other values on the right and left side of the vehicle (difference by 30%) caused differences in wheelbase between the left and right sides of the vehicle as well as differences in wheelbase between the front and rear axles (by as much as over 1o), and a shift of wheels of the front axle and a shift of wheels of the rear axle.

Bibliography

1. J. J. Aguilar, M. Sanz, D. Guillomía, M. Lope, I. Bueno, International J. Adv. Man. T. **30**, 1174 (2006)
2. R. Anderson, *Division of Production and Materials Engineering* (2007)
3. M. Berlemann, A. Matthes, Journal of Policy Modeling **36**, 313 (2014)
4. H. Chen, X. Gong, Y. F. Hu, Q. F. Liu, B. Z. Gao, H. Y. Guo, Acta Autom. Sinica **39** (2013)
5. J. E. Duffy, *Auto Body Repair Technology; Canada by Nelsin Education* (2014)
6. J. Gonera, J. Napiórkowski, A. Olszewski, M. Janulin, Studies & Proc. of Pol. As. of Kn. **79**, 34 (2016)
7. A. Habibovic, J. Davidsson, Accident Analysis and Prevention **49**, 493 (2012)
8. H. Huang, C. Li, Q. Zeng, Accident Analysis and Prevention **86**, 129 (2016)
9. H. S. Kim, H. J. Kim, B. Bongsoo Son, Accident Analysis and Prevention **38**, 981 (2006)
10. L. Laflamme, M. Vaez, M. Hasselberg, A. Kullgren, Safety Science **43**, 1 (2005)
11. W. A. Livesey, A. Robinson, *The repair of vehicle bodies* (2006)
12. R. Michalski, J. Gonera, Studia i Materiały Pol. Stow. Zarz. Wied. **47**, 169 (2011)
13. B. Raatz, *Nowoczesne technologie pomiarów i napraw karoserii pow.* (2005)
14. A. Tobota, *Naprawy powypadkowe nadwozi, a bezpieczeństwo* (1997)
15. Dziennik Ustaw RP Poz. 407 z dnia 25 marca 2013 r.
16. Dziennik Ustaw RP Poz. 776 z dnia 21 kwietnia 2015 r.
17. Dziennik Ustaw RP Poz. 1475 z dnia 24 sierpnia 2015 r.
18. Rozporządzenie Komisji (UE) nr 214/2014 z dnia 25 lutego 2014 r.

Statistical assessment of injection aparatus of diesel

Wiktor Kupraszewicz[1,*]

[1]Stanisława Staszic State Higher School of Vocational Education in 64-920 Pila, Poland

Abstract. The issue of estimating technological condition of internal combustion engine gathers special meaning mostly in using stage, in consideration of economic and ecological conditions. Appliances which are used now a days and which diagnose condition of subjects such as compression - ignition engines, are limited only to the measurements. This article discusses problems of methods diagnosis with special regard of fuel apparatus.

1 Introduction

The condition of injection apparatus of the ICE primarily determines its correct operation. Based on many studies of Diesel engines, it was concluded that the damage initiated by maladjustment starts first in the fuel supply system, especially in the high pressure systems of the injection apparatus [8, 9, 10]. Every case of failure of the injection pump or the injector causes engine revolution drop which leads to the faster use and serious failures. Therefore, even more important in diagnosing Diesel engines are installation-free methods consisting in the early identification of inoperable elements of injection apparatus.

The low repeatability of fuel pressure parameters in the injection conduits (Fig. 1) being a cycle of complex, independent wave phenomena, seriously complicates the diagnosing with comparative method and requires high perception from the operator which often exceeds his capabilities [3, 8, 9, 10].

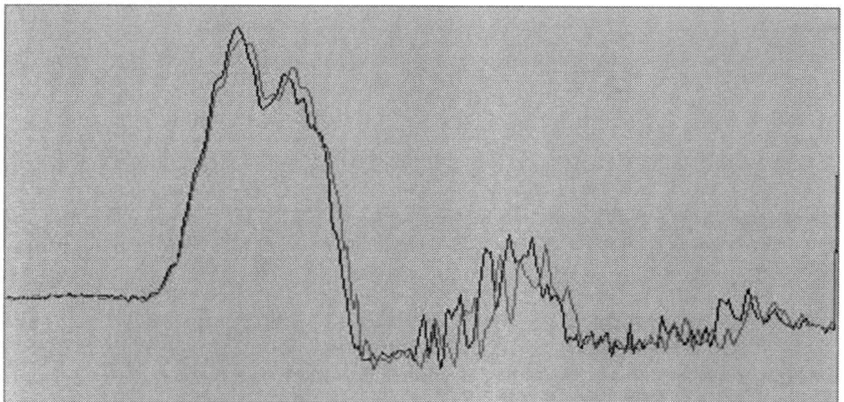

Fig.1. Two fuel pressure oscillograms in injection conduit measured in different time intervals for working apparatus.

In order to identify maladjustment and failure of engine systems in the operation of UTD-20 engines (high repair cost) as early as possible, an attempt was made to fit the existing diagnostic tool (Diesel engine tester), developed at the Department of Working Machines and Vehicles, Technical and Agricultural University in Bydgoszcz, into the operator support sys-tem – advisory system.

*Corresponding author: wkupraszewicz@pwsz.pila.pl

2 Concept of advisory system

The advisory system [2, 5, 7, 11] to support Diesel engine tester's operator was provided based on the skeleton system. The skeleton system "Exsys Developer v8.0" by MultiLogic operating under Windows 9x was chosen to develop the advisory system prototype. The system has built-in mechanisms to create knowledge databases which is represented as rules.

The architecture of the system (Fig. 2) includes the following basic elements:

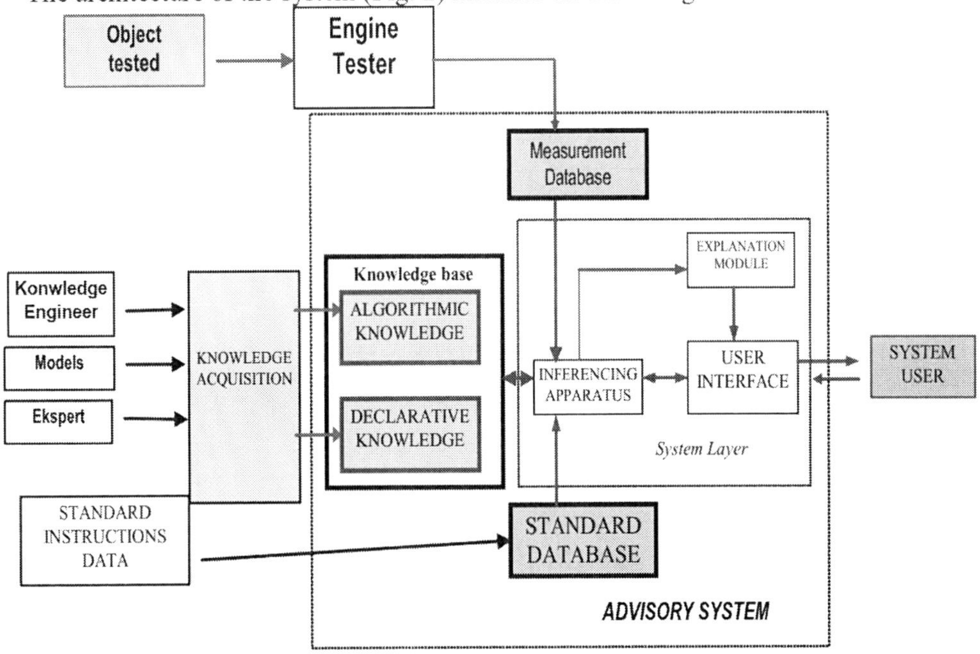

Fig. 2. Architecture of advisory system supporting operator's decisions based on tester measurement [5]

- Knowledge database – its source is knowledge from experts, operation instructions and UTD-20 engine tests – recorded as inferencing rules;
- Database – set of operation characteristics such as: fuel injection angle, crank-shaft rotary rate on idle run, electric voltage and accumulator amperage during start-up, model fuel pressure oscillograms in injection conduit and the data from operation tests and instructions being fixed data and measured data entered by the operator during engine test being variable data;
- Inferencing procedures – defined as dialogue control algorithm, saved in the sys-tem and updated by the knowledge engineer;
- Explanations – inform on the inferencing strategy, allow to explain during knowledge database updating why a given solution is chosen;
- Dialogue control procedures – input / output procedures to formulate questions and give answers by the operator, and provide the solution as a report;
- Modules to extend and modify database.

Given the complex nature of the phenomena accompanying the fuel pumping and injection, the neuronal computer vision classifier was abandoned. Few tests conducted within identification of injection apparatus with the use of artificial neuronal network (despite more frequent uses in the technical diagnostics) confirm certain limitations in their use in this case. These limitations refer to, among others, the type of network, its computer form and calculation capacities.

Since the skeleton used for design has no tool in the form of, for example, artificial neuronal network, which can be used to identify the pressure oscillograms (most often identification of the injection apparatus is done by the tester operator), the following oscillogram solution was proposed (Fig. 3) consisting in assigning the oscillogram monitoring to the specific class.

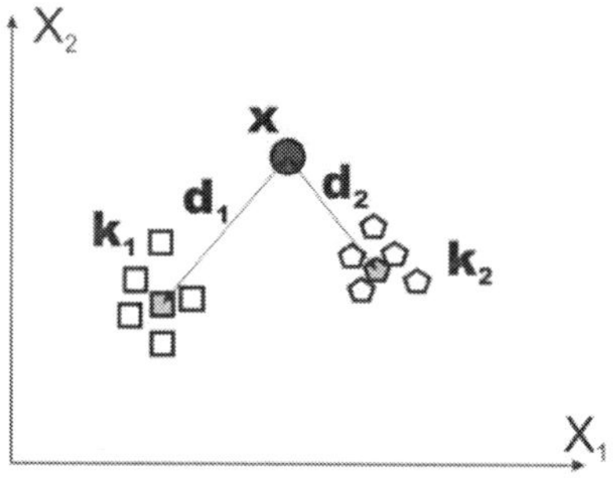

Fig. 3. Classification of oscillogram observations according to nearest neighbour [1].

The system user saves the pressure oscillograms in text files compatible with ASCII. A file is imported into MS Excel, where a recorded value is classified into one of the classes: class 1 – engine injection apparatus is good; class 2 – injection apparatus not working, injector spring damage; class 3 – injection apparatus not working, leaking injector nozzle. Based on the identified and classified recorded value into class (number 1,2, ... m) and placing it into the advisory system, a finding is generated as a result of identification of the injection apparatus.

3 Statistical decision-making criterion

Statistical decision-making criterion is based on Bayes recognition algorithm [1,5]. Let |P, where P, the dimension of characteristics vector is the space of observation. Let M be the space of classes and (to simplify) the decision space. For the given (or estimated) probabilities, observation (image) x from given class m appears, that is P1, P2, ... PM, and for the given estimated distributions of density M, conditional probabilities Q(x|1), Q(x|2), ... Q(x|M) (Fig. 4.), the statistical recognition consists in assigning to a random image x = = $(x_1, x_2, ..., x_p ..., x_M) \in \chi^P$; the decision on affinity to one of the classes.

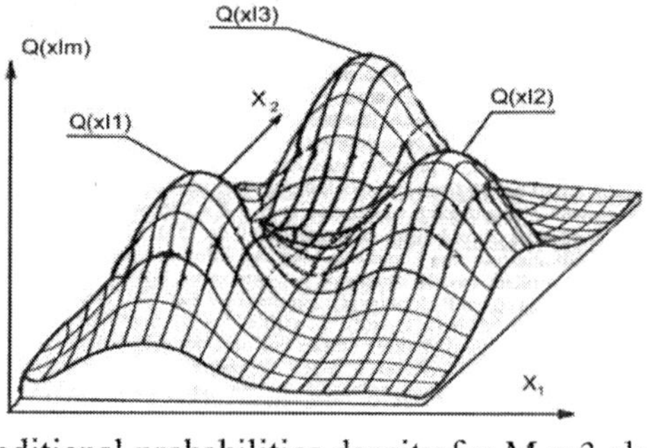

Fig.4. Distribution of conditional probabilities density for M = 3 classes.

Function Ψ(x) for which we call image x ∈ m as decision rule.
Introducing the concept of medium risk R(Ψ):

$$R(\Psi) = R_m(x) = \sum_{l=1}^{M} c_{m,l} P\,(l/x), \quad (1)$$

where: cm,l – element of loss matrix which defines the loss value resulting from classifying an observation (image) from class 1 as belonging to class m.
Since the probability P(l| x) with which we recognize the image, represents class m, is related to the density of conditional probabilities, Bayes dependence:

$$P(m|\, x) = \frac{Q(x/m)P_m}{\sum_{l=1}^{M} Q(x/l)P_l} \;;\; m = 1, 2, \dots, M \tag{2}$$

And considering the fact that for a given image x the denominator of dependence on P(x| m) is fixed, the dependence for medium risk can be brought down to the form:

$$R_m(x) = \sum_{l=1}^{M} c_{m,l} Q(x/l) P_l \,. \tag{3}$$

Therefore, Bayes recognition rule $\Psi(x)$ has the following form:

$$\Psi(x) = m \quad \text{i.e.. } x \in m \tag{4}$$

if:

$$R_m(x) < R_1(x);\ \text{for} \bigwedge_{k \neq l};\ k = 1, 2, \dots, M. \tag{5}$$

The situation where we know the values of all distributions is ideal and virtually does not occur. Many recognition systems are based on the simple idea to assign an observation to the nearest class. The non-parametric recognition algorithms has specific relations between two observations (images) or between an observation and its standard of a given class, known as similarity functions. They are the measure to provide the decision rule. The similarity functions may occur as a distance function (known as distance in short) or a proximity function. The most common algorithms using this classification method include:
Nearest neighbor method (used as a classifier in the form of macro);
K-nearest neighbors method (a modification of the first one)
NN algorithm (Nearest Neighbor) acts based on assigning a new observation x to the already known learning sequence classes (Fig.5.)

																					Przypisane		
11	19	11	0	5	13	7	1	-5	-10	-11	-15	-21	-21	-12	-5	-6	-20	-20	3	12	1		
13	9	8	4	9	15	9	-1	-1	2	-3	-15	-25	-30	-45	-54	-41	-32	-30	-22	-8	1		
10	19	11	-1	5	12	6	2	-4	-10	-11	-15	-21	-21	-12	-5	-6	-20	-20	3	12	1		
12	15	6	5	8	14	10	-1	-3	2	-4	-15	-25	-34	-46	-54	-41	-32	-30	-22	-8	1		
-23	-12	-7	-6	-7	-7	-8	-11	-8	-4	-3	-6	-13	-15	-14	-14	-16	-14	-7	-15	-26	2		
-13	-6	0	-4	-8	-8	-7	-10	-12	-12	-9	-8	-23	-20	-12	-17	-17	-10	-11	-16	-6	2		
-13	-6	0	-4	-8	-8	-7	-10	-12	-12	-9	-8	-23	-20	-12	-17	-17	-10	-11	-16	-6	2		
Wektory średnich w poszczególnych klasach																							
Klasa 1																					Klasa 2		
12	14	8,3	2,7	7,3	14	8,33	0	-2,67	-2	-6	-15	-24	-28	-34	-38	-29	-28	-27	-14	-1	-16	-8	-
Macierze kowariancji w poszczególnych klasach																							
Klasa 1																					Klasa 2		
1	0	0	0	0	0	0	0	0	0	0	0	0	0	0	0	0	0	0	0	0	1	0	
0	1	0	0	0	0	0	0	0	0	0	0	0	0	0	0	0	0	0	0	0	0	1	

Fig.5. Learning set composed of three model observations and recognized new observation (classified as class 1)

In the process of proper recognition, a new observation (image) appears whose assignment to a specific class is unknown $\{x\} \in \chi P$. The procedure of NN algorithm performs calculation of the similarity function (usually the distance) between all observations (images) of the learning sequence and the unknown (new) observation x:

$$FP(x, W_{m,i}) \text{ for } m = 1, 2, \dots, M,\ i = 1, 2, \dots, I_m \tag{6}$$

where Wm,i = x(m,i).

Upon calculation of all similarity functions, the lowest one is looked for (in case of distance) and as a classifier's decision, the name (cod) or class number to which the image of learning sequence belonged is given and which is the nearest in the sense of function of similarity to the recognized image:

$$\Psi(x) = m \text{ i.e. } x \in m \tag{7}$$

if:

$$FP(x, W_{m,l}) < FP(x, W_{l,l}) \tag{8}$$

$$\text{for } \bigwedge_{m \neq l};\ m, l = 1, 2, \dots, M. \tag{9}$$

The nearest method despite considerable recognition time is characterized by high accuracy of recognition of 87.4 % [3, 6, 7, 10]. Considering such criteria as: type of image (400 vectors Fig.1.), the calculation capacity of the equipment and available programs, recognition time and accuracy, a macro has been developed using the built-in Visual Basic module for MS Excel.

4 Image recognition algorithm

In the proposed solution as a similarity function, from many used in minimum-distance algorithms, the Euclidean distance was applied:

$$d_{ij} = \sqrt{\sum_{k=1}^{p} \left(x_{ik} - x_{jk}\right)^2} \tag{10}$$

Recognition algorithm and assignment of a new observation to a class is shown in Fig. 6. Preparation of the learning set was done during tests of engines in the test room and consisted in recording oscillograms for engines after the main repair with a working fuel ap-paratus. Thus, a data matrix X_1 belonging to class 1 was created. Next, by introducing further maladjustments and damage, recording of oscillograms for non-working apparatus was con-ducted, creating matrix X_2, X_3 , etc. which corresponded to specific damage responsible for such courses. The whole learning set, based on which new observations were qualified, in its simplest version, has four classes: C1 – working injection apparatus, C_2 – improper adjustment of injector spring, C_3 – leaky atomizer nozzle, C_4 – other damage (Fig. 5 shows classification for two classes, fitness and unfitness for use of the injection apparatus).

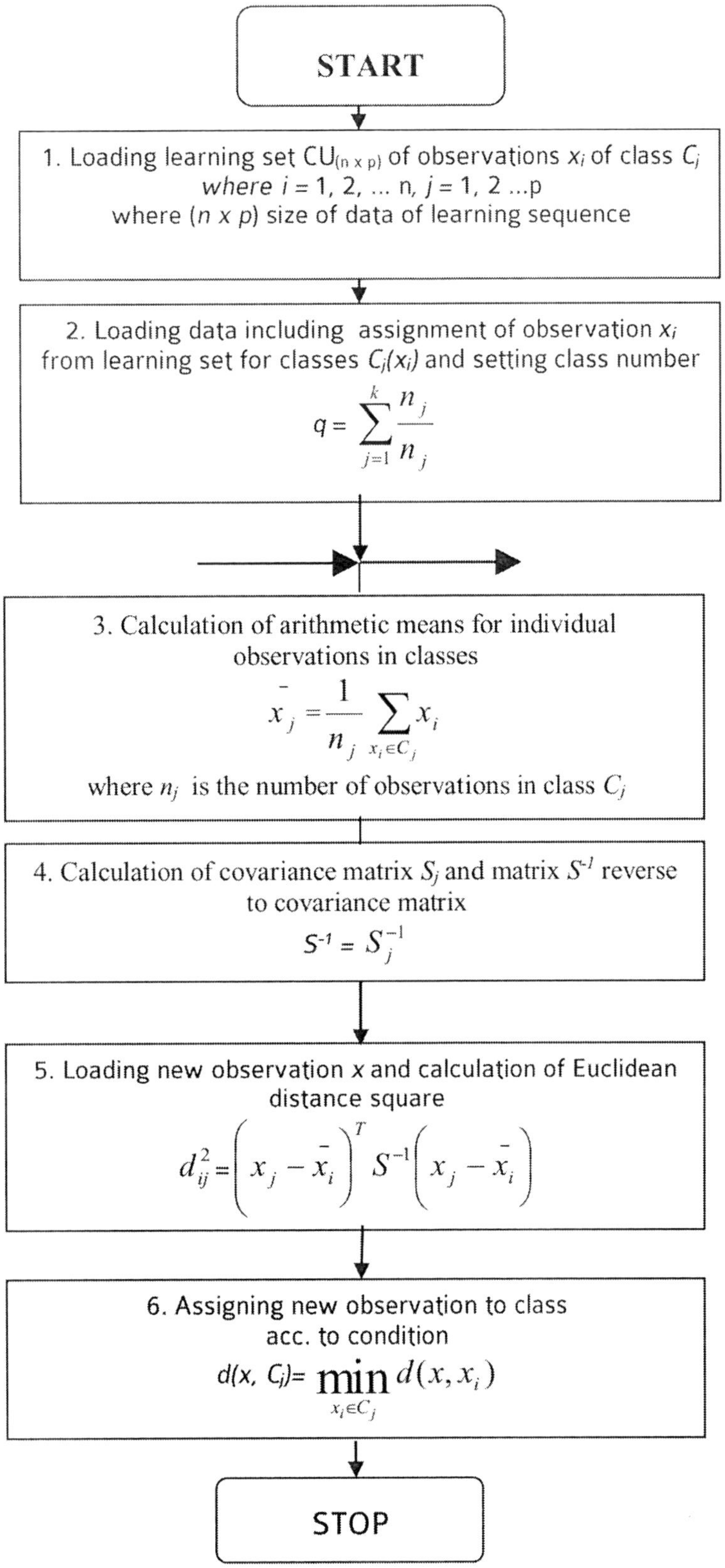

Fig.6. Algorithm of assigning new observation to class from learning set [5]

5 Summary

Only in few cases, the designers of expert systems created on skeleton systems are able to limit themselves to the tools proposed by the given skeleton system. When the created advisory system must be able to solve specialist issues with high complexity, it is necessary to refer to other specialist solutions.

The rather well known and discussed method for observation of fuel pressure oscillograms in the injection conduit is limited during the engine test to classification of the technical condition based on operator's knowledge and experience.

The proposed solution for recognition of the injection apparatus by recognizing the condition of fuel injection high pressure systems and equipping the advisory system with them, will allow to eliminate often subjective opinions on the technical condition of engines fitted with injection apparatus.

References

1. Basztura Cz.: Jak rozmawiać z komputerem ludzkim głosem?, WPN Format, Wrocław (1993)
2. Cholewa W., Pedrycz W.: Systemy doradcze, Skrypt uczelniany Politechniki Ś lą - skiej nr 1447, Gliwice (1987)
3. Jankowski M.: Ocena wraż liwoś ci diagnostycznej sygnału pulsacji ciś nienia aparatury wtryskowej. Praca doktorska. Bydgoszcz, (1997)
4. Kramarenko G.; Techniczna eksploatacja samochodów. WKŁ. Warszawa, (1989)
5. Kupraszewicz W.; Ż ółtowski B.: Wykorzystanie układu szkieletowego w diagnostyce silnika. Materiały konferencyjne AIRDIAG (1999)
6. Ostasiewicz S.; Statystyczne metody analizy danych; WAE Wrocław, (1999)
7. Tadeusiewicz R., Wszołek W., Izworski A.: Zastosowanie sieci neuronowej oraz rozpoznawania obrazów w diagnostyce mowy zdeformowanej, Materiały konferencyjne, Mię dzynarodowy Kon-gres Diagnostyki Technicznej KDT'96, Gdań sk (1996)
8. Tylicki H.: Diagnozowanie układu wtrysku paliwa silników spalinowych, Materiały konferencyjne, Diagnostyka'99, KMR i P, ATR Bydgoszcz (1999)
9. Ż ółtowski B,: Podstawy diagnostyki maszyn. ATR Bydgoszcz (1996)
10. Ż ółtowski B,: Diagnostyka silnika wysokoprę ż nego. Instytut Technologii Eksploatacji, Radom, (1995)
11. Ż ółtowski B, Ć wik Z,: Leksykon Diagnostyki Technicznej. ATR; Bydgoszcz, (1996)

The influence of the applied window function and sampling frequency on the assessment accuracy of the turbine engine in flight control condition based on its ground tests

Karol Golak[1,*], *Paweł* Lindstedt[2]

[1]Faculty of Mechanical Engineering, Bialystok University of Technology, Wiejska 45C street, 15-351 Białystok, Poland
[2]Air Force Institute of Technology, Księcia Bolesława 6 street, 01-494 Warszawa, Poland

Abstract. The article presents the basis of the method allowing to assess the state of control condition of a turbine jet engine during the plane flight when the input signal to the engine is unpredictable interference signal „*z*". Method is based only on signals acquired during engine ground test when the input signal is „*w*" delivered by technical support using the engine control lever (DSS). The control state of the engine both from the set value „*w*" and „*z*" is expressed by the quotient of its cross power spectral densities (in frequency domain). The way of transition from the frequency domain to the time domain was presented, it allows to obtain step responses of engine from „*z*" and „*w*", greatly facilitating the interpretation of the obtained results.. The analysis of the influence of choosing types of the window function and signals sampling frequency on the accuracy of the proposed model has been carried out.

1 Introduction

The aviation engine is used during the flight when, apart from the „*w*" signal set by pilot using the engine control lever (DSS), it is also affected by unpredictable interference signal „*z*" delivered from the environment. Engine is adjusted on the basis of signals recorded during its ground tests, when the input signal is only the signal „*w*" from the DSS, and the signal „*z*" has low power and are constant so it is often overlooked. This may lead to a situation where the aircraft properly adjusted during ground tests from the „*w*" signal may react inappropriately to the „*z*" interference signals occurring during the flight. It may lead to an undesirable change in the rotational speed (so-called vibrations and fluctuations of rotation), and even its stalling. It is also known that improving the quality of the engine from „*w*" (on the ground) causes deterioration of the quality of the engine from „*z*" (in flight) [1, 2, 3, 5, 8].

2 Theoretical basis

The simplified diagram of the aircraft jet engine rotational speed automatic control system jet engine can be reduced to a standard scheme of the automatic control system (Fig. 1.). To derive a model describing the assessment of the control quality of a jet engine with input „*z*" (during a flight) using only signals measured with input „*w*" during its standard ground tests, it is necessary to identify the transfer functions of individual elements of the automatic control system.

* Corresponding author: k.golak@pb.edu.pl

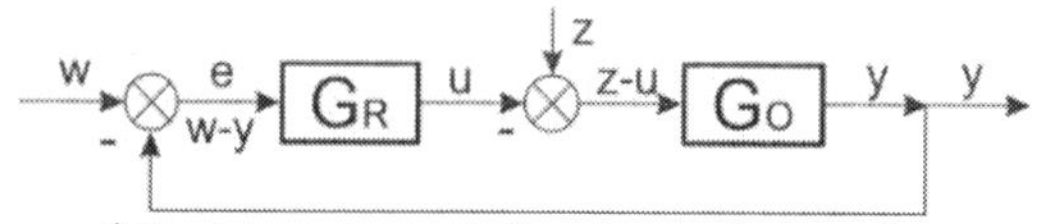

Fig. 1. Simplified diagram of the jet engine rotational speed control system, where: G_R - controller transfer function, Go - object transfer function, w - input signal (setpoint using the engine control lever), z - input signal (interference), y - output signal (rotational speed), $e=w-y$ - input signal to the controller, u- output signal from the controller (control signal). [3, 4]

Using the diagram from Fig. 1. and presenting the object and controller transfer functions in the form with cross power spectral density (S_{xy}) it is possible to determine the following models: H_W (when $z = 0$) describing the operation of the engine during the ground test and H_Z (when $w = 0$) describing the operation of the engine during the flight. It takes the form: [4, 7]

$$H_W = \frac{S_{y_w u} S_{ue}}{S_{y_w u} S_{ue} + S_{uu} S_{ee}} \tag{1}$$

$$H_Z = H_W \cdot \frac{1}{G_R} = \frac{S_{y_w u} S_{ee}}{S_{y_w u} S_{ue} + S_{uu} S_{ee}} \tag{2}$$

Thus, by using H_W transfer function, the H_Z transfer function can be determined. It is therefore possible to determine the characteristics of the engine with input „z" (on the fly) based only on the signals registered during the engine ground tests when the input signal is „w".

3 Verification of the turbine engine control quality assessment method

The work of a turbine jet engine during ground tests (with the „w" input) and during the flight (with the input interference signal „z") is described by the model shown in Fig. 1. This model for the simulation needs can be presented in the form of two separate models responsible for work from just one of inputs (two SISO models). Fig. 2a shows the model with input „w", whereas Fig. 2b shows the model with input „z".

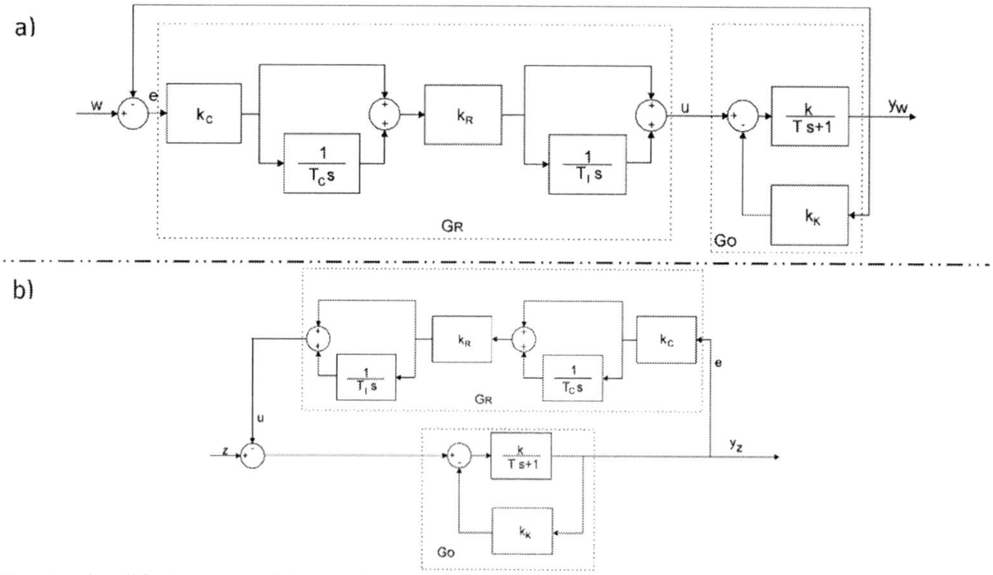

Fig. 2. Simplified model of the engine speed control system, where: a) when input $z = 0$ (during ground tests), b) when input $w = 0$ (during the flight), y_W - system response when the input signal is

"*w*" delivered to the controller, y_Z- system response when input signal is delivered directly on the object, T_i, T, T_c - these are the time constants of the controller, the object and the sensor; k_C, k_R, k, k_K – gain for the sensor, controller, object and object compensator. [2, 5, 6, 9, 10]

For selected parameters: [2, 4, 9, 10, 11, 12]:

$T=1,7$ $k=0,45$, $k_K=0,075$

$k_R=7$ $T_I=2$

$k_C=0,15$ $T_C=0.35$

Transfer functions of the object and controller from Fig. 2. take the form:

$$G_O(s) = \frac{0,45}{1,7s + 1,034} \tag{3}$$

$$G_R(s) = \frac{0,3675s^2 + 1,234s + 0,525}{0,35s^2} \tag{4}$$

Thus, transfer functions of the system from „*w*" and from „*z*" take the form:

$$H_W(s) = \frac{0,3307s^2 + 1,11s + 0,4725}{1,19s^3 + 1,054s^2 + 1,11s + 0,4725} \tag{5}$$

$$H_W(s) \cdot \frac{1}{G_R(s)} = \frac{0,315s^2}{1,19s^3 + 1,054s^2 + 1,11s + 0,4725} = H_Z(s) \tag{6}$$

Transfer functions $H_W(s)$ and $H_Z(s)$ can be written in the frequency domain by substituting $s=j\omega$. The real parts of these Transfer functions for both the „*w*" and „*z*" inputs take the form:

$$\mathrm{Re}_W(\omega) = \frac{(27*(-47050*\omega^4 + 27983*\omega^2 + 10800))}{(1849600*\omega^6 - 1999655*\omega^4 + 308961*\omega^2 + 291600)} \tag{7}$$

$$\mathrm{Re}_Z(\omega) = \frac{(1800*\omega^2*(241*\omega^2 - 108))}{(1849600*\omega^6 - 1999655*\omega^4 + 308961*\omega^2 + 291600)} \tag{8}$$

In order to prove the correctness of the transformation from (Eq 1. and 2) a simulation of the both models from Fig. 2 (described also by $H_W(s)$ (Eq. 5) and $H_Z(s)$ Eq. 6) was carried out for step and Dirac delta function as input arguments.

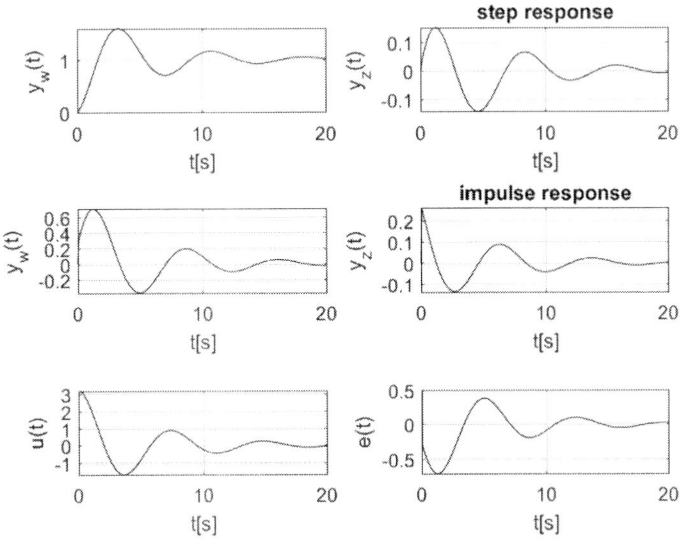

Fig. 3. Step and impulse response of the transfer functions $H_W(s)$ and $H_Z(s)$.

Using the signals obtained as results of impulse response from Fig. 3. graphs of real part of transfer functions were made in three variants (Fig. 4.):

a) directly from the For Eq. 7 and Eq. 8

b) $Re_{2W}(\omega) = real(\frac{S_{y_w w}}{S_{ww}})$ and $Re_{1Z}(\omega) = real(\frac{S_{y_Z z}}{S_{zz}})$ - using only the input and output signals for describing the H_W and H_Z models by power spectral density;

c) $Re_{2W}(\omega) = real(\frac{S_{y_w u} S_{ue}}{S_{y_w u} S_{ue} + S_{uu} S_{ee}})$ and $Re_{2Z}(\omega) = real(\frac{S_{y_w u} S_{ee}}{S_{y_w u} S_{ue} + S_{uu} S_{ee}})$ - using only the signals from impulse response of $H_W(s)$ (Eq. 5): $y_w(t)$, $e(t)$ and $u(t)$.

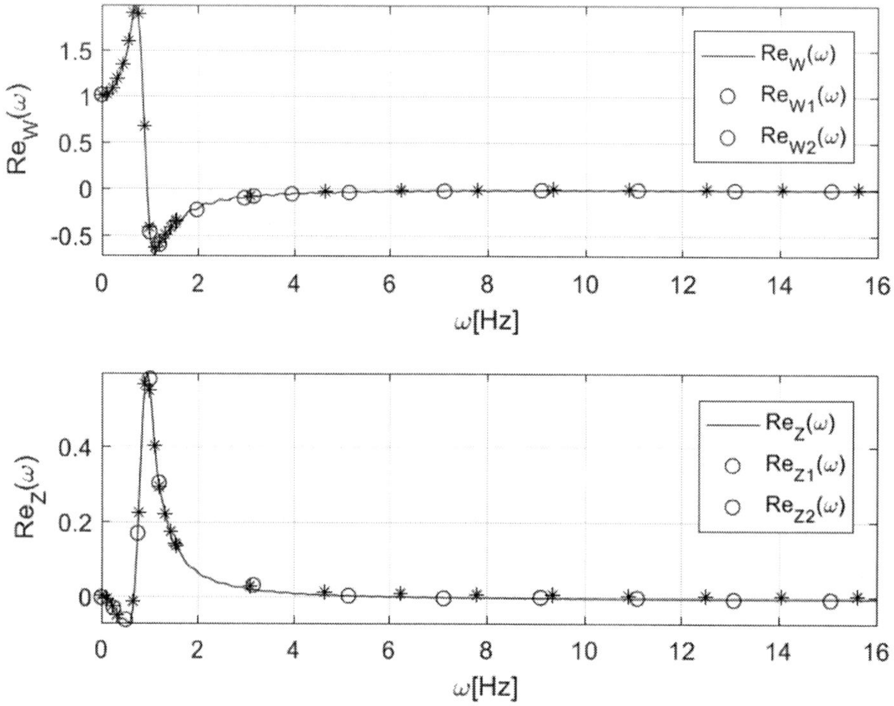

Fig. 4. Comparison of real part of spectral transfer functions obtained from responses on the impulse signal.

In order to transfer from the frequency domain to the time domain, based on the values of the real parts of the spectral transfer functions from impulse response (Fig. 4.) and the transformation from Eq. 9 step responses of the $H_Z(s)$ in time domain were obtained [3].

$$y(t_i) = \sum_{n=1}^{n=\infty} \frac{2}{\pi} \cdot \frac{\sin(\omega_n \cdot t_i)}{\omega_n} \mathrm{Re}(\omega) \cdot \Delta\omega \qquad (9)$$

The obtained results were compared with $y_z(t)$ which is a step response of the $H_Z(s)$ (Eq. 6 - Fig. 3) are presented in Fig. 5.

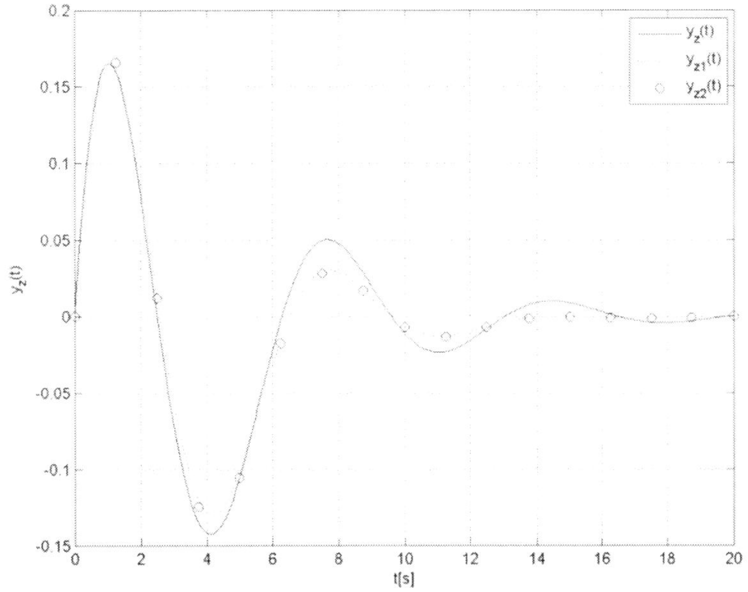

Fig. 5. A comparison of the step responses, where: $y_z(t)$ is from Eq. 6 – Fig. 3., $y_{z1}(t)$ from $Re_{Z1}(\omega)$ and Eq. 9, $y_{z2}(t)$ for $Re_{Z2}(\omega)$ and Eq. 9.

As shown in Fig. 5, all three step response variants are very similar. This proves the correctness of the proposed method for assessing the state of engine regulation in flight (from „*z*") based only on signals registered during the ground test from „*w*".

The accuracy of the obtained results is strongly influenced by two parameters: $\Delta\omega$ - the sampling frequency, and the type of used window function (used for auto and cross correlation necessary to obtain the power spectral density functions S_{yx}). Their impact is shown below.

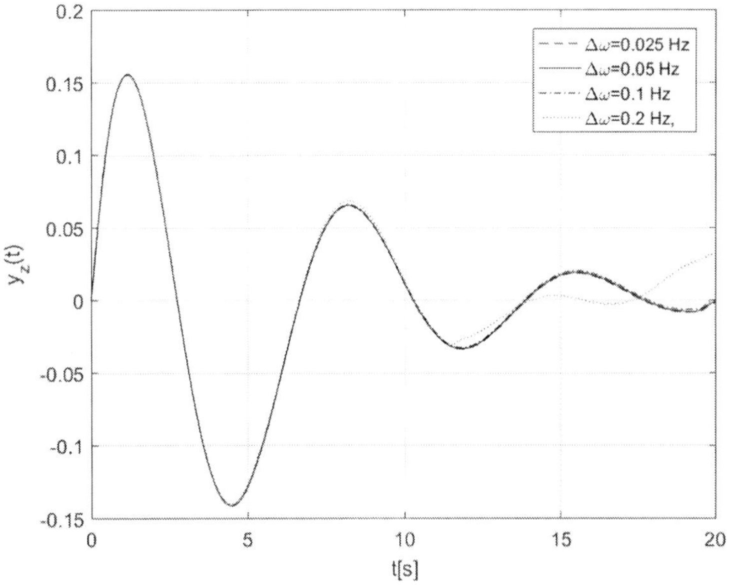

Fig. 6. The impact of the sampling frequency $\Delta\omega$ on the $y_z(t)$ step response.

Fig. 6 shows the influence of the sampling frequency $\Delta\omega$ on the obtained step response. The waveforms for the four sampling frequencies were calculated. The obtained results show a small influence of the sampling rate in the tested range, only $\Delta\omega$ = 0.2 Hz shows clear deviations from the signal waveform.

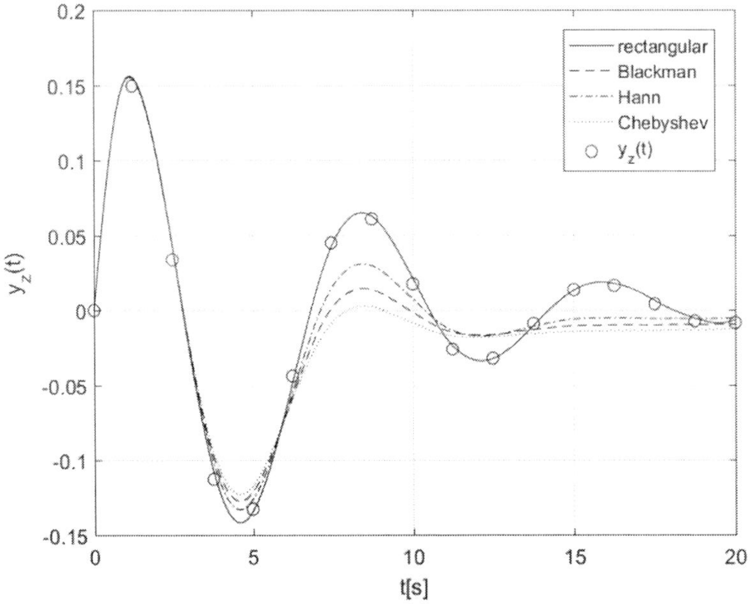

Fig. 7. The impact of the used window function on the $y_z(t)$ step response.

Fig. 7 shows the influence of the window functions used. The calculations were carried out for four selected window functions: rectangular, Hanning, Blackman and Chebyshev and compared to the answer obtained directly from Hz(s) (Eq. 6.). Results show that the best approximation is obtained for a rectangular window. However, for further research it is recommended to use the Hanning window due to the significant noise in signals obtained during ground tests.

Summary

According to the conducted simulation test, as a result of multiplication of $H_W(s)$ (transfer function describing engine during the ground tests) with $1/G_R(s)$ (inverse of the controller transfer function also from ground tests) could be obtained the $H_Z(s)$ (which describes the engine during the flight). This fact has been confirmed in the frequency domain using the power spectral density of signals y_w, u, e - by comparing the real parts of spectral transfer functions(Fig. 4.).

By using the real parts of transfer functions ($Re_{Z1}(\omega)$ and $Re_{Z2}(\omega)$) obtained from the impulse response and using the transition from the frequency domain to the time domain, the step responses of the model could obtained (Fig. 5.). These responses are very similar to the responses obtained directly from the $H_Z(s)$ transmittance (Eq. 6) shown in Fig. 5. Summing up, it is proved that there is a real possibility to assess the engine control quality during the flight based only on signals registered during its ground tests.

The engine characteristics obtained directly from its ground tests will allow the assessment of the impact of engine control changes on its operation. This will allow to propose a more appropriate way to control the engine after the ground test (without the need for flight test), which will reduce the risk associated with the inspection flight carried out after each repair and refurbishment.

This paper was supported by the research project No. S/WM/2/2017 and financed by the Ministry of Science and Higher Education.

References

1. W. Balicki, S. Szczeciń ski, *Diagnostyka lotniczych silników turbinowych. Zastosowanie symulacyjnych modeli silników do optymalizacji zbioru parametrów diagnostycznych.* Biblioteka Naukowa Instytutu Lotnictwa (2001)
2. W. A. Bodner, *Automatyka silników lotniczych*, Wyd. MON (1958).
3. K. Golak, *Parametryczna ocena turbinowego silnika odrzutowego w locie na podstawie jego prób naziemnych* (PhD thesis, 2016)
4. K. Golak, P. Lindstedt, R. Grą dzki, *Studies of the jet engine control quality based on its response to the disturbance inflicted on the object, designated from its response to the set point inflicted to the controller*, In: Safety and Reliability: Methodology and Applications - Proceedings of the European Safety and Reliability Conference, 137- 140 (2014)
5. R. Grądzki, P. Lindstedt, K. Golak, *Premises of evaluation of the technical object suitability with including the quality of its maintenance and operation, and their initial conditions*. In: Safety and Reliability: Methodology and Applications - Proceedings of the European Safety and Reliability Conference, chapter 44; pp. 319 - 326 (2015)
6. K. Kurman, *Teoria regulacji podstawy analiza projektowanie*, WNT, (1975)
7. P. Lindstedt, K. Golak: *Premises of parametrical assessment of turbojet engine in flight regulation condition during ground test*, Acta Mechanica et Automatica, **6**, no. 1 (2012)
8. Lindstedt P.: *Possibilities of assessment of the potential of Aircraft Engines*, Solid State Phenomena, **147-149/2009**, Trans Tech Publications, (2009)
9. M. Orkisz, *Simulation analysis of the influence of variation in some selected design and control parameters on the acceleration time of a turbojet engine*, Engineering Transactions, **39**, 3-4 (1991|)
10. W. I. Pawlak, *Computer simulation of transient processes in a turbojet engine, with special attention to amplitudes of thermal shocks in some selected fault models of operations*, The Archive of Mechanical Engineering, **LIV**, no 3 (2007)
11. S. Szczeciń ski, *Napę d samolotu I-22 IRYDA*, WPTiL **3/1995** (1995)
12. А. А. Шевяков, *Автоматика авиационных и ракетных силовых установок*, Машиностроение (1970)

The impact of a vehicle braking system state on safe driving – part one

Piotr Aleksandrowicz[1,*]

[1]Machine Maintenance Department, Institute of Machinery Operation and Transport, University of Science and Technology, Al. prof. S. Kaliskiego 7, 85-796, Bydgoszcz, Poland

Abstract. The state of a vehicle braking system is one of the key factors determining the safety of the driver, passengers and other road users. During operation of a vehicle, the elements of its braking system undergo the process of wear and they are exposed to damage. Deterioration of the braking system efficiency has a direct impact on the collision speed. Thus, diagnosing a braking system during a vehicle operation is of key importance. Therefore, the authors of this article have undertaken the task to identify the influence of the state of a braking system particular element on the collision speed and kinetic energy of the impact. The obtained results of simulation calculations have proven a significant role of a braking system diagnosing in elimination of vehicles whose braking systems do not function properly as using inefficient brakes poses a significant threat to road traffic users and vehicles.

1 Introduction

Road traffic involves a risk of accidents, collisions and vehicle failures. For example, work [1] deals with the risk of failure and the impact of prevention repairs on the risk reduction. However, elements and systems of a vehicle undergo natural wear affecting the quality of transport processes, which is described in works [2, 3]. An increase in transport volume and people's mobility increases these phenomena. The number of cars in Poland is constantly growing. In 2017, the number of registered motor vehicles was 29 149 178. As many 32 760 accidents were reported to the police, in which 2 831 people were killed and 39 466 were injured. Whereas, 436 469 collisions were reported to the police. No significant improvement for neither of these categories had been reported until 2016 (http://statystyka.policja.pl).

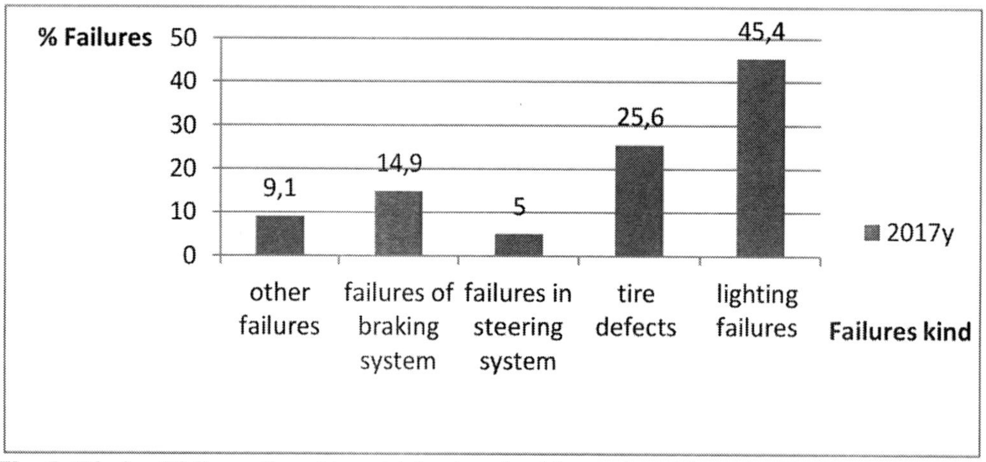

Fig. 1. Kinds of failures found in vehicles involved in road accidents in 2017.

* Corresponding author: paleksandrowicz@utp.edu.pl

The main cause of road accidents is improper behavior of road users, though some accidents are caused by poor technical condition of a vehicle. Failures which were found in post-accident vehicle technical state diagnosing included lighting, tires and braking system. They account for 14.9 % of all the failures found in the diagnosed vehicles.

Below, there are kinds of failures reported for vehicles which were involved in accidents in 2017 (http://statystyka.policja.pl).

The possibility to reduce speed at any time of driving or pull up the vehicle, regardless of the weather conditions and time of operation, is the key factor of road traffic safety. Especially that, dangerous situations involve the necessity to brake immediately. To perform this task, a driver needs to use the vehicle braking system. Thus, a braking system belongs to the most important structural systems of a vehicle and, therefore, it needs to meet several requirements. The most important ones were formulated in regulations R13 or R13-H ECE as well as in the EU directive 98/12/WE [4].

2 Braking system of a vehicle

The process of braking involves changing the vehicle kinetic energy into, most frequently, heat and it is advisable to increase the motion resistance. Once the driver starts pulling up the braking force, acting at the contact of the vehicle wheels with the road surface, reduces its speed. Currently, vehicles are equipped with hydraulic braking systems. In this type of systems, transmission of braking force is performed by means of a braking fluid.

2.1 Design of two-circuit braking system

Fig. 2 shows a scheme of a typical two circuit disc braking system. Such a system includes: braking pump (1), correctors of braking force distribution (2), supporting device (3) brake pedal (4), front wheel brake (5) rear wheel brake (6) (http://zssplus.pl/publikacje/publikacje45.htm).

However, Fig. 3 shows a scheme of a disc brake with two expanders in Dunlop system. This kind of braking system includes: expander (1), brake disc (2), brake pads 3 and 4), body of clamp (5) sealing ring (6) (http://zssplus.pl/publikacje/publikacje45.htm).

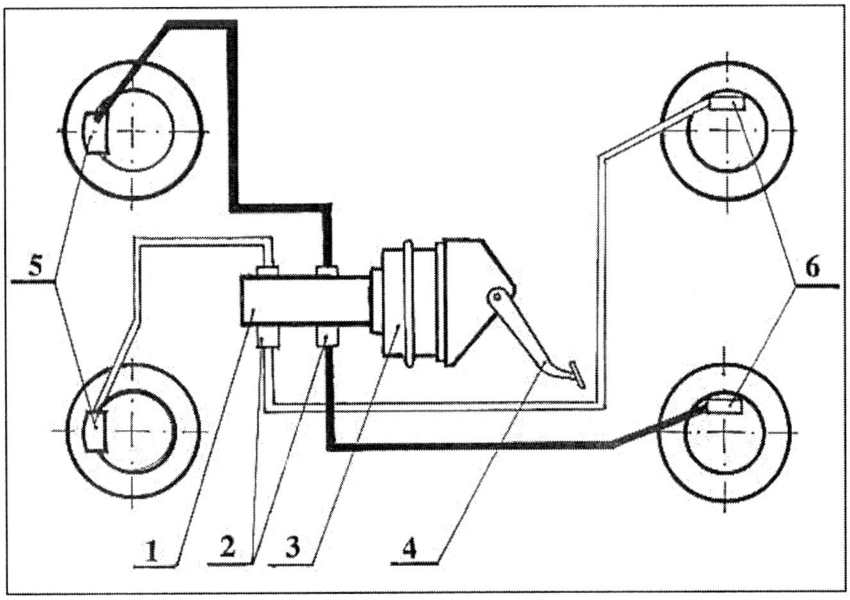

Fig. 2. Scheme of a two two-circuit disc braking system.

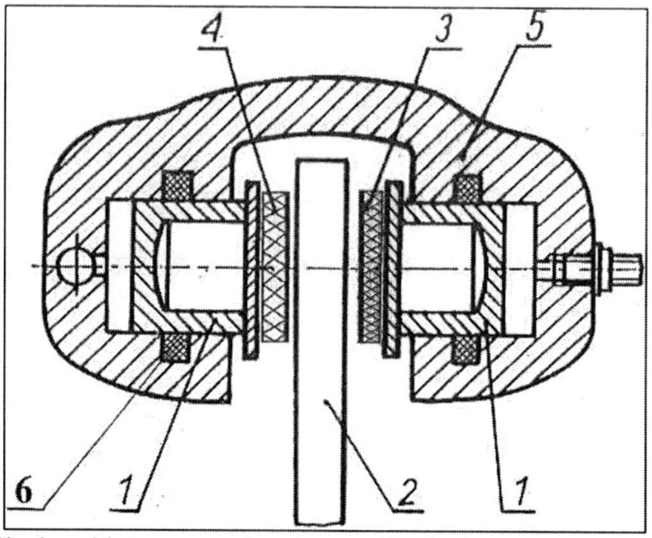

Fig. 3. Scheme of brake with two expanders (system Dunlop).

2.2 Design requirements for braking systems

The major design requirements for braking systems are supposed to provide the possibility of [5]:

- blocking the vehicle wheels upon pressing the brake pedal 490 N, on dry asphalt (surface with adhesion coefficient, at least μ=0.7),
- stroke of the brake pedal from the initial position to wheel blockage, not be higher than 130 mm,
- uniform increase of braking force and its proportionality to the brake pedal pressure,
- uniform distribution of braking force onto the left and the right wheels of the vehicle according to established proportionality of braking force of the front axle wheels,
- possibility of operation with multiple use in short periods of time, regardless of atmospheric conditions,
- durability,
- long periods between necessary regulations.

2.3 Forces acting on the vehicle while braking

In order to establish obtainable braking delay, the force system basic conditions are used, which are presented in Fig. 4.

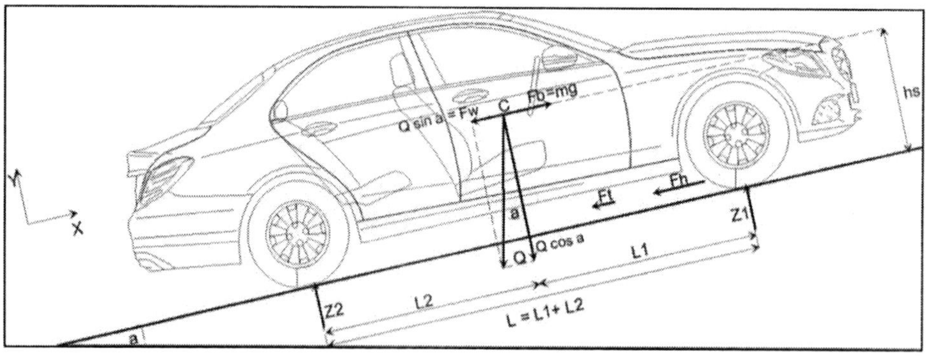

Fig. 4. System of forces acting on the vehicle while braking.

Conditions of forces balance can be expressed by formulas:

$$F_h - F_b + F_t + F_w = 0 \quad (1)$$

$$Z_1 L_1 - F_b h_s + h_s Q\sin\alpha - L_2 Q\cos\alpha = 0 \quad (2)$$

where:
F_b – inertia force,
F_h – braking force,
α – elevation angle,
F_t – rolling resistance force,
F_w – elevation reistance force,
h_s – height of mass center location,
Z_1 and Z_2 – wheel reaction normal forces,
L_1 and L_2 – distance of the mass center from the wheel axle,
$Q = m \cdot g$ – m - vehicle curb weight, g - acceleration of gravity.

3 Parameters the research object and the movement environment

The research object is a passenger vehicle Mercedes class C type W205. The most important technical data of the vehicle and parameters of the movement environment accepted for simulation calculations are presented in table 1.

Table 1. Technical data of the object of the research and parameters of the movement environment.

Parameter	Value
Vehicle curb weight	1 430 kg
Driver mass	68 kg
Length/Width/ Height of vehicle	4.69 m /1.81 m /1.44 m
Wheel base 1-2	2.84 m
Braking efficiency of vehicle	10.2 kN
ABS	Yes
Size of vehicle tires	205/60 R16
Dry asphalt road	Adhesion coefficient: adhesive friction μ_1=0.8 slip friction μ_2=0.75
Rolling resistance coefficient	0.015
Road inclination	α=2 °

4 Case study – own research

Using data from works [4, 5] the friction coefficient of usable brake pads was accepted to be 0.38 for simulation calculations (Fig. 4a), whereas, for totally worn brake pads it was 0.24 (Fig. 4b) which being converted into the braking system efficiency, were, respectively, 100 % and 63 %. For simulation, the vehicle initial speed was 25 m·s^{-1}. Simulation calculations were performed with the use of V-SIM4 code. This program is used in reconstruction of road accidents, but it is also applied for verification of damage by SDC method, which is described in work [6,7] and for calculation of the impact parameters identified from the crash a test [8]. The obtained calculation results are shown in Fig. 5a-j and an video (https://youtu.be/rxxkmEsE6O8).

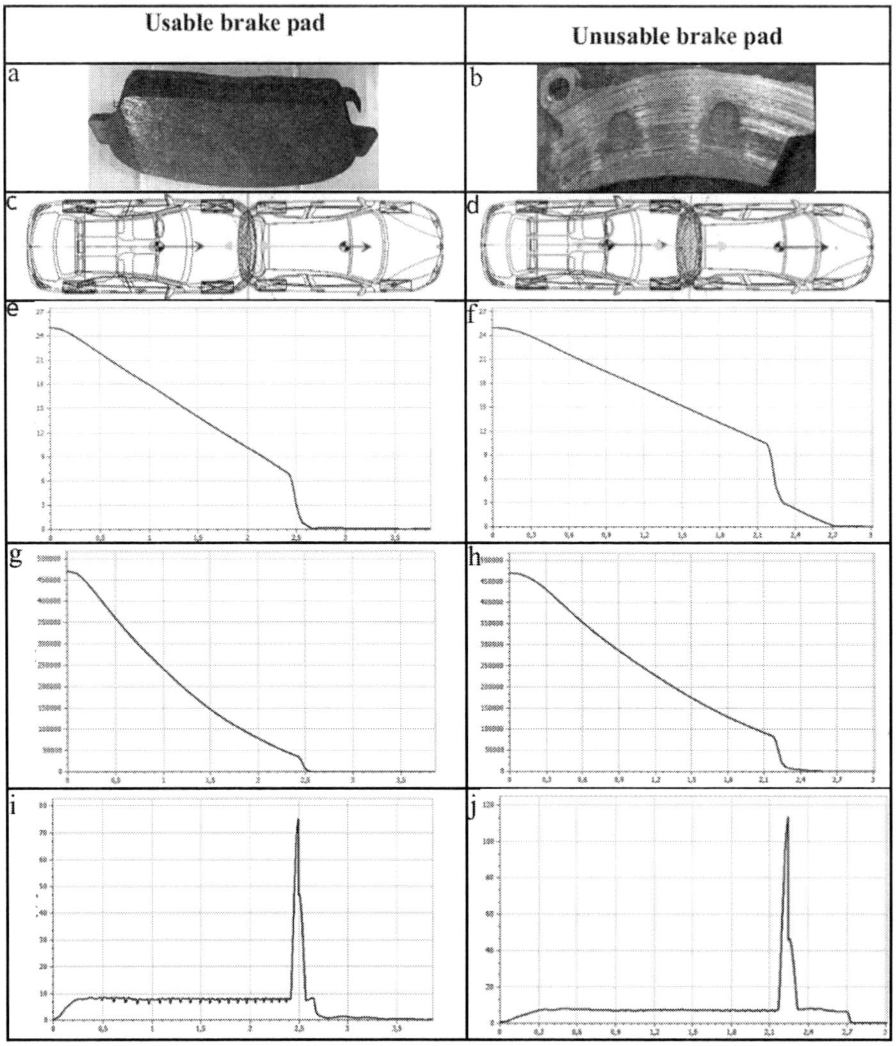

Fig. 5. Usable (a) and unusable brake pads (b), depth of car body deformation for usable (c) and unusable brake pads (d), impact speeds for usable (e) and unusable brake pads (f), kinetic energy of vehicle impact for usable (g) and unusable brake pads (h), delay of vehicle braking for usable (i) and unusable brake pads (j).

5 Conclusions

Simulation calculations have shown that worn out brake pads have an influence on:

- impact with collision speed higher than 3.6 $m \cdot s^{-1}$,
- increase in depth of car bodies penetration during impact by 0.20 m,
- impact with kinetic energy higher by 47 195 kJ,
- decrease in delay resulting in prolongation of the braking distance by 1.8 m.

The obtained results show the importance of brake pads diagnosing to eliminate from operation vehicles whose friction elements of braking system indicate limit wear or close to limit wear.

References

1. L. Knopik, K. Migawa. *Optimal age-replacement policy for non-repairable technical objects with warranty*. Eksploatacja i Niezawodność – Maintenance and Reliability, vol. 19, no. 2, pp. 172 -178, (2017)
2. Ł, Muślewski. *Evaluation Method of Transportation Systems Operation Quality*. Polish Journal of Environmental Studies. vol. 18, No. 2A, Hard Olsztyn, (2009)
3. Ł. Muślewski, M.Woropay. *Theoretical grounds to evaluate of the transport system operation*. Maritime Transportation and Exploitation of Ocean and Coastal Resources – GuedesSoares, Garbatov& Fonseca, Taylor & Francis Group, London (2005)
4. J. Wicher. *Motor vehicles and road traffic safety*. WKŁ Warsaw. pp. 126-127, (2012)
5. H. Tylicki, B. Żółtowski. *Terra- Technology Operation and Maintenance of Selected Systems of Vehicles*. University Publisher PWSZ, Piła, pp. 80 (2005)
6. P. Aleksandrowicz. *Analysis of vehicle collisions with the SDC method*. 23rd International Conference Engineering Mechanics, 15-18.05.2017 Svratka, Czech Republic, pp. 78-81, (2017)
7. P. Aleksandrowicz. *Verification of airbag activation during collision of vehicles with the use of SDC method*. 24th International Conference Engineering Mechanics, 14-17.05.2018 Svratka, Czech Republic, pp. 17-20, (2018)
8. R. Kostek, P. Aleksandrowicz. *Effect of contact parameters on the pattern of vehicle collisions with a round pillar*. 23rd International Conference Engineering Mechanics, 15-18.05.2017 Svratka, Czech Republic, pp. 490-493, (2017)

The assessment of the impact of the chosen exploational conditions of hydraulic arrangement on the working liquid condition

Joanna Wilczarska[1,*], *Ewa* Kuliś[1], *Marcin* Łukasiewicz[1], *Łukasz* Fornal[2], *Natalia* Dluhunovych[3]

[1]Department of Vehicle Engineering, UTP Bydgoszcz, Poland
[2]Stanisława Staszic State Higher School of Vocational Education in Pila, Poland
[3]Faculty of Programming Computer and Telecommunication System, Khmelnytsky National University, Ukraine

Abstract. Hydraulic systems in modern vehicles and agricultural machines have many components. In order to achieve high efficiency of the system, the system requires an accurate filtration system that allows you to reduce the amount of contaminants. Therefore, an important aspect is to determine the content of impurities in the working liquid, so we can diagnose and then predict the durability of the entire hydraulic system. This paper presents selected methods and means of state assessment, as well as the methodology for testing the working fluid. The results of experimental studies conducted in accordance with the developed methodology have been presented, which made it possible to obtain an appropriate analysis of results and draw appropriate conclusions. The results of the research confirmed the validity of the research activities undertaken using the passive experiment.

1 Introduction

Operating conditions may affect the functioning of hydraulic systems in vehicles and working machines. The quality of use, maintenance and storage influence this aspects. Currently, most vehicles and working machines have extensive hydraulic systems. This has a direct effect on the amount of working liquid in the hydraulic system. The costs associated with the replacement of the hydraulic energy medium are large. Modern hydraulic systems require high efficiency with relatively small dimensions. That is why manufacturers of hydraulic system components, to meet the customer's needs they perform elements with smaller and smaller fits of cooperating elements, which is associated with better quality of the working liquid and its purity. In addition to the very precise system of filtration, this warrants the need to determine changes in selected properties of working fluids - even in the aspect of their use [4,14].

1.1 Analysis of methods and means for assessing the state of the working fluid

For the functioning of hydraulic systems, the condition of the working liquid is very important. During operating the hydraulic system, the quality of the oil generally deteriorates – a phenomenon known as degradation. It was determined that 70-80% of failures of hydraulic systems result from the contamination of the working medium. Therefore, it is necessary to control the parameters determining the hydraulic factor, which have a serious impact on the change of physical properties, but also utility properties [2,3,6].

* Corresponding author: asiulazol@utp.edu.pl

The most commonly used methods for assessing the working fluid state include:

a) organoleptic methods - these are the methods that evaluate the state of the working medium by means of human senses,

b) instrument methods (the most frequently used instruments are [5]: - hydrometer for determining the density of liquids, laboratory scale, viscometer),

c) laboratory methods - these are the methods that use appropriate chemical reactions that allow you to determine the appropriate states of the working fluid. Examples of chemical reactions include [7,8]: - distillation; - crystallization; - sorption, etc.

Depending on the tested parameter of the working fluid state, the following measures can be determined:

a) color and transparency [8,9,10]:

- dark color indicates aging of the agent during operation and too long time of use, additionally, milky precipitation indicates the water in the working medium,

- Lovibond method - it functions by applying patterns that can be applied to each other to create a palette of colors ;

- Stammer type colorimeter - using the standard filters, the illuminated liquid column is compared until the same color as the standard is obtained.

b) means for determining the content of solid contaminants [2,9]:

- microscopic methods consist of filtering a proper amount of working liquid through a specific membrane filter with a grid, in order to calculate the amount of solid particles. It allows to determine their quantity, and to group the particles in terms of size, shape and color;

- colmatage meter – it is a device based on the relationship between the pressure difference of the working liquid and the size of the gap, the flow rate of the liquid and the content of particles with dimensions greater than the dimensions of the gap. They are used in precision systems in which continuous monitoring of the working fluid condition is required;

- particle counters are equipped with photoelectric sensors to determine the degree of contamination from the flowing liquid. In addition, the analyzed data is presented in the appropriate purity class according to ISO standards 4406:1999, NAS 1638 or SAE AS 4059. This type of devices allow quick and continuous monitoring of changes occurring in the hydraulic system.

c) means for determining the water content [9]:

- Karl Fischer's method consists in mixing the right amount of oil and two reagents in a closed chamber. The hydrogen produced from the water in oil allows to determine its content;

- Dean-Stark method relies on the distillation of water from the working medium to the receiver, which has a measuring scale to read;

- capacitive transducers enable not only continuous measurement of water content in oil, but also its temperature. Saturated water affects the sensor signal, which allows obtaining measurement results;

d) the kinematic viscosity determined by means of the capillary viscometer consists of measuring the time of flow of the working fluid from one capillary reservoir to the other ;

e) means to determine the acid number and the basic number [11]:

- titration of the weighed amount of the test liquid in a well-defined solvent with an appropriate indicator;

- comparison of the measuring potential of the indicator electrode to the non-polarizing electrode immersed in the tested working liquid;

f) means to determine the content of elements [9,12]:

- spectrometry generally involves examining the electron spectra of chemical compounds in a liquid working molecule, followed by comparison with the standard spectra.

- the activation analysis consists of measuring the radiation of the nuclear isotope formed in the sample. This allows to determine the amount of traces of metallic elements on the surface of the working liquid;

- the colorimetry is used to determine the concentration of the liquid based on the intensity of the color of the chemical compounds as compared to the standard factor. Photocell allows to determine content of some elements and organic substances;

- polarography is based on the registration of changes in the intensity of current flowing through the electrolytic vessel depending on the voltage on the mercury electrode or a constant. It is used to determine the content of elements such as: sulfur, nitrogen, oxygen, etc.

g) the ability to separate water is carried out under static conditions that are relevant to the operating conditions of the spray liquid;

h) the tendency to foam is determined by the volume of foam. The test sample is placed in a measuring cylinder and aerated for a suitable time. Then, in the meantime, the volume amount of foam is determined [5,9,10];

i) the electric breakdown resistance consists in determining the voltage at which the electric spark passes through the liquid sample under test [8,10] ;

j) ignition temperature - the principle of determining this parameter consists of heating up the tested oil until the generated vapours mixed with air create a combustible or explosive mixture. In this methods we used apparatus with an open crucible (e.g. Cleveland) or a closed crucible (e.g. Martens-Pensky) [13];

k) coking residues - the parameter is determined by two methods, e.g. Conradson and Ramsbottom. Both methods have a similar principle of operation, - where the liquid at high temperature, without access to air, transforms into a residue in the form of coke. The mass is weighed and expressed in % in relation to the mass of the medium to be treated [5,7,8,9].

2 Methodology for assessing the working fluid condition

2.1 Determining the operating conditions of hydraulic systems

According to the definition, the operating conditions of technical facilities are divided into terms of use, maintenance and storage.

Due to the conditions of use of the hydraulic system, conditions [11,12] can be distinguished as: light, medium and heavy.

Due to the operating conditions of the hydraulic system, it is possible to distinguish between the following types [13]:

- extreme - these are the activities performed as a last resort such as, supplying the system with any working fluid and performing repairs of the largest damage to the object directly affecting the start-up of this device system;

- correct - these are the activities that allow to maintain the fitness condition of the task system, while all other elements remain without significant service;

- proper - these are the activities supporting the entire facility in the state of fitness in terms of task and function. This type of technical object is ready for any random events regardless of time and place.

Due to the method of storage of the hydraulic system, the following forms can be distinguished:

- bad - the system is exposed directly to weather conditions throughout its lifetime.

- good - the system is secured appropriately. The whole object is located in a drained room, ensuring the limitation of all or a large part of the atmospheric factors that affect its aging.

2.2 Set of working liquid parameters determination

Following the discussion above, the following features of the state, i.e. solid impurities content in the working liquid and water content in the working liquid, require further investigation.

The parameter describing the amount of impurities in the tested hydraulic oil will be the parameter referring to the oil purity classes determined using the instrument method.

The parameter describing the amount of water in the tested hydraulic oil will be the percentage of water content.

The scope of the assessment of the condition of the spray liquid, in accordance with specific sets of state characteristics and diagnostic parameters, will refer to oil contamination with solid particles and water content. The following conditions for the assessment of the status of the subject status were adopted by measuring the determined parameters [14]:

a) Oil samples for testing should show the actual state of the entire working medium. Samples for testing should be taken directly during the system operation or when the system stops, within a time not exceeding 15 minutes (the oil temperature should correspond to the recommended temperature specified by the equipment manufacturer).

b) You should take samples from the same place, which allows better interpretation of the results and eliminate measurement error.

c) Ensure cleanliness while collecting the sample and the vessel itself, which should be clean and dry.

d) The method of sampling is determined by the appropriate standard.

e) If the system does not allow to take a sample from the pipe, it can be taken from the tank using a suitable device.

f) The appropriate frequency of sampling should be considered.

The purpose of the assessment of condition of the working liquid is to determine the water content and solid contaminants in the hydraulic oil.

a) The purity class parameter should be used to determine the solid impurity content. Due to the minimization of the dimensions of the elements and to increase the efficiency of the system, it performs tribological systems with smaller and smaller fitting tolerances, which translates directly into the purity of the oil used. Therefore, it is required to deliver and introduce clean liquid to the cleaned system, as well as effective filtration and control of the cleanliness of the liquid during the operation of the system. Thanks to the cleanliness class, you can determine the quality of the hydraulic system operating fluid. The most commonly used standards for determining the purity classes are ISO 4406: 1999 and NAS 1638. It could be determined their number depending on the size and amount of working liquid.

b) The percentage of water content should be used to determine the water content in the spray liquid. An important point is the saturation point of the oil, in which the oil can not take more water and is determined by the value of 100%. Value up to 100% - suggests dissolved water, i.e. a small amount of water is absorbed into the chemical composition of the spray liquid, which allows the drainage of water from the system without any consequences. However, the value above 100% - suggests undissolved water, the amount of which exceeds the value of water that it can absorb in its composition, which is visible in the form of water precipitates (emulsions). Most currently used working fluids in modern hydraulic systems have a saturation level below 50%. This value prevents the corrosion of system components and the loss of oil properties. [1,3,5].

3 Evaluation of the impact of selected operating conditions of the hydraulic system on the working fluid condition

3.1 Research object

The object of the current research was hydraulic oil used in three hydraulic systems of agricultural machines. Hydraulic oils from 3 agricultural machines, namely 2 farm tractors fed with the same oil grade and a telehandler fed with a second oil grade were used for the tests. The first test object was oil from the Claas Axion 830 agricultural tractor hydraulic system from 2015 year production (fig. .1). The tractor has combined transmission and hydraulics arragements. The object is used on a farm only in transportation activities. Another object of the research was hydraulic oil from the Claas Xerion 5000 agricultural tractor from 2012 year production (fig. 2). The tractor has a separate hydraulic system, which means that hydraulic oil is used only in this system. The object works on a farm only in field works.

Fig. 1. Agricultural tractor Claas Axion 830.

Fig. 2. Agricultural tractor Claas Xerion 5000.

In the presented Claas farm tractors, hydraulic oil CLAAS AGRIHYD HVLP-D46 is used. It is a mineral hydraulic oil with inhibitors of oxidation and corrosion, anti-wear additives and viscosity additives, with the following parameters:

density at 15°C	-	0,874 kg/dm^3;
kinematic viscosity at 40°C	-	45 mm^2/s (cSt);
kinematic viscosity at 100°C	-	8,4 mm^2/s (cSt);
viscosity index	-	166;
flow temperature	-	33°C.

Another object of the study was the hydraulic oil from the JCB 535-95 telehandler from 2009 (fig. 3). The loader has a separate hydraulic system. The object is used on farms only for animal production and alongside other milk production machines .

Fig. 3. Telescopic loader JCB 535-95

The JCB telescopic loader uses JCB hydraulic oil: JCB HYDRAULIC FLUID HP 46. It is a mineral hydraulic oil based on zinc technology and on a refined soluble oil base and the following parameters:

density at 15°C - 0,872 kg/dm^3;
kinematic viscosity at 40°C - 46 mm^2/s (cSt);
kinematic viscosity at 100°C - 7,9 mm^2/s (cSt);
viscosity index - 143;
flow temperature - 39°C.

3.2 Research tools

For the tests, the measuring instrument HYDAC FCU 1000 was used. The apparatus ensured the determination of the solid contamination content, the percentage water content and the temperature in the essence of the mineral working liquid. It is a portable device that, with the help of appropriate hydraulic connections, enables its use in the control and supply circuit of the hydraulic system [15]. Technical specifications of the device:

- particle counter (optical method) determining the size and amount of solid particles;
- capacitive-resistive transducer defining the percentage of water and temperature of the liquid;
- intended for hydraulic oils 10 … 350 mm2/s/16 … 1622 Sus;
- determination of cleanliness classes according to the following standards: ISO 4406:1999; NAS 1638 and SAE AS 4059;
- accuracy of measurement by code +/- 0,5 ISO;
- supply voltage 24V DC;
- working pressure: 45bar/max. 650 psi (without adapter); 345 bar/max. 5000 psi (with adapter); 0 bar (from the tank);
- pump for automatic control of the working fluid flow [15].

3.3 Workplace tests and analysis of test results

The tests were carried out on farms, in the region of Złotów and in the Pila, from August 2016 to June 2017. Operating conditions were deliberately varied to allow for differentiated results. In addition, each object has only one operator. At the beginning of the assessment of the impact of selected operating conditions of the hydraulic system on the condition of the working fluid, messages were obtained about hydraulic oils used in the facilities, i.e. their detailed specification. It was determined that samples for testing hydraulic oils would be collected using sterile, disposable surgical instruments every 50 hours. Samples were collected using laboratory syringes and using a suitable hydraulic conduit. Then, 10 measurements were taken from each sample under laboratory conditions (at 20°C), to determine the oil cleanliness class according to ISO 4406: 1999 and NAS 1638, as well as the percentage water content. After the tests, the oil tested was returned for re-use of the hydraulic system - according to the instructions of HYDAC FCU 1000 [10,14]. Table 1 presents the general operating conditions under which agricultural vehicles and machines were used, maintained and stored at the time of sampling.

Table 1. Information on the operational conditions of vehicles and agricultural machinery.

Operating conditions:	Claas Axion 830	Claas Xerion 5000	JCB 535-95
use	medium	medium	heavy
maintenance	propper	propper	correct
storage	good	dood	bad

The tables and graphs below present the results of research related to the number of solid particles and the percentage of water in oil. In addition, all values related to the liquid purity class have been included.

Table 2. Results of hydraulic tests of agricultural tractor Claas Axion 830 in terms of the condition of the spray liquid, and the course of work - Pearson's correlation coefficient.

No	Workflow [mth]	Average amount of particles > 4μm according to the ISO standard	Average amount of particles > 2 μm according to the NAS standard	Average percentage of water content [%]	
1	3020	6000000	4498696	21,2	
2	3069	6000000	5278212	20,7	
3	3122	6000000	3745344	20,6	
4	3171	6000000	4653944	20,0	
5	3219	3000000	3032132	19,8	
6	3268	3000000	2899232	19,9	
7	3320	3000000	4379032	21,2	
8	3372	3000000	3554868	23,6	
9	3417	3000000	5681520	25,5	
10	3470	3000000	5604796	28,3	
Pearson's correlation coefficient		Dependence between the course of the object's operation and the amount of solid particles> 2 μm according to the NAS standard		.20	Weak correlation
		The dependence between the course of the object's operation and the percentage of water content.		.75	Very high correlation

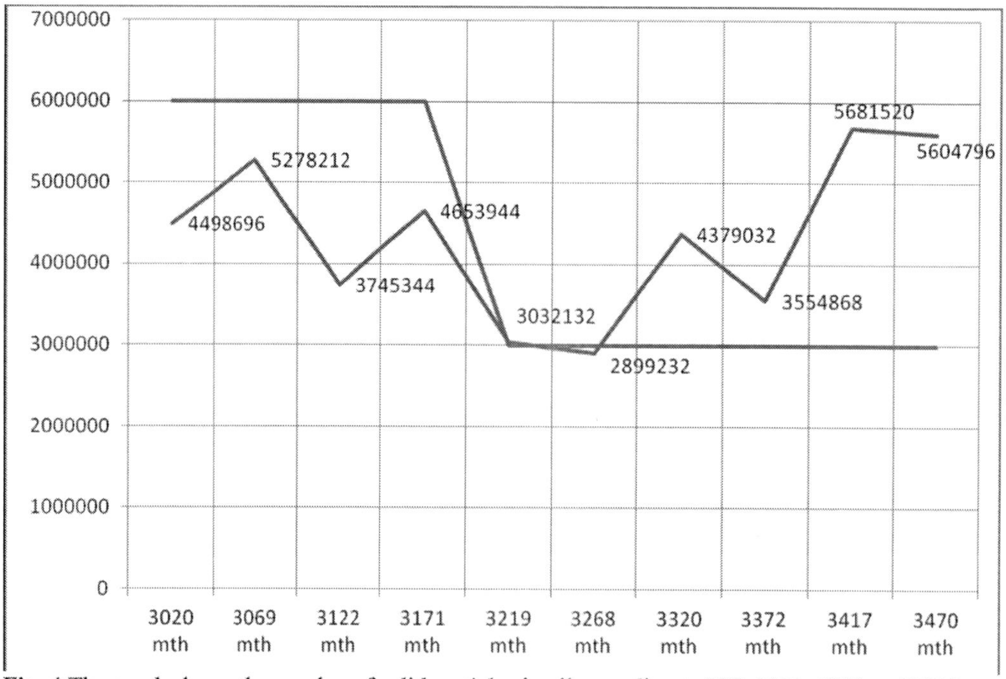

Fig. 4 The graph shows the number of solid particles in oil according to ISO 4406: 1999 and NAS 1638 in relation to the work progress of the Claas Axion 830 farm tractor. Blue line - Average amount of particles > 4μm according to the ISO standard. Red line - summed average number of solid particles above 2 mm in size according to NAS standard.

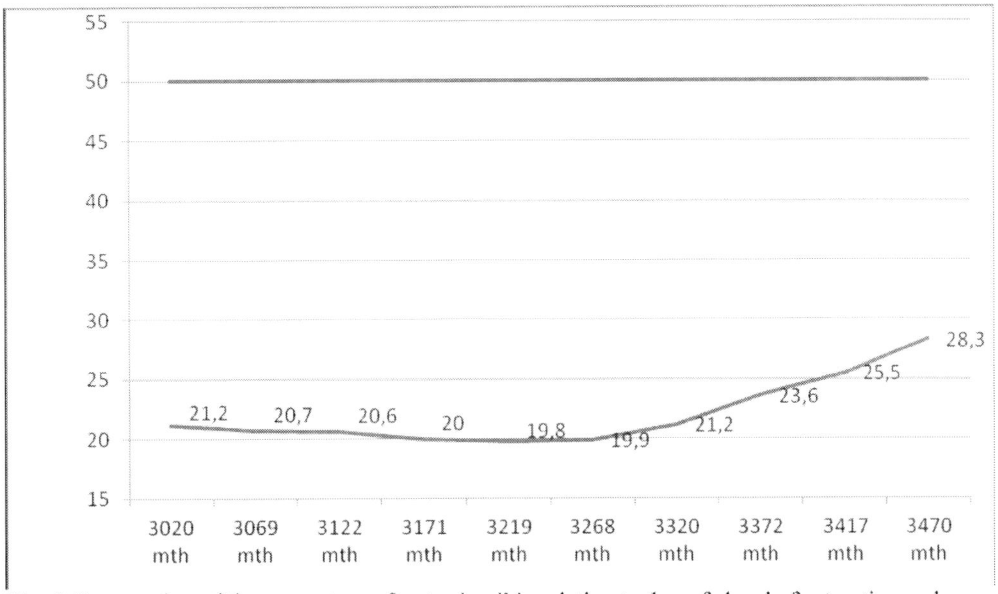

Fig. 5. Presentation of the percentage of water in oil in relation to the safe level of saturation and work progress of the agricultural tractor Claas Axion 830. Blue line - average percentage of water content [%]. Red line - safe level of water saturation in oil (less than 50%).

Table 3. Results of hydraulic tests of agricultural tractor Claas Xerion 5000 in terms of the condition of the spray liquid, and the course of work - Pearson's correlation coefficient.

<table>
<tr><th>No</th><th>Workflow [mth]</th><th>Average amount of particles > 4µm according to the ISO standard</th><th>Average amount of particles > 2 µm according to the NAS standard</th><th colspan="2">Average percentage of water content [%]</th></tr>
<tr><td>1</td><td>1509</td><td>750000</td><td>287478</td><td colspan="2">22,2</td></tr>
<tr><td>2</td><td>1561</td><td>750000</td><td>486868</td><td colspan="2">24,5</td></tr>
<tr><td>3</td><td>1610</td><td>1500000</td><td>771804</td><td colspan="2">28,7</td></tr>
<tr><td>4</td><td>1662</td><td>1500000</td><td>1679772</td><td colspan="2">32,6</td></tr>
<tr><td>5</td><td>1708</td><td>1500000</td><td>1860880</td><td colspan="2">37,0</td></tr>
<tr><td>6</td><td>1758</td><td>3000000</td><td>2375376</td><td colspan="2">40,7</td></tr>
<tr><td>7</td><td>1812</td><td>3000000</td><td>2515554</td><td colspan="2">42,8</td></tr>
<tr><td>8</td><td>1864</td><td>1500000</td><td>1988732</td><td colspan="2">43,2</td></tr>
<tr><td>9</td><td>1910</td><td>750000</td><td>1896948</td><td colspan="2">44,4</td></tr>
<tr><td>10</td><td>1960</td><td>1500000</td><td>2310776</td><td colspan="2">45,6</td></tr>
<tr><td colspan="2" rowspan="2">Pearson's correlation coefficient</td><td colspan="2">Dependence between the course of the object's operation and the amount of solid particles> 2 µm according to the NAS standard</td><td>.84</td><td>Very high correlation</td></tr>
<tr><td colspan="2">The dependence between the course of the object's operation and the percentage of water content.</td><td>.97</td><td>Almost fully correlated</td></tr>
</table>

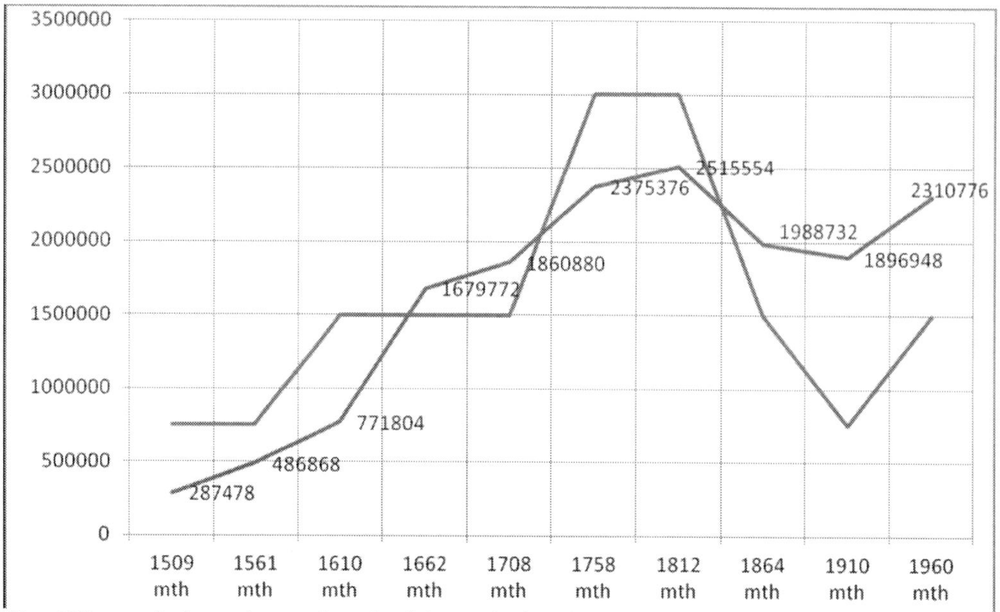

Fig. 6 The graph shows the number of solid particles in oil according to ISO 4406: 1999 and NAS 1638 in relation to the work progress of the Claas Axion 5000 farm tractor. Blue line - Average amount of particles > 4µm according to the ISO standard. Red line - summed average number of solid particles above 2 mm in size according to NAS standard..

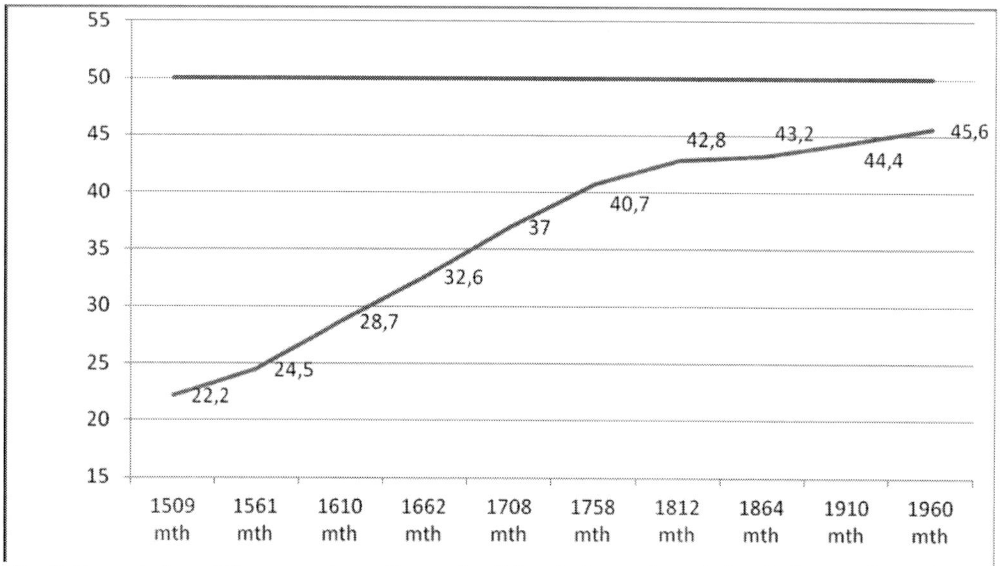

Fig. 7. Presentation of the percentage of water in oil in relation to the safe level of saturation and work progress of the agricultural tractor Claas Axion 5000. Blue line - average percentage of water content [%]. Red line - safe level of water saturation in oil (less than 50%).

Table 4. Analysis of the test results of the JCB 535-95 telehandler in terms of the condition of the spray liquid, and the course of work - Pearson's correlation coefficient.

No	**Workflow [mth]**	**Average amount of particles > 4μm according to the ISO standard**	**Average amount of particles > 2 μm according to the NAS standard**	**Average percentage of water content [%]**	
1	4500	750000	646230	22,1	
2	4550	750000	712874	25,1	
3	4601	750000	842048	29,3	
4	4649	750000	926912	34,2	
5	4701	1500000	1422176	40,1	
6	4751	1500000	1714204	48,5	
7	4803	3000000	2259152	61,7	
8	4854	3000000	1824648	69,8	
9	4901	3000000	2187576	78,3	
10	4952	3000000	2335984	84,1	
Pearson's correlation coefficient		Dependence between the course of the object's operation and the amount of solid particles> 2 μm according to the NAS standard		.96	Almost fully correlated
		The dependence between the course of the object's operation and the percentage of water content.		.99	Almost fully correlated

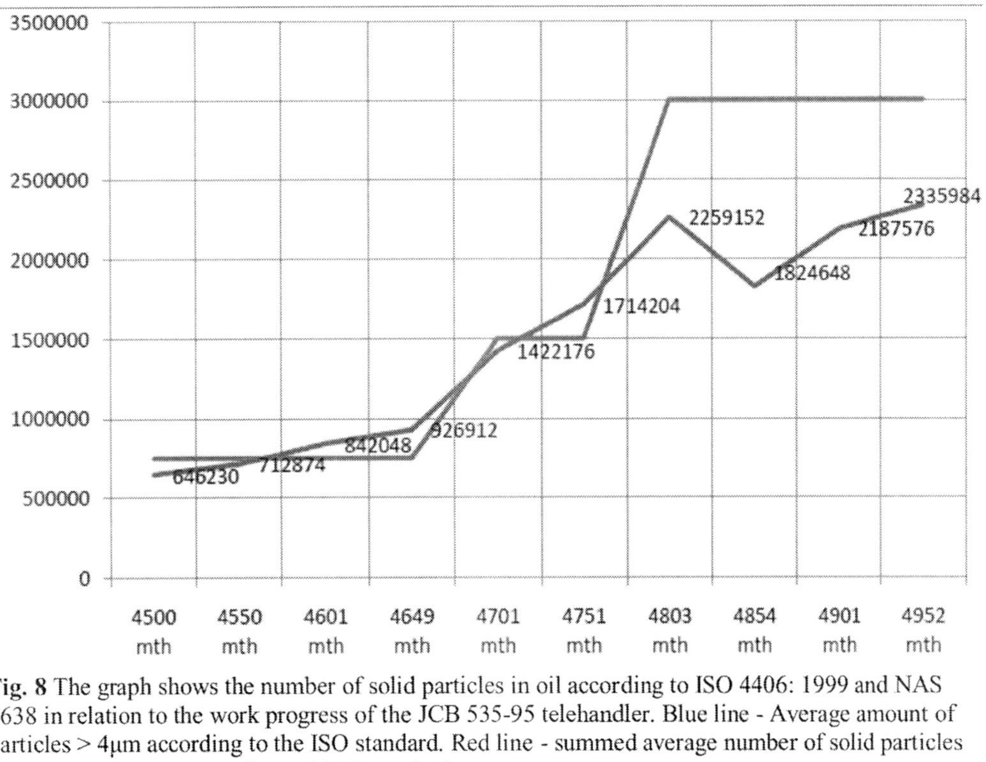

Fig. 8 The graph shows the number of solid particles in oil according to ISO 4406: 1999 and NAS 1638 in relation to the work progress of the JCB 535-95 telehandler. Blue line - Average amount of particles > 4μm according to the ISO standard. Red line - summed average number of solid particles above 2 mm in size according to NAS standard.

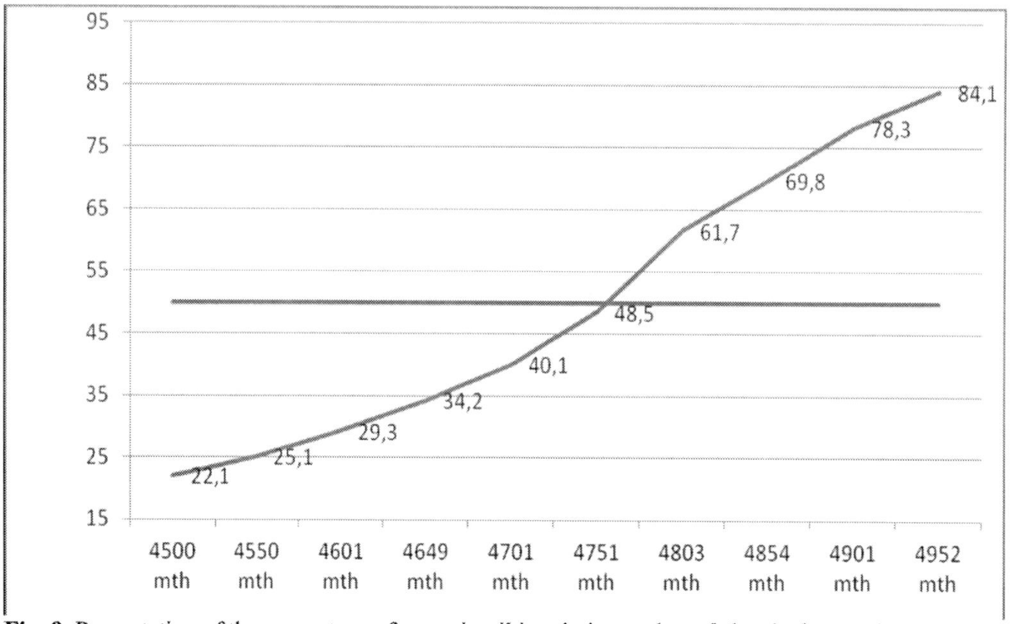

Fig. 9. Presentation of the percentage of water in oil in relation to the safe level of saturation and work progress of the JCB 535-95 telehandler. Blue line - average percentage of water content [%]. Red line - safe level of water saturation in oil (less than 50%).

4 Conclusions

The measurements for each sample indicate that they were made correctly, which confirms the value of the standard deviation within the limits of unity. Also the number of measurements allows for correct interpretation of the results for each sample.

The values of particulate pollution and the percentage of water content are different for each object. This is mainly due to the operating conditions and the construction and specifications of the hydraulic system. The Claas Axion 830 agricultural tractor was operated under favorable operating conditions, but due to the complex hydraulic system with the transmission system, a significant impact on the greater amount of particulate matter contamination was observed compared to the other hydraulic systems.

The Claas Xerion 5000 agricultural tractor has tougher conditions of use compared to the Claas Axion 830 tractor, but its hydraulic oil supplies only one hydraulic system, which translates into results related to the amount of particulate matter.

Hydraulic system of the JCB 535-95 telehandler has a large amount of contaminants, especially the percentage of the water content, due to poor operating conditions. The water content also results from the working conditions of the machine, because the system works in the environment with a lot of live animals every day, where the humidity is high and there are significant temperature differences. For example, during the winter period, the hydraulic system works in a building where the temperature is within 20°C, then the agricultural machine leaves the building where the temperature reaches -15°C, which significantly affects the condensation of the water on the walls, e.g. hydraulic tank. However, the fact that the hydraulic system has comparable amounts of contamination with solid particles to a similar system of the Claas Xerion 5000 farm tractor suggests that the hydraulic systems exploitation conditions have a large influence on their amount. By altering the exploitation conditions of the hydraulic systems, one can reduce the amount of solid particles, which in turn will impact the life of the system.

Thanks to Pearson's correlation coefficient, it has been proven that in most cases the amount of particulate contamination and the percentage of water content is also important for the system operation. The only case where the correlation between the amount of particulate matter and the course of work was weak the hydraulic system of the Claas Axion 830 farm tractor. This effect might have been influenced by the fact that the hydraulic oil also fed the transmission system, which shook the test results of this object.

References

1. T. Bujak, *Równokołowe 520 KM.* Rolniczy Przegląd Techniczny, nr 1/2015, Agencja Promocji Rolnictwa i Agrobiznesu APRA Sp. z o.o., Bydgoszcz 2015.
2. P. Drogosz, *Ocena zużycia oleju hydraulicznego w urządzeniach pracujących w małych przedsiębiorstwach,.* Hydraulika i Pneumatyka, nr 3/2009, Oficyna Wydawnicza SIMP, Wrocław 2009.
3. K. Klarecki, *Możliwości zwiększania sprawności napędów hydraulicznych.* Hydraulika i Pneumatyka, nr 5/2014, Oficyna Wydawnicza SIMP, Wrocław 2014.
4. Kostek R., Landowski B., Muślewski Ł.: Simulation of rolling bearing vibration in diagnostics. Journal of Vibroengineering 2015, volume 17, issue 8, pp. 4268-4278, ISSN 1392-8716.
5. G. Kotnis, *Budowa i eksploatacja układów hydraulicznych w maszynach.* wyd. KaBe, Krosno 2011.
6. M.Matuszewski., T.Mikołoajczyk, D.YU.Pimenov, M.Styp-Rekowski, *Influence of structure isotropy of machined surface on the wear process.* International Journal of Advanced Manufacturing Technology, 2017.
7. P. Rybacki, *Oleje do silnika, przekładni i hydrauliki.* Top Agrar Polska, nr 7/2011, Polskie Wydawnictwo Rolnicze Sp. z o.o., Poznań 2011.
8. W. Sieradzki, M. Jastrzębski M, *Wpływ zanieczyszczeń stałych i wody na pracę urządzeń górniczych.* Hydraulika i Pneumatyka, nr 2/ 2013, Oficyna Wydawnicza SIMP, Wrocław 2013.
9. A. Skrobacki, A. Ekielski: *Pojazdy i ciągniki rolnicze.* Wydawnictwo „Wieś Jutra",Warszawa 2012.
10. Z. Szydelski, *Pojazdy samochodowe. Napęd i sterowanie hydrauliczne.* Wydawnictwa Komunikacji i Łączności, Warszawa 1999.
11. The Drive & Control Company: *Hydrostatyczny napęd wentylatora firmy Bosch Rexroth w pojazdach i maszynach samobieżnych.* Hydraulika i Pneumatyka, nr 2/2014, Oficyna Wydawnicza SIMP, Wrocław 2014.
12. The Drive & Control Company: *Nowoczesne pompy hydrauliczne o stałej i zmiennej wydajności.* Hydraulika i Pneumatyka, nr 2/2015, Oficyna Wydawnicza SIMP, Wrocław 2015.
13. Z. Trela, M*onitorowanie stanu oleju w diagnozowaniu układów hydraulicznych.* Hydraulika i Pneumatyka, nr 1/2009, Oficyna Wydawnicza SIMP, Wrocław 2009.
14. W. Zwierzycki *Oleje, paliwa i smary dla motoryzacji i przemysłu.* Wydawnictwo i Zakład Poligrafii Instytutu Technologii Eksploatacji w Radomiu, Gorlicach 2001.
15. https://www.hydac.com (15.03.2017).

The use of thermography in the diagnosis of ship piston internal combustion engines

Jan Monieta[1,**]

[1]EDP, Maritime University of Szczecin, Wały Chrobrego 1–2, 70-500 Szczecin, Poland

Abstract. The intensity of infrared radiation emitted by objects depends mainly on their temperature. One of the diagnostic signals may be the temperature field. In infrared thermography, this quantity is used as an indicator of the technical condition of marine objects. The article presents an overview of the use of infrared thermography for the diagnosis mainly of marine piston floating objects and various types of reciprocating internal combustion engines as well as examples of own research results. A general introduction to infrared thermography and common procedures for temperature measurement and non-destructive testing are presented. Experimental research was carried out both in laboratory conditions and in the operating conditions of sea-going vessels. Experimental studies consisted of the presentation of photographs of the same objects made in visible light and the use of infrared thermography. The same objects were also compared, but for different cylinders of the tested internal combustion engines as well as for the up state and fault state. The characteristics of the temperature values at selected points were taken depending on the engine load along with the approximation mathematical models of these dependencies.

1 Introduction

Infrared thermography is a technique for producing an image of the invisible to eyes infrared light emitted by objects due to their thermal condition. A typical a thermography camera produces a live video picture of heat radiation intensities. More sophisticated cameras can actually measure the temperature values of any object or surface in the field-of-view and produce colour images that make interpretation of thermal patterns easier. An image produced by an infrared (IR) camera is called a thermogram [1].

The properties of measuring heat transmission from objects using the thermography camera are more than an interesting novelty. An advance in IR sensor production has helped the infrared viewing technology to be adopted as a cost-effective, non-invasive measuring method. Advanced optics and sophisticated software interfaces continue to add to the variety of Thermal IR cameras.

However, the most powerful means of thermal energy transfer is radiation, which moves at the speed of light. Infrared energy is emitted by all materials above 0 K. Infrared radiation is part of the electromagnetic spectrum and occupies frequencies between visible light and radio waves. The infrared part of the spectrum spans wavelengths from 0.7 μm to 1000 μm [1]. Within this wave band, only frequencies of 0.7 μm to 20 μm are used for practical, everyday temperature measurement.

The three methods by which heat flows from one object to another are: radiation, convection and conduction. These three phenomena in most situations they occur together. The invention of the thermal imaging camera has caused a number of changes in the field of temperature measurement. The measurement method bases its operation on the phenomena of the objects' impact relative to their surrounding space. The use of

* Corresponding author: j.monieta@am.szczecin.pl

electromagnetic radiation reception in the low energy range of this radiation has contributed to the development of new directions in exact sciences.

2 An examples of the use of thermography in the diagnostics of marine combustion engines

The intensity of the infrared radiation emitted by objects is mainly a function of their temperature [2]. In infrared thermography, this feature is used for multiple purposes: as a health indicator in medicine, process monitoring, manufacturing inspection, and non-destructive testing, as a sign of malfunction in mechanical and electrical maintenance or as the indicator of heat loss in objects. In paper presents a review of infrared thermography focused on two applications [3]: temperature measurement and non-destructive testing for the general introduction to infrared thermography and the common procedures for temperature measurement and non-destructive testing are presented.

A ship classification and certification body Lloyd's Register, made the prediction, that the near future, mechanical machinery onboard vessels will also benefit from thermal imaging to repaired, especially as a pre-docking strategy to identify and target equipment and systems which need maintenance as well as to eliminate necessary works [1]. Commercial ships are objects for thermographic inspections with big machinery, vast installations, extended electronic systems that can hardly be surveyed by visual inspections. According to the International Convention for the Safety of Life at Sea (SOLAS), the maximum surface temperature of machinery, parts and components in a vessel's engine room should not rise above 220°C [1]. The majority of engine room fires are caused by ruptured fuel or oil pipes which eventually spray fuel or oil on adjacent hot surfaces [1, 4]. IR thermography has the advantage of being a noncontact inspection and measurement tool able to display and store exact temperature values as well as visual evidence. When something happens to the engines or to other vital installations, inspection tool as an infrared camera can be used. Thermography is used in marine applications for engineering inspection of electrical and mechanical systems of ships. IR thermography can play a major role in maintaining the safety and reliability of marine floating objects where catastrophic failure could lead to the loss of a vessel or its crew.

The thermal imaging camera has the receiving function here and is a signal transducer. Advanced thermal imaging cameras are characterized by high resolution of the image being processed and recorded the image on a storage medium. Saving images or video recordings can then be played back to analyse the course of changes in temperature areas [5]. Thermal imaging cameras have recently been the object of interest of many specialists dealing with diagnostics [5–7].

In the modern internal combustion engines it strives to reduce the toxicity of exhaust gases, fuel economy and how best cohesive properties. In the operation of the engine of internal combustion important element is the ongoing control of its technical state. The essence of activity, directed is on the maximum use of the time of the engine and its diagnosis. This makes it possible to diagnose the consumption of failure. There are many methods of diagnostics. One of them is the thermography diagnosis for maintenance. In the article presents technical state assessment of selected components of the engine of internal combustion of crank train and oil supply system based on the obtained IR image [6].

Recently, thermography has proved to be a powerful tool to deepen the knowledge about the chemical and physical processes that take place in automotive engine. 2D infrared (IR) chemiluminescence measurements have been performed in an optical Common Rail self-ignition engine in order to investigate the different processes that regulate the pollutant emission formation [8]. The time evolution of the injected fuel, of the flames, and of the burned gases has been analyzed. Infrared images have been detected in the spectrum from 1.5 mu m to 5 mu m. It has allowed obtaining more information with respect to the visible imaging and it has been possible to detect the fuel vapour and the burned gas motion in the combustion chamber for a longer period with respect to visible range. CO_2 evolution in the available volume during several combustion phases has been detected: during the inlet phase, during the combustion phase, in the burned gases cloud that moves toward the outlet valves. The results obtained suggest IR diagnostics as useful tool for engine control.

Infrared thermography has become the powerful tool for basic and applied scientific research and for the application in maritime affairs. As a prognosis maintenance tool [9], IR thermography has the ability to identify problems before they occur. It is especially helpful for trouble shooting potential electrical overloads, worn or bad circuit breakers. IR thermography can also be used to detect bad bearings of shafts or any application where heat detection would be beneficial. This paper has basic principles of IR thermography are presented and examples of the tool application in maritime affairs are given [1].

One of the benefits of using thermal imaging is the ability to predict possible the fire hazards due to increases in temperature in controlled areas in a particular marine object. The work proposes the possibility of applying new electronic and computer technology as part of the ship's fire detection system, such as the use of computer vision, using existing marine thermography systems and installing thermal imaging IR cameras on the same system [5]. There has been proposed communication between the thermal imaging system and the fire detection and central alarm system of the ship. The visual analysis of certain areas on board the ship and the related facilities inside it, with the addition of certain software applications into the existing video technology system, makes is the fire alarm system. With the forecast and send early warnings to the ship's central fire alarm system, thereby contributing to improved safety with to equipment, the ship as a whole, cargo and human lives. The spreads widely very fastly in high temperatures and becomes the cause of great damage to ship equipment [10], often resulting in the loss of the ship. Videotapes from thermal cameras can be used for risk analysis and the prevention of future ship fire incidents. It is possible the fast detection of temperature changes very early warning.

The article presented non destruction method and results of using thermal method in diagnostic of marine combustion engines operation of fishing fleet vessel [7]. The article presents a mathematical model for determining the temperature distribution by means of thermovision [5]. Knowledge of this model makes it possible to analyse the influence of external factors [11] on the accuracy of temperature determination by thermovision.

3 Principles of Infrared Thermography

The basic dependence used to determine the temperature is the thermovision method is basic equation of radiation. Effective radiation power of the thermographic camera that reaches the lens from the surface of the object being examined consists of the following components [12]:

- the effective radiation power emitted by the tested object,
- radiation of the background reflected from the object,
- radiation of your own atmosphere.

The hotter an object becomes the more infrared energy it emits. The Stefan-Boltzmann law, states that the total energy radiated per unit surface area of a black body in unit time E_{BB} is directly proportional to the fourth power of the black body absolute temperature T:

$$E_{BB} = \sigma \cdot T^4 \text{ [W/m}^2\text{]} \tag{1}$$

The constant of proportionality σ, called the Stefan-Boltzmann constant is non-fundamental in the sense that it derives from other known constants of nature.

When the object is at thermal equilibrium, the amount of absorption will equal the amount of emission. The Kirchoff's law determined for any material, the emissivity and absorptance of the body are equal at any specified temperature and wavelength [3]:

$$\varepsilon_\lambda = a_\lambda \tag{2}$$

where:
a – coefficient of absorptance,
λ – wavelength.

When the infrared energy radiated by real object reaches another body, a portion of the energy received will be absorbed a, a portion will be reflected r and, if the body is not opaque, a portion t_r will be transmitted. The sum of the individual parts is added up to the initial value of radiation which left the source:

$$a + r + t_r = 1 \tag{3}$$

The total radiation received by the camera W_t comes from three sources: the emission of the target objects E_o, the emission of the surroundings and reflected by the object E_r and the emission of the atmosphere E_a. It can be expressed as equation:

$$W_{to} = E_o + E_r + E_a \tag{4}$$

The total strength of the radiation can be expressed with the formula FLIR Systems:

$$W_{to} = \varepsilon_o \tau_o W_o + (1 - \varepsilon_o)\tau W_r + (1 - \tau_a) W_a \tag{5}$$

where:
$\varepsilon_o \tau_o W_o$ – is the emission of radiation from the object,
ε_o – the emission of the object,
τ_a – the atmospheric release,
$(1 - \varepsilon_o)\tau W_r$ – the reflection from the environment,
$(1 - \varepsilon_o)$ – the reflection of the object,
$\tau_a W_r$ – the temperature of the environment;
$(1 - \tau_a) W_a$ – the emission of radiation from the atmosphere, $(1 - \tau_a)$ is the emission of the atmosphere and W_a is the temperature of the atmosphere.

The basic equation of radiation for the complexity of the measurement scene can be saved:

$$W_{to} = \tau_a \varepsilon_o \sigma T_o^4 + \tau_a (1 - \varepsilon_o) \sigma T_r^4 + (1 - \tau_a) \sigma T_a^4 \ \text{[W]} \tag{6}$$

Hence the measured temperature of object is equal:

$$T_o = \sqrt[4]{\frac{W_{to} - (1 - \varepsilon_o)\tau_a \sigma T_r^4 + (1 - \tau_a)\sigma(1 - \tau_a)\sigma T_a^4}{\varepsilon_o T_a \sigma}} \tag{7}$$

In order to solve equation (6), the following parameters must be supplied: the emissivity of the object, the reflected temperature T_r, the transmittance of the atmosphere and the temperature of the atmosphere T_a. The transmittance of the atmosphere is estimated using the distance from the object to the camera and the relative humidity. This value is very close to one. The temperature of the atmosphere is obtained using a common thermometer. However, as the emittance of the atmosphere is very close to zero (1 = a), this parameter has little influence on the temperature measurement. The emissivity of the object and the reflected temperature has but the very high influence on the temperature measurement.

4 Methodology and research objects

The object of thermal imaging studies were the combustion engines driving generators of 3 types and one main drive. The research was carried out both in the laboratory conditions of the Maritime University of Szczecin and in the conditions of operation of sea-going vessels. Different types of FIR cameras were used for the tests. First, the measuring track was tested by comparing the temperature readings with thermometers and the thermovision camera. Thermal imaging was preceded by the configuration of the camera settings to the conditions that prevailed during the tests, which is why the parameters of the reference conditions were measured: ambient temperature, atmospheric pressure and relative humidity.

5 Sample investigations results

The measurement path was calibrated by measuring the temperature with contactless and contactless devices inside the muffle furnace for calibrating thermoelectric thermometers and a thermal imaging camera (Fig. 1). The same tests were carried out in the conditions of sea-going vessels. Differences in indications were insignificant.

a) b)

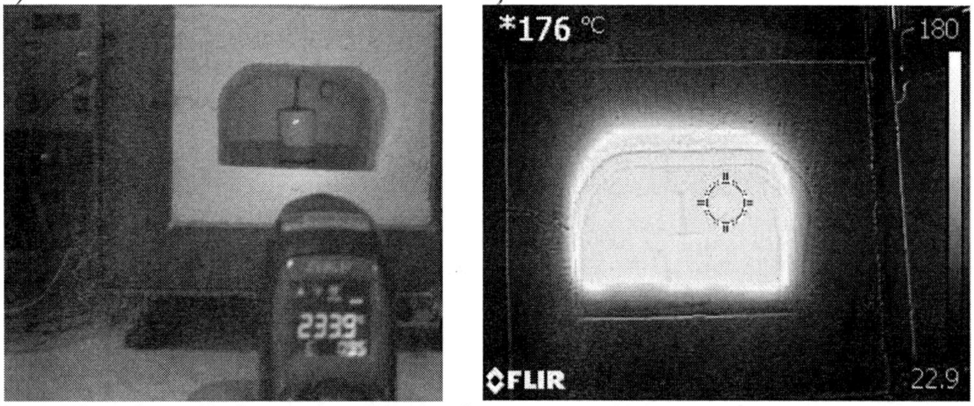

Fig. 1. View of a muffle furnace for calibrating thermometers with measurement using a contactless thermometer a) and a photo taken with a thermal imaging camera b).

Fig. 2 presents a global view of a functioning engine driving the generator in the compartment of the ship's power plant and the temperature field from the infrared camera. Fig. 3, 4 and 5 show images in visible light and infrared from the thermovision camera of such elements as an injection pump, body of injector valve and head cover together with the temperature values.

a) b)

Fig. 2. Photograph in visible light of the engine with the generator (a) and thermographic image b).

a) b)

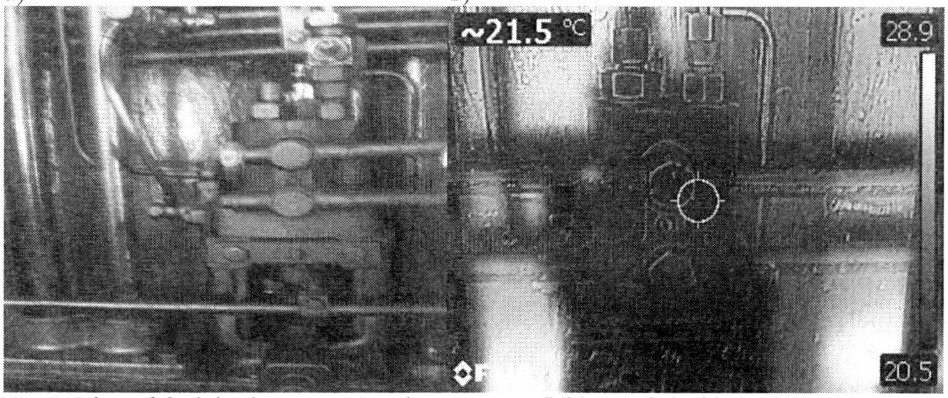

Fig. 3. View of the injection pump a) and temperature field recorded with the thermal imaging camera b).

a) b)

Fig. 4. Photograph in the visible light of the internal combustion engine head without cover a) and the temperature field from the infrared camera.

a) b)

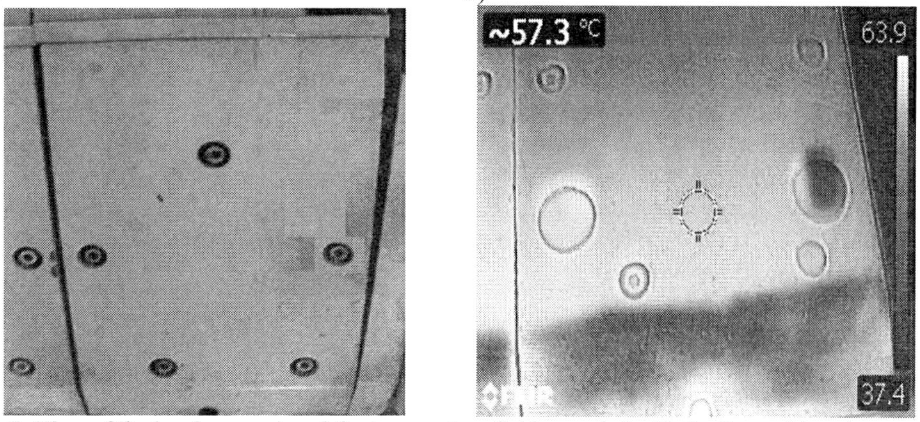

Fig. 5. View of the head cover a) and the temperature field recorded with the thermal imaging camera b).

The tests were carried out for various cylinders of combustion engines, with different loads and for different technical conditions. In contrast, Fig. 6 shows dependencies the temperature of the injector bodies depending on the engine load. The curve for the third cylinder deviating from the others results from the inability state of the injection subsystem.

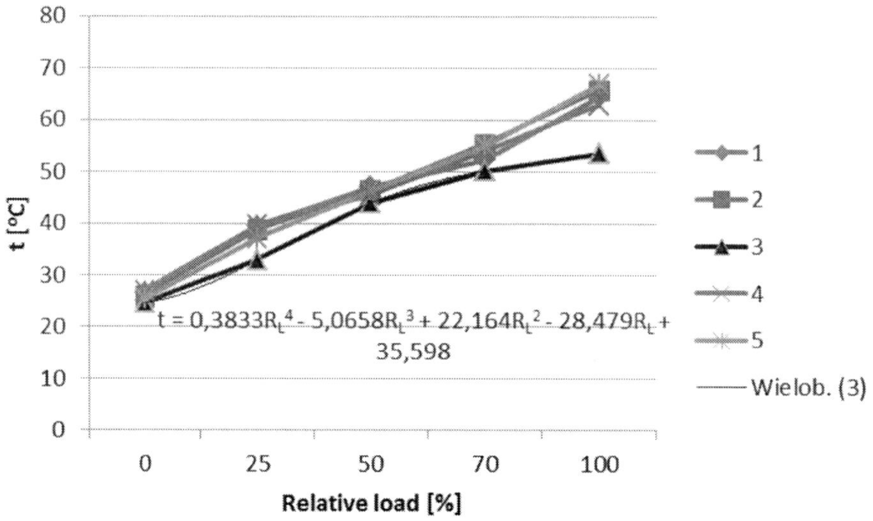

Fig. 6. The influence of engine load on temperature values on the injector body for five cylinders of an internal combustion engine.

This technique is based on the polynomial fit of each pixel time history from the given thermographic image sequence. The evolution of the temperature is adjusted to an n degree polynomial for input value x, what described in equation [13]:

$$T(t) = a_n x^n + a_{n-1} x^{n-1} + \ldots + a_1 x + a_0 \tag{8}$$

This paper presents an example of the polynomial model for the temperature dependence of the injector body function on the relative load R_L of the combustion engine for one of the curves of Fig. 6.

$$t = 0{,}3833R_L^4 - 5{,}0658R_L^3 + 22{,}164R_L^2 - 28{,}479R_L + 35{,}598 \tag{9}$$

6 Conclusions

Diagnostics of ship engine facilities using thermovision and thermography are non-invasive, which is likely to be recognized by classification societies. Application is possible for objects with different temperature fields. The measurement process itself takes place without any prior preparation and quickly.

The benefits of using thermovision imaging are the ability to predict possible fire hazards caused by growth of temperature and catastrophic failures in controlled areas in specific marine objects or systems.

The article presents the results of preliminary experimental research. Photographs in visible light and images from infrared camera of the surface elements of marine engines were placed.

Further research should demonstrate the usefulness of selected diagnostic symptoms, check the possibility of detecting specific damages, determine the measurement errors and combine the monitoring with the alarm system of the ship's actuator.

The results so far have to be considered as encouraging.

References

1. V. Pašagić, M. Mužević, D. Kelenc, Brodogradnja **59**, 2, pp. 123–130 (2008)

2. C. Meola, *Origin and Theory of Infrared Thermography. In Infrared Thermography*: Recent Advances and Future Trends; Meola, C., Ed.; Bentham Science: New York, NY, USA (2012).
3. R. Usamentiaga, P. Venegas, J. Guerediaga, L. Vega, J. Molleda and F. G. Bulnes, Sensors, **14**, pp. 12305–12348 (2014)
4. J. Monieta, Scient. Journ. of the Maritim. Univ. of Szczecin, **47**, 119, pp. 36–41 (2016)
5. M. Bistrović, P. Ristov, D. Komorčec, Scient. Journ. of the Marit. Univ. of Szczecin, **50**, 122, pp. 23–29 (2017)
6. P. Jarmużek, W. Sawczuk, Log. **3**, pp. 2557–2562 (2014)
7. O. Klyus, C. Behrend, P. Rajewski, M. Szczepanek, O. Bezyukov, A. Kardanov, Scient. Journ. of the Maritim. Univ. of Szczecin, **35**, 107, pp. 36–41 (2013)
8. E. Mancaruso, L. Sequino, M.B Vaglieco, 16th Fot. AEIT Italian Conf. on Phot. Technol. Loc., Naples, Italy, 12–14 (2014)
9. B. Żółtowski, C. Cempel, *Engineering of diagnostics machines* (in Polish), Polish Society of Technical Diagnostics, Warsaw, Bydgoszcz, Radom (2004).
10. J. Monieta, Tribol. 4, pp. 55–63 (2017)
11. C. Jermak, R. Majchrowski, J. Monieta, Combust. Eng. **146**, 3, pp. 71–72, cd. pp. 1–8 (2011)
12. Z. Rymarczyk, M. Dzierzgowski, M. Strzeszewski, *The use of thermography for a qualitative assessment of thermal protection of buildings* in: "Energy-efficient shaping of the internal environment" (work done by Z. Popiołek). (in Polish), Silesian University of Technology, Gliwice (2005)
13. A.V. Nguyen, N.J. Cohen, H. Lipman, C.M. Brown, N.A. Molinari, W.L Jackson, H. Kirking, P. Szymanowski, T.W. Wilson, B.A. Salhi, et al, Infect. Dis., **16**, pp. 1710–1717 (2010).

Acknowledgements

This research outcome has been achieved under the research project "Ecological and economic aspects of the operation of selected elements of marine power systems" No. 1/S/IESO/2014 financed from a subsidy of the Ministry of Science and Higher Education for statutory activities.

Analysis of the vehicle operation process including diagnostic information

Mirosław Siergiejczyk[1], *Adam* Rosiński[2,*], and *Marcin* Wawrzyński[1]

[1]Warsaw University of Technology, Faculty of Transport, 00-662 Warsaw, Koszykowa 75, Poland
[2]Military University of Technology, Faculty of Electronics, 00-908 Warsaw, gen. W. Urbanowicza 2, Poland

Abstract. Modern telecommunications techniques allow for continuous transmission of information from vehicles to dispatch centres during the operation process. The information included in this transmission may contain diagnostic data that make it possible to monitor the state of responsible components of the vehicle construction. The article presents the issues related to the analysis of the vehicle operation process including information received with the use of a CAN bus. The model of the vehicle operation process was presented, and the diagnosis state was taken into account. The use of transport telematics systems and intelligent transport systems in transport allow for the rolling stock profiling at the next stage. Owing to the rolling stock profiling, operational databases powered by a telematics interface are created. This, in turn, will allow to carry out the operational application process along with further optimisation of the operation process.

1 Introduction

Modern telecommunications techniques allow for continuous transmission of information from vehicles to dispatch centres during the operation process [1, 2]. The information included in this transmission may contain diagnostic data that make it possible to monitor the state of responsible components of the vehicle construction. It is encouraged by the fact that the control of individual actuators of the vehicle equipment is carried out with the use of on-board communication buses, the messages of which constitute a medium of information on specific load parameters of the devices. The information techniques, in turn, based on specialised IT equipment, and the use of advanced methods for information processing allow to observe changes in the size of diagnostic symptoms with a minimum delay, and often in real time [3, 4].

This situation makes it possible to modify the applied vehicle maintenance strategies including the transport telematics technical capabilities in the area of information transmission and processing [5-7]. The article analyses the vehicle operation process including diagnostic information. The reliability and operating approach was presented [8-10], and the diagnosis state was taken into account. The use of transport telematics systems and intelligent transport systems [11-14] in transport allow for the rolling stock profiling at the next stage. Owing to the rolling stock profiling, operational databases powered by the telematics interface are created [15].

* Corresponding author: adro@wt.pw.edu.pl

2 Use of CAN buses of vehicles in obtaining diagnostic information

On-board vehicle data exchange systems evolve with the development of means of transport. They constitute specialised structures created from scratch in order to fulfil specific tasks, which are posed by modern control systems [16].

In the automotive industry, the development of an engine operation control unit, ECU (Electronic Control Unit), can be considered as the beginning of the creation of a protocol capable of reliable and safe exchange of data between sensors and actuators. It was the moment of the birth of bus systems in vehicles. One of the most common is the CAN bus [17, 18]. From the very beginning, the main objective of the CAN standard designers was to ensure reliability of operation while simplifying the vehicle's electric bundle. The reconciliation of contradictory expectations had to result in the search for a solution simultaneously on several layers of the protocol stack.

In the CAN standard, a differential signal transmission system was used. It increases the resistance to interference. The induced voltage acts similarly on both wires maintaining a mutual voltage relation, which allows the receiver to correctly read the right signal value, even under strong interference conditions. As a consequence, the redundancy of the CAN bus physical layer planned by the constructors brings another advantage in the form of resistance to damage to one of two bus wires. As a result, the system can function on one wire by closing the circuit with the use of connection to the vehicle mass at both ends, on the transmitter and receiver sides. In such a situation, in most vehicles, the driver will not be even informed about partial unfitness of the system, because there are no current consequences for operation. Another important feature of CAN is the orientation of transmission to the message, which, in short, means that there is no need to address the devices themselves, but only to give a distinguishing feature to a type of the message transmitted by any device connected to the bus. This feature has a double implication. On the one hand, it provides the solution flexibility, expected in multi-series production of many versions of the same vehicle, but on the other hand, it opens the door to Cyber Security threats [17, 18].

The current generations of the bus systems are focused on new functionalities related to the maintenance of real-time systems, such as fully mechatronic steering systems, however, the vehicle's IT framework includes the buses based on a proven and reliable CAN solution.

The use of the CAN buses in vehicles makes it possible to obtain numerous diagnostic information. Modern vehicles have many buses that exchange data with each other. The vehicle telematics equipment capable of reporting the events marked as essential for operation sends data via radio to the processing centre. These data can be used to determine the technical condition of the vehicle and its reliability and operation analysis.

3 Reliability and operation analysis of the vehicle including the diagnosis process

The process of the vehicle operation is always accompanied by degradation processes. They consist in a change of the state towards deterioration of its functional properties [19]. The used vehicle is always subject to degradation (even due to wear), only the intensity of these changes can be different. A significant role is played by external factors basically related to an unfavourable environmental impact, to which the vehicle has to adapt its operation, many times, at the expense of the increased rate of degradation. The vehicle response to unfavourable external conditions leading to maintaining the same level of efficiency of operation usually results in an increase in the rate of degradation.

In order to prevent, stop or reduce the technical degradation processes of vehicles and restore their service potential, various operation strategies are used [20, 21]. The following main maintenance strategies can be distinguished:

- repair strategy after damage (*Corrective Maintenance*),
- on-condition maintenance strategy (*On-condition Maintenance*),
- time based maintenance strategy (*TBM- Time Based Maintenance*).

In the vehicle operation process, an important role is played by monitoring. Monitoring is an activity based on continuous observation of changes in the object condition [22]. It is a kind of supervision, the concept of which includes activities involving, in general, discrete (at certain time intervals) or continuous observations.

The monitoring process uses such diagnostic signals, the observation of which does not interfere with the object functioning, and in particular, its operation process.

By carrying out the vehicle functioning analysis, it can be stated that the relationships taking place in it, in reliability and operation terms, can be visualised as shown in Fig. 1.

The state of usability S_Z is a state, in which the vehicle functions properly. The state of partial usability S_{ZC} is a state, in which the vehicle is partially usable, but it carries out the tasks that it was assigned to. The state of diagnosis S_{D1} is a state, in which the processes related to diagnosing of the vehicle state are conducted, but it is usable. The state of diagnosis S_{D2} is a state, in which the processes related to diagnosing of the vehicle state are conducted, but it is partially usable. The maintenance state Q_O is a state, in which the activities necessary to restore the usability state of the vehicle are carried out. The state of unfitness Q_N is a state, in which the vehicle does not implement the tasks that it was assigned to.

The vehicle being in the state of usability S_Z performs periodic diagnoses. It is illustrated by the transition to the state of diagnosis S_{D1} with the intensity λ_{ZD1}. The diagnosis time is determined by the intensity of transition μ_{D1Z}. This process takes place during the stop, and it makes it possible to obtain diagnostic information, which is collected from the on-board telematics systems of the vehicle. Being in the state of usability S_Z, it is possible to transit to the state of partial usability S_{ZC} with the intensity λ_{ZZC} in case of exceeding the limit value for the state of usability by the diagnostic symptom. By being in the state of partial usability S_{ZC}, the periodic diagnoses are performed. It is illustrated by the transition to the state of diagnosis S_{D2} with the intensity λ_{ZCD2}. Being in the state of diagnosis S_{D2}, a decision (based on diagnostic information) was made whether the vehicle, which is partially usable, is still operated, or renovation is necessary. In the first case, there is a transition to the state of partial S_{ZC} with the intensity μ_{D2ZC}. In the second case, there is a transition to the state of renovation S_O with the intensity λ_{D2O}. Being in the state of renovation, the vehicle operational potential is restored, and at the same time, there is a transition to the state of usability S_Z with the intensity μ_{OZ}. In the system, the transitions to the state of unfitness S_N from the state of usability S_Z (with the intensity λ_{ZN}) or from the state of partial usability S_{ZC} (with the intensity λ_{ZCN}) are possible. The mentioned situations occur when the diagnostic symptom reaches the value characteristic for the state of unfitness. Being in the state of unfitness S_N, there is a transition to the state of renovation S_O with the intensity λ_{NO} in order to restore the operational potential.

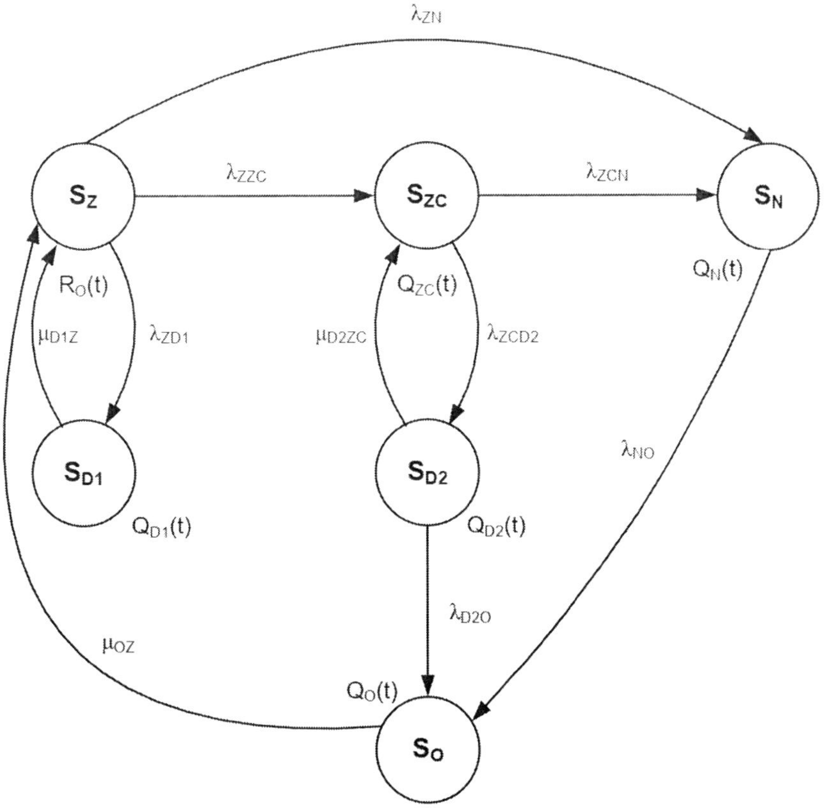

Fig. 1. Relationships during the vehicle operation.

Markings in Fig.:
$R_O(t)$ – probability function of the system's staying in the state of usability
$Q_{ZC}(t)$ – probability function of the system's staying in the state of partial usability,
$Q_{D1}(t)$ – probability function in the system's staying in the state of diagnosis 1,
$Q_{D2}(t)$ – probability function of the system's staying in the state of diagnosis 2,
$Q_O(t)$ – probability function of the system's staying in the state of maintenance,
$Q_N(t)$ – probability function of the system's staying in the state of unfitness,
λ_{ZZC} – intensity of transitions from the state of usability to the state of partial usability,
λ_{ZN} – intensity of transitions from the state of usability to the state of unfitness,
λ_{ZD1} – intensity of transitions from the state of usability to the state of diagnosis 1,
μ_{D1Z} – intensity of transitions from the state of diagnosis 1 to the state of usability,
λ_{ZCD2} – intensity of transitions from the state of partial usability to the state of diagnosis 2,
μ_{D2ZC} – intensity of transitions from the state of diagnosis 2 to the state of partial usability,
λ_{ZCN} – intensity of transitions from the state of partial usability to the state of unfitness,
λ_{D2O} – intensity of transitions from the state of diagnosis 2 to the state of maintenance,
λ_{NO} – intensity of transitions from the state of unfitness to the state of maintenance,
μ_{OZ} – intensity of transitions from the state of maintenance to the state of usability.

The system presented in Fig. 1 can be described by the following Chapman–Kolmogorov equations:

$$R_0'(t) = -\lambda_{ZZC} \cdot R_0(t) - \lambda_{ZN} \cdot R_0(t) - \lambda_{ZD1} \cdot R_0(t) + \mu_{D1Z} \cdot Q_{D1}(t) + \mu_{OZ} \cdot Q_O(t)$$

$$Q_{D1}'(t) = \lambda_{ZD1} \cdot R_0(t) - \mu_{D1Z} \cdot Q_{D1}(t)$$

$$Q_{ZC}'(t) = \lambda_{ZZC} \cdot R_0(t) - \lambda_{ZCN} \cdot Q_{ZC}(t) - \lambda_{ZCD2} \cdot Q_{ZC}(t) + \mu_{D2ZC} \cdot Q_{D2}(t)$$

(1)

$$Q'_{D2}(t) = \lambda_{ZCD2} \cdot Q_{ZC}(t) - \mu_{D2ZC} \cdot Q_{D2}(t) - \lambda_{D2O} \cdot Q_{D2}(t)$$

$$Q'_{N}(t) = \lambda_{ZCN} \cdot Q_{ZC}(t) + \lambda_{ZN} \cdot R_0(t) - \lambda_{NO} \cdot Q_N(t)$$

$$Q'_{O}(t) = \lambda_{D2O} \cdot Q_{D2}(t) + \lambda_{NO} \cdot Q_N(t) - \mu_{OZ} \cdot Q_O(t)$$

Assuming the baseline conditions:

$$R_0(t) = 1$$
$$Q_{D1}(t) = Q_{D2}(t) = Q_{ZC}(t) = Q_N(t) = Q_O(t) = 0 \tag{2}$$

and applying the Laplace transform, we obtain the following system of linear equation:

$$s \cdot R_0^*(s) - 1 = -\lambda_{ZZC} \cdot R_0^*(s) - \lambda_{ZN} \cdot R_0^*(s) - \lambda_{ZD1} \cdot R_0^*(s) + \mu_{D1Z} \cdot Q_{D1}^*(s) + \mu_{OZ} \cdot Q_O^*(s)$$

$$s \cdot Q_{D1}^*(s) = \lambda_{ZD1} \cdot R_0^*(s) - \mu_{D1Z} \cdot Q_{D1}^*(s)$$

$$s \cdot Q_{ZC}^*(s) = \lambda_{ZZC} \cdot R_0^*(s) - \lambda_{ZCN} \cdot Q_{ZC}^*(s) - \lambda_{ZCD2} \cdot Q_{ZC}^*(s) + \mu_{D2ZC} \cdot Q_{D2}^*(s)$$

$$s \cdot Q_{D2}^*(s) = \lambda_{ZCD2} \cdot Q_{ZC}^*(s) - \mu_{D2ZC} \cdot Q_{D2}^*(s) - \lambda_{D2O} \cdot Q_{D2}^*(s) \tag{3}$$

$$s \cdot Q_{N}^*(s) = \lambda_{ZCN} \cdot Q_{ZC}^*(s) + \lambda_{ZN} \cdot R_0^*(s) - \lambda_{NO} \cdot Q_N^*(s)$$

$$s \cdot Q_{O}^*(s) = \lambda_{D2O} \cdot Q_{D2}^*(s) + \lambda_{NO} \cdot Q_N^*(s) - \mu_{OZ} \cdot Q_O^*(s)$$

The probabilities of the system's staying in the distinguished functional states from the symbolic (Laplace's) perspective have the following form:

$$R_0^*(s) = \frac{b \cdot e \cdot f \cdot (c \cdot d - \lambda_{ZCD2} \cdot \mu_{D2ZC})}{\begin{matrix} a \cdot b \cdot c \cdot d \cdot e \cdot f - c \cdot d \cdot e \cdot f \cdot \lambda_{ZD1} \cdot \mu_{D1Z} - a \cdot b \cdot e \cdot f \cdot \lambda_{ZCD2} \cdot \mu_{D2ZC} + \\ + e \cdot f \cdot \lambda_{ZCD2} \cdot \lambda_{ZD1} \cdot \mu_{D1Z} \cdot \mu_{D2ZC} - b \cdot c \cdot d \cdot \lambda_{NO} \cdot \lambda_{ZN} \cdot \mu_{OZ} - \\ - b \cdot e \cdot \lambda_{D2O} \cdot \lambda_{ZCD2} \cdot \lambda_{ZZC} \cdot \mu_{OZ} - b \cdot d \cdot \lambda_{NO} \cdot \lambda_{ZCN} \cdot \lambda_{ZZC} \cdot \mu_{OZ} + \\ + b \cdot \lambda_{NO} \cdot \lambda_{ZCD2} \cdot \lambda_{ZN} \cdot \mu_{D2ZC} \cdot \mu_{OZ} \end{matrix}} \tag{4}$$

$$Q_{D1}^*(s) = \frac{e \cdot f \cdot \lambda_{ZD1} \cdot (c \cdot d - \lambda_{ZCD2} \cdot \mu_{D2ZC})}{\begin{matrix} a \cdot b \cdot c \cdot d \cdot e \cdot f - c \cdot d \cdot e \cdot f \cdot \lambda_{ZD1} \cdot \mu_{D1Z} - a \cdot b \cdot e \cdot f \cdot \lambda_{ZCD2} \cdot \mu_{D2ZC} + \\ + e \cdot f \cdot \lambda_{ZCD2} \cdot \lambda_{ZD1} \cdot \mu_{D1Z} \cdot \mu_{D2ZC} - b \cdot c \cdot d \cdot \lambda_{NO} \cdot \lambda_{ZN} \cdot \mu_{OZ} - \\ - b \cdot e \cdot \lambda_{D2O} \cdot \lambda_{ZCD2} \cdot \lambda_{ZZC} \cdot \mu_{OZ} - b \cdot d \cdot \lambda_{NO} \cdot \lambda_{ZCN} \cdot \lambda_{ZZC} \cdot \mu_{OZ} + \\ + b \cdot \lambda_{NO} \cdot \lambda_{ZCD2} \cdot \lambda_{ZN} \cdot \mu_{D2ZC} \cdot \mu_{OZ} \end{matrix}} \tag{5}$$

$$Q_{ZC}^*(s) = \frac{b \cdot d \cdot e \cdot f \cdot \lambda_{ZZC}}{\begin{matrix} a \cdot b \cdot c \cdot d \cdot e \cdot f - c \cdot d \cdot e \cdot f \cdot \lambda_{ZD1} \cdot \mu_{D1Z} - a \cdot b \cdot e \cdot f \cdot \lambda_{ZCD2} \cdot \mu_{D2ZC} + \\ + e \cdot f \cdot \lambda_{ZCD2} \cdot \lambda_{ZD1} \cdot \mu_{D1Z} \cdot \mu_{D2ZC} - b \cdot c \cdot d \cdot \lambda_{NO} \cdot \lambda_{ZN} \cdot \mu_{OZ} - \\ - b \cdot e \cdot \lambda_{D2O} \cdot \lambda_{ZCD2} \cdot \lambda_{ZZC} \cdot \mu_{OZ} - b \cdot d \cdot \lambda_{NO} \cdot \lambda_{ZCN} \cdot \lambda_{ZZC} \cdot \mu_{OZ} + \\ + b \cdot \lambda_{NO} \cdot \lambda_{ZCD2} \cdot \lambda_{ZN} \cdot \mu_{D2ZC} \cdot \mu_{OZ} \end{matrix}} \tag{6}$$

$$Q_{D2}^{*}(s)=\frac{b\cdot e\cdot f\cdot\lambda_{ZCD2}\cdot\lambda_{ZZC}}{\begin{gathered}a\cdot b\cdot c\cdot d\cdot e\cdot f-c\cdot d\cdot e\cdot f\cdot\lambda_{ZD1}\cdot\mu_{D1Z}-a\cdot b\cdot e\cdot f\cdot\lambda_{ZCD2}\cdot\mu_{D2ZC}+\\+e\cdot f\cdot\lambda_{ZCD2}\cdot\lambda_{ZD1}\cdot\mu_{D1Z}\cdot\mu_{D2ZC}-b\cdot c\cdot d\cdot\lambda_{NO}\cdot\lambda_{ZN}\cdot\mu_{OZ}-\\-b\cdot e\cdot\lambda_{D2O}\cdot\lambda_{ZCD2}\cdot\lambda_{ZZC}\cdot\mu_{OZ}-b\cdot d\cdot\lambda_{NO}\cdot\lambda_{ZCN}\cdot\lambda_{ZZC}\cdot\mu_{OZ}+\\+b\cdot\lambda_{NO}\cdot\lambda_{ZCD2}\cdot\lambda_{ZN}\cdot\mu_{D2ZC}\cdot\mu_{OZ}\end{gathered}} \quad (7)$$

$$Q_{N}^{*}(s)=\frac{b\cdot f\cdot(c\cdot d\cdot\lambda_{ZN}+d\cdot\lambda_{ZCN}\cdot\lambda_{ZZC}-\lambda_{ZCD2}\cdot\lambda_{ZN}\cdot\mu_{D2ZC})}{\begin{gathered}a\cdot b\cdot c\cdot d\cdot e\cdot f-c\cdot d\cdot e\cdot f\cdot\lambda_{ZD1}\cdot\mu_{D1Z}-a\cdot b\cdot e\cdot f\cdot\lambda_{ZCD2}\cdot\mu_{D2ZC}+\\+e\cdot f\cdot\lambda_{ZCD2}\cdot\lambda_{ZD1}\cdot\mu_{D1Z}\cdot\mu_{D2ZC}-b\cdot c\cdot d\cdot\lambda_{NO}\cdot\lambda_{ZN}\cdot\mu_{OZ}-\\-b\cdot e\cdot\lambda_{D2O}\cdot\lambda_{ZCD2}\cdot\lambda_{ZZC}\cdot\mu_{OZ}-b\cdot d\cdot\lambda_{NO}\cdot\lambda_{ZCN}\cdot\lambda_{ZZC}\cdot\mu_{OZ}+\\+b\cdot\lambda_{NO}\cdot\lambda_{ZCD2}\cdot\lambda_{ZN}\cdot\mu_{D2ZC}\cdot\mu_{OZ}\end{gathered}} \quad (8)$$

$$Q_{O}^{*}(s)=\frac{b\cdot\begin{pmatrix}c\cdot d\cdot\lambda_{NO}\cdot\lambda_{ZN}+e\cdot\lambda_{D2O}\cdot\lambda_{ZCD2}\cdot\lambda_{ZZC}+\\+d\cdot\lambda_{NO}\cdot\lambda_{ZCN}\cdot\lambda_{ZZC}-\lambda_{NO}\cdot\lambda_{ZCD2}\cdot\lambda_{ZN}\cdot\mu_{D2ZC}\end{pmatrix}}{\begin{gathered}a\cdot b\cdot c\cdot d\cdot e\cdot f-c\cdot d\cdot e\cdot f\cdot\lambda_{ZD1}\cdot\mu_{D1Z}-a\cdot b\cdot e\cdot f\cdot\lambda_{ZCD2}\cdot\mu_{D2ZC}+\\+e\cdot f\cdot\lambda_{ZCD2}\cdot\lambda_{ZD1}\cdot\mu_{D1Z}\cdot\mu_{D2ZC}-b\cdot c\cdot d\cdot\lambda_{NO}\cdot\lambda_{ZN}\cdot\mu_{OZ}-\\-b\cdot e\cdot\lambda_{D2O}\cdot\lambda_{ZCD2}\cdot\lambda_{ZZC}\cdot\mu_{OZ}-b\cdot d\cdot\lambda_{NO}\cdot\lambda_{ZCN}\cdot\lambda_{ZZC}\cdot\mu_{OZ}+\\+b\cdot\lambda_{NO}\cdot\lambda_{ZCD2}\cdot\lambda_{ZN}\cdot\mu_{D2ZC}\cdot\mu_{OZ}\end{gathered}} \quad (9)$$

where:

$$a=s+\lambda_{ZZC}+\lambda_{ZN}+\lambda_{ZD1}$$

$$b=s+\mu_{D1Z}$$

$$c=s+\lambda_{ZCN}+\lambda_{ZCD2}$$

$$d=s+\mu_{D2ZC}+\lambda_{D2O}$$

$$e=s+\lambda_{NO}$$

$$f=s+\mu_{OZ}$$

The solution of the above set of equations in the field of time is the next step of the analysis and it is not discussed here.

The obtained relationships allow to determine the probabilities of staying of the vehicle operation system, including data collected from the on-board telematics systems of the vehicles, in the states of: usability S_Z (4), diagnosis S_{D1} (5), partial usability S_{ZC} (6), diagnosis S_{D2} (7), unfitness S_N (8) and maintenance S_O (9).

4 Conclusion

Modern vehicles are equipped with a large number of electronic devices that form on-board IT networks. Almost all the systems that initially consisted of autonomous devices are currently connected with each other via bus systems. As a result, the electric bundles were simplified consequently reducing the vehicle wiring mass, but above all, it was allowed to exchange information between any elements of the system or even between separate systems. Thus, the data transferred in the buses are used for the current control of the processes of many systems, but they also have information potential that can be used to rationalise the maintenance of the vehicle fleet.

The article presents the reliability and operation analysis of the vehicle, while the diagnosis process was taken into account. It made it possible to determine the relationships that allow to define the values of probabilities of the vehicles' staying in the distinguished states of operation. In further studies, the authors plan to use operational data received via telematics interface in order to develop a database. It will allow for profiling the rolling stock which consists in distinguishing individual vehicles or their collections due to their functional properties. This, in turn, will allow to carry out the operational application process along with further optimisation of the operation process. It leads to the rolling stock operation specification that provides the opportunity to rationalise the operation and maintenance.

References

1. J. Paś, *Operation of electronic transportation systems* (Publishing House University of Technology and Humanities in Radom, Radom, 2015)
2. A. Rosiński, *Modelling the maintenance process of transport telematics systems* (Publishing House Warsaw University of Technology, Warsaw, 2015)
3. T. Dąbrowski, M. Bednarek, K. Fokow, M. Wiśnios, The method of threshold-comparative diagnosing insensitive on disturbances of diagnostic signals, Przegląd Elektrotechniczny, **11A**, pp. 93-97 (2012).
4. F. Losurdo, I. Dileo, M. Siergiejczyk, K. Krzykowska, M. Krzykowski, Innovation in the ICT infrastructure as a key factor in enhancing road safety: a multi-sectoral approach, Proceedings 25th International Conference on Systems Engineering ICSEng 2017, IEEE Computer Society Conference Publishing Services (CPS), Las Vegas, Nevada, USA, pp. 157–162 (2017). doi: 10.1109/ICSEng.2017.69
5. M. Stawowy, Model for information quality determination of teleinformation systems of transport, in: "Proceedings of the European Safety and Reliability Conference ESREL 2014", CRC Press/Balkema, pp. 1909–1914 (2015)
6. M. Rychlicki, Z. Kasprzyk, Increasing performance of SMS based information systems, in „Proceedings of the Ninth International Conference Dependability and Complex Systems DepCoS-RELCOMEX", given as the monographic publishing series – „Advances in intelligent systems and computing", vol. 286, Springer, pp. 373–382 (2014). doi: 10.1007/978-3-319-07013-1_36
7. Z. Kasprzyk, M. Rychlicki, Analysis of physical layer model of WLAN 802.11g data transmission protocol in wireless networks used by telematic systems, in „Proceedings of the Ninth International Conference Dependability and Complex Systems DepCoS-RELCOMEX", given as the monographic publishing series – „Advances in intelligent systems and computing", vol. 286, Springer, pp. 265-274 (2014)
8. J. Paś, A. Rosiński, Selected issues regarding the reliability-operational assessment of electronic transport systems with regard to electromagnetic interference, Eksploatacja i Niezawodnosc – Maintenance and Reliability, **19(3)**, pp. 375–381 (2017). doi: 10.17531/ein.2017.3.8
9. M. Siergiejczyk, J. Paś, A. Rosiński, Issue of reliability–exploitation evaluation of electronic transport systems used in the railway environment with consideration of electromagnetic interference, IET Intelligent Transport Systems, **10(9)**, pp. 587–593 (2016). doi: 10.1049/iet-its.2015.0183
10. M. Siergiejczyk, K. Krzykowska, A. Rosiński, L.A. Grieco, Reliability and viewpoints of selected ITS system, in: Proceedings 25th International Conference on Systems Engineering ICSEng 2017, IEEE, Conference Publishing Services (CPS), pp. 141–146 (2017). doi: 10.1109/ICSEng.2017.68
11. B. Kalupova, I. Hlavon, Intelligent transport systems in the management of road transportation, Open Engineering, **6(1)**, pp. 492–497 (2016). doi: 10.1515/eng-2016-0062
12. J. Młyńczak, T. Hejczyk, B. Wszołek, A. Gałuszka, D. Surma, R. Ogaza, R. Burdzik, Passenger safety and information module in intelligent integrated traffic management system, Vibroengineering PROCEDIA, **6**, pp. 234–237 (2015)

13. L. Janusova, S. Cicmancova, Improving safety of transportation by using inteligent transportation systems, 9th International Scientific Conference Transbaltica 2015, Procedia Engineering 134, pp. 14–22 (2016). doi: 10.1016/j.proeng.2016.01.031
14. U. Ehlers, E. Ryeng, E. McCormack, F. Khan, S. Ehlers, Assessing the safety effects of cooperative intelligent transport systems: A bowtie analysis approach, Accident Analysis & Prevention, **99**, pp. 125 – 141 (2017). doi: 10.1016/j.aap.2016.11.014
15. P. Dersin, Fleet Maintenance as a Dynamic Decision Process, European Safety and Reliability Conference ESREL 2017
16. M. Dziubinski. A. Drozd, M. Adamiec, E. Siemionek, Electromagnetic interference in electrical systems of motor vehicles, in: Scientific Conference On Automotive Vehicles And Combustion Engines (KONMOT 2016), Book Series: IOP Conference Series-Materials Science and Engineering, vol. 148, (2016). DOI: 10.1088/1757-899X/148/1/012036.
17. R. Curie, Developments in Car Hacking, SANS Institute (2015)
18. R. Curie, Hacking the CAN Bus: Basic manipulation of modern Automobile Through CAN Bus Reverse Engineering, SANS Institute (2017)
19. L. Będkowski, T. Dąbrowski, *The basis of exploitation, part II: The basis of exploational reliability* (Publishing House MUT, Warsaw, 2006)
20. A. Jodejko-Pietruczuk, S. Werbińska-Wojciechowska, Analysis of maintenance models' parameters estimation for technical systems with delay time, Eksploatacja i Niezawodność – Maintenance and Reliability, **16(2)**, pp. 288–294 (2014)
21. C. Dibsdale, Holistic Prognostics, in Through-life Engineering Services, Springer, New York (2015)
22. T. Dąbrowski, *Diagnosis of anthropotechnical systems in potential and effect terms* (Publishing House MUT, Warsaw, 2001)

Wheel rim state assessment using modal and geometrical parameters

Arkadiusz Rychlik[1,*]

[1]University of Warmia and Mazury in Olsztyn, Faculty of Technical Sciences, Department of Vehicle and Machine Construction and Operation, Oczapowskiego 11, 10-719 Olsztyn, Poland

Abstract. This paper presents the concept of a non-invasive method for assessing the wheel rim state to determine its technical status. The research procedure includes an analysis of the wheel rim frequency spectrum in a trial at a diagnostic station, which involves applying vibration forces with specific parameters on the wheel rim. The distribution of the wheel rim's natural frequencies in rotation angle and unbalanced value with its positioning and radial run-out and axial run-out for the wheel rim bead in relation to the wheel disc centring hole on the internal and external wheel rim surfaces, are established based on the analysis and calculations. These parameters also serve to determine the condition of joints (fusion-welded, pressure-welded, etc.) and to identify the cracks or loss of integrity in the wheel rim structure. This method can be applied for tests of both new wheel rims during the quality control and used rims used for passenger cars, special vehicles and slow-running vehicles.

1 Introduction

Cold-formed sections are the most commonly used structural elements in engineering and technology, for instance, in simple constructions for manometric connectors on pipelines, photovoltaic or solar panel installation systems and traffic signs or banner ad holders. Examples of formed sections also include more complex and critical elements, e.g. plane wings and propeller blades or wheel rims.

Since wheel rims play a crucial role for the driver's safety and vehicle's operation, they should be periodically assessed during both the manufacturing process and operation. However, this approach is only followed in motorsports as the driver's and audience's safety is most important. In professional car and motor racing teams, qualified and experienced technicians perform visual checks of the complete wheels before they are assembled on vehicles or Formula One cars. The teams monitor the total number of driving hours for individual wheel rims and check their geometrical parameters before each assemblage. They use ultrasound and penetration methods to detect cracks and structural changes that are invisible to the naked eye. A wheel rim with any sign of fatigue or damage is scrapped and replaced with a new one. In fact, technical referees during Indy Car or F1 races require the teams to comply with strict standards for new wheel rims and to monitor the period in which the wheel rims are in operation. According to the technical regulations of such associations, used wheel rims should be scrapped every 500 miles (800 km). A similar rule applies to road races although the limitations to the period in operations are not that restrictive.

Unfortunately, such rigorous control procedures are uncommon among users and in the legal regulations on the vehicles involved in standard road traffic. For these type of vehicles, the condition of the wheel rim and the wheel is assessed during seasonal type replacement in tyre service points where a technician basically visually assesses the degree of whole wheel run-out and rarely the wheel rim alone while balancing the whole wheel.

* Corresponding author: arkadiusz.rychlik@uwm.edu.pl

Importantly, assembling and disassembling of wheel rims is not beneficial since this procedure negatively mainly affects the wheel disc by provoking accelerated wear and tear of screw holes and alignment hole.

It should be noted that the wheel disc and hoop fatigue phenomenon results from a combination of cyclicity and intensity of load to which the wheel rims are exposed. While small loadings are repeated thousands of times (cycles), substantial loads accelerate the fatigue process and the number of cycles till the structure is damaged (cracked) is significantly reduced [1].

All-wheel rims are subject to major deformations when a vehicle takes sharp turns. During standard road traffic or off-road driving, the magnitude of wheel rim and hoop deformation is small. Since modern tyres have almost the same operational parameters as the former racing tyres, each turn taken at the limits of wheel grip and stability of a vehicle's driving track results in substantially greater deformations of the wheel rim fragments. Furthermore, wheel spins (applying motor power allowance in relation to the wheel grip), driving over the edges of hard objects (kerbs, ruts, faults), off-road driving and the operation of vehicles at their maximum speed and load (the phenomena of bodywork movement dynamics) should all be considered, as all of this translates into additional forces acting on the wheel rim. When a vehicle moves in this way, the wheel rims often reach temperature levels that are uncommon for standard operational conditions since they contribute to dissipating heat generated during braking (frictional parts of the braking unit) and for electric drive vehicles – also from cooling of the drive system situated inside the wheel rim [2], which is very challenging for the wheel rims of such wheels. These constant cycles of changes, from environmental temperature, dynamics of longitudinal and cross-sectional loads to operational conditions and the goal of reducing the mass of wheel rims, may accelerate and limit the life durability of a wheel rim.

The methods of monitoring the structures and objects to identify and predict their conditions have become a major research area that is called NDT (Non-Destructive Testing). Damage to the structure and elements of machines can be prevented by detecting the cracks early, both those resulting from fatigue, with different non-destructive testing methods. Standard non-destructive testing techniques, such as penetration tests, magnetic methods, ultrasound, etc., have their intrinsic limitations and are often expensive and provide inconclusive feedback on the condition. During the next decade, a number of alternative methods based on identifying the nature and parameters of frequencies may become an effective, fast and handy diagnostic tool to detect the lack of structural integrity in machine parts and design systems.

It is commonly believed that the vibration theory is correlated with the modal parameter of a system, i.e. with natural frequencies, damping and types of vibrations. In other words, this is a physical system composed of the physical properties of a construction (mass, stiffness, and damping). These modal parameters are homogeneous systems described with differential equations of a model's physical motion that are expressed in relation to its mass, damping and stiffness, acceleration, velocity and displacement. Therefore, all changes in the modal parameters are conditioned by the physical features of the modelled object, which is due to damage.

Of these methods and diagnostic tools, the following should be mentioned: identification of dynamic characteristics such as natural frequencies (NF) [3], mode shapes and their derivative parameters [4], time domain vibration response [5-8] frequency response functions [9], power spectral density [10], transmissibility functions [11], coherence functions [12], correlation functions [13], etc., which serves to identify the condition of an investigated object based on emitted vibration signals.

This paper presents a comprehensive diagnostic method for identifying the condition of wheel rims not only in their modal (wheel rim structure parameters) or dynamic (assessment of unbalance and its positioning) aspects, but also with regards to the geometry (distribution of radial run-out and axial run-out for the wheel rim bead in relation to the wheel disc centering hole). The proposed methodology can be used to determine the optimal diagnostic signals and features that, after disassembling, help to identify the types and nature of damage to the `1wheel rims that occurred during operation, whereas in the future it will be applied to diagnose the complete wheel on the vehicle with the principles of the SHM (Structural Health Monitoring) philosophy.

2 Methods and experimental

In general, the method used to determine the condition of wheel rims consists of identifying the nature of natural frequencies (NF) with respect to the time and frequency for which the selected features of signal are determined and compared to the general reference model typical of a given type and model of wheel rim.

The analyses of damage to the wheel disc revealed that the phenomenon of inadequate join penetration between the wheel disc and the hoop results in a lack of connection between one or several wheel disc arms with the hoop. For new wheel discs, this issue is mainly an effect of incorrect welding parameters during the manufacturing process. It is emphasised that this type of non-operational state is difficult to detect and does not disqualify the wheel rim from operation; the wheel rim is most often damaged when it is overloaded or during extreme operating conditions. A similar phenomenon is also found for the wheel rims with a specific mileage. Loss of quality in the connections between the wheel disc and the hoop may be caused by a fatigue crack or an overload crack, a change in the structure's properties due to different phenomena, such as repairs (e.g. levelling), corrosion of fusion weld or materials of the wheel disc or hoop. The geometry of the wheel rims in relation to the wheel disc opening is very important for the safety and durability of the entire wheel operation.

At the initial stage, the objective was to develop a non-invasive method of assessing the condition of disc wheel rims after their manufacturing process. Verification tests of the developed method and the diagnostic station confirmed the efficacy and usability of the proposed methodology and, thus, it was also applied for the used wheel rims, i.e. ones that were already in operation.

2.1 Research objective

According to the developed method, the objective of diagnostic studies was to identify the following wheel rim parameters:

- loss of integrity or structural cracks that are assessed based on changes in stiffness and/or wheel rim damping (however, the damping phenomenon is not covered in this paper) for fusion-welded, pressure-welded, glued or riveted connections made of steel, steel cast or aluminium;
- geometrical parameters, such as radial run-out and axial run-out for the wheel rim bead in relation to the wheel rim cantering hole;
- dynamic parameters: balance quality factor, i.e. identifying the heaviest and lightest side of the wheel rim and its marking.

2.2 Subject of research

The studies were carried out on car and working machine wheel rims of discs, in different dimensions and of varied types, both new and already in operation. The wheel discs were all joint with the hoop via the arms, i.e. with the so-called "sequential bonding".

2.3 Diagnostic station and equipment

The nature of the ZRTOK diagnostic station, together with all major components, is presented in Figure 1. This station was constructed based on a serial balancing machine for vehicle wheels.

The identification of wheel rim modal parameters involves analysing the vibration signal recorded by the measuring sensors (4) situated on the shaft keeps frame (2). The measuring process begins with mounting a tested wheel rim (1) on the centring holder and fixing it with the pressure nut (3), similarly to the case of wheel mounting on a wheel balancing machine. When in such position, the wheel rim is exposed to vibration input (surge force) via a jackhammer (6) or via a mechanical vibration exciter. The vibration-exciting strokes are located on the edge of the wheel rim, at its lowest point, perpendicularly to the hoop. This process is repeated cyclically, every 90° of the hoop turn, to identify the damping and stiffness parameters for the wheel disc. This measurement generates four amplitude-time runs for each position of the wheel rim (0°, 90°, 180°, 270°) that are determined by the angle of rotation sensor (5). The angle of rotation for the wheel rim during the next measurement may have any value and mainly depends on the number of wheel disc arms connected with the wheel hoop.

Fig.1. The components of the ZRTOK diagnostic station: 1- an investigated wheel rim; 2 – frame with the shaft to assemble a wheel rim; 3 – pressure nut for the wheel rim; 4 – vibration acceleration sensors (C1 – the front one, C2 – the rear one); 5 – angle of rotation sensor (encoder); 6 – vibration exciter; 7 – wheel rim bead radial run-out and axial run-out measuring headers. The shaft rotation drive system and head positioning system are not presented in the drawing.

The value of rotation angle for the shaft on which the wheel rim is mounted is referred to as the hoop valve hole. When vibrations are excited and registered, the hoop is immobilized (it does not turn). Based on the analysis of time runs recorded by two sensors for four positions of the wheel rims, a damping factor is determined and following a FFT analysis, an vibration frequency spectrum is generated for each position and used to analyse the modal parameters of the hoop, among others.

Measurements of the wheel rim's geometrical parameters (as the fifth measurement) is run at the same hoop positioning as for the modal parameter testing, yet before the measurement is initiated, the automated system will move the measuring sensors (7) to the wheel rim bead and the drive unit then performs several full wheel rim turns. Meanwhile, the sensors record the length parameters recorded for the axial (A) and radial (R) plane and for the internal (IN) and external (OUT) wheel rim bead in relation to the wheel rotation angle. This approach serves to determine the maximum and minimum axial and radial run-out for the internal and external wheel rim bead. The results of these measurements and calculations may be reflected on the wheel rim by putting a spot or other appropriate marking.

In a similar way, using the phase angle method, the "light" and "heavy" side of the wheel rim is determined. The only difference is the application of acceleration sensors and a shaft rotation angle sensor. The total time of all the measurements regardless of the type or wheel rim dimension is approx. 20 s, not including the mounting and removal of wheel rim on or off the shaft of the diagnostic station.

The ZRTOK measuring system is equipped with two piezoelectric acceleration sensors (ICP 100 series – 100mV/g) and a shaft position sensor – encoder with 8-bit resolution (1,40°). Four OP-200-CW-10/80 eddy-current tracks are used as the axial run-out and radial run-out sensor, while the Impulse Force Hammer model: 086D05 (PCB) serves as a vibration exciter.

The measuring process of signal and working apparatus movements (i.e. measurement induction, wheel rim turns, positioning of wheel rim geometry measuring head) is automated and controlled by the NI USB 6343 card from the LabVIEW level.

3 Results and discussion

3.1 Course of test trials

The test trials of the wheel rim were conducted according to the following schedule: when the wheel rim was installed on the station shaft in relation to the valve hole position, a hammer stroke on the wheel rim was manually induced (the stroke force: 100N ±10N). Preliminary experiments confirmed the repeatability of diagnostic signals regardless of the sample of rim fixing, providing there was no damage to a wheel disc cantering hole and the procedure of mounting was complied with the requirements of the balancing machine manufacturer. Together with a hammer motion (the stroke), a procedure of recording the diagnostic signals was initiated, which lasts approximately 2 seconds at 2 kHz sampling frequency per channel for each angular position of the wheel rim.

3.2 Trial results and analysis

The time runs and acceleration frequency runs were generated during the trials and the values were quantitatively analysed. The selected results of these analyses are depicted below.

Based on the time runs for each of the measurements, the damping factor was determined and its value and change in relation to the angle of rotation for the wheel rim is a measure of wheel rim structure damping in the identified directions. The logarithmic decrement is defined as the ratio of absolute values for two consecutive extreme deflections. A change in this parameter may result from such issues as a lack of symmetry of the hoop due to its damage (e.g. deformation) or a lack of integrity of wheel rim material structure or the connection between the wheel disc and the hoop. The issue of damping decrement in the identification of the technical condition of wheel rims will not be further presented here due to the restrictions as to the length of this article.

The next stage of the analysis involves analysing the time runs with the Fourier transform. Figure 2 presents the frequency runs for the individual wheel rim rotation angles of a new 5Jx13H2 (material light alloy) wheel rim, together with the selected frequencies characteristic of the diagnostic station and the tested rim, recorded by the C1 sensor (Fig.1).

The analysis of the runs presented on the vibration spectrum allows distinguishing two organized groups of peaks. The first one is associated with the design of the ZRTOK diagnostic station and the other with the tested wheel rim. The spectrum visualized in

Figure 2 features the resonant frequency for the station shaft on which the tested wheel rim was mounted (labelled: NF of Shaft ZRTOK) and vibration frequency of the frame on which the shaft of positioned (NF of Frame ZRTOK) and their next harmonics. The following frequencies are associated with the tested wheel rim: natural frequencies of the complete wheel rim (NF of WHEEL RIM), natural frequencies of the hoop (HF of HOOP (Sound of Hoop)) and its next harmonics at high frequencies (not visible in the plot).

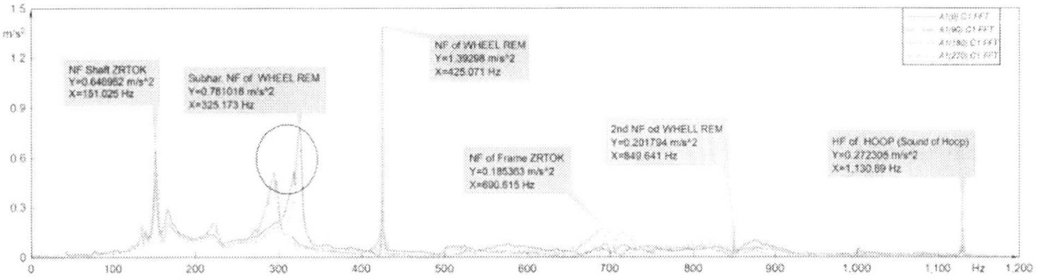

Fig. 2. The vibration amplitude spectra for the new 5Jx13H2 wheel rim, recorded on the ZRTOK diagnostic station; the selected frequencies characteristic of the station and the tested wheel rim are emphasized for the following four measurement angles: 0º, 90º, 180º, 270º.

When comparing the obtained frequency spectrums for the particular measurement angles of the tested wheel rim (Fig. 2), it could be concluded that the spectrums are similar (they overlap each-other) in the characteristic points (PEAK). This is confirmed by a comparative analysis of both runs, based of which it was found that the max relative error in the analysed characteristic points does not exceed 0.5% for the frequency values, and it does not exceed 15% for the vibration amplitude. Such a large relative error of the vibration amplitude resulted from a manual method of vibration excitation.

In order to differentiate the vibration form of the station and the tested wheel rim, a time frequency spectrum of the 5Jx13H2 wheel rim (the same, whose vibration spectrum was presented in Fig. 2) was used (Fig. 3). The frequencies related to the diagnostic station are characterised by a high damping of vibration amplitudes. On the other hand, the vibration amplitudes at the frequencies related to the tested wheel rim, are much longer due to significantly lower damping of material and structure of the wheel rim compared to the damping of the station's construction. During the preliminary studies of the method and the diagnostic station, a detailed identification of these frequencies (mod), with both experimental method using the vibration signal and the acoustic signal and the obtained study results were confirmed with the analytic and simulation method.

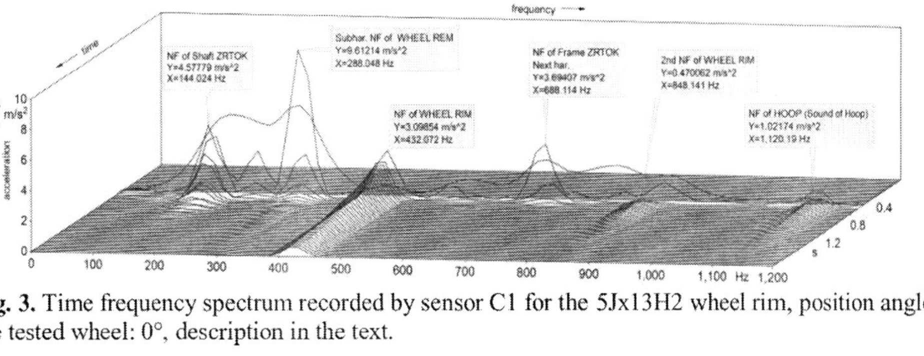

Fig. 3. Time frequency spectrum recorded by sensor C1 for the 5Jx13H2 wheel rim, position angle of the tested wheel: 0°, description in the text.

For the discussed method, the nature of evaluation and identification of the condition of a tested wheel rim is based on comparing the recorded time and frequency runs to the angle of rotation for a tested wheel rim during measurements. If the wheel rim has similar geometric and structural properties and a quality connection between the wheel disc and the hoop at each of the investigated directions, the runs are similar to a wheel disc

Based on an analysis of the study results and identification of diagnostic signals the following sets of wheel rim unfitness symptoms were distinguished at the ZRTOK diagnostic station:

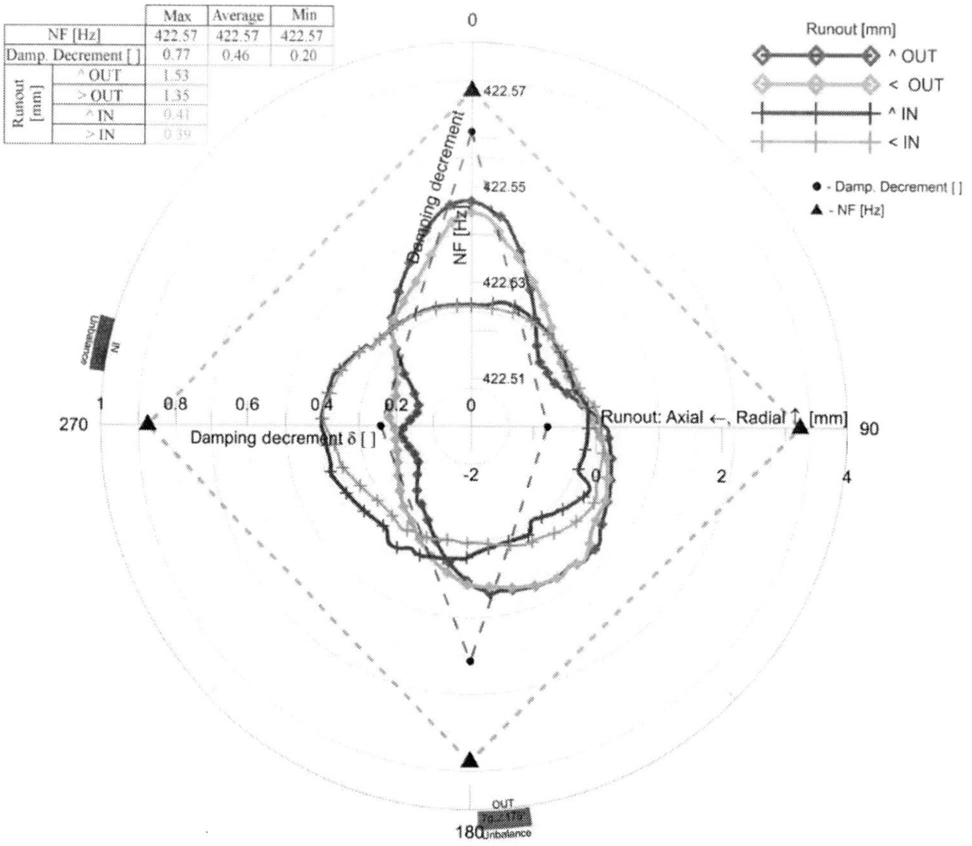

Fig. 4. The graphical report of the 5Jx13H2 wheel rim test trials (manufacturing date: 1993), generated by the ZRTOK system.

Spectral asymmetry – understood as the value of standard deviation or relative error of NF and the vibration spectrum in the characteristic points in relation to subsequent measurements (0°, 90°,180° and 270°) for the same wheel rim. Examples of damage escribed by this feature: rim material crack, faulty connection of a rim with a disc, etc.

Damping decrement asymmetry – defined as a "high" value of the damping decrement average (δ) and a "low" value of damping decrement standard deviation ($SD_δ$) for subsequent measurements of a tested wheel rim. Examples of damage: wheel rim structure discontinuity, varied directional stiffness of a wheel rim.

Faulty mounting on the ZRTOK station shaft or a damage to a centring hole of the wheel disc – described as a "low" value of an average damping decrement δ and a "low" value of $SD_δ$, for subsequent measurements of a tested wheel rim as well as a "low" value of the amplitude of the recorded vibrations.

High frequency spectral asymmetry (within the frequencies of rim sounds – sound of Hoop) – understood as a formation of a series of neighbouring peaks (organised amplitudes) at the wheel hoop sound frequencies. This phenomenon is caused by discontinuity of a butt weld structure of the wheel hoop (e.g. cracking during spin-forming or operation resulting from a lack of re-melting during wheel welding).

Geometric asymmetry – exceeding the permissible value of radial and axial run-out of the internal (IN) and external (OUT) side of the wheel rim bead relative to a centring hole of a wheel disc with marking in respect of the norms requirements.

Good balancing – identifying the correction mass value in g (grams) reduced to the diameter of wheel rim necessary to dynamically balance a wheel rim for each of the sides including marking in respect of the norms requirements.

Permissible values and parameter limit values for conditions: NF, damping decrement, run-out value or good balancing are individualised and connected with a mass, dimension, kind and type of a tested wheel rim and the requirements imposed by a producer or a technical norm.

A two-state classification of conditions (fit, unfit), identical for all the symptoms with an unconditional differentiation of conditions was used in the developed diagnostic model of wheel rim technical condition identification. Figure 4 presents a graphical report (spider diagrams) of a 5Jx13H2 wheel rim test trials (the wheel rime is made of light alloy, manufacturing date: 1993) that was generated by the ZRTOK system at the end of the test trial procedure.

The analysis of the results also reveals a constant value of natural frequencies (NF) for each measured angle of the wheel rim. For the geometry and damping factor, asymmetry becomes obvious. According to the standards, the value of axial run-out and radial run-out of the tested wheel rim on the OUT side eliminates the wheel rim from further operation. This geometry asymmetry of the external (OUT) side of the wheel rim includes the distribution of the damping factor resulting from other damping of the wheel rim at 0-180° and 90-270° directions.

4 Conclusions

Based on the test trials and the analysis of discussed method and diagnostic station, the following conclusions are presented:

- The discussed methodology and the diagnostic station is applied both for identifying the unfitness of wheel rims, such as a lack of design integrity or structural integrity of the wheel rim caused by a crack, corrosion or mechanical wear and other issues making the wheel rims unfit for use, for instance, unbalance and geometrical asymmetry of wheel rim;
- The assessment of amplitude-frequency runs for the subsequent angle measurements of the wheel rim allows identifying the condition of connection between the wheel disc and the hoop. Significant spread of NF values for the subsequent measurements is the main feature of the signal indicating this type of damage. However, it should be considered that a similar symptom is caused by substantial asymmetry of the wheel rim or a lack of structural integrity of the wheel rim.
- The condition of centring the hole of a wheel rim has a substantial impact on measurements of modal parameters; all defects, damages (cracks) and additional centring sleeves (often included in wheel rims made of light alloys) significantly limit the accuracy of measurements and the quality of results analyses, therefore for the purpose of the presented diagnostic method, an identification procedure for such a condition was developed in the form of a faulty mounting or damage to the centring hole of the tested rim;
- The increment of measuring angle (number of measurements) for the individual diagnostic signals can be reduced to 20° to perform more precise analyses of the wheel rim condition. A smaller value of increments for the wheel rim rotation angle than the proposed one does not impact the quality or informative nature of the recorded values;

– The discussed research methodology can be applied for identifying the modal parameters of new wheel rims (verification of the manufacturing process quality – the method is used in the quality control system for the end-products in the disc wheel rims manufacturer Polkar Warmia) and wheel rims that are already in operation in order to identify their condition, e.g. during a seasonal tyre exchange in a tyre exchange station.

5 Acknowledgments

This research was conducted under NCBR project no. POIR.01.01.01-00-1746/15 by POLKAR Warmia Sp. z o., ul. Fabryczna 21, 11-040 Dobre Miasto.

6 References

1. J. K. Sinha, M. I. Friswell, S. Edwards, *Simplified models for the location of cracks in beam structures using measured vibration data*. Journal of Sound and Vibration **251(1)**, p. 13-38. (2002). DOI: 10.1006/jsvi.2001.3978
2. Matej Biček, Gorazd Gotovac, Damijan Miljavec, Samo Zupan. *Mechanical failure mode causes of in-wheel motors. Strojniški vestnik* - Journal of Mechanical Engineering, Vol. **61**, p. 74-85. (2015). DOI:10.5545/sv-jme.2014.2022
3. K. He and W.D. Zhu, *Structural damage detection using changes in natural frequencies: theory and applications*. Journal of Physics: Conference Series, Vol. **305**, Issue 1, p. 012054. (2011)
4. N. M. M. Maia, J. M. M. Silva, E. A. M. Almas, *Damage detection in structures; from mode shape of frequency response function methods. Mechanical Systems and Signal Processing*, Vol. **17**, Issue 3, p. 489–498. (2003)
5. J. Cattarius, D. J. Inman, *Time domain analysis for damage detection in smart structures, Mechanical Systems and Signal Processing*, Vol. **11**, Issue 3, p. 409–423. (1997)
6. T. Kuroiwa, H. Iemura, *Vibration based damage detection using time series analysis*, The 14th world conference on earthquake engineering, Beijing, China, October 12-17. (2008)
7. A. Rychlik. *Detection of structural damage in vibroacoustic analysis*. Journal of KONES Powertrain and Transport, Vol. **23**, No. 1, p. 279-288. (2016).
8. A. Rychlik, *A method for evaluating the technical condition of wheel rims in slow moving vehicles based on modal parameters*. DIAGNOSTYKA, Vol. **17**, No. 2, p: 65-70, (2016)
9. N. Guo, Z. Yan , Y. Jia , L. Wang. *Model updating using correlation analysis of strain frequency response function,* Mechanical Systems and Signal Processing, Vols. **70-71**, p. 284–299. (2016)
10. Z. D. Zheng, Z. R. Lu, W. H. Chen, J. K. Liu., (2014). *Structural damage identification based on power spectral density sensitivity analysis of dynamic responses*, Computers & Structures, Vol. **146**, p. 176–184. DOI: 10.1016/j.compstruc.2014.10.011
11. D. Zhu, Yi X., Y. Wang. *Sensitivity analysis of transmissibility functions for structural damage detection*, Proceedings of SPIE 7983, (2011) 79832M-1, DOI: 10.1117/12.879867
12. D. D. Rizos, S. D. Fassois, Z. P. Marioli-Riga, A. N. Karanika, *Vibration-based skin damage statistical detection and restoration assessment in a stiffened aircraft panel*. Mechanical Systems and Signal Processing, Vol. **22**, Issue 2, p. 315-337. (2008)

13. M. Zhang, R. Schmidt, *Sensitivity analysis of an auto-correlation-function-based damage index and its application in structural damage detection*. Journal of Sound and Vibration, Vol. **333**, Issue 26, p. 7352-7363. (2014)

Autocalibration of gyroscope based two-step regulator for autonomous vehicles.

Piotr Wasilewski[1], *Krystian* Chojnowski[1], *Rafał* Grądzki[1*]

[1]Bialystok University of Technology, Faculty of Mechanical Engineering, , Wiejska 45C, Bialystok 15-351, Poland

Abstract. In this article was presented an auto calibration algorithm based on a two-step regulator and gyroscope readings. Unable to use PID, as the PWM signal causes the motors to overheat, it was decided to introduce a two-step algorithm. This type of controller reaches the setpoint much faster, though an error is present. Gyroscope is used to sample angular speed and then integrate these values to get the angle. To calibrate the offsets the robot performs a calibration run during which it measures and averages the error on each gyroscope axis while no movement is present. Afterwards it makes a characteristics of braking angle and angular speed. Interpolated polynomial from 13 samples allows to designate the point in time where the braking has to start to achieve the desired angle. The calibration aims for adjusting the parameters of movement depending on the surface it runs on. Additional check is made using encoders to estimate the accuracy of gyroscope output.

1 Introduction

Positioning of a wheeled mobile vehicle is a complex subject. It can be done using plentiful of sensors like accelerometer, gyroscope, encoders or GPS. None of them can calculate position perfectly, that is why for path planning it is necessary to use different types of sensors. Path planning using accelerometer and gyroscope only is difficult due to errors they might produce. It is often done using those two sensors along with magnetometer mounted in the same structure called IMU (inertial measurement unit). Measurements need to be adjusted using reference points that have a fixed, well known position.. GPS has a poor accuracy and cannot be used for indoor positioning. This environment requires good calibration of other sensors.

Usage of gyroscope and accelerometer is burdened with errors resulting from integrating, noise or environmental factors like centrifugal force. Mounting of the sensors is also important and done improperly can affect the readings producing offsets. Encoders often used in mobile vehicles, on the other hand, are sensitive to any kind of slips and cannot detect movements parallel to the axis of the wheels. LIDAR sensor can also be implemented to measure the distance from obstacles. As for static environment they can be treated as reference points.

Mega Sumo robots are specific types of two wheeled platforms. On the steel ring the movement resistances are very high, though the inertia is low. This allows to drive the robot at high velocities on relatively small area. The resistances cause the robot to stop instantly after turning the motors off, and start at minimum 30% of PWM duty. However low PWM duty cycle causes the motor to draw higher current resulting in faster heat buildup. Its coreless design results in lower inductance than traditional core-based motors. The current rises and falls much faster therefore introducing high current spikes and high instantaneous power dissipation in windings. Considering this problem it was decided to use a two state regulator in order to reduce the heating and maintain maximum performance. PID and its derivative

* Corresponding author: r.gradzki@pb.edu.pl

forms are not suitable as their output covers the whole PWM range increasing the motor heating. Moreover that kind of regulators reaches the target angle slower than the one implemented in this project. The error resulting from the two-step controller is always present, however an error of $\pm 10\%$ can be treated as insignificant, taking under consideration the high velocities and accelerations occurring while moving.

Positioning the platform is complex due to the lack of any kind of markers with known constant position [1,2]. Walls or other obstacles can be used as reference points, though their coordinates are not exactly known. Different techniques of combining odometry models and gyroscope/accelerometer are used in order to decrease the influence of errors [4]. To reduce IMU sensor error Kalman filter can be applied [3]. It fuses data from different sensors in order to compensate the error resulting from the design of each sensor. Kalman filter requires detailed knowledge about the object, as well as high computing power. Indoor path planning requires proper calibration of sensors built in IMU. Without the calibration they introduce an error basing on their positioning. The article presents a way of calibrating the gyroscope and using its readings to rotate by a set angle.

Calibration of sensors mentioned above is complex subject. It requires plenty of tests and calculations. Because of that the focus of the article is on rotating the robot by a desired angle only. The rotation is done using two-step regulator which requires previous readings to predict the angle the robot will turn after control signal to stop. To get the proper readings offset calculations has to be made. The whole calibration process is based on the gyroscope alone. It can be burdened with error in long-term due to gyroscope's angle drift. Encoders are used only as a comparison of angle calculated from gyroscope readings, and to check if there are slips during the movement.

2 Theory

As it was decided to use two-step regulator there are only two drive states - 0 and 100% velocity. In theory the idea was to designate the braking distance from 12 different values of angular velocity read from the gyroscope. Knowing the velocity the robot has to turn with in order to stop at setpoint angle allows to turn off the motors at appropriate moment. Braking distance is a recurrent value dependent on the actual angular velocity, due to the low inertia of the whole system. The value of angular acceleration slightly varies when the voltage on the batteries drops, that is why the curve of degrees per second and angle has to be designated again when the voltage drops significantly.

For the proper readings of gyroscope, calibration of the offsets has to be made. It can be caused by soldering the sensor, internal stresses or by change in ambient temperature. Removing offsets is done by subtraction of measured and averaged values while standing still on the drive surface. After removing offsets any reading on gyroscope is supposed to be equivalent to actual movement, and should be read 0 without the movement.

Braking angle based on angular velocity is useful for the two-step regulator. It allows to predict the angle the vehicle will turn with motors off at given angular velocity. Surely it is impossible to measure braking angle at all the possible points, so the braking curve is fitted with least-square method designated polynomial using several measurements presented on equation 1. The fit is not perfect but it can average differences in angle due to delays between reading and control signal and can be successfully used for the test. Another useful property of the polynomial as fitting curve is that it is easier to calculate using on-board microcontroller without so much computing power.

Polynomial is fitted by solving Matrix equation using Gaussian elimination [6].

$$\begin{bmatrix} N & \sum_{i=1}^{N} x_i & \cdots & \sum_{i=1}^{N} x_i^k \\ \sum_{i=1}^{N} x_i & \sum_{i=1}^{N} x_i^2 & \cdots & \sum_{i=1}^{N} x_i^{k+1} \\ \vdots & \vdots & \ddots & \vdots \\ \sum_{i=1}^{N} x_i^k & \sum_{i=1}^{N} x_i^{k+1} & \cdots & \sum_{i=1}^{N} x_i^{2k} \end{bmatrix} \begin{bmatrix} a_0 \\ a_1 \\ \vdots \\ a_k \end{bmatrix} = \begin{bmatrix} \sum_{i=1}^{N} y_i \\ \sum_{i=1}^{N} x_i y_i \\ \vdots \\ \sum_{i=1}^{N} x_i^k y_i \end{bmatrix} \quad (1)$$

In equation 1, k is polynomial degree and N is number of measured points. Increasing number of points allows using higher polynomial degree or getting better fit. Calculating a set of 6 equations takes approximately 1,5ms on the microcontroller using single-precision floating-points operations. The precision of the calculations can be improved by using double-precision type but it requires a lot more calculating time due to lack of hardware double-precision unit in microcontroller that was used.

Having the braking angle from angular velocity curve allows creating two-step regulator. The regulator has three inputs: current angle change, angular velocity and targeted angle. When current angle added to braking angle calculated for current angular velocity is higher than targeted angle, the motors are turned off and vehicle starts braking.

Using the regulator a difference between set and actual angle can be calculated. It can be represented as absolute error or relative error. Absolute error gives the difference presented in degrees. This method allows to assess the regulator and apply that as error zone for regulator. Relative error tells the percentage of the angle over- or undershoot for the angle. The difference between both the errors is that same absolute error is different relatively to set point. It is crucial to asses if the regulator can be used for every angle.

3 Test platform

The platform is two wheeled autonomous robot with a fulcrum on the front and two 40mm diameter wheels on the back. Magnets placed on the chassis give the robot additional downforce due to steel surface at which it operates. It results in compression of the rubber on the wheels up to 1.5mm depending on the surface it is placed on. Spacing between wheels' centers is 128 mm. The platform reaches high velocities of up to 3.6m/s, which means it is capable of crossing the 1,54m diameter Dohyo in about 400ms. High forces present while accelerating, braking and rotating require high performance sensors able to measure them.

ICM-20601 is the 3-axis 4000 dps (degrees per second) gyroscope that has been used in the tests. Built in low-pass filter with cutoff frequency 176 Hz gives more stable readings. Calibration of the gyroscope before each test is needed to overcome the problem of inaccurate data without the movement.

AS5045 quadrature encoders for the assessment of the gyroscope's readings have been placed on the wheel axis. Placing of the encoders results in precision decrease, however it became resistant to the backlash of the gears. With this setup we can assume the encoders show the actual distance traveled by each wheel.

All readings are recorded onto micro SD card present on the mainboard of the robot. Samples are saved with 1 kHz frequency basing on the "data ready" interrupts from the gyroscope. Higher frequencies caused problems with synchronization and saving data, leading to delays that are unacceptable, as the system is time-based. Subsequently the recorded data is analyzed with Matlab software to compare it with the computing made with on board microcontroller.

Due to extreme accelerations and velocities occurring in MegaSumo robot operations it is difficult to adjust the regulators and the error that may be present is not significant. However it is then easier to apply these kinds of calibration in other autonomous vehicles working in less harsh conditions.

4 Calibration process

Mega Sumo robots run on a heavy steel ring usually placed on the floor. The difference in placement of both the ring and the robot itself can affect the sensors' readings. Depending on the surface angle there is a need of changing the default gyroscope offsets to get a reliable output. To calibrate the robot we introduced a single program, performing a few separate tests. After executing it the platform should be calibrated well enough to make the set angles with 10% tolerance. However this implementation is susceptible to any changes of the environment it was calibrated in, as it assumes no distortion of the surface it runs on. The calibration has to be performed before every run when the conditions change significantly. Such change can be produced due to the progressive voltage drop on the batteries that could affect the accelerations and maximal angular speed.

Figure 1 shows the control signal and robot's angular velocity. The first vertical red line is start indicator, the other is stop signal. After selecting stop the robot goes further and stops with a significant delay. In this article the dependence of breaking angle from a set angular velocity is used to estimate where the robot is going to stop.

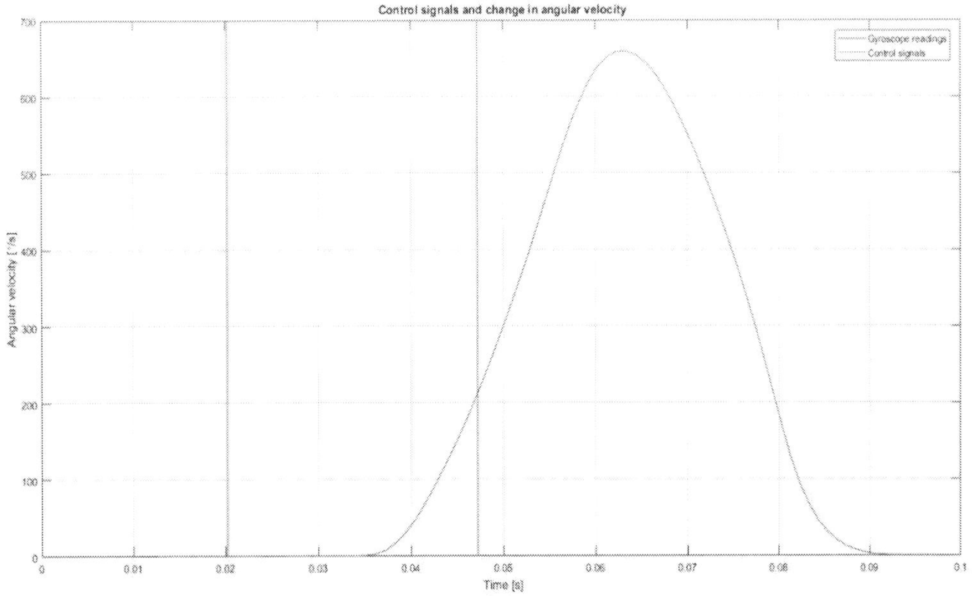

Fig. 1. Control signal and change in angular velocity.

At first the robot is placed on the ring by the operator. After start signal it performs four measurements of the gyroscope offsets at different places on the ring. During each test the robot goes straight and then rotates by 90°. It samples the values present on each axis between the tests while no movement is present. After fourth run the sampled data is averaged and calculated values are added to each axis.

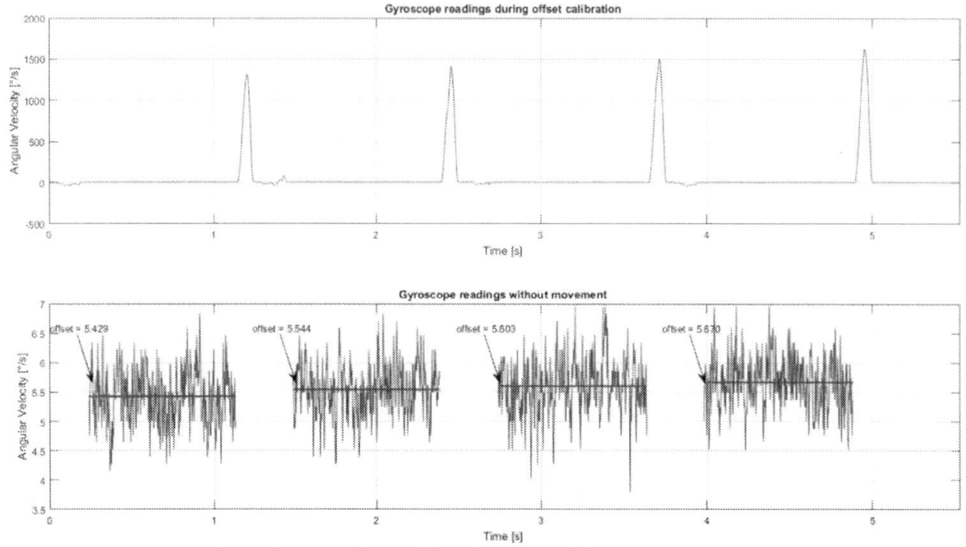

Fig. 2. Gyroscope readings during offset calibration and without movement.

Figure 2 presents gyroscope readings during the calibration of offsets. First one shows angular velocity for the whole test to show the magnitude of changes during movement. Second one presents oscillations caused by noise without movement. Readings change in range of about ± 1 dps but its average is not changing significantly for the whole time of the test. The offset calculated during the test is 5.5615 dps and this is the value that is subtracted from the readings during further tests.

Second test is performed to gain information about the braking angle from the set angular speed. The robot is rotated from 200 up to 2300 dps with a step of 200 dps, and half a second delay between tests, to be sure the robot has stopped. The value of 2400 dps becomes unreachable after a few tests when the voltage on the Li-po batteries drops, that is why a safe value of 2300 dps is used. With the zero dps point where surely angle has to be zero there are 13 samples from which the fitted polynomial coefficients are calculated.

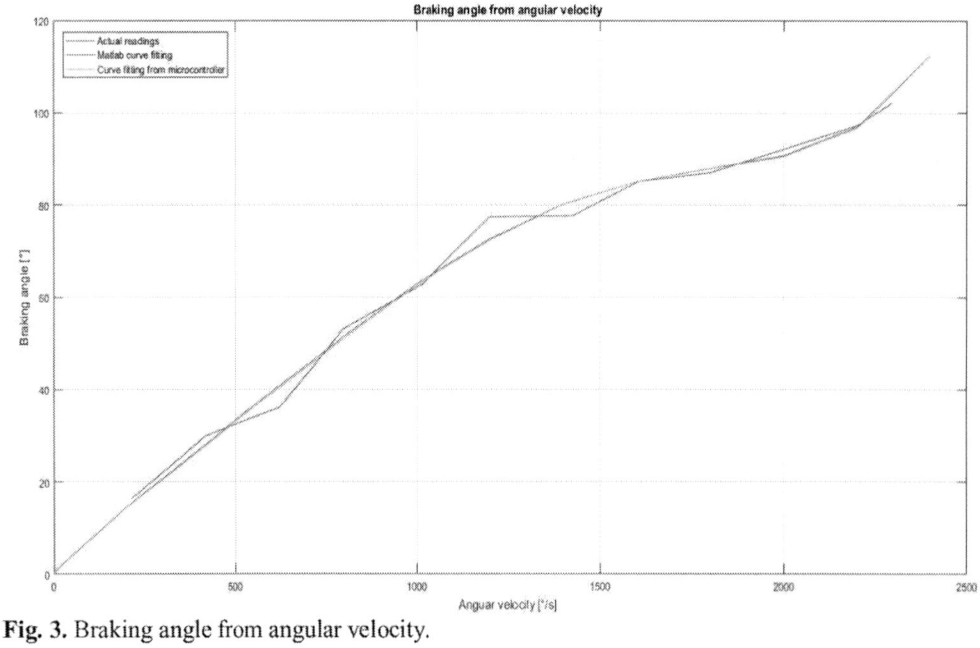

Fig. 3. Braking angle from angular velocity.

Designated polynomial is used to predict when the robot has to start braking to stop at setpoint angle. This method gives about ±10% accuracy. Figure 3 presents actual curve from readings during second test and fitted polynomials calculated by microcontroller as well as one calculated with higher precision in Matlab software. Both polynomials were calculated from the same gyroscope readings yet the points differ insignificantly due to difference of floating point precision. It is difficult to distinguish one from another which confirms the correctness of the method used on microcontroller. The difference in precision is also unimportant for this kind of calculations.

Last test is performed to check if the polynomial was accurately calculated and if it can be used for two-step regulator. The robot makes 18 tests - turn every 10° starting from 10° up to 180°. During each test the actual angle is compared to the setpoint angle and an error is appointed. If the error is for consecutive tests greater than 10% the robot notifies the operator it needs another run to gain new data.

At the end the polynomial coefficients are saved onto micro SD card in order to read the calculated values at another run.

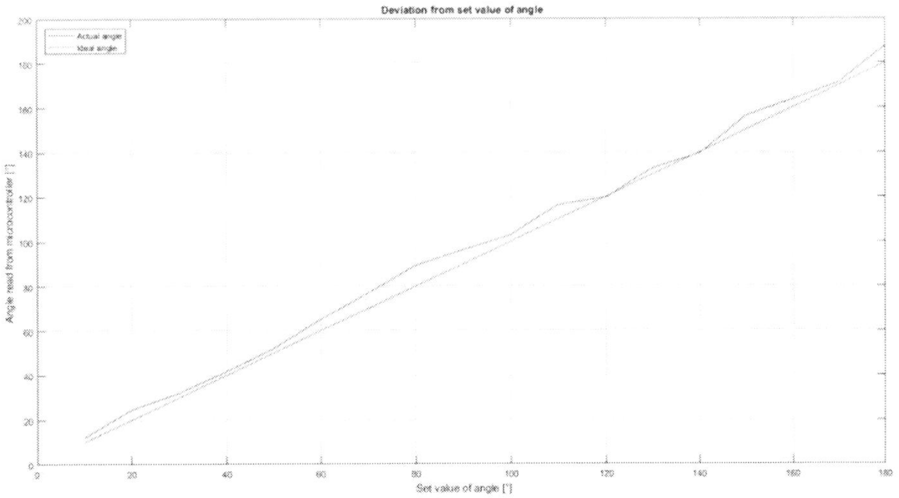

Fig. 4. Deviation from set value of angle.

Figure 4 presents actual angle by which the robot turned using two-step regulator. Both characteristics have similar shape yet the actual value is greater than ideal by few degrees. It is clear that it always overshoots the angle. It implies that the regulator should have an error zone which would decrease the difference. There is no reading showing significant changes in the shape of the characteristics which confirms the accuracy of the fitting in figure 3. All the actual readings and set angles are shown in Table 1.

Table 1. Set angle compared to actual angle read from the vehicle.

Set angle	10°	20°	30°	40°	50°	60°	70°	80°	90°
Read angle	11.96°	24.61°	32.09°	41.68°	52.03°	64.99°	77.06°	89.24°	96.34°

Set angle	100°	110°	120°	130°	140°	150°	160°	170°	180°
Read angle	102.92°	116.26°	119.66°	132.76°	139.56°	156.19°	163.64°	171.25°	187.78°

To assess the significance of the overshooting the angle the absolute and relative error is shown on figure 5. The absolute error gives the difference in degrees which can represent the accuracy of the regulator but cannot assess if it can be used for the specific set angle. This assessment can only be done by relative error.

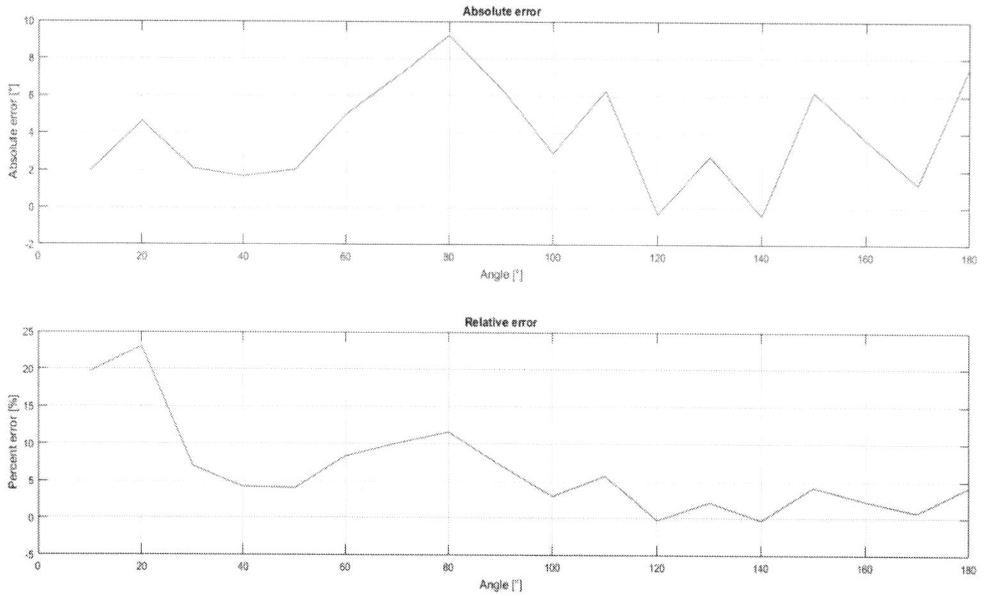

Fig. 5. Absolute and relative error.

The absolute error during test is in the interval of -0.44° up to 9.24° as presented in table 2. It shows that during the test actual angle was mostly greater than set angle. Introducing error zone as mean of the absolute error would decrease the value of errors with the same magnitude of absolute errors which is 9.68°. The relative error could help assess the accuracy of the test for each angle better than absolute error, e.g for set angle of 10° absolute error is less than 2° which gives relative error of 20%. Relative error for most of the readings is less than 10%. Adding the error zone could also decrease the relative error. The highest errors recorded are for the smallest set angles. This implies that two-step regulator is not useful for small angles. The error for set angle of 80° is higher than 10% but it could be treated as an error caused by slight delay in control signal with high angular velocity.

Table 2. Set angle compared to relative error and absolute error for each angle made.

Set angle	10°	20°	30°	40°	50°	60°	70°	80°	90°
Absolute error	1.96°	4.61°	2.09°	1.68°	2.03°	4.99°	7.06°	9.24°	6.34°
Relative error	19.59%	23.04%	6.98%	4.20%	4.06%	8.32%	10.08%	11.55%	7.04%

Set angle	100°	110°	120°	130°	140°	150°	160°	170°	180°
Absolute error	2.92°	6.26°	-0.34°	2.76°	-0.44°	6.19°	3.64°	1.25°	7.78°
Relative error	2.92%	5.69%	-0.28%	2.12%	-0.31%	4.12%	2.27%	0.73%	4.32%

What is important to check if there are no slips during the movement. While testing it turned out there are no slips, so the encoders were the reference for the angle integrated from gyroscope data. As the encoders are directly connected with wheels they are a more reliable source of the distance travelled by the platform. A comparison of encoder data and gyroscope readings presented on figure 6 showed a delay between these two. The gyroscope seems to detect the angle later and output a lower angle, which could be an effect of flexibility of mounting. The change in angle between the two ways of calculating is insignificantly different. The change in gyroscope based angle can be caused by not detecting slight changes in angles while encoder based angle detects it.

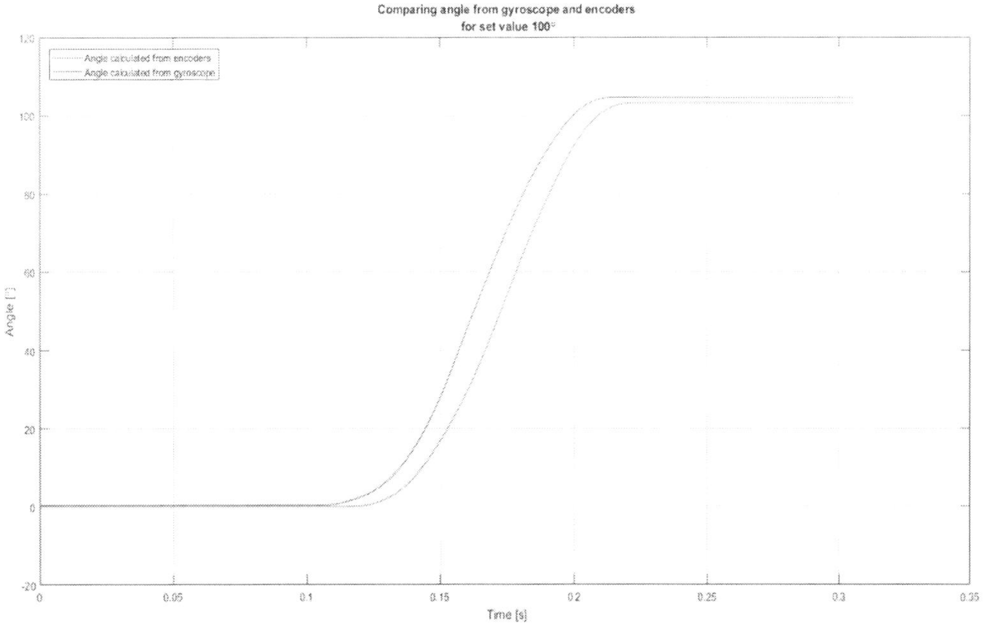

Fig. 6. Comparison of angle from gyroscope and encoders for a value of 100°.

Figure 7 and table 3 presents differences between actual angle and angle designated from the polynomial. For lower angles both characteristics are similar. Angle done before reaching angular velocity corresponding to proper angle braking does not exceed 10° during test up to 60° of set angle. Tests for higher angles do not show high angular velocity changes which means that the time to reach those angles does not differ significantly. Reaching higher angles than testes is unnecessary to test due to possible rotation in both directions.

Table 3. Comparison of a set angle, angular velocity, theoretical and read braking angle.

Set angle	10°	20°	30°	40°	50°	60°	70°	80°	90°
Angular velocity	139,8	288,4	405,5	583,2	690,1	842,2	947,8	1027,3	1205
Braking angle	11,68	23,34	29,90	37,48	45,96	56,56	66,30	74,17	78,13
Calculated braking angle	10,62	20,16	27,27	37,96	44,40	53,46	59,54	63,89	72,63
Set angle	100°	110°	120°	130°	140°	150°	160°	170°	180°
Angular velocity	1281,3	1417,9	1609,6	1750,6	1855,4	1913,4	2017,8	2070,8	2113
Braking angle	80,14	87,31	86,23	91,07	88,84	96,78	95,71	94,00	101,69
Calculated braking angle	75,84	80,61	85,18	87,24	88,44	89,13	90,77	91,96	93,16

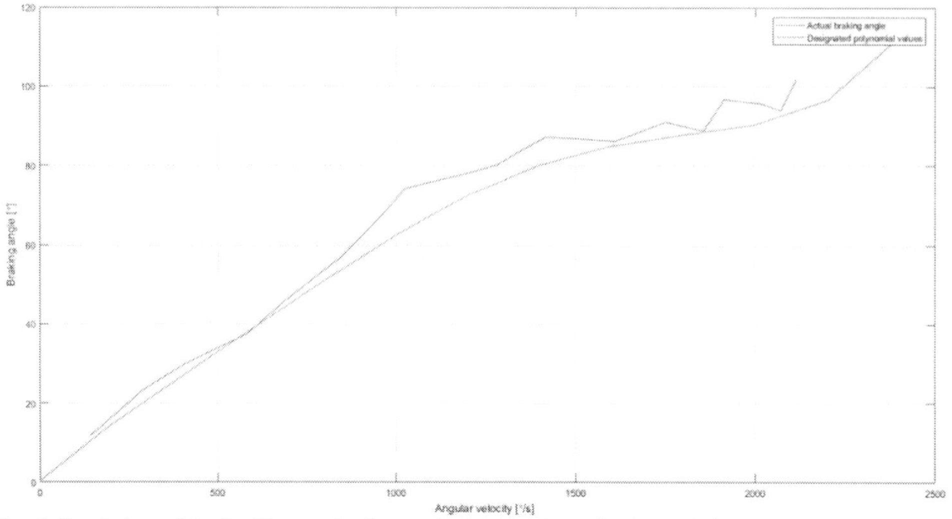

Fig. 7. Deviation of the braking angle from calculated value of polynomial.

5 Conclusions

Research conducted using the created algorithm allows to calibrate the two wheeled vehicle. The main idea was to achieve a relatively low error while rotating the robot to a set angle. Two-step regulator using gyroscope gave satisfying results, as other regulators were not taken under consideration due to physical limitations. Even though the platform is specific type of a two wheeled robot, the algorithm can be successfully adapted in other mobile vehicles. High accelerations and velocities reached by Mega Sumo robot required a wide range sensors which tend to be less accurate. Usage of a smaller scale gyroscope may result in a better accuracy of movement.

Presented solution of a two-step regulator is suitable for angles bigger than 30°. Lower angles have a big error due angle error that is present in all the readings. While rotating by a small angle the drive time has to be really short and as a result it tends to overshoot the desired angle. High accelerations and frequency of the gyroscope readings can be cause to the errors. Time between the readings is around 1 ms and in that time angular velocity can change significantly causing the regulator to overshoot.

Absolute error present in the readings can be taken into account while building regulator as an error zone at which the robot should begin braking. Using that does not change the outcome of error occurring it just allows for it to be closer to the set angle. Delays between control signals and movement cause problems in proper controlling of the vehicle. They cannot be reduced because they are the property of the mechanical set of the vehicle. Given the time sensitivity for the whole system it is important to be aware of the delays and introduce them in the control algorithms.

Using polynomial as the curve fitting is the most efficient way to be used on microcontrollers. It can be easily designated and calculated with low computing power of microcontroller. Other function used as curve fitting may present more accurate data but are more complex.

Presented two-step regulator cannot produce ideal results but for the operation condition of the testing platform slight change in the angle is accepted. The error during the turn can be later compensated during forward motion. Using only gyroscope in long term can produce errors due to angle drift which can be compensated using sensor fusion. Progressive voltage drop is one of the problems that the operator has to face with. Lower voltage equals to lower acceleration which leads to distortion of the fitted curve. It causes changes in braking angle for given angular velocity. Voltage drop can also be problematic during calibration making it difficult to reach given angular velocity to produce polynomial.

The calibration process can be expanded by adapting the accelerometer for translational movements. Later it could be used for inverse kinematics and full path planning algorithm. For the proper use of path planning algorithm it is necessary to calibrate all the possible movement types (straight forward motion, curved motion and rotation in place) and transition between them. Autonomous vehicle will then be able to change the control signals earlier which results in proper achievement of the set points.

This paper was supported by the research project No. S/WM/1/2016 and financed by the Ministry of Science and Higher Education.

References

1. W. Lv, Y. Kang, J. Qin, *Indoor Localization for Skid-Steering Mobile Robot by Fusing Encoder, Gyroscope, and Magnetometer*, IEEE Transactions on Systems, Man, and Cybernetics: Systems, pp. 1-13 (2017), DOI: 10.1109/TSMC.2017.2701353
2. H. Chung, C. Hou, Y. Chen, *Indoor Intelligent Mobile Robot Localization Using Fuzzy Compensation and Kalman Filter to Fuse the Data of Gyroscope and Magnetometer,* IEEE Transactions on Industrial Electronics, **62**, Issue:10, pp. 6436-6447 (2015), DOI: 10.1109/TIE.2015.2416692
3. H. Shah, K. Mehta, S. Gandhi, *Autonomous Navigation of 3 Wheel Robots using Rotary Encoders and Gyroscope*, International Conference on Computational Intelligence and Communication Networks (CICN), 14-16 November 2014 Bhopal, India, DOI: 10.1109/CICN.2014.244
4. N. Houshangi, F. Azizi, *Mobile robot position determination using data integration of odometry and gyroscope*, World Automation Congress (WAC), July 24-26 2006, Budapest, Hungary, DOI: 10.1109/WAC.2006.375994
5. S. Hong, G. Zhao, X. Wang, X. Zhong, *A Method to Coordinate Balance and Speed Control of a Two-wheeled Robot*, 29th Chinese Control And Decision Conference (CCDC), 28-30 May 2017, Chongqing, China, DOI: 10.1109/CCDC.2017.7978203
6. L. Yi, W. Y. Ping, *The Realization of Polynomial Fit between EMF and Temperature Changes Based on Least Squares Fit,* International Conference on Electronic & Mechanical Engineering and Information Technology (EMEIT), 12-14 August 2011, Harbin, China, DOI: 10.1109/EMEIT.2011.6022967

The theoretical basis of the method of optimizing parameters of the propulsion multipurpose wheeled vehicles

Valery Guskov[1,*], *Tatiana* Mikulic[2], *Victoria* Pavlova[3], and *Alexey* Sochnev[1]

[1]Belarusian National Technical University, Department "Tractors", 220013 Minsk, Nezavisimosti, 65, Belarus
[2]Belarusian National Technical University, Department "Details of machines", 220013 Minsk, Nezavisimosti, 65, Belarus
[3]Belarusian National Technical University, Department "Transport systems and technologies", 220013 Minsk, Nezavisimosti, 65, Belarus

Abstract. In the design of the wheel machine there are problems of selecting the parameters of its engine at a given load on the wheel or determining the optimal normal load at a given wheel parameters. The method of optimization of parameters of a wheel and its loading proceeding from an integral criterion of efficiency is offered. As a criterion, such an indicator as the efficiency of the driving wheel is taken. The article offers mathematical dependences for determining the efficiency, which take into account the wheel slipping, the resistance force due to the soil crumpling and the formation of a track, as well as the tangential force of traction realized by the wheel.

1 Introduction

Questions related to the study of traction qualities of wheeled vehicles are considered in many works [1-3]. It should be noted that for the machine as a whole, traction and dynamic qualities are usually evaluated by the traction efficiency and dynamic factor [3-4].
However, each of the wheels operates under certain conditions in terms of vertical load, torque input, driving conditions. In addition, for example, for tractors, the wheels of the front and rear axles have different sizes. In this regard, of particular interest is the assessment of the contribution of each individual wheel to the overall traction dynamics of a multi-axle wheeled vehicle.

Traction properties of the wheel depend on a large number of parameters. These include design parameters, geometric dimensions, and operating conditions [1-2].

When assessing the efficiency of the wheel in the traction mode during mathematical modeling, as well as in experimental studies, it is advisable to use some integral index, taking into account the variety of factors acting on the wheel. This will allow the design stage and research to choose the optimal mode of operation of the wheel or to choose the appropriate tire for certain operating conditions of the wheel machine. The paper proposes to use as such an indicator the coefficient of performance, which takes into account the wheel slipping, the force of resistance due to the crushing of the soil by the driver and the formation of a track, as well as the tangential force of traction realized by the wheel.

2 Main part

As mentioned above, the design of the wheel machine raises questions about the choice of parameters of its engine at a given load on the wheel or determining the optimal normal load at a given wheel parameters.

* Corresponding author: POVAREKHO@bntu.by

The variety of structural and layout solutions chassis of wheeled vehicles (Figures 1, 2) complicates the selection of tires and load modes of the wheels of individual bridges.

Fig. 1. Some layouts wheeled chassis.

Fig. 2. The circuit chassis of wheeled tractors and tractor units.

Layout solutions may differ in the number of wheel axle, the number of driving and driven, controlled and not controlled bridges, the nature of their location along the base of the machine. Double and built trolleys can be used. All this leads to the fact that the load on the wheels and the conditions of their movement are different. Therefore, the choice of the wheel scheme of the chassis, the nature of the load being transported, the load distribution along the cargo platform must be made from the condition of ensuring the greatest efficiency of the wheels of the drive axles in terms of the best implementation of traction and dynamic qualities. All this will improve the traction quality of the machine, fuel efficiency and performance in General.

Depending on the type of torque supply to the drive wheels (differential or locked), driving conditions (straight or cornering), the design features of the transmission wheel can operate in traction or brake modes. In this paper, the wheel movement in the traction mode is considered.

In General, the wheel slip characterizes the speed loss due to the displacement of the instantaneous center of speeds relative to the contact spot and is calculated by the expression:

$$\delta = \frac{\omega \cdot r_k - v}{\omega \cdot r_k} = 1 - \frac{v}{\omega \cdot r_k}, \quad (1)$$

where ω – the angular velocity of the wheel; r_k – the kinematic radius of the wheel; v – the linear velocity of the wheel center.

The settlement scheme of movement of the drive wheel taking into account influencing forces is shown in Figure 3.

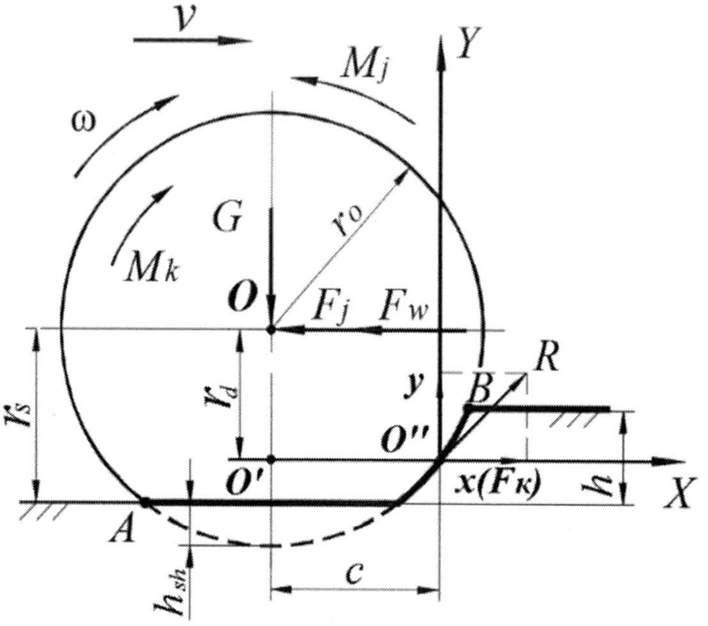

Fig. 3. Diagram of forces and moments applied to the drive wheel moving along a horizontal ground surface in a steady state.

In this figure, the following designations are accepted: r_o, r_s, r_d – nominal, static and dynamic wheel radiuses; c – coordinate of the point O'' of application of the reaction of a soil surface; M_k – torque applied to the center O of the wheel; h и h_{sh} – rut depth and deformation of the tire under normal load G.

The analysis of the driving wheel working processes during its movement allowed us to make an assumption that the coefficient of efficiency of the driving wheel can be used as a complex indicator of the efficiency of the traction properties.

The efficiency of the drive wheel η_k is determined according to the expression:

$$\eta_k = (1 - \delta) \cdot \left(1 - \frac{F_f}{F_k}\right), \quad (2)$$

where F_f - the strength of the resistance due to the soil crumpling and the formation of a track; F_k - tangential thrust force (Figure 3).

The authors propose a method for optimizing the parameters of the wheel and its load, when as a criterion of efficiency (objective function) takes the coefficient of efficiency of the driving wheel η_k.

To determine the parameters included in the formula for determining the efficiency of the wheel, a calculation scheme was used (Figure 3). On the basis of the D'Alembert principle, the equations of equilibrium of forces and moments applied to the driving wheel moving along the horizontal ground surface in an unsteady mode were compiled:

$$\begin{cases} \Sigma X = x - F_j - F_w = 0 \\ \Sigma Y = y - G = 0 \\ \Sigma M_0 = y \cdot c + x \cdot r_d + M_j - M_k = 0 \end{cases}, \quad (3)$$

where M_k – torque applied to drive wheel; G – normal load acting on the wheel; R – the reaction of soil; x и y – the horizontal and vertical components of reaction R; F_j и M_j – force and moment of inertia; F_w – the force of air resistance.

Taking into account the designations given in Figure 3, the force of resistance to movement due to the crumpling of the ground by the wheel and the formation of a track is equal to:

$$F_f = y \cdot \frac{M_k}{r_d} = G \cdot \frac{c}{r_d} = f \cdot G, \tag{4}$$

where f – coefficient of resistance to movement.

The reaction x represents the driving force, which is denoted as F_k.

To determine the values included in the system of equations (1), Professor V.V. Guskov [6] developed the theoretical basis of the interaction of the driving wheel with the ground surface. They are based on the modern provisions of the theory of soil mechanics in the application of dynamic loads and the dependence of the resistance of the soil compression and shear, proposed by Professor V.V. Katsygin [7].

In particular, the compressive stress of the soil is determined by the formula:

$$\sigma = \sigma_0 \cdot th\left(\frac{k}{\sigma_0} \cdot h\right), \tag{5}$$

where σ_0 – soil bearing capacity; k – the coefficient of volumetric collapse of the soil; h – the depth of immersion of the stamp.

Shear stress arising from the deformation of the soil:

$$\tau = f_{sk} \cdot q_x \cdot \left(1 + \frac{f_n}{ch\frac{\Delta}{k_\tau}}\right) \cdot th\frac{\Delta}{k_\tau}, \tag{6}$$

where q_x – wheel pressure on the ground; f_{sk} – the coefficient of sliding friction; f_n – the coefficient of static friction; k_τ – the coefficient of deformation of the soil; Δ – shear deformation.

The graphic image of the dependence of the ground compression stresses is shown in Figure 4.

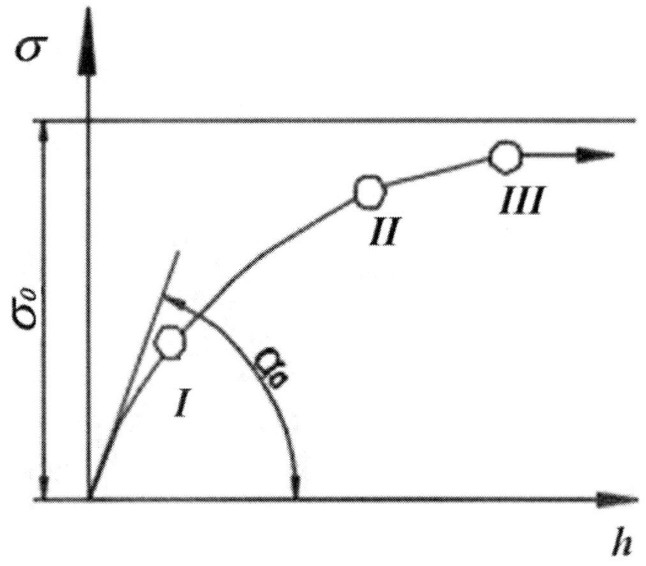

Fig. 4. Compression stress-strain dependence($k = tg\alpha$).

Figure 4 shows that there are three areas of this dependence: the first section reflects the elastic deformation; the second – plastic; the third – the flow of the soil.

The graphic image of the dependence of shear stresses arising from the deformation of the soil is shown in Figure 5.

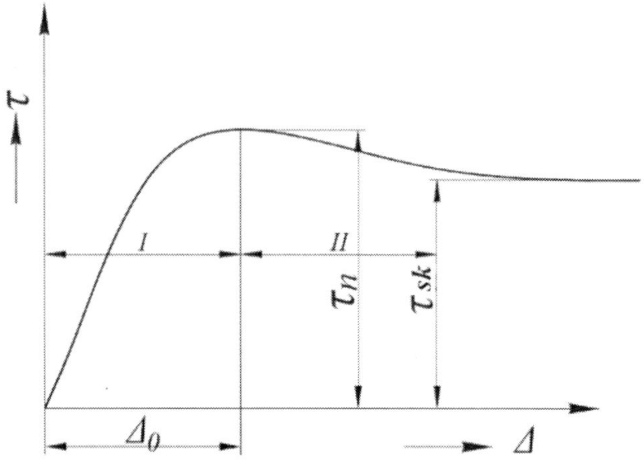

Fig. 5. The dependence of shear stress against deformation.

Figure 5 shows that shear stresses reach a maximum at some strain Δ_0 and then decrease. This phenomenon is explained by the fact that on the site I the soil is compacted (static friction), and on the site II – is shifted (sliding friction).

On the basis of the provisions of soil mechanics obtained mathematical expressions for determining the power parameters acting on the wheel. These indicators are proposed to determine according to the developed algorithm of the interaction of the drive wheel with the ground surface, proposed by Professor V.V. Guskov [6].

Resistance to movement:

$$F_f = \int_0^{h_0} b \cdot \sigma_0 \cdot th\left[\frac{k}{4 \cdot b \cdot \sigma_0} \cdot D_{pr} \cdot ln\left(\frac{D_{pr}-h}{D_{pr}-h_0}\right)\right] dh, \quad (7)$$

where b – wheel width; D_{pr} – led wheel diameter; h_0 – the soil deformation for the corresponding vertical loads.

Vertical load, leading to soil deformation on the value of h_0:

$$G = \int_0^{h_0} \frac{b \cdot \sigma_0 \cdot (D_{pr} - 2 \cdot h)}{2 \cdot \sqrt{D_{pr} \cdot h - h^2}} \cdot th\left[\frac{k}{4 \cdot b \cdot \sigma_0} \cdot D_{pr} \cdot ln\left(\frac{D_{pr}-h}{D_{pr}-h_0}\right)\right] dh. \quad (8)$$

Tangential traction force developed by the wheel:

$$F_k = \int_0^{L_{pr}} \frac{2 \cdot b \cdot f_{sk} \cdot q_x}{L_{pr} \cdot \delta_x} \cdot \left(1 + \frac{f_n}{ch\frac{\delta_x \cdot L_{pr}}{k_\tau}}\right) \cdot th\frac{\delta_x \cdot L_{pr}}{k_\tau} dL, \quad (9)$$

where L_{pr} – the length of the bearing surface of the wheel.

The following assumptions were made in obtaining these equations.

1. The wheel moves under the action of the applied torque M_k in a steady state on a horizontal surface.

2. When considering the process of interaction of the driving wheel with the ground surface, mechanical characteristics of soils are used to determine which mathematical dependences proposed by Professor V. V. Katsygin are used [7].

3. The nominal diameter D_0 of the real wheel is replaced by the reduced diameter D_{pr} of the hard wheel according to the dependence [5]:

$$D_{pr} = D_0 + \frac{h_{sh}}{h} \cdot (D_0 - 2 \cdot h - h_{sh}). \quad (10)$$

4. The actual length of the reference surface is calculated by the formula:

$$L_{pr} = \frac{D_0}{2} \cdot arctg\sqrt{\frac{D_{pr} \cdot h - h^2}{\frac{D_{pr}}{2} - h}} + \sqrt{D_{pr} \cdot h}. \quad (11)$$

Thus, on the basis of the proposed mathematical models, the assessment of the optimal conditions for the operation of the wheel machine engine can be solved in two directions.

1. The optimal vertical load on the wheel is determined for its specified design parameters, i.e. the equation of the form is solved:

$$\frac{d\eta_k}{dG} \to 0. \quad (12)$$

2. The design parameters of the wheel at a given load are determined:

$$\frac{\partial \eta_k}{\partial (D_0, b_0, p_w, \dots)} \to 0, \quad (13)$$

where p_w – the air pressure in the tire.

To verify the proposed method, an example of solving the problem of optimizing the normal G load on the drive wheel (bus 580/70R42) with dimensions is given: D_0=1,9 m; b_0=0,714 m a pressure of p_w =0,2 MPa, moving along the stubble of a loam of normal humidity (w=16%). The soil has the following physical and mechanical properties: f_{sk}=0,76; f_n=0,79; k_T=0,04 m; σ_0=1,58·10^6 N/m^2.

Figure 6 shows the results of the calculation of the traction-chain properties of the specified tire depending on the normal load.

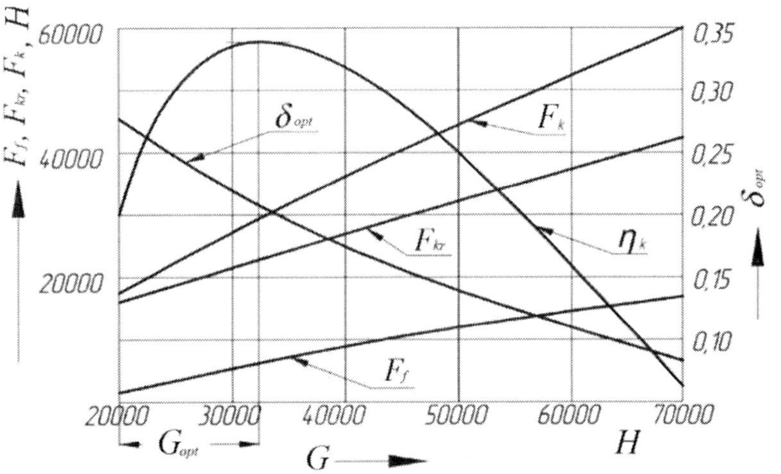

Fig. 6. Traction properties of the tires 580/70R42 depending on the vertical load.

In this figure F_{kr} - hook effort. The hook force is the useful force realized by the wheel and is determined by:

$$F_{kr} = F_k - F_f \quad (14)$$

As can be seen from the presented graphic dependences, with the growth of the vertical load on the wheel, there is an increase in the tangential thrust force F_k, the force of resistance to the movement of the wheel F_f and, accordingly, the realized hook load F_{kr}. The nature of these parameters changes is close to linear.

The dependence of the efficiency of the wheel η_k has a pronounced maximum corresponding to the vertical load G_{opt}=32.5 kN.

3 Conclusion

As a result of the research the following results were obtained.

1. The proposed method of optimization when used as a criterion for the efficiency of the traction coefficient of efficiency of the wheel allows to determine the rational parameters of the designed wheel drive with high efficiency and efficiency.

2. This method was introduced into the practice of designing the perspective wheeled tractors of the family "Belarus" Minsk tractor plant.

References

1. E.D. Lvov, *Tractor Theory* (Moscow: Mashgiz, 1960)
2. A.A. Khachaturov, *The Dynamics of the system road – tyre –vehicle - driver* (Moscow, Mashinostroenie, 1976)
3. K.Popp, W.Schiehlen, *Ground Vehicle Dynamics* (Springer, 2010)
4. A.A. Pulungan, *Design of four-wheel drive wheeled vehicles* **1-2** (Moscow, MGTU, 2000)
5. M.G. Becker, *Introduction to the theory of terrain-machine systems* (Moscow: Mashinostroenie, 1973)
6. V.V. Guskov, *Optimal parameters of agricultural tractors* (Moscow: Mashinostroenie, 1966)
7. V.V. Katsygin, Some aspects of deformations of soils. "Issues of agricultural mechanics", **13**, 117 ...185 (Minsk, Selkhozgiz, 1964)

Methodology for measuring car traction parameters

Bronisław Kolator[1,*], and *Michał* Janulin[1]

[1]University of Warmia and Mazury in Olsztyn, 10-719 Olsztyn, Poland

Abstract. The article presents methodology for testing the traction parameters of the vehicle on the laboratory stand. The tests have been carried out using MAHA LPS 3000 chassis dynamometer. Power output can be measured in two ways, at steady state or during acceleration. Both have benefits as well as drawbacks. Car manufacturers are required to use steady state tests for certification of power output. It has an advantage as it is a well defined test that is easy to repeat and verify. The problem with results measured during acceleration is that the acceleration itself has a considerable effect on the result.

1 Introduction

Engine Performance, declared by the automotive manufacturer, is related to a specific test procedure where the engine is separately mounted and tested in an engine dynamometer. Chassis dynamometer testing analyses the main factors affecting a vehicle in order to closely reproduce them in a testing facility. It provides complete guidance on how these tests should be performed, including the requirements affecting the test cells themselves so that they can be fully optimized [1].

SAE Recommended Practice J1540 specifies test procedures to map transmission efficiency and parasitic losses in a manual transmission. The procedure comprises two parts. The first compares input and output torque over a range of speed to determine efficiency. The second measures parasitic losses at zero input torque over a range of speed. As specified in J1540, efficiency of transmissions is routinely measured on a test-stand under steady torque and speed. While such testing is useful to compare different transmissions, it is unclear whether the "in-use" efficiency of a given transmission is the same as that measured on the stand [2].

Kołodziej et al presents way of conduct of automobiles' diagnostics testing on uniaxial chasses dynamometer [5]. Chassis dynamometers are very popular to run some quick tests for installed power and check out the chassis and drivetrain [4]. The identification of traction and speed characteristics as well as of fuel economy of motor vehicles in road conditions has been presented Sergey N Krivtsov et al [3].

Rototest Research Institute defines Powertrain Performance as the performance available to the wheels (measured at the wheel hubs). A simple description is the engine performance minus the drivetrain losses [6].

This has historically always been the way to do it and has the benefit of a controlled environment during the test. However, the modern cars of today have increased the integration of the whole powertrain to a degree where it has made it in practice impossible to test an engine separately.

2 Test method and measuring

Power output can be measured in two ways, at steady state or during acceleration. Both have benefits as well as drawbacks. Car manufacturers are required to use steady state tests

* Corresponding author: bronislaw.kolator@uwm.edu.pl

for certification of power output. It has an advantage as it is a well defined test that is easy to repeat and verify.

The simplest way of testing is to let the engine / car accelerate against a fixed inertia load. Simple in one way most often means drawbacks in other ways. In this case it has numerous drawbacks, which affects the results. Also you are limited to just one type of test. A more demanding test for the dyno is to accelerate against the road load. Then the dyno has to simulate running on the road. This can be good when tuning a specific car. However the results will be affected by the vehicle set-up. Changing the weight of the car will produce different power readings even if it is the same engine.

The chassis dynamometer LPS 3000 is available in various versions for performance testing of cars [7]. Depending on the version, wheel power from 260 kW to 520 kW with a max. test speed of 260km/h can be tested. The dyno load simulation is done with an eddy-current brake. The LPS 3000 enables engine power measurements to be made on cars with Otto and diesel engines. Testing of four wheel drive vehicles is possible if the LPS 3000 is equipped with the appropriate roller set and the corresponding control electronics. Measurement programs, this dynamometer provides an exact simulation of defined road load conditions. Fuel consumption measurement and exhaust analysis, as well as serial monitoring of vehicles can be carried out. The following operation modes can be selected for conducting the various measurement tasks (Fig. 1):

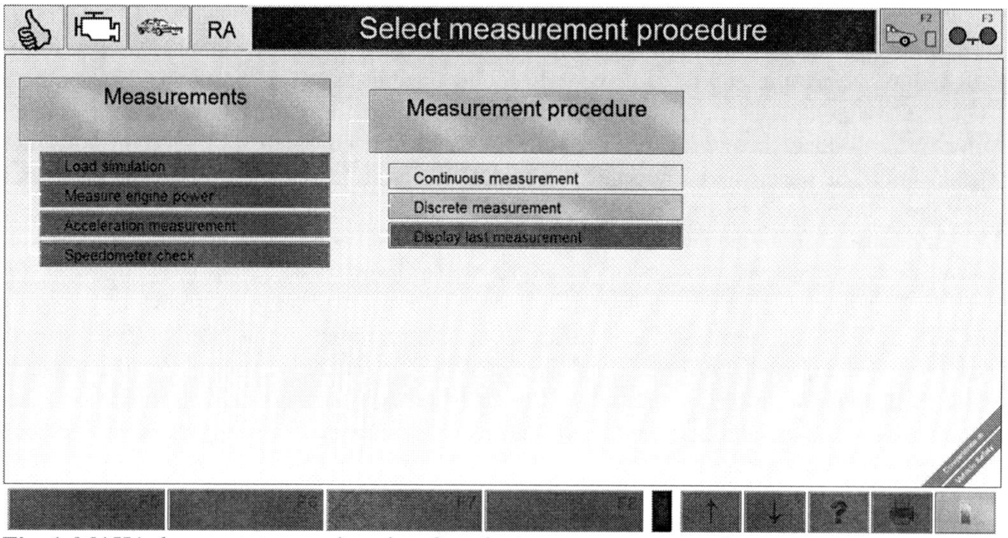

Fig. 1. MAHA dynamometer monitor view for select measurement procedure test.

- Load Simulation - The user can pre-select from various load conditions which make possible the determination of e.g. a specific engine diagnosis or vehicle climbing power. Depending upon need, various load conditions can be simulated such as constant traction, a constant speed, a constant RPM or a driving simulation.
- Measure Engine Power - Use this menu item to determine the vehicle's engine power (continuous and discrete measurement).
- Extrapolation of the Engine Power - Extrapolation of the engine power based on DIN 70020, EWG 80/1269, ISO 1585, JIS D 1001, SAE J 1349.
- Flexibility Test - The engine flexibility can be checked using this menu item. Once the flexibility test has been called up the vehicle weight and road load must be entered as in the road load simulation test.
- Speedometer Test - The vehicle's speedometer can be tested using this menu item. The speed can be tested as well.

- Road Load Adaption - Various dyno coefficients can be determined using this menu item. The iterative coefficient adaption can be done based on ECE or SAE J2264 standard.
- Lug-Down Test - Use this menu item to do a diesel exhaust measurement under full load.

The max. power is determined during a continous measurement and the engine power is calculated according to DIN 70020, EWG 80/1269, ISO 1585, SAE J1349, depending on which setting has been selected in category "Power Correction". The curves for wheel, drag and engine performance are displayed graphically. If air pressure and temperature have been taken into consideration (measured) the engine performance is displayed as standard performance.

Discrete power measurement makes it possible to approach pre-defined points and hold them for a specified time as a function of the speed or RPM (Fig. 2). These target points must be set before testing begins. The Start- and End speeds or RPM as well as the test increments and hold time are entered by the inspector. Hold time is defined as the length of time which the pre-set approached RPM and/or speed is held constant. The hold time assures a stable test point so that for example enough time is left for a fuel consumption test to be done. The Start-Hold time eases the regulation procedure. More accurate test results are supplied for turbo-charged engines, whose charger often has a delayed reaction.

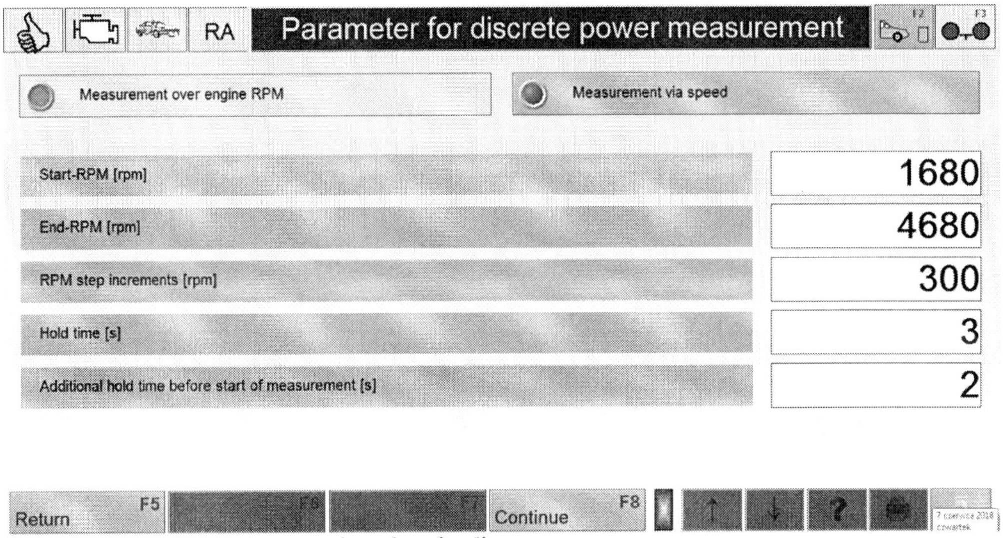

Fig. 2. MAHA dynamometer monitor view for discrete power measurement test.

You are free to use colour illustrations for the online version of the proceedings but any print version will be printed in black and white unless special arrangements have been made with the conference organiser. Please check whether or not this is the case. If the print version will be black and white only, you should check your figure captions carefully and remove any reference to colour in the illustration and text. In addition, some colour figures will degrade or suffer loss of information when converted to black and white, and this should be taken into account when preparing them. These test instructions serve to set out the basic conditions for wheel power measurement in writing and are to provide a basis for reproducible measurements [8].

3 Test results

The problem with results measured during acceleration is that the acceleration itself has a considerable effect on the result. It is therefore important that the rate of acceleration is defined and controlled within tight tolerances (Fig. 3). Only then can the test be repeated and verified.

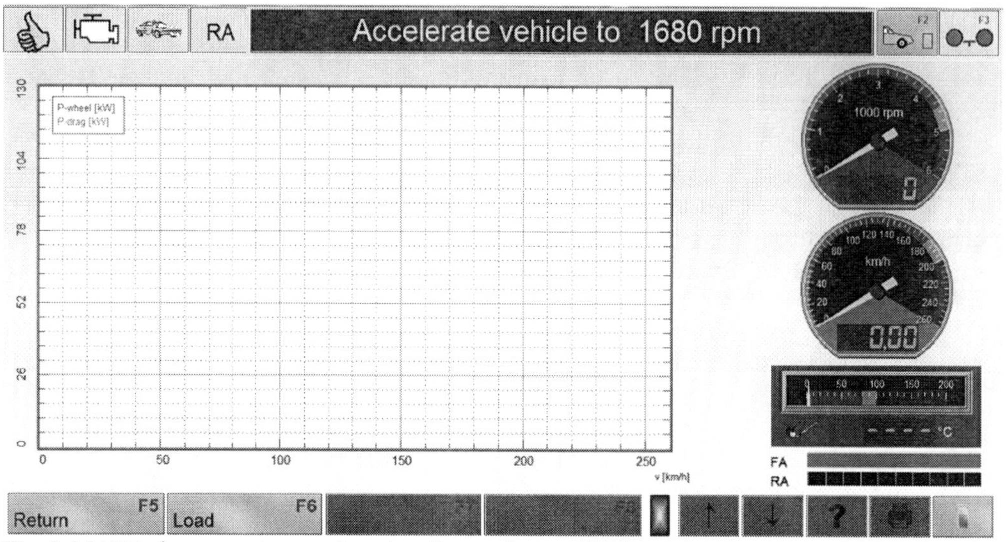

Fig. 3. MAHA dynamometer monitor view for accelerate vehicle.

This is a comparison of steady state vs acceleration at a steady rate. The acceleration is started at 1680 rpm and stopped at 4680 rpm. The main difference is caused by the inertia of the drive train but also the fact that the engine is running in a different way when it is accelerating compared to steady state.

The influence is even more evident in the following figures. Engine power continuous measurement (Fig. 4) and discrete measurement (Fig.5). As the steady rate increases full torque is reached later and the torque available is decreased. The inertia of the drive train is reducing the available torque directly related to the acceleration rate. However it is obvious that the difference is related to other variables as well, especially if you compare the fastest steady rate to the others. Sample waveforms of the recorded values have been presented in Table 1 continuous measurement and in Table 2 discrete measurement.

Table 1. Parameters values of the engine during continuous measurement tests on the chassis dynamometer.

Measurement date: 14.05.2018 (8:59)

Measurement data table

n [rpm]	v [km/h]	P_{Wheel} [kW]	P_{Eng} [kW]	P_{Norm} [kW]	M_{Morm} [Nm]
1800	53,9	4,3	9,8	9,8	51,9
2100	62,9	12,4	18,9	18,9	85,9
2400	71,8	18,9	26,6	26,5	105,5
2700	80,8	26,0	34,9	34,8	123,2
3000	89,8	38,0	48,1	48,0	152,9
3300	98,8	40,7	52,2	52,1	150,8
3600	107,7	42,7	55,7	55,6	147,4
3900	116,7	44,4	59,0	58,9	144,1
4200	125,7	45,3	61,8	61,7	140,2
4500	134,7	44,8	63,2	63,1	133,9

Minimum value Maximum value

Measurement date 14.05.2018 (8:59) Page 1

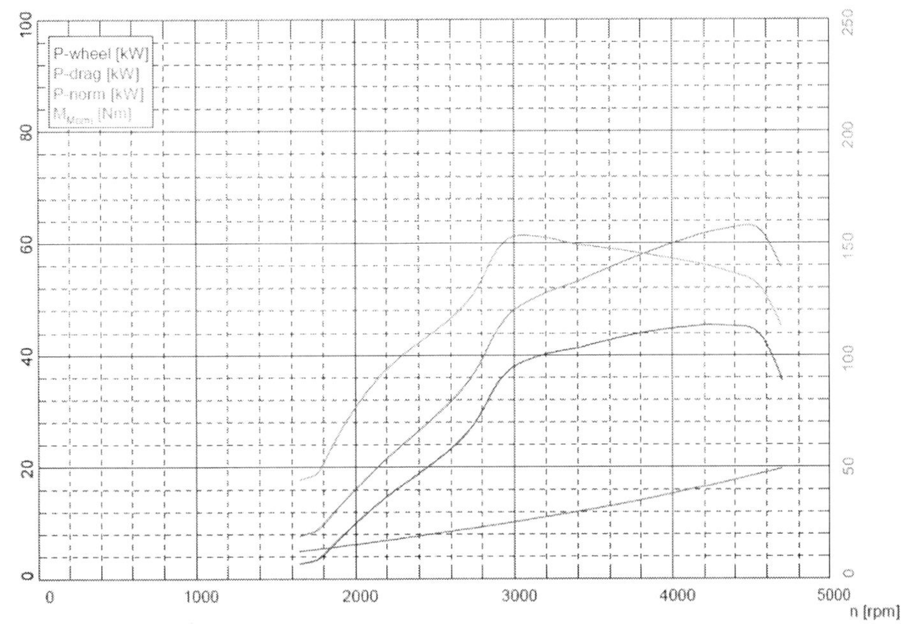

Power data					
Corrected power 1)	P_{Norm}	63,1 kW	/	85,8	BHP
Engine power	P_{Eng}	63,2 kW	/	86,0	BHP
Wheel power	P_{Wheel}	45,0 kW	/	61,2	BHP
Drag power	P_{Drag}	18,2 kW	/	24,8	BHP
Max. power at		4485 rpm	/	134,1	km/h
Torque 1)	M_{Morm}	153,3 Nm			
Max. Torque at		3045 rpm	/	91,1	km/h
Max. attained RPM		4695 rpm	/	140,6	km/h

Ambient data			
Ambient temperature	$T_{Ambient}$	23,3	°C
Intake air temperature	$T_{Intake\ air}$	18,6	°C
Relative humidity	H_{Air}	43,8	%
Air pressure	p_{Air}	1011,6	hPa
Steam pressure	p_{Steam}	12,5	hPa
Oil temperature	T_{Oil}	----,-	°C
Fuel temperature	T_{Fuel}	----,-	°C

Fig. 4. Characteristics of power and torque for engine continuous measurement.

Table 2. Parameters values of the engine during discrete measurement tests on the chassis dynamometer.

Measurement date: 14.05.2018 (9:13)

Measurement data table

n [rpm]	v [km/h]	P_{Wheel} [kW]	P_{Eng} [kW]	P_{Norm} [kW]	M_{Morm} [Nm]
1682	50,3	20,1	24,8	24,8	140,7
2003	59,9	26,1	31,8	31,9	151,9
2336	69,9	30,8	37,6	37,6	153,9
2671	79,9	35,2	43,3	43,3	154,8
3004	89,9	38,6	48,0	48,0	152,6
3339	99,9	41,8	52,6	52,6	150,5
3673	109,9	44,1	56,5	56,5	147,0
4008	120,0	46,1	60,1	60,1	143,3
4343	130,0	47,1	62,5	62,5	137,5
4673	139,9	35,2	52,4	52,5	107,2

Minimum value Maximum value

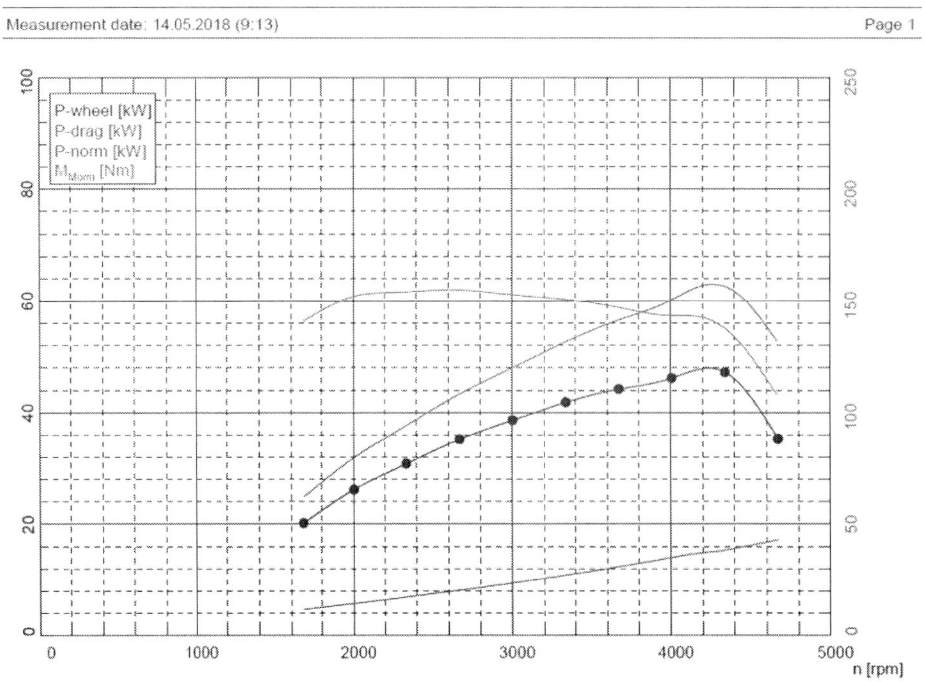

Power data				Ambient data			
Corrected power 1)	P_{Norm}	62,9 kW	/ 85,6 BHP	Ambient temperature	$T_{Ambient}$	24,2	°C
Engine power	P_{Eng}	62,9 kW	/ 85,5 BHP	Intake air temperature	$T_{Intake\ air}$	20,0	°C
Wheel power	P_{Wheel}	47,9 kW	/ 65,1 BHP	Relative humidity	H_{Air}	40,3	%
Drag power	P_{Drag}	15,0 kW	/ 20,4 BHP	Air pressure	p_{Air}	1011,6	hPa
Max. power at		4260 rpm	/ 127,4 km/h	Steam pressure	p_{Steam}	12,2	hPa
Torque 1)	M_{Norm}	154,8 Nm		Oil temperature	T_{Oil}	---,-	°C
Max. Torque at		2630 rpm	/ 78,8 km/h	Fuel temperature	T_{Fuel}	---,-	°C
Max. attained RPM		4670 rpm	/ 139,8 km/h				

Fig. 5. Characteristics of power and torque for engine discrete measurement.

3 Conclusion

Compare the continuous measurement results of vehicles with manually shifted transmission which are based on the standard conditions and have been determined in the nominal engine power range with the target curves. In the nominal torque range discrete measurement is precise. The load dependent losses are included in the adaptation factors of the chassis dynamometer altogether.

Engine combinations tested on a chassis dyno can be done rapidly and data can be evaluated quickly as well. However reliable test data is dependent on many things that any engine builder or vehicle owner should acknowledge and be willing to accept so that they each can learn from the experiences.

References

1. E. Galindo, D. Blanco, C. J. Brace, E. Chappell, R. Burke, *Chassis Dynamometer Testing* (SAE, R-452, 2017)
2. S. Rengarajan et al, *In-Situ Measurement of Transmission Efficiency in Vehicles*, SAE Technical Paper 2017-01-1095, 2017, https://doi.org/10.4271/2017-01-1095
3. S. N. Krivtsov et al 2016 IOP Conf. Ser.: Mater. Sci. Eng. 142 012101
4. Bettes H.: *A quick view of chassis dynamometer testing*, AERA Newsletter, 2010
5. Kołodziej K., Skrzyniowski A.: Badania diagnostyczne samochodów na hamowni podwoziowej AutoDyn 30, Cracow University of Technology, Cracow, 2012

6. http://rototest-research.eu/index.php?DN=39&Menu=38&ID=11
7. The Chassis Dynamometer LPS 3000, Operating Instructions
8. Test Instructions for Engine Power Measurement Chassis Dynamometer: MAHA LPS 3000

Modelling of the grain materials 'shredding process for the purposes of the beater shredders' designing

Andrzej Bochat[1], and *Marcin* Zastempowski[1*]

[1]UTP University of Science and Technology in Bydgoszcz, Faculty of Mechanical Engineering, Al. prof. S. Kaliskiego 7, 85-796 Bydgoszcz, Poland

Abstract. Within the frames of the study's performance, the modeling attempt of the gran materials' shredding with the beater shredder was taken up. Modelling was conducted for two constructions of working shredder's assemblies, that is for a classical and a new one, developed by the study's authors. The classical construction is characterised by the fact, that it has beaters assembled to the rotor of a cuboid shape. However, the new construction, has beaters in the shape of a circular section. It results from the conducted shredding process' analysis, that the shape of the beaters has a crucial impact on the shredding output's increase at simultaneous decrease of the process' energy consumption. The developed kinematics and energy consumption models of the shredding process may be used at the phase of the beater shredders' designing for grain materials.

1 Introduction

Shredding is the process of the material's splitting into smaller particles with the use of machines' working parts, overcoming the cohesion forces of the material's particles. Materials' shredding is the effect of influencing such processes as: splitting, crushing, grinding or breaking by the materials particles' hitting. While selecting of the shredding method, there have to be considered the shredded material's mechanical properties, selecting such a manner of the working element's operating as to obtain the required shredding degree at possibly the lowest destructive stresses of the shredded material. In the professional literature, the process of shredding is considered most often in three fundamental perspectives, that is productivity, energy consumption and the shredding degree. The subject matter of the gran materials' shredding is being widely entered into in the available literature, both with the reference to the process' effectiveness as well as new constructions of the shredders' working assemblies [1-4].

In the economy (mining industry, chemical industry, food processing industry), the following types of shredders are commonly used:

- beater shredders,
- crushing mills,
- radial plate grinding mills,
- roller mills.

Quick designing of that type of machines requires, among the others existence of appropriate mathematical models describing the shredding processes conducted by the.

Within the frames of the study's performance, developing of the models for the classical and innovative beater shredder's construction was undertaken. The innovative construction of the shredder is original and patented.

* Corresponding author: zastemp@utp.edu.pl

2 The essence of the classical and innovative construction of the beater shredder

The fundamental elements of a typical beater shredder are the rotor with self-aligning or stiffy mounted beaters, a shredder's sieve or sieves, a shredding plate or plates and a supporting structure with a power transmission system. In beater shredders, the material shifts from the charging hopper to the working area, in which there is a rotor with beaters. The supplied material reaches a considerable speed (40 – 110 m/s) under the influence of the beaters' strokes and moves on the circular path hitting a sieve (or sieves) and one another. As a result of the percussive forces' effect, the shredded material remains in the shredding chamber as long, till it gets the particles' dimensions smaller than the holes in sieves. The shredded material, having reached appropriate geometrical features, comes through outside the shredder's chamber through the holes in the sieves.

A flaw and an inconvenience of the known, classical design solutions of the beater shredders is their low throughput as compared to the energy consumption [5-9]. It is determined most of all b a rotor's construction, where the beaters are of the shape of rectangular plates. As an effect of that, under the impact of the beaters' strokes, the material's particles start to move along the paths similar to a circle. They form a thin layer oscillating on the shredding chamber's internal perimeter, what results in the fact that in spite of sometimes sufficient shredding degree, the material is circling still for a long time before it passes through the holes in the sieves. In figure 1 the classical rotor of a beater shredder in the isometric view is presented.

Fig. 1. The isometric view of a classical beater shredder's rotor with rectangular beaters (α=0°), designed in the Solid Works programme.

The essence of the new construction consists in the fact, that the shredder's working assembly consists of a disk rotor mounted on a shaft on which the self-aligning beaters are assembled. The beaters are of plates' shape in the form of a circular section of the angle of flare of at least 350, while the beater's fastening opening is on the symmetry axis of the circular section close to its arc bed. Such a construction of a beater shredder's rotor resuls in the fact, that the shredded material's particles, hit by the beaters, do not move along the circular path and do not form a rotaring ring, but they move approximately radially with reference to the sieves and they immediately hit them. It results in quicker material's passing through the holes in the sieves. In figure 2 there is presented a new construction of a beater shredder's rotor in isometric view. The presented new solution of a beater shredder's rotor is the subject matter of the patent of the Republic of Poland no 173497 (Bochat A.).

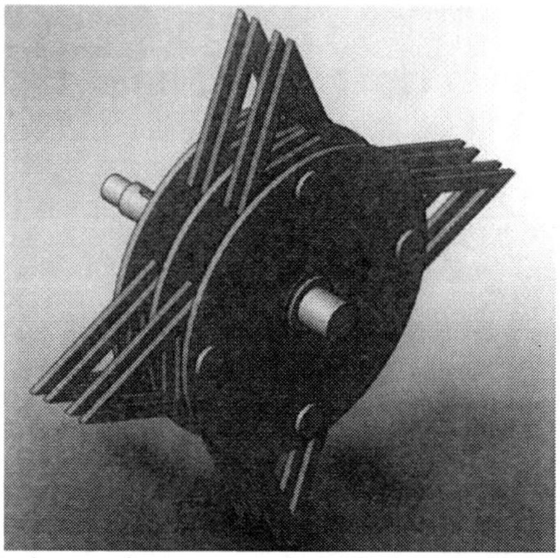

Fig. 2. The isometric view of the innovative construction of a beater shredder's rotor with beaters in the shape of a circular section (α=45°), designed in the Solid Works programme.

Shredding of the grain material is most of all conducted with mechanical methods, and on the process' effectiveness [10] there decide the physiomechanical material's properties [11] as well as the constructional and exploitation shredders' features. The following may be counted among the others as the constructional and exploitation features: shape, dimensions and construction of the shredder's working assembly, kinematic and dynamic process' parameters [12], level of wear and tear as well as resistance to mechanical destruction of working elements. Resistance to mechanical destructions is determined, among the others, by conducting studies on the strength of materials from which the component elements of the working assembly have been made of [13-16], while the wear and tear may be forecasted based on modelling of the machines elements' conditions [17].

3 Modelling of the kinematics of the grain materials' shredding process

In the beater shredder, tha grain materia lis shredded by a beater's hitting the matrial and the material's hitting the sieves. The shredded material passes through the shredding chambers outside through the sieve. The grain material which is in the working chamber, rotates till it reaches the required shredding degree. The degree of the material's shredding depends on the circumferential speed of the beaters, the shape of the beaters, the angle of contact and the size of the holes in the sieves. The efficiency W of the shredded may be described with the dependence:

$$W = \frac{m}{t}, \tag{1}$$

where:

m - mass of the shredded material,
t - time of the material's staying in the shredder's working chamber.

The time of the grain material's staying in the working chamber, depends on the multiplicity of circulation k and the average speed of the circulating grain mass. The average value has been assumed to be the resultant speed of the shredded material's particle in the beater's middle length:

$$k = \frac{v_w\, t}{2\,\pi\, r} = \frac{m\, v_w}{W\, 2\,\pi\, r}, \tag{2}$$

where:

k - multiplicity of the material's circulation in the working chamber,
v_w - resultant speed of the circulating grain, falling on the beater's active middle length,
r - distance of the shredder's rotor axis of rotation measured to the middle of the beater's active length

The speed of the moving grains in the working chamber, depends on the rotor's rotational speed and the applied constructional solution of the beaters. The figure 3 presents the distribution of the grain particle's speed, which is in the beater shredder's working chamber considering the beaters' shape. In case of shredding the grains with the use of the beaters of the shape of the rectangular prism (fig. 3a), then the normal speed vector which equals to the peripheral tangential speed of the shredded particle equals to zero, and the normal speed vector is equal to the peripheral speed, falling on the half of the active beater's length.

So:

$$v_n = \omega r, \tag{3}$$

$$v_n = v_r, \tag{4}$$

where:

v_n - normal speed of the shredded material's particle,
v_r - peripheral beater's speed measured to the half of the active beater's length,
ω - angular speed of the beater's rotor.

Application of a beater in the shape of the perpendicular prism results in the fact, that the shredded material oscillates in the shredding chamber, forming an oscillating ring. Inconvenience of such a solution is the decrease of the shredding process' efficiency, and the increase of the dusty fraction's content [5].

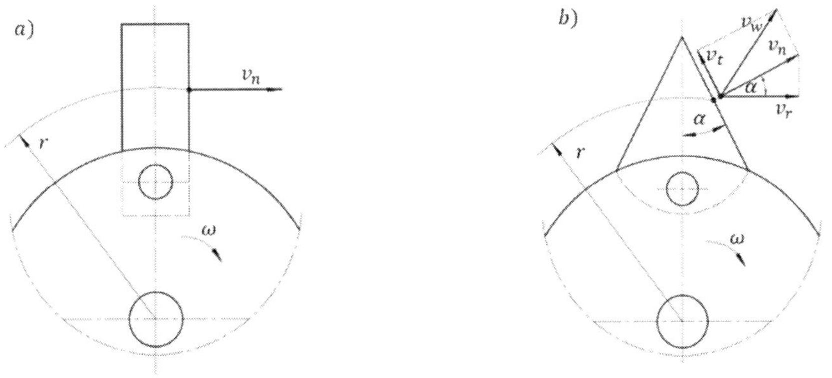

Fig. 3. Diagram of the grain speed's distribution at the moment of hitting the beater: a) the beater in the shape of a rectangular prism, b) the beater in the shape of a circular section.

In case of the gran material's shredding with a working assembly with beaters in the shape of a circular section (fig. 3b), the normal speed of the shredded particle equals to the relation of the peripheral beater's speed to the beater's angle cosine. The direction of the resultant speed's value of the shredded particle depends on the rotor's speed and the angle between the front beater's surface and its symmetry axis:

$$v_r = \omega\, r, \tag{5}$$

$$v_n = \frac{v_r}{cos\alpha} = \frac{\omega\, r}{cos\alpha}, \tag{6}$$

$$v_w = \frac{v_n}{cos\alpha} = \frac{\omega\, r}{cos^2\alpha}, \tag{7}$$

where:

α - angle of the beater.

Applying in the shredder's rotor of the beater in the shape of a circular section results in the fact, that the shredded particles do not form an oscillating ring, but they collapse at a bigger angle on the sieve's surface and are disposed of outside the working chamber (fig. 4a and 4b).

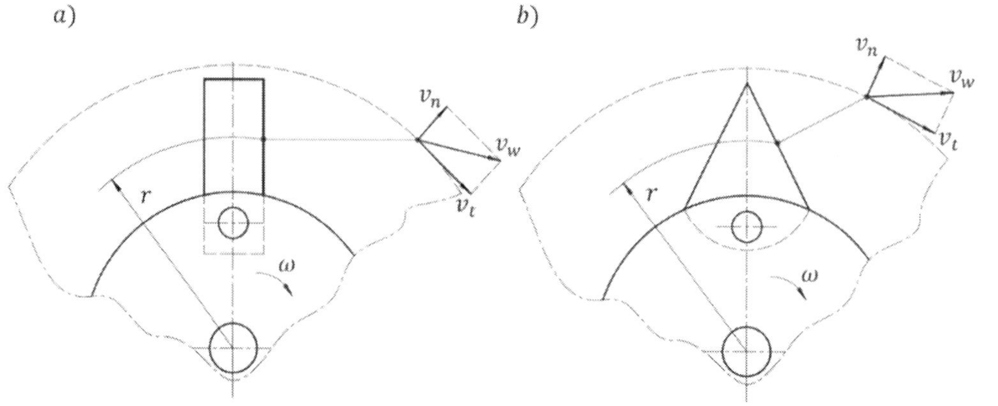

Fig. 4. Diagram of the gran speed's distribution while hitting the sieve: a) the beater in the shape of a rectangular prism, b) the beater in the shape of a circular section.

4 Modelling of the shredding's energy consumption in the beater shredder

Energy consumption of the beater shredders depends on their features and design parameters as well as on the shredded material's properties. The summary energy E_c necessary for the shredding process' conducting, is the sum of the shredding energy and the energy of the running light:

$$E_c = E_r + E_l, \quad (8)$$

where:
E_r - shredding power rozdrabniania,
E_l - power of the shredder's running light.

Te shredded material's power may be determined based on the dependence:

$$E_r = E_c - E_l, \quad (9)$$

whereas

$$E_c = E_j\, m, \quad (10)$$

where:
E_j - unit power of shredding,
m - mass of the shredded grain material.

The unit energy necessary for conducting of the material's splitting, is the function of power intake by the bear shredder's working assembly with respect to the volume of the shredded material. Considering of this process with respect to the mass unit, the formula fot the unit energy is received:

$$E_j = \frac{N}{W}, \quad (11)$$

where:
N - intake of power by the shredder's working assembly,
W - efficiency of the shredder.

The shredding process in the beater shredder is the most intensive on the rotor's circumference, so in order to calculate the power demand it may be assumed, that the material's particle having hit the beater, changes its speed for $\Delta\omega$, where the change of this speed is proportional to the resultant speed of the shredded material's particle on the beater's mid length. The power needed for the material's shredding is the product of the shredded particle's resultant speed and of the shredding power. However, the power of the running lights' resistance is the function of the turning moment and the angular speed in respect of time needed for its shredding.

So:

$$N_c = F\,v_w, \tag{12}$$

$$N_o = \frac{J\,\omega^2}{2t}, \tag{13}$$

where:

N_c - power necessary for the material's shredding,

F - resisting force of the the particles' shredding on the rotor's circumference,

N_o - the shredder's power of running light,

J - moment of inertia of the shredder's working assembly.

Here it should be pointed out, thay one of the inconveniences of the shredding process is the fact, that the shredded material, in spite of reaching the demanded shredding degree, oscillates in the shredder's working space. Due to that, there may be singled out the useful and unessential hits, which result in linger material's staying in the working chamber. The experimental data point out, that shredding of the grain materials takes place most often at the hitting speed of 40 m/s and 30-40 hits. Assuming, that in time t there shall be shreaded material of the mass m, the number of total hits n resulting in the material's shredding, is equal to:

$$n = z\,\frac{\omega}{2\pi}t, \tag{14}$$

where:

z - number of the beaters.

Mass of the material circulating in the shredder's chamber, has been determined based on the design features of the shredding chamber :

$$m = 2\,\pi\,r\,L\,h_m\,\rho\,c_m, \tag{15}$$

where:

L - width of the working assembly (rotor),

h_m - thickness of the circulating material's layer,

ρ - the shredded material's density,

c_m - material's concentration in the shredding chamber.

So, following the transformations, the shredding energy's equation assumes the form:

$$E_r = E_j\,m - \frac{J\omega^2}{2} = z\frac{\omega}{2\pi}t\,m\,v_w^2\,B - \frac{J\omega^2}{2} = z\,r\,L\,h_m\,\rho\,c_m\,t\,m\,v_w^2\,B - \frac{J\omega^2}{2}, \tag{16}$$

where:

B - proportionality coefficient resulting from the fact, that the change of the speed $\Delta\omega$ is proportional to the beaters' peripheral speed.

5 Summary

From the conducted analysis of the shredding process it results, thay the shape of the beaters has a fundamental impact on the energy consumption of the shredding process conducted with the beater shredder. The resultant speed and the direction of the shredded material's flight, determine the time of its staying in the shredder's chamber and the process' energy consumption. Losses of energy resulting from the grains' friction without the effect of their shredding as well as the losses of the kinematic and potential energy connected with the grains' deformation, are a big inconvenience of the shredding process. However, most of these losses are hard to be assessed to be described mathematically. Due to the above, the developed model should be experimentally verified at the time of gran materials' shredding of different physio-mechanical properties. The developed models of thee shredding process kinematics and energy-consumption, following their positive verification, may be used at the stahe of the the beater shredders' designing.

References

1. A. Tomporowski, J. Flizikowski, A. Al-Zubiedy, Przemysl Chemiczny, **97**, 2 (2018)
2. A. Tomporowski, J. Flizikowski, W. Kruszelnicka, Przemysl Chemiczny, **96**, 8 (2017)
3. M. Zastempowski, A. Bochat, *6th International Conference on Trends in Agricultural Engineering (TAE)*,: Czech Univ Life Sci, Prague, 726-735 (2016)
4. M. Zastempowski, S. Borowski, J. Kaszkowiak, *5th International Conference on Trends in Agricultural Engineering,* Czech Univ Life Sci, Prague, 673-676 (2013)
5. A. Bochat, L. Wesołowski, M. Zastempowski, Transactions of ASABE, **58**,3, (2015)
6. M. Chwiej. Roczniki Nauk Rolniczych, **76**, 9-19 (1986)
7. P.C. Kaptur, K. Schoenert, D. Furstenau, International Journal of Mineral Processing, **29**, 221-233 (1990)
8. S. Obidziński, R. Hejft, Zeszyty Naukowe Politechniki Opolskiej, **60**, 185-193 (2000)
9. M. Opielak, H. Komsta, Zeszyty Naukowe Politechniki Opolskiej, **60**, 209-217 (2000)
10. A. Tomporowski, ski, Eksploatacja i Niezawodnosc-Maintenance and Reliability, **2**, 150 (2012)
11. W. Keska, L. Gierz, *69th International Conference on Agricultural Engineering* LAND TECHNIK AgEng. Hannover, GERMANY, 459-464, (2011)
12. A. Bochat, M. Zastempowski, Transactions of the ASABE, **56**, 4 (2013)
13. B. Ligaj, G. Szala, Polish Maritime Research, **17**, 1 (2010)
14. T. Piątkowski, J. Sempruch, T. Tomaszewski,Transactions of the Canadian Society for Mechanical Engineering, **38**, 1 (2014)
15. T. Tomaszewski, P. Strzelecki, J. Sempruch, 22nd International Conference on Engineering Mechanics, Svratka, Czech Republic (2016)
16. G. Szala, B. Ligaj, International Journal of Fatigue, 91, 39-49 (2016)
17. L. Knopik, K. Migawa, Eksploatacja i Niezawodnosc – Maintenance and Reliability, **20**, 1 (2018)

Analysis of assessment criteria for selected systems of transport means operation

Łukasz Muślewski[1*], *Leszek* Knopik[2], *Bogdan* Landowski[3] and *Oleh* Polishchuk[4]

[1] University of Science and Technology in Bydgoszcz, Faculty of Mechanical Engineering, Al. prof. S. Kaliskiego 7, 85-796 Bydgoszcz, Poland
[2] University of Science and Technology in Bydgoszcz, Faculty of Managenent, Al. prof. S. Kaliskiego 7,85-796 Bydgoszcz, Poland
[3] University of Science and Technology in Bydgoszcz, Faculty of Mechanical Engineering, Al. prof. S. Kaliskiego 7, 85-796 Bydgoszcz, Poland
[4] Khmelnitsky National University, Instytuts'ka Street, 11, Khmel'nyts'kyi, Khmel'nyts'ka oblast, 29000, Ukraine

Abstract. The problems discussed in the study are connected with evaluation of complex technical systems functioning, in particular, transport systems. It was assumed that evaluation of their functioning depends on fulfilment degree of selected criteria. Therefore, it is important to determine a set of criteria including their type, number and importance. Since the research object is a public city transport system considered to be a sociotechnical system of the type: human (driver) –machine (vehicle) and the environment <H-M-E>, the criteria to be used for the assessment must include behavior of humans, operation of transport means, and the environmental impact. Thus, selection and determination of importance of significant, time variable, measurable and independent characteristics whose values, in a given time moment or a given time interval, define fulfillment degree of the criteria provide the basis for evaluation of such systems functioning. The quality of technical systems functioning in time is assessed on the basis of the criteria fulfilment degree or comparison and classification of different systems of the same type. It needs to be emphasized that the choice of optimal methods for selection of relevant criteria and determination of their impact on the analyzed system functioning is the research subject of study.

1 Introduction

The choice and specification of assessment criteria for complex operating systems, including the analysed transport ones, from the point of view of their functioning quality evaluation, is an important research problem. A criterion is defined as fulfilment degree of assumed requirements. At the same time, selection of the most important assessment criteria, whose fulfilment degree are values of independent and measurable features describing a given system, was carried out.

The notion of a transport system quality covers a set of its features expressed by means of numerical values, in a given time t or fixed time interval (t+Δt), determining the fulfillment degree of the specified criteria (its users requirements) [1].

In connection, with the above, in order to build an assessment model for a given system, it is necessary to establish the number and significance of the assessment criteria to be used for determination of characteristics describing the system in terms of its functioning quality evaluation. The assessment process involves referring each criteria from set X to the

* Corresponding author: lukasz.muslewski@utp.edu.pl

characteristics they describe and determining, on this basis, whether and to what extent the criteria have been fulfilled in given time t (during assessment).

Having in mind adequacy of the constructed resultant models, it needs to be emphasized that identification and determination of the assessment criteria significance is an important issue in diagnosis and analysis of the system functioning. Therefore, this study uses selected methods for evaluation of the criteria significance and analyzes them in terms of resultant approach to the set of the most important criteria.

A research method and a resultant model for transport systems operation quality assessment has been developed on the basis of literature and the author's own study The developed method enables assessment and comparison of operation quality of different transport systems of the same type. It is also one of rational diagnostic factors of the processes carried out within the discussed systems, depending on value changes of the characteristics describing actions of operators, controlled by them objects, and the environmental impact. For this purpose, metrics to be used for transport system operation quality assessment, have been developed. Values of the metrics described by dependence (1), are determined basing on the values of significant features describing the system, taking into account the values of weights attributed to particular features. A random process is defined for the analyzed system:

$$Z_x(t) = \sum_{i=1}^{p} \alpha_i X_i(t) \tag{1}$$

where:

$\alpha_i \geq 0, \sum_{i=1}^{p} \alpha_i = 1$

It should be emphasized that determination of a set of the most significant assessment criteria to be used for identification of significant characteristics consistent with the accepted criteria and establishing their significance is an important task for determination of a given system functioning quality assessment [2-3].

Further, in the next section of this paper, methods applied in the process of selection of the most significant criteria to be used for development of a resultant model for functioning quality assessment of a transport means operation system, will be presented.

2 Research object

The research object is a real, complex, transport system, providing passenger transport services in an urban agglomeration and suburban areas. This is a socio-technical system of the type < H-M-E> (Human –Machine – environment) in which the operation quality assessment is carried out in dependence on changes of values of features describing behavior of operators, controlled by them technical objects and the environmental impact [4].

The basic goal of a transport system operation is providing passenger transport services in a given area, in a given quantity and in a given time. Providing the systems with the required operation quality including efficiency, reliability, safety and economy is the main task in the process of their operating [1, 5-10].

The considered system consists of major subsystems such as: management subsystem, subsystem ensuring operation continuity (providing vehicles with usability, diagnostic, fuel supply), and executive one consisting of elementary subsystems of the type <H – TO> (driver – bus) and the environment as a cooperating system which has a direct impact on the technical system and its subsystems functioning.

3 Analysis of research survey results

A research survey was carried out whose goal was to determine a set of the most important criteria to be accepted for evaluation of the transport system operation quality. It was conducted by a group of experts, including workers employed in the considered transport system, and academic workers who conducted the research on optimization of the processes involved in the system functioning.

On the basis of a survey conducted by the study authors, a set of sixteen criteria, which were assessed by respondents, was determined. These criteria included: safety, efficiency, availability, accessibility, ergonomics, environment friendliness, usability, atmospheric factors, damageability, esthetics, informativeness, timeliness, reliability, transport task accomplishment time, external factors and punctuality.

The statistical group (respondents) were users of the analyzed system, different in terms of gender, age, education and occupation. The user was a statistical unit, and the number of respondents was N=100. Significance of the evaluated criteria, expressed according to a grading scale $\{0,1,...,10\}$, was the variability criterion.

It was assumed that the significance threshold for a criterion to be accepted was 5 points [2], according to the applied grading scale.

On the basis of the performed survey research, mean values of grades given to particular criteria, were determined. On the basis of the obtained survey results, a statistical analysis was made and histograms of distributions of grades obtained for particular criteria were determined. In order to check the correlation between particular criteria and find out whether the analyzed set of criteria in redundant, a test of the correlation coefficient significance was performed [1, 2].

According to the correlation coefficient significance test it can be said that particular criteria are strongly correlated. On the basis, of the results of analysis it can be said that the obtained criteria make up a redundant set. This means that some of the considered criteria can be neglected, with slight loss of information conveyed by the criterion, as the remaining, strongly correlated criteria include this information.

An analysis of statistical tests results, especially of the arithmetic mean being the most effective, not burdened estimator of an unknown expected value, proves that users of the analyzed transportation system are of the opinion that the most significant criterion is the system safety (9.35) and timeliness (8,96), which were given the maximal number of pints by more than half of the respondents. Besides, values of the variability coefficient for these criteria, within the analyzed set, are the lowest, respectively: 9.99% and 10.2%. This means that the responses given by the respondents on the subject of the above criteria are the least diversified. The criteria of the service performance time and reliability were given 8 points on the average, which also makes them significant and they need to be accepted in the process of the considered system functioning quality assessment.

However, the least significant, from the point of view of the surveyed respondents, were: informativeness and esthetics which were graded from 3.91 to 4.53 points. It means that from the point of view of the operation quality assessment, these criteria are of small significance and can be omitted. Moreover, the differences that occur in the analyzed set of criteria are found to be the highest for the same features, which reflects the users' lower precision while providing their values.

4 Principal Component Analysis

In this study, a description of methodological assumptions has been made, and the principal component analysis has been applied to analyze sets of grades, determined on the basis of research conducted in a real transportation system.

A survey usually contains a big number of variables. However, the questions in a questionnaire are related to each other and the survey output is unnecessarily extended. On the one hand, we try to describe the obtained data in the most complete way. On the other hand, we are limited by the survey costs. While analyzing the survey, different problems can be encountered. One of them is the question how to make an optimal reduction of the set of variables without significant loss of information. Popular statistical multidimensional methods do not provide ideal statistical procedures for the best possible choice of a subset of variables. One of such methods is the principal component analysis. However, application of this analysis requires fulfillment of certain assumptions.

This method is carried out on measurable variables (quantitative), though according to literature, it is possible to use it for variables of order type. The examined variables should be linearly dependent on each other and the correlation between features should be measured by Pearson coefficient. If the analyzed variables are not related to each other, then application of the principal component method is not advisable.

Literature provides [4, 11] that when all coefficients of correlation are smaller than 0.3, application of the principal component method is not effective. The higher correlation coefficients, the more justified application of this method.

At the beginning of the statistical analysis it is necessary to he use the test of Bartlett [12]. Bartlett test answers the question whether all correlation coefficients are equal to zero. Application of the method of principal component analysis requires a test with an adequate quantity. Literature [13] suggests that if the correlations are strong, it is enough for the statistical test to have the quantity equal to 50. An assumption about the distribution normality is not necessary for a description of relations between the variables. However, when statistical tests are used to define significance of the components, the assumption about multi-dimensional normality of the studied features distribution is necessary.

The analyzed data sets are numerical matrixes with dimensions n x p, where n stands for the number of surveys, p denotes the number of criteria. A matrix of correlation coefficients is determined for a given matrix X(n x p). Matrix of correlation coefficients undergoes Bartlett test which decides about advisability of application of the principal component analysis. If R denotes matrix of correlation coefficients, then Bartlett test involves verification of a statistical hypothesis form:

$$H_0 : R = I, \quad (2)$$

where:
I – is a unit matrix of dimension p x p.

Hypothesis Ho means that all coefficients of correlation contained in matrix R are equal to zero. Test statistics for this hypothesis has the form:

$$U = -(n-1-\frac{2p+5}{6})\sum_{i=1}^{p}\ln\lambda_i \quad (3)$$

where:
p= number of variables,
n – number of tests,
λ_i – i-th value of matrix R.

It is assumed that proper values are arranged in a non-ascending order, which means:

$$\lambda_1 \geq \lambda_2 \geq \ldots \geq \lambda_p, \quad (4)$$

Statistics U has, with the assumption that hypothesis Ho is true, chi-square distribution with p(p-1) freedom degrees. The result of Bartlett test is presented in table 1.

Table 1. Bartlett test result.

Value of statistics U	p-value
256.69	0.000001

The analysis of test results of Ho hypothesis: R=I shows that, for a very low p-value, it is necessary to reject the null hypothesis.

This means that it is justified to use the principal component method.

The principal component analysis enables determination of linear transformation of the form:

$$Z = A\,X, \tag{5}$$

where:
A is a matrix of p x p dimension linear transformation,
X is a column matrix of $X^T = [X_1, X_2, \ldots, X_p]^T$.
Z is a column matrix containing dependent variables $Z_1, Z_2, \ldots, Z_p$ called components.
The principal component analysis determines the first row vector of matrix A in such a way that component Z_1 has the maximal variance with the form:

$$\sum_{i=1}^{p} a_{1i}^2 = 1 \tag{6}$$

Next, the second component is determined so that variance of variable Z_2 will be maximal for proper limits.

One of the main reasons for using the principal component analysis is verification of hypothesis of the form:

$$H_0 : \lambda_{k+1} = \lambda_{k+2} = \ldots = \lambda_p, \tag{7}$$

In relations to the alternative hypothesis H_1 : not all $\lambda_{k+1}, \lambda_{k+2}, \ldots, \lambda_p$ are equal. Testing statistics for H_0 hypothesis has the form:

$$\chi^2 = -(n-k)\left[\sum_{j=1}^{p} \ln \lambda_j - q \ln \frac{1}{q} \sum_{i=k+1}^{p} \lambda_i\right], \tag{8}$$

where:
$\chi 2$ has distribution,
$\chi 2$ z df = q(q+1)/2 – 1, q = n – k degrees of freedom for the assumption that the hypothesis is true.

Statistical hypothesis described by dependence (7) was successively verified for k=0,1,2,…,p–2. The analysis of data contained in table 2 shows that all the proper values differ significantly statistically, for significance level $p < 0.033$.

Table 2. Results of testing.

Lp	Proper values	p – value
1	4.161	0.000
2	2.783	0.000
3	1.563	0.000
4	1.181	0.000
5	0.957	0.000
6	0.892	0.000
7	0.778	0.000
8	0651	0.000
9	0.603	0.001
10	0.49	0.015
11	0.415	0.032
12	0.4	0.020
13	0.359	0.025
14	0.343	0.033
15	0.212	0.039
16	0.111	

Proper values included in table 2 have non-ascending order. A chart of successive proper values, in dependence on the proper value number, in an ordered sequence, described by dependence (4), is often used in graphical presentations of the principal component analysis. Such a chart is called an avalanche. In order to place all the sequences in one chart, they need to be normalized by dividing each proper value by the prime (maximal). Chart of avalanche is presented in figure 1.

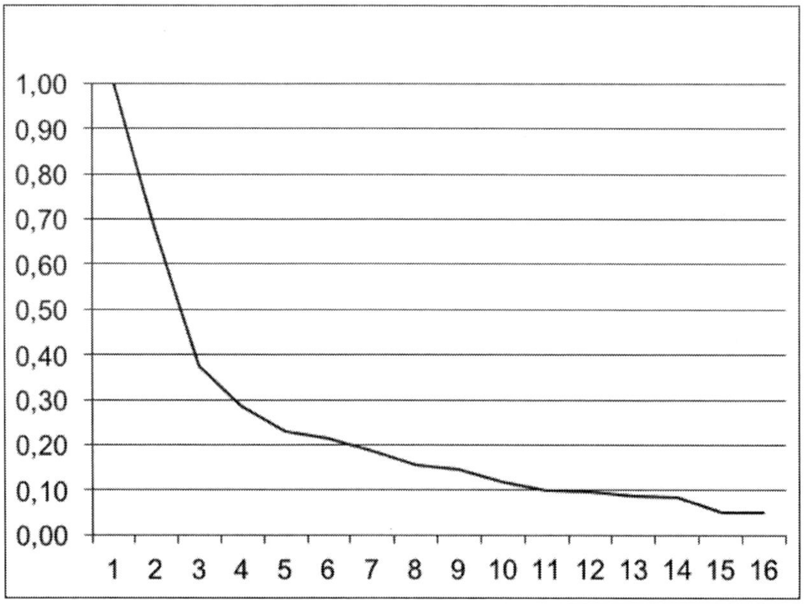

Fig. 1. Chart of avalanche for the analyzed data set.

An analysis of the above chart shows that it gets stabilized along with the growth of the proper value number. However, the stabilization degree is different which has been shown before in statistical tests.

Whereas, an analysis of the assessment results obtained from set (N=100) shows that all proper values are statistically different. On this basis, it can be said that the considered set of criteria should not be reduced to their subsystems. Distinct differentiation in the speed of the avalanche curves stabilization, confirms advisability of further study on the differences between the criteria and their statistical dimensionality.

5 Application of Analytic Hierarchy Process (AHP)

In this study a set of 16 criteria is defined according to scale from 0-10, where grade 10 refers to the most significant criterion. In order, to determine the resultant weight of a particular criterion, a table with comparison of criteria pairs (each with each), has been presented, according to AHP method [14]. The number of the verse and column of each element denotes the number of the compared criteria, whereas, a comparison between criterion i and j is, according to the reasearch, a difference in significance degree between criterion i and j, in the element of verse i and column j.

$$a_{ij} = si_i - si_j \tag{9}$$

where:

a_{ij} - denotes value in element i and column j of the comparison table,
si_j - significance degree defined in the surveys for criterion j,
si_i - significance degree defined in the surveys for criterion i.

The calculated values change in the range from -10 to 10, and according to the applied method, they should change within the range from 6 to 6 [15]. Thus, normalization was carried out on elements of the tables, according to dependence (10), and the results were rounded off to integral values.

$$an_{ij} = round(0.6 \cdot a_{ij}) \tag{10}$$

where:

an_{ij} - denotes the value in verse element i and column j after normalization.

Comparisons of criteria with each other were not taken into account in further calculations. With such an assumption, a matrix of coefficients and absolute terms describing the system of normal equations was created for the comparison tables. These matrixes are of the following form:

$$X = \begin{bmatrix} il_1 & -il_{12} & \Lambda & -il_{1n} \\ -il_{21} & il_2 & \Lambda & -il_{2n} \\ M & M & M & M \\ -il_{n1} & -il_{n2} & \Lambda & il_{nn} \end{bmatrix} \quad Y = \begin{bmatrix} \sum_{j=1}^{n} an_{1j} \\ \sum_{j=1}^{n} an_{2j} \\ M \\ \sum_{j=1}^{n} an_{nj} \end{bmatrix} \tag{11}$$

where:

il_{ij} - denotes the number of grades in elements of verse i and column j for all the tables,

il_i - denotes the number of grades in elements of verse i for all the tables.

According to AHP, the obtained systems are undetermined. In order, to solve them the weight of one of the criteria with value is assumed to be 0, and its corresponding verse is removed. Acceptance of zero value results also in removing an adequate column. For the purpose, of calculation, the weight value of criterion 1 was accepted to be zero (also analogical operations were performed with acceptance of the last criterion weight to be zero, obtaining an equivalent solution).

The reduced systems of equations were solved with the use of 'scilab' program. In this way, these solutions have been received. The obtained solutions can be normalized, by value 10 being added. Assessment of the criteria significance was performed according to the SMART method. This assessments can be interpreted according to the table 3.

Table 3. Interpretation of criterion assessment.

Criterion assessment	Interpretation
10	Extremely significant
9	Very significant
8	Significant
7	Medium significant
6	Little significant
5	Very little significant
4	Insignificant

For calculations of a transport system functioning quality it is better to use the weight of criteria established for the AHP method as their values approach to 1, which reflects relative degrees of the criteria significance in relation to each other. For this purpose, a transformation was made, according to the dependence described in work [16-18]:

$$c_j = \frac{2^{w_j}}{\sum_{i=1}^{n} 2^{w_i}} \tag{12}$$

where:

c_j - denotes AHP assessment of criterion j,

w_j - denotes solution of a normal equation system for criterion j,

n - number of criteria.

In order to analyze the obtained results, a relative weight drop was calculated for each criterion, according to the dependence:

$$wz_{j,j+1} = \frac{c_{j+1} - c_j}{c_{j+1}} \tag{13}$$

where:

$wz_{j,j+1}$ denotes a relative drop in weight of criterion j, in relations to j+1.

Whereas, it seems that a better method for reduction of the criteria number is to take into consideration only three criteria whose weight is larger than the half of weight of the most significant criterion. The considered criteria are presented in the tables in italics.

Table 4. Results of surveys concerning qualitative criteria.

Criterion number	AHP weight	Relative drop in weight
1	*0.19938*	
15	*0.12398*	0.37817
11	0.08378	0.32423
14	0.06982	0.16659
4	0.06817	0.02368
3	0.06611	0.03015
2	0.05692	0.13895
7	0.05225	0.08220
13	0.04564	0.12642
16	0.04038	0.11525
6	0.03923	0.02846
10	0.03699	0.05694
5	0.03150	0.14859
12	0.03100	0.01575
9	0.02806	0.09483
8	0.02671	0.04818

In further research on the subject, of a given transport system operation quality, the determined criteria will provide basis for its assessment whereas, their fulfillment degree will be crucial for the analyzed transport system operation quality.

In connection, with the above, for further research on the analyzed transport system operation quality, the following criteria have been accepted: safety, efficiency, availability, ergonomics, punctuality, damageability and reliability.

6 Conclusions

In this study, methods for selection and determination of significance of criteria necessary to carry out the assessment process of the research object functioning quality, have been presented. On the basis, of the obtained results, it can be said that, each particular, method yielded divergent data - criteria sets with different significance.

It should be noted that sets of the most significant criteria have been distinguished on the basis, of applied statistical analysis and the AHP method. However, the results obtained with the use the PCA method show that all eigenvalues are statistically different. On this basis, one can conclude that the considered set of criteria should not be reduced to their subsets. A distinct differentiation of the avalanche curve stabilization speeds (Fig. 1) shows that it is advisable to carry out further research on particular criteria and analyze their statistical dimensionality.

However, it is possible to determine one subset of significant criteria, representative for this study, by analyzing particular results and comparing them with each other.

Thus, on the basis of the obtained results, it was possible to identify from a set of 16 criteria, the most significant ones: safety, efficiency, availability, atmospheric factors, timeliness, reliability, external factors and punctuality.

In connection with the above, it should be emphasized that in order to evaluate usability of methods it is advisable to compare different ones, as using only one method does not guarantee that it is the most suitable one, and the obtained results are the most optimal from the research point of view.

References

1. M. Woropay, Ł. Muślewski, Quality in terms of system. ITeE-PIB, Radom (2005)
2. Ł. Muślewski, Evaluation Method of Transport Systems Operation Quality. Polish Journal of Environmental Studies. Vol. **18**, No. 2A, Hard Olsztyn, Olsztyn (2009)
3. E. Mańka, M. Słomion, M. Matuszewski, Constructional features of ropes in functional units of mining shaft hoist. Acta Mechanica et Automatica, Vol. **12**, No. 1, pp. 66-71 (2018)
4. B.G. Tabachnik, L. Fidell, Using Multivariate Statistics. Haper & Row, New York (1996)
5. B. Landowski, D. Perczyński, Ł. Muślewski, P. Kolber, Economic aspects of a city transport means purchase. Proceedings of 58th International Conference of Machine Design Departments - ICMD 2017, Publisher: Czech University of Life Sciences Prague, Czech Republic (2017)
6. B. Landowski, Ł. Muślewski, Decision model of an operation and maintenance process of city buses, Proceedings of 58th International Conference of Machine Design Departments - ICMD 2017, Publisher: Czech University of Life Sciences Prague, Czech Republic (2017)
7. B. Landowski, Ł. Muślewski, Numerical simulation of stochastic process as a model of technical object state changes. Engineering Mechanics 2018, 24nd International Conference, may 14 – 17, 2018, Svratka, Czech Republic, Book of full texts, Institute of Theoretical and Appiled Mechanics of the Czech Academy of Sciences, Prague (2018)
8. P. Aleksandrowicz, Analysis of vehicle collisions with the SDC method. 23rd International Conference Engineering Mechanics, 15-18.05.2017 Svratka, Czech Republic (2017)
9. M. Matuszewski, T. Mikołajczyk, D. Yu. Pimenov, M. Styp-Rekowski, Influence of structure isotropy of machined surface on the wear process. International Journal of Advanced Manufacturing Technology, Vol. **88**, Iss. 9, pp. 2477-2483 (2017)
10. M. Styp-Rekowski, E. Mańka, M. Matuszewski, M. Madej, D. Ozimina, Tribological problems in shaft hoist ropes wear process. Industrial Lubrication and Tribology, Vol. **67**, Iss. 1, pp. 47-51 (2015)
11. B.G. Tabachnik, L. Fidell L, Computer – Assisted Research Design and Analysis. Allyn & Bacon, Boston (2001)
12. M.S. Bartlett, A note on the multiplying factors for various chi square approximations. Journal of the Royal Statistical Society **16**, series B (1954)
13. A.L. Comrey, A First Course in Factor Analysis. Academic Press, New York (1973)
14. F.A. Lootsma, Fuzzy logic for planning and decision making. Kluwer Academic Publishers, Dordrecht (1997)
15. F.A. Lootsma, A Model for the Relative Importance of the Criteria in the Multiplicative AHP and SMART. European Journal of Operational Research, Vol. **94** (1996)
16. Ł. Muślewkski, M. Pająk, Fuzzy interpretation of systems operation evaluation. Journal of KONES - European Science Society of Powertrain and Transport Publication, Vol. **13**, No. 3, Warsaw (2006)
17. Ł. Muślewski, The implementation of the fuzzy logic elements in the area of the transport system operation quality assessment. Maritime Industry, Ocean Engineering and Coastal Resources. Volume 1, Maritime Transportation. Taylor & Francis Group, Balkema – Proceedings and Monographs in Engineering, Water and Earth Sciences, Carlos Guedes Soares, Peter N. Kolev (eds), London/Leiden/New York/Philadelphia/Singapore (2007)

18. M. Pająk, Fuzzy Modelling of Cardinal Features of a Complex Technical System. Risk, Reliability and Safety: Innovating Theory and Practice. CRC Press Taylor &Francis Group, Glasgow (2016)

Diagnostics in ballast water management

Marcin Bielski[1], *Jacek* Wachowicz[2], *Ryszard* Bielski[3,*] , *Arkadiusz* Adamczyk[4], *Hendra* Jantanata[5]

[1] Shiprepair Yard Nauta SA, Newbuildings Division Gdańsk, Poland
[2] UTP University of Science and Technology, Faculty of Management, Bydgoszcz, Poland
[3] Gdańsk University of Technology, Faculty of Management and Economics, Gdańsk, Poland
[4] Polish Naval Academy, Faculty of Mechanical and Electrical Engineering, Gdynia, Poland,
[5] LGT Indonesia, Batam, Riau Island, Indonesia

Abstract. Diagnostics in ballast water management is the main remedy to protect against the threat of spreading invasive species that can be carried in ships' ballast tanks. This phenomenon is getting better known and understood. New and more effective methods of preventing threats related to this are being developed. Procedures are created to ensure environmental safety in the face of ever-increasing transport by sea. The article characterizes both the background of the problem, basic diagnostic medics used in ballast water management, as well as procedures and basic techniques used to ensure the safety of the natural environment..

1 BACKROUND

Nowadays most of cargo is being transported by sea. It may be not the fastest but for sur the cheapest solution. Vessels transport three to five billion tonnes of ballast water in one year period. Maritime scientist assumed that every given day, up to 10000 different marine species are released outside their natural habitats. We often consider which is the most economical way, but should that be the only one factor? We should not forget about the slogan "SAFETY FIRST". Hearing that we consider human and other living species safety, cargo and equipment safety. Awareness of growing risk that some species may be transported in ballast water to a brand new area, where due to lack of natural predators treating them they may dominate or even eliminate others. This threat, not only potential, has huge influence on marine environment.

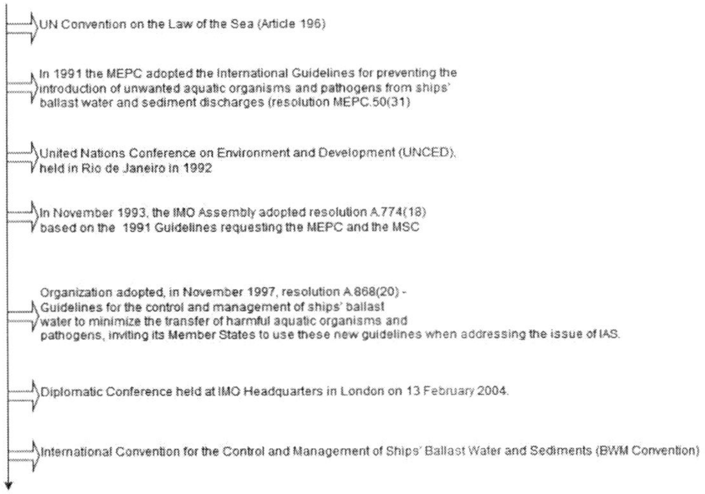

Fig. 1. Legislation evolution.

* Corresponding author: rysbiels@pg.edu.pl

One of the legislation pointis Ballast Water Management Awareness. This subject is related to water transport form one to another region. It came together with steel-hulled vessels more than 100 years ago. Ballast operation is needed (due to its impact on vessel stabilisation) for safe travelling around seas and oceans. The first recognition of the problem of transportation of alien species is dated for 1903, when scientist discovered mass occurrence of the Asian phytoplankton algae Odontella in the North Sea.

MPEC – IMO Marine Environment Protection Committee is the Organization that put effort to establish rules that may prevent such cases of spreading unwanted organisms. The negotiation period can be counted in years. Global maritime community analyzed all pros and cos and decided that ballast water management is entering into force.

Diagram of legislation evolution is presented on Fig. 1.

Now we can find more than 25 regional authorities that announced their own requirements. As the main are treated: IMO and United States Coast Guard. However case study based on new technology implementation, should be divided into two groups: new buildings and retrofits. The thesis that it should be split into two case studies is related to practice installation conditioning. At the first quarter of the 2017 USCG revealed their final rule implementation schedule. The table 1 presents main dates and requirements

Table 1. USCG implementation schedule. [4]

	Total ballast capacity	Date of vessel launching	Compliance date
New buildings	All	On or after 1 Dec 2013	On delivery
Retrofits	< 1500 m^3	Before 1 Dec 2013	First scheduled dry-docking after 1 Jan 2016
	$1500m^3$ – $5000m^3$		First scheduled dry-docking 1 Jan 2014
	> $5000m^3$		First scheduled dry-docking after 1 Jan 2016

Ship-owners must consider their operation area, because if they have clients in United States of America they have to install proper system for ballast water treatment. Having on mind strict restrictions USCG provides two additional compliance alternatives:

- Exclusive use of ballast water from a public water system in United States.
- Discharge of all ballast water into an onshore facility or another vessel, barge for treatment purposes.

2 BALLAST WATER DIAGNOSTICS AND ITS ENFORCEMENT

Ballast water management is governed by regulations which stipulate limits for: quantity of organisms as well as concentration of indicator microbes that are allowed to be discharged in ballast water. Units dedicated to ballast water treatment must show that they meet these limits not only during type approval, but also during whole product life cycle.

The maritime industry require standardization. BWM convention define international standards for ballast water treatment in Regulation D-2. According to diagnostics, which show measured levels different actions may be required. This may be divided into three consecutive levels, where different actions should be undertaken. This may be presented as on Fig. 2.

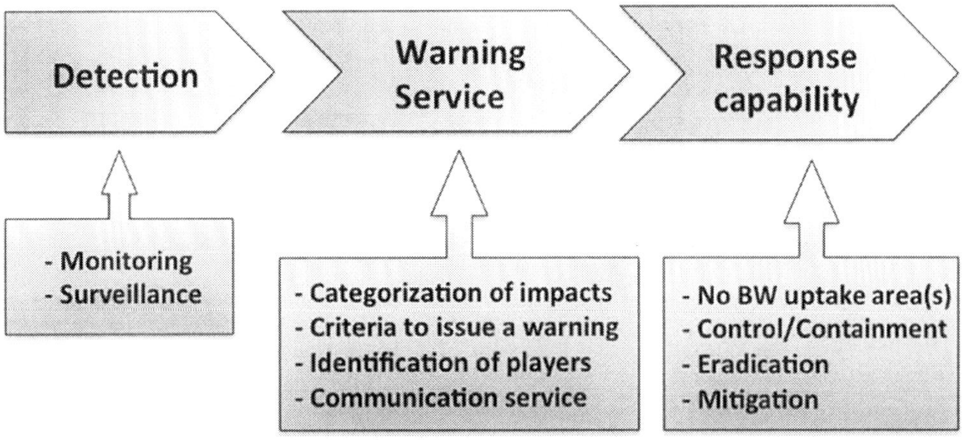

Fig. 1. Main components of an Early Warning System for Non-indigenous Species and Ballast Water Management. BW = Ballast Water. [6]

2 DETECTION

First step, in all cases requires measurement. This is very difficult issue in case of ballast water in ships, as they may only be probed – and outcomes may vary significantly due to ambient conditions, as presented on Fig. 6. Different probing techniqes may also be sensitive to various factors. Basic, simplest method is microscopic evaluation, which diagram is presented on Fig. 3. More advanced technique is HR-LOPC, which diagram is presented on Fig. 4 The FlowCAM technique is the first one that may execute tests in different classes of samples. Its diagram is presented on Fig. 5. A broad overview of existing specific methods can be found in [3].

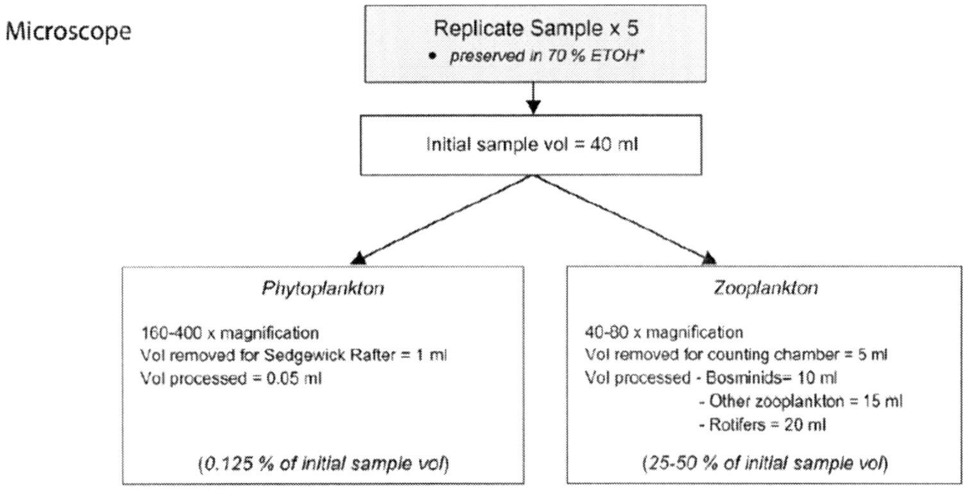

Fig. 3. Diagram of microscopic evaluation. [5]

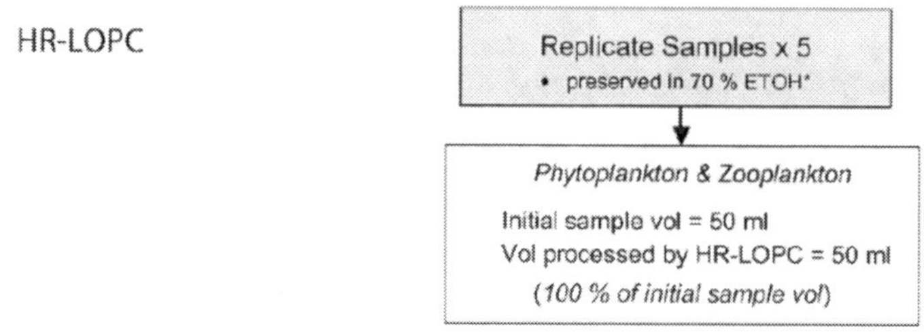

Fig. 4. Diagram of HR-LOPC evaluation. [5]

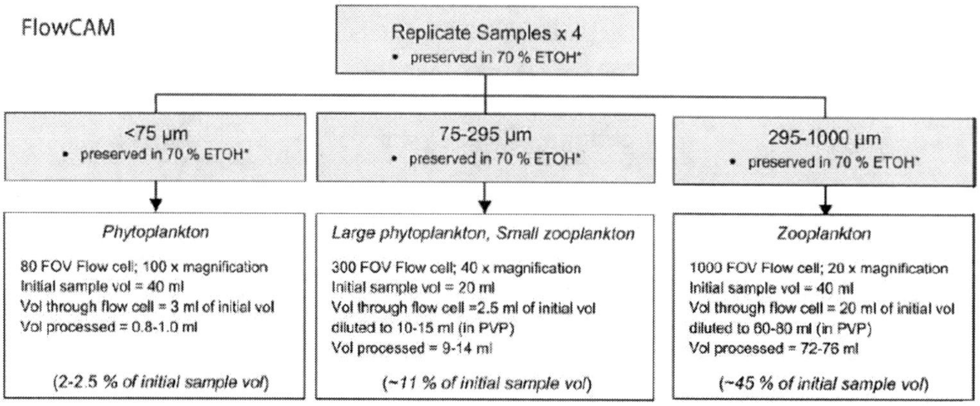

Fig. 5. Diagram of FlowCAM eevaluation. [5]

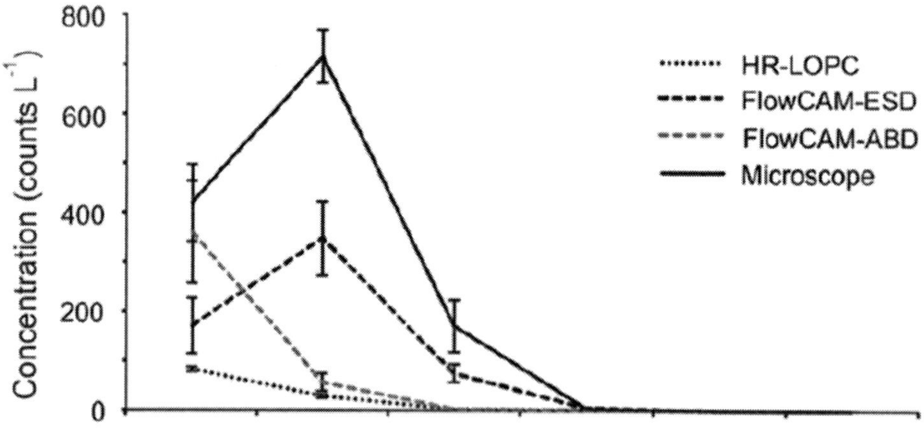

Fig. 6. Concentration plots for particles N 30 μm measured by the HR-LOPC, FlowCAM (organism only) ESD and ABD measurements, and the microscope. [5]

Differences in method sensitivity resulted in spliting sampling methods into targeted ones (which are highly sensitive to specific spicies, but omit possible other ones, even if they had been very dangerous) and into community profilling (which are less accurate, but give responses to a far higher spectrum of potential threats). Their processes may be comared and presented as on Fig 7

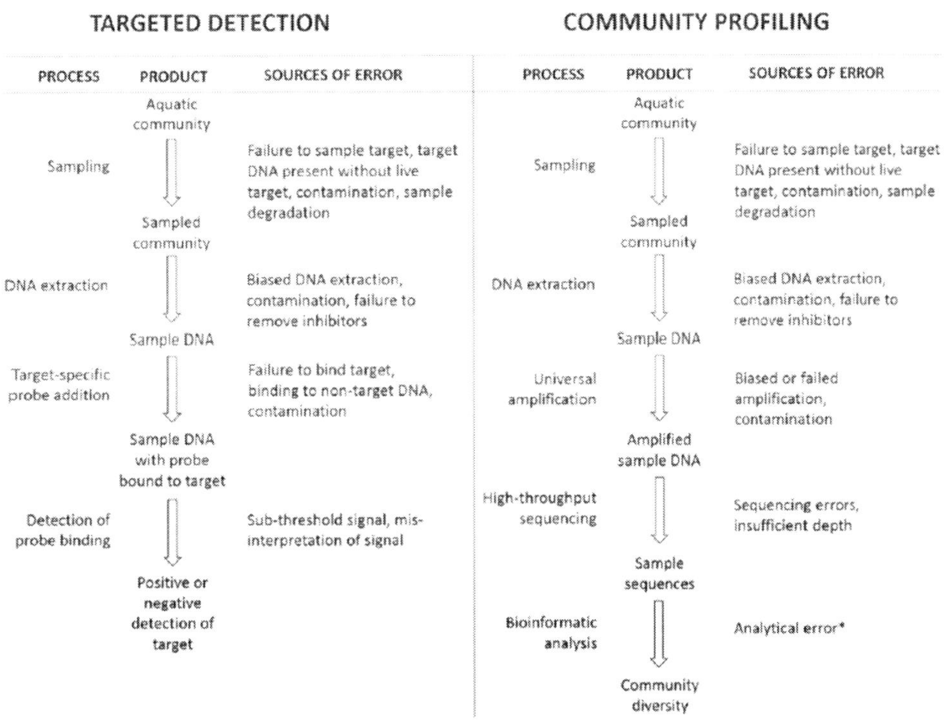

Fig. 7. Workflow for targeted probe-based detection (left) and HTS-based community profiling (right), including potential sources of error introduced at each process step. False positive and false negative errors can derive frommultiple steps in each process. *Encompasses a broad range of error sources including, but not limited to, poor filtering of reads, failure to remove chimeras, and inaccuracies in taxonomic assignment.. [3]

Relatively new class of methods used in ballast water diagnostics are genetic methods. They offer brand new approach and possibilities, due to their extremely distinctive properties, as in comparison to traditional, morphological methods. The may be compared as on Fig. 8.

Criterion	Genetic methods	Traditional morphological methods
Sensitivity	HIGH	low
Specificity	HIGH	low
Ability to identify sub-adult or partial specimens	HIGH	low
Ability to identify cryptic taxa	HIGH	low
Quantification	low	HIGH
Opportunity for passive surveillance	HIGH	low
Affordability of up-front costs	low	HIGH
Affordability per sample	HIGH*	low
Speed of analytical turnaround	HIGH	low
False negative avoidance	HIGH	low
False positive avoidance	low	HIGH

Fig. 8. Genetic vs. morphological methods. [3]

3 WARNING SERVICE AND RESPONSE CAPABILITY

Results of ballast water testing testing may force necessity of further actions. They may have different directions – towards vessel, other vessels, or authorities. Simple graph of possible warning actions is presented on Fig. 9.

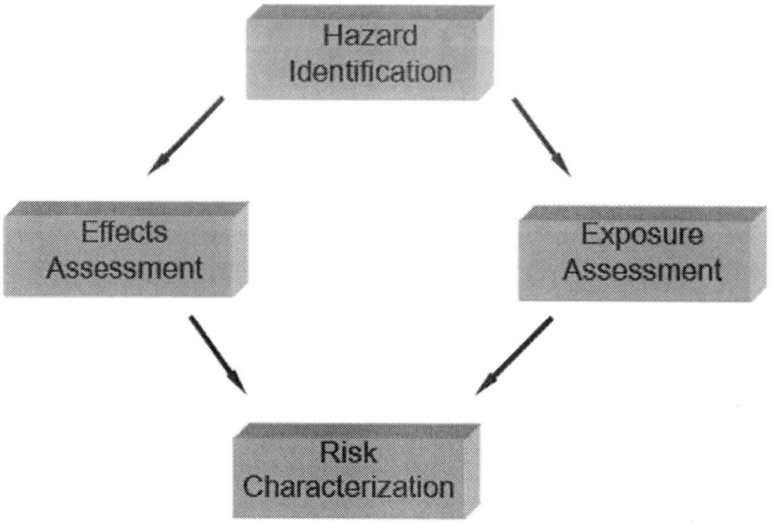

Fig. 9. Risk assessment approach. Four-step risk assessment approach. [1]

Group of organisms	Counts per volume
⩾50 μm	<10 per m^3
⩾10 μm < 50 μm	<10 per mL
Vibrio cholerae (O1 and O139)	<1 cfu per 100 mL or <1 cfu per 1 g (wet weight) zooplankton samples
Escherichia coli	<250 cfu per 100 mL
Intestinal Enterococci	<100 cfu per 100 mL

Fig. 10. The BWM Convention ballast water performance standard (Regulation D-2) for maximum limits of viable organisms per defined volume of discharged ballast water. [7]

Finding an excessive amount of chemicals or phyto- or zooplankton may require undertaking standard preventive actions to protect the local environment against the intrusion of alien species. The most widespread among them are presented in Fig. 10. Decision graph for additional actions may consist of many basic steps. It may be presented as on Fig. 11.

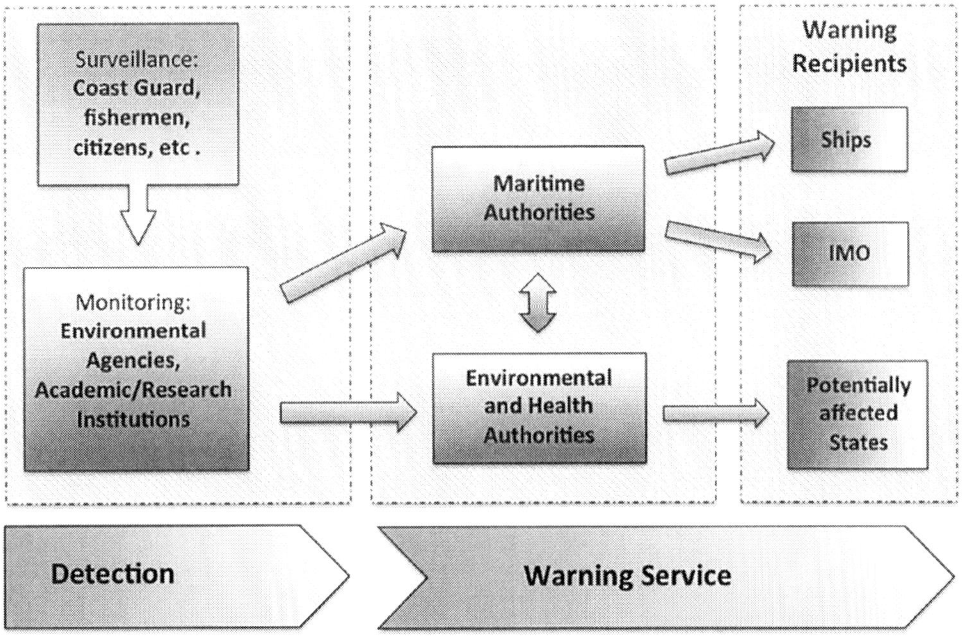

Fig. 11. Institutions, authorities and stakeholders involved in an Early Warning System for Non-indigenous Species and Ballast Water Management. [6]

Depending on hazard different actions may be required. They are illustrated on following Figures.

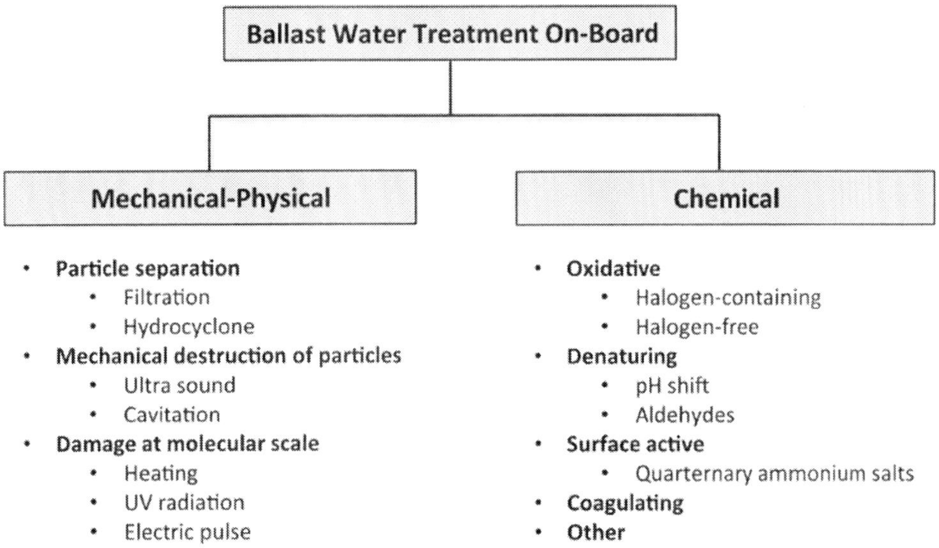

Fig. 12. Overview of ballast water treatment measures on board a ship. [7]

Most frequently used techniques reducing biohazards are shown on Fig. 13

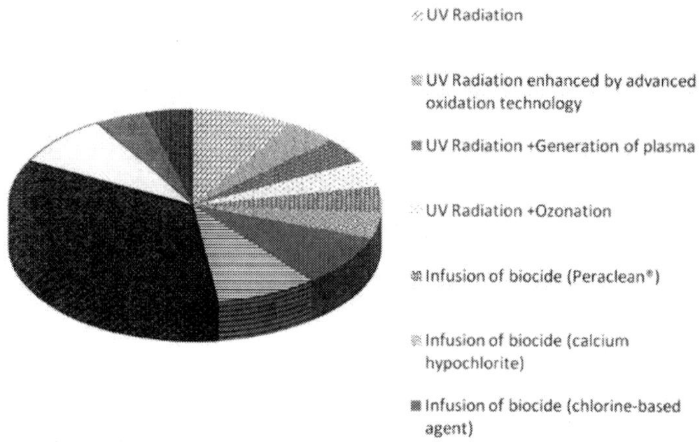

Fig. 13. Techniques of reducing biohazards. [2]

There is always a risk of injuries to human crew; some standard security procedures are shown on Fig. 14.

Activity	Exposure	Quantification (IMO, 2012)
Crew		
Handling of chemicals	Dermal/inhalation (leakages, spills)	100 mg (0.1 mL)/container
Starting of BWMS	Type-specific	Case by case
BW treatment	Type-specific	Case by case
Ballasting/routine deck work	Inhalation (BW tank exhaust)	1 h d ^{1} over 6 mth ^{1}
De-ballasting	Dermal/inhalation (spray drift)	Not considered
Tank cleaning	Dermal (whole body)/inhalation	8 h d ^{1} on 5 d wk ^{1}; 1 wk y ^{1}
Other maintenance work	Type-specific	Case by case
Malfunctions, accidents	Inhalation/dermal	Not considered
Port state control		
Sampling (compliance control)	Inhalation/dermal (hands)	2 h d ^{1} on 5 d wk ^{1}; 45 wk y ^{1}
Tank inspection	Inhalation	3 h d ^{1} on 1 d mth ^{1}
General public		
Swimming/recreational activities	Inhalation/dermal/oral	5 h d ^{1} on 14 d y ^{1}
Seafood consumption	Oral	~200 g d ^{1}

Fig. 14. Scenarios and pathways for human exposure to chemicals from BWMS. [7]

In serious cases it may be necessery to inform other vessels. Such decision diagram is shown on Fig. 15.

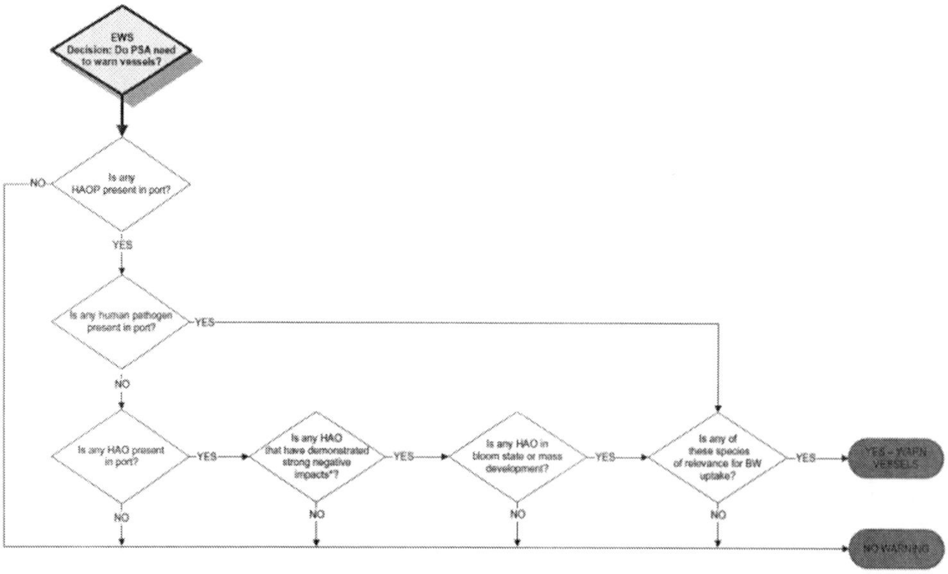

* impacts on human health, environment, economy.

Fig. 15. Decision Support System model for an Early Warning System: warnings to vessels. EWS

=Early Warning System, PSA = Port State Authority, HAOP =Harmful Aquatic Organisms and Pathogens, HAO =Harmful Aquatic Organisms. [6]

3 CONCLUSIONS

Diagnostics in ballast water management is the main remedy to protect against the threat of spreading invasive species that can be carried in ships' ballast tanks. This phenomenon is getting better known and understood. However, it is still a problem widely known only to a small group of specialists. This article presents a comprehensive picture of the problem, characterizing the basic methods and procedures. It has a cognitive character, as well as allowing for a deeper look at the problems occurring as a result of the unintentional transport of living organisms in ballast water tanks.

References

1. S. Baner, B. Werschkun, T. Höfer, Assessing the risk of ballast water treatment to human health, Regulatory Toxicology and Pharmacology, Vol 62, Iss 3, April 2012 (2012)
2. J. Culin, B. Musta, Environmental risks associated with ballast water management systems that create disinfection by-products (DBPs), Ocean & Coastal Management 105 (2015)
3. J. A. Darling, R. M. Frederick, Nucleic acids-based tools for ballast water surveillance, monitoring, and research, Journal of Sea Research, 133 (2018)
4. IMO/USCG D2 implementation scheme, DNV GL, https://www.dnvgl.de/maritime/ballast-water-management-imo-uscg.html accessed 2018-05-10
5. J. Kydd, H. Rajakaruna, E. Briski, S. Bailey, Examination of a high resolution laser optical plankton counter and FlowCAM for measuring plankton concentration and size, Journal of Sea Research, 133 (2018)
6. E. Magalettia, F. Garaventab, M. Davidc, L. Castriotad, R. Krause, G. M. Lunaf, C. Silvestria, C. Forteg, M. Bastianinih, M. Falautanod, T. Maggiod, G. Raka, S. Gollaschi ; Developing and testing an Early Warning System for Non Indigenous Species and Ballast Water Management, Sea Research, 133 (2018)
7. B. Werschkuna, S. B. Oihane, C. Basurkob, M. David, F. Fuhrd, S. Gollasche, T. Grummtf, M. Haarich, A. N. Jhah, S Kacan, A. Kehrer, J. Lindersk, E. Mesbahi, D. Pughiuc, S. D. Richardsonn, B. Schwarz-Schulz, A. Shaho, N. Theobald, T. Höfer; Emerging risks from ballast water treatment: The run-up to the International Ballast Water Management Convention, Chemosphere, Vol. 112, October 2014 (2014)

Traceable calibration of automatic weighing instruments operating in dynamic mode

Rafał Sarnecki[1,*] , *Wojciech* Wiśniewski[1] , *Wiktor* Ślusarski [1] and *Piotr* Wiłkojć[1]

[1]Central Office of Measures, Laboratory of Mass, Elektoralna 2 Warsaw, Poland

Abstract. This article describes calibration method of automatic weighing instruments for measuring vehicles in motion. These measuring instruments in future will be widely used to determine the axle loads, if applicable the axle-group and vehicle mass of road vehicles when the vehicles are weighed in motion. For several years, some activities have been carried out on the weight control system that determines the mass of vehicles moving at higher speeds, called HS-WIM. The implementation of such systems will affect both the improvement of road safety and indirectly limit the number of road users moving by overloaded vehicles.

1 Introduction

Automatic instruments for weighing vehicles in motion (WIM) [1], are one or two-bridging measuring instruments used to determine the axle load, axle group load and total vehicle mass while the vehicle is in motion,. These instruments are more effective and efficient in the long term than non-automatic weighing instruments .The development of automatic instruments emphasizes the need to confirm their metrological quality by calibrating and determining their measurement uncertainty. Calibration of such instruments also is carried on demand of different companies due to environment protection procedures (continuous totalizing automatic weighing instruments - belt weighers - used to control CO_2).

2 Calibration method

In order to realize a calibration, the influence of various factors on the measured value should be taken into account. Such factors include the rate of operation of the instrument, type of material and nominal value of the testload.

The object of the calibration is the indication provided by the instrument in response to an applied load. The results are expressed in units of mass. The uncertainty of measurement depends significantly on properties of the calibrated instrument itself, the control instrument, the characteristics of the test loads, not only on the equipment of the calibrating laboratory.

In agreement with the client, the calibrating laboratory determines anticipated value of the uncertainty of measurement, which is appropriate in view of the use of the instrument and in view of the cost of the calibration.

Any such procedure must include, for a limited number of test loads, the determination of the error of indication and of the uncertainty of measurement assigned to these errors. The calibration should be processed as closely as possible to the weighing operations that are routinely being performed by the user.

The procedure may further include rules how to derive from the results advice to the user of the instrument with regard to the errors, and assigned uncertainty of measurement, of indications which may occur under normal conditions of use of the instrument.

Calibration of the complete WIM instrument is performed in the conditions of its use in a location and at the rate of operation requested by the client.

* Corresponding author: rafal.sarnecki@gum.gov.pl

Before the application of the test procedure, extreme caution and safety in the dock entrance road and the weighing area should be guaranteed and provided.

When starting to calibrate, a test load should be applied to the instrument under specified conditions. Also, the total weight and axle load of the test vehicles constituting test loads should be determined, and the indication errors of WIM instrument should be calculated. The last step is to estimate the measurement uncertainty.

The weighing range is given by the values Max and Min. For Axle load, the Min=1000 [kg] and Max =20000 [kg]. In case of the mass od vehicle, Min=3500 and

Max $\leq$ 48000 or as specified by national legislation.

Calibration of the complete WIM instrument is performed in the conditions of its use in a location agreed with the client. Before the application of the test procedure, extreme caution and safety in the dock entrance road and the weighing area should be guaranteed and provided.

2.1 Determining the axle load Ax_{ref} of control vehicle

Terminology and equation used in this article are based on:

- OIML R 76-1 [4]
- OIML R 134[1]
- EURAMET Calibration Guide No. 18 [2]
- OIML R 111-1[3]

The mass Ax_{refi} determined two axle vehicle Perform at least 10 load static measurements of Ax_i for the setting direction of the vehicle. The loads shall be determined in practice, at least two different loads: vehicle unloaded and the most loaded with standard weights or ballast.

Calculates the average value Ax_i for each axis of a 2-axle vehicle representing the value $Ax_{ref.}$:

$$Ax_{refi} = Ax_i = \frac{\sum_{1}^{10} A_{xi}}{10} \tag{1}$$

where: Ax_{is} is value of the mass of the control vehicle axle load during the biaxial static weighing. Calculation of the total mass of the vehicle is realized by VM_i:

$$VM_i = VM = \sum Ax_i \tag{2}$$

2.2 Method of calibration the WIM instrument of the dynamic loading

A complete WIM instrument is checked using control vehicle and other measuring equipment at rated operating conditions of WIM instrument, according to its characteristics. The mass of control vehicle should be such that it can be as close as possible to the maximum measuring range of the WIM instrument. The mass of control vehicle may be less adapted to customer needs. This should be agreed before starting the measuring procedure. Vehicles carrying liquids or other materials, which means harder-bones moves during the weighing may be used as a control vehicle only in duly substantiated cases (eg. customer needing, special application of WIM instrument). Speed of vehicles should be constant, without acceleration or braking, and should include the full speed range, including the minimum speed V_{min}, and the typical speed on that road. If necessary (eg. according to the customer needs), the speed can be different (higher or lower than V_{max}), but only with the acceptance of the customer and printing such information on the calibration certificate. For each of vehicle control, and for each set of the weight and speed and specified driving directions, takes at least $n = 5$ passages:

- 3 runs down over the center of the load receptor,
- 1 ride left side of the load receptor,
- 1 ride right side of the load receptor.

3 MEASUREMENT RESULTS

3.1 Test of WIM instrument with 2-axles control vehicle dynamic loading

Determines the error of axle E_{Axi} and vehicle total mass E_{VMi} from every passage of the individual vehicle and:

$$E_{Axi} = Ax_i - Ax_{refi} \tag{3}$$

$$E_{VMi} = VM_i - VM_{ref} \tag{4}$$

4 UNCERTAINTY BUDGET

4.1 Standard uncertainty of discrete values

The error is equal to difference between the value indicated by the instrumentin the certain entity and the reference load[2]:

$$E = I - m_{ref} \tag{5}$$

The equation of uncertainty has the form:

$$u^2(E) = u^2(I) + u^2(m_{ref}) \tag{6}$$

In case of automatic weighing instruments, reference loads replaced by a value of control instrument and the formula for E calculation is:

$$E = I - R_K \tag{7}$$

where R_K is value from the control instrument and

$$R_K = I_K - E_K \tag{8}$$

where E_K is the error of the control instrument which means

$$E = I - (I_K - E_K) \tag{9}$$

Then the equation of uncertainty is:

$$u^2(E) = u^2(I) + u^2(R_K) \tag{10}$$

where $u(I)$ is the uncertainty of the indication and
$u(R_K)$ is the uncertainty of the value from the control instrument

4.2 Standard uncertainty of indication

$$I = I_L + \delta I_{digL} + \delta I_{rep} - I_0 - \delta I_{dig0} \tag{11}$$

All these correction have expected value of zero. Their standard uncertainties are:

δI_{dig0} stands for the rounding error at zero loading

$$u(\delta I_{dig0}) = \frac{d_0}{2\sqrt{3}} \tag{12}$$

or

$$u(\delta I_{dig0}) = \frac{d_T}{2\sqrt{3}} \tag{13}$$

δI_{digL} stands for the rounding error of indication at a load

$$u(\delta I_{digL}) = \frac{d_I}{2\sqrt{3}} \tag{14}$$

or

$$u(\delta I_{digL}) = \frac{d_T}{2\sqrt{3}} \tag{15}$$

δI_{rep} stands for the error of repeatability

$$u(\delta I_{rep}) = s(I_j) \tag{16}$$

Standard deviation calculation is carried out form a series of calculated errors for each entity (gross mass or axle load).

4.3 Standard uncertainty of a value from the control instrument u(R$_K$)

$$u(R_K)^2 = u^2_{opak} + u^2_{d0} + u^2_{dI} + u^2_{EI} \tag{17}$$

where $u(R_K)$ is equal to the uncertainty of the instrument in use. The uncertainty of the control instrument value u_{EI} for the given load (the value is obtained from the control instrument calibration certificate) is expanded by an uncertainty of indication.

The value of the uncertainty taken from a calibration certificate is valid only for a control instrument intended for weighing the gross mass of a reference vehicle.

The uncertainty of a control instrument for weighing axle load shall be obtained as follows.

$$u_h^2 = \left(\frac{mpe}{\sqrt{3}}\right)^2 + (0{,}005\, I)^2 \tag{18}$$

where: mpe is the maximum permissible error for the given load in service;

0,005 is the value resulting from external effects;

I is the indication.

The individual results of measurement are considered to be as correlated quantities (coefficient: $r = 1$) and summing the uncertainties of individual axles shall be as follows:

$$u_C = u_1 + u_2 + \ldots + u_n \tag{19}$$

The expanded uncertainty:

$$U = 2 \times u_C \tag{20}$$

4.4 Combined standard uncertainty

$$u(E_I)^2 = u^2_{opak} + u^2_{d0} + u^2_{dI} + (u^2_{opak} + u^2_{d0} + u^2_{dI} + u^2_{EI} + u(m_T)^2) \tag{21}$$

where items in brackets represent an uncertainty of the control instrument.

5 Implementation of HS- WIM

In last year Central Office of Measures intensive effort to establish legal metrology control of high speed weighing in motion. Roads are being damaged by overweight trucks. Too heavy trucks also are dangerous because of longer way of reducing their speed. To reduce wear of road surface and increase safety, Poland decided to focus effort on implementing high-speed weighing of road vehicles subject to legal metrology. To achieve this, it is necessary to implement devices enabling the imposition of penalties on road users destroying the surface by moving overloaded vehicles, for legal metrological control. The implementation of such systems will improve both road safety and indirectly limit the number of road users moving overloaded vehicles. In Poland currently over hundred HS-WIM instruments work as "pre-selection", which means they measure load and Road and Traffic Security Guard must repeat measurment on legally approved non-automatic weighing instruments (NAWIs). Drivers know when road inspectors are in work and too heavy trucks avoid control waiting for officers to end their work so the best solution is automatic 24h/day system which controls loads the same way like it is speed measuring.

5.1 The method of carrying out the measurement

There were vehicles: an empty bus, a bus with a 500 kg standard, a truck with a weight of 3000 kg and a weight of 39,000 kg. The vehicles passed through two measuring stations, turned back and passed them again. The instruments installed on the gantry identified the vehicle and measured its length. The vehicle was passing through transducers installed in the roadway.

The measurement was repeated 10 times. The time of the vehicle's passage through the transducers was also monitored. The results were recorded in the system that was available for the Inspectorate of Road Transport.

Fig1. and Fig2. Show error points for each measurement. Testing different types of load sensors shows that is possible to build automatic load measurment system. Accuracy of systems which are now used is 3-5% while overweight of trucks vary between 10 and 20 %, so for road inspection it is efficient accuracy.

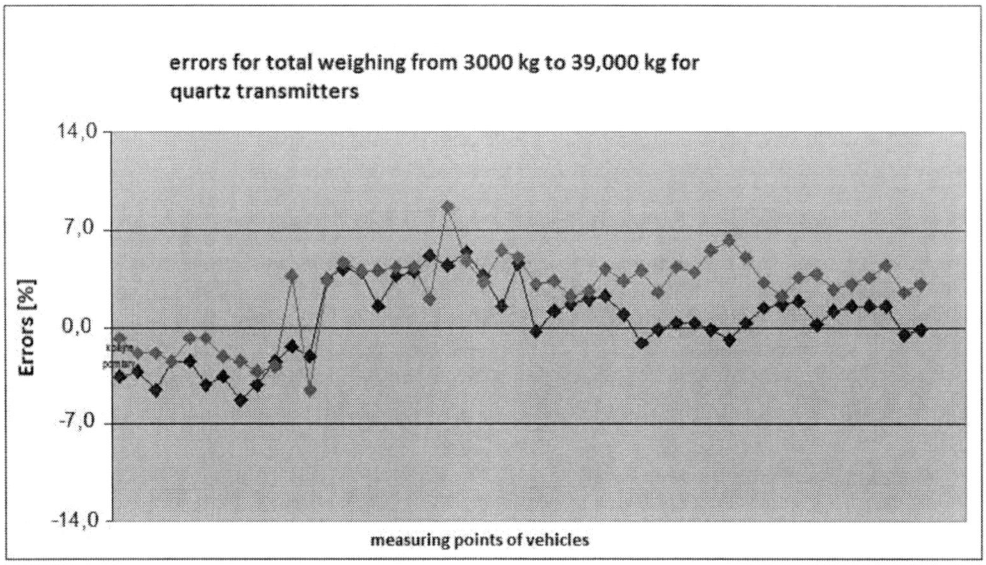

Fig1. Errors for total weight for quart transmitters.

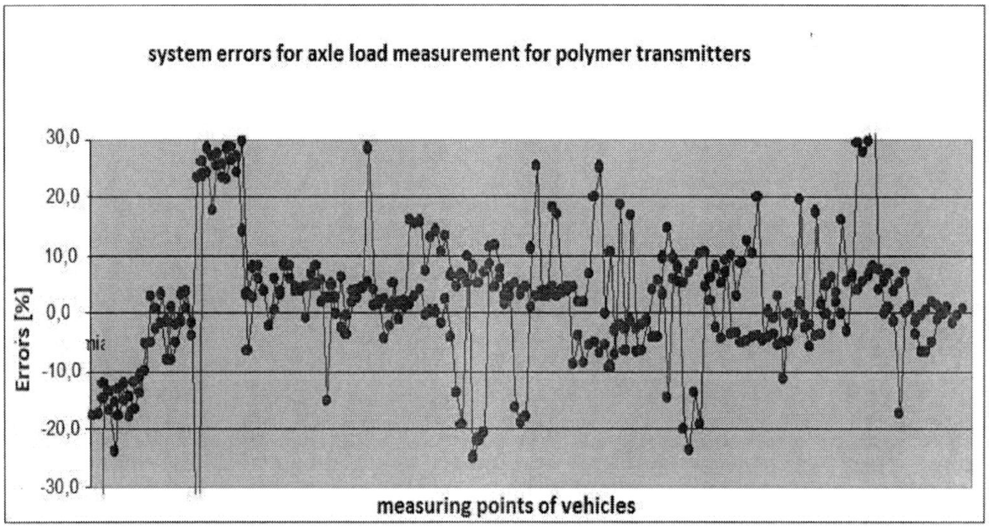

Fig2. Errors for total weight for polymer transmitters.

5.2 Plans for testing laboratory of HS-WIM

Central Office of Measures plans to build a complex that allows testing all devices in this type of instruments.

The complex will include:

- test road
- laboratory of speed measurement

- warehouses
- laboratory of testing electrical interference
- laboratory of total mass measurement

Full complex is shown on Fig3.

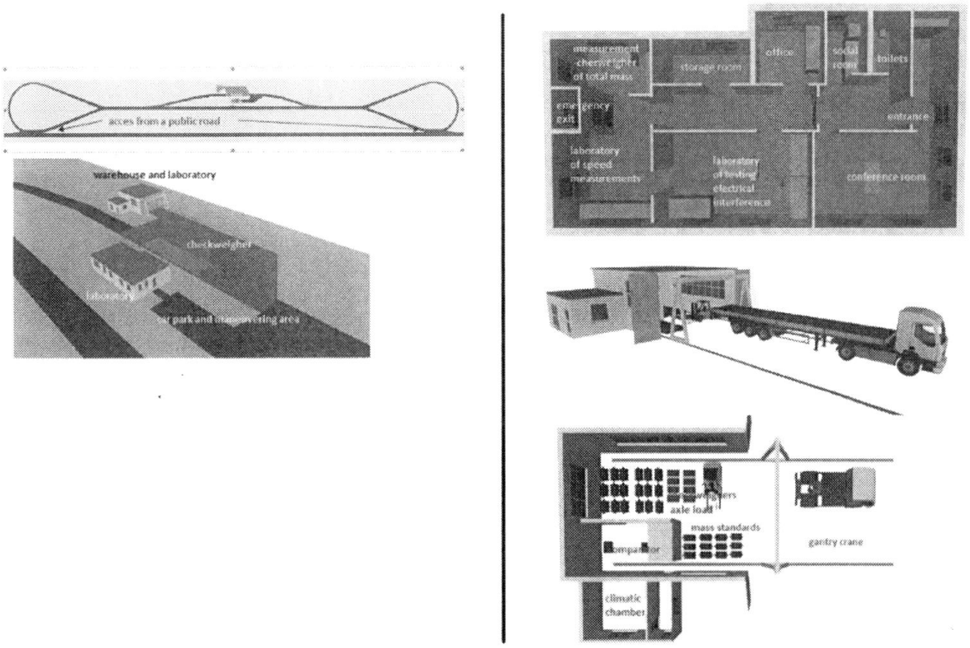

Fig3. Laboratory complex for testing HS-WIM.

6 Conclusion

The method of calibration of automatic instruments described in the article will improve weighing process control in different industries such as production pre-packed products and transport. The calibration laboratories be accredited by accreditation bodies based on the harmonised guidance. In the case of AWI, the uncertainty of measurement depends to a large extent on the properties of the calibrated instrument itself, the characteristics of test loads, the control instrument, and not only the other possibilities of the calibration laboratory.

Several methods for determination of the reference mass are available however, it is necessary to look for a suitable compromise between the accuracy and simplicity of the selected method. Additionally, a density of the test load does not need to be known without significant influence to the measurement uncertainty. Significant progress has been made in creating the guide for automatic balances.Mass Laboratory in Central Office of Measures wants to convince state to build laboratory which is necessary to carry further tests and carry type approval in future. Building such testing laboratory will cost a lot of money but implementing working HS-WIM system should prevent much bigger cost of damaging roads.

References

1. OIML R 134-1, Automatic instruments for weighing road vehicles in motion and measuring axle loads. Part 1: Metrological and Technical Requirements – Tests (2006)
2. EURAMET Calibration Guide No. 18, *Guidelines on the Calibration of Non-Automatic Weighing Instruments*, Version 4.0 (1/205)
3. OIML R 111-1, Weights of classes E_1, E_2, F_1, F_2, M_1, M_{1-2}, M_2, M_{2-3} and M_3. Part 1: Metrological and Technical Requirements, (2004)

4. *OIML R 76-1 , Non-automatic weighing instruments. Part 1: Metrological and Technical Requirements – Tests* 206)

Diagnostics of the bainite transformation mechanism and the effect of normalizing and tempering on the hardness and microstructure of the new Cr-Mo-V-Ti steel for operation at elevated temperature

Zdzisław Ławrynowicz[1,*]

[1]University of Science and Technology UTP, Mechanical Engineering Faculty, Department of Materials Science and Engineering, Av. Kaliskiego 7, 85-796 Bydgoszcz, Poland

Abstract. The Cr-Mo-V-Ti based low alloy steels are widely used in thermal power plants because of their ability to withstand elevated temperatures and high pressure under continuous service. In the present work conventional heat treatment like normalizing and tempering of the alloys has been performed. The material used in this study was the laboratory prepared experimental low alloy Cr-Mo-V-Ti steel. Samples were austenitized at 980°C for 0.5 hour air cooled and tempered at 500, 550, 600, 650, 700 and 750°C for 1 hour. Mechanism of bainite transformation has been studied in Fe-C-Cr-Mo-V-Ti steel using high speed dilatometry. These experimental data indicate that bainitic ferrite forms by a displacive transformation mechanism, but soon afterwards, excess of carbon is partitioned into the residual austenite. The changes observed in the microstructure of the steel tempered at the higher temperature, i.e. 750°C were more advanced than those observed at the temperature of 500°C. Performed microstructural investigations have shown that the degradation of the microstructure of the examined steel was mostly connected with the processes of recovery and polygonization of the matrix, disappearance of lath bainitic microstructure and the growth of the carbides. The magnitude of these changes depended on the temperature of tempering.

1 Introduction

The first Cr-Mo steels were used for conventional power-generation applications in the 1920s. The 2¼Cr-1Mo (nominally Fe-2.25Cr-1.0 Mo-0.3Si-0.45Mn-0.12C) steel, designated by ASTM as Grade 22, was introduced in the 1940s and is still widely used today. The next generation of heat-resistant steels, in addition to increased chromium, involved primarily the addition of the carbide formers vanadium and niobium to chemical compositions (i.e. T22 steel) to add precipitate strengthening and in some cases, a small molybdenum addition was made for further solid-solution strengthening [1,2,3]. Typically Cr-Mo steels are intended for boiler tubes, wire pairs, parts of boilers and steam turbines, components and other equipment operating at temperatures up to 580°C [4].

The efficiency of power plants could be improved by enhancing the steam parameter. At present, heat-resistant steels for the high-steam parameter of 650°C are being developed. This has put heat-resistant steels such as T/P91, T/P92 and E211 out of consideration because of the loss of the microstructure stability during service at the high temperature [5-7]. More advanced steels should be developed to meet this requirement. It is well accepted in heat-resistant steels that highly stable microstructure will produce excellent creep strength. The precipitates are basically $M_{23}C_6$ and MX, and the MX-type carbides show much better stability than the $M_{23}C_6$ type carbide. In order to achieve microstructure

* Corresponding author: lawry@utp.edu.pl

with high stability, stable precipitates such as MX-type carbides are expected in heat resistant steels [8-11].

The aim of the present investigations was to study the mechanism of bainite transformation and the influence of tempering temperature on microstructure previously normalized steel, the new experimental Cr-Mo-V-Ti heat-resistant steel which contains more alloying contents relative to T22 (2¼Cr-1Mo) and other high-quality heat-resistant steels.

2 Material and experimental procedures

The chemical composition of the experimental Cr-Mo-V-Ti steel (denoted as 15HM2VT) is listed in Table 1 and the critical A_1, A_3, B_S and M_S temperatures are shown in Table 2.

Table 1. Chemical compositions of the 15HM2VT steel used in this study. All concentrations are given in wt.% .

Alloy	C	Si	Cr	Mn	Mo	V	Ti	P	S	Al
15HM2VT	0.15	0.24	0.84	0.92	2.6	0.24	0.12	0.022	0.019	0.06

Table 2. A_1, A_3, B_S and M_S temperatures.

Alloy	Temperature, °C				Experimentally determined ***		Thermodynamically calculated			
	A_{c_1}	A_{c_3}	A_{r_1}	A_{r_3}	M_S	B_S	M_S*	B_S*	M_S**	B_S**
15HM2VT	807	969	730	896	453	-	423	555	479	601

* Assuming full austenitization (1200°C) ensuring complete dissolving of carbides in austenite
** Assuming partial dissolving of carbides in austenite (1000°C)
*** After austenitization at 1000°C/10min.

In order to establish the A_1 and A_3 temperatures (Table 2) dilatometric analysis was carried out on a Leitz-Wetzlar vacuum dilatometer using a 20 mm long by 3.5 mm diameter specimens. Adamel Lhomargy LK-02 high-speed dilatometer was used to establish change of length ($\Delta L/L$) during bainitic transformation and the M_S temperatures (Table 2). In order to ensure rapid quenching (300 Ks^{-1}) the specimens were 12 mm in length and 1.0 mm in diameter.

Optical microscopy was used to examine etched structures. Microscopic images were recorded using the inverted metallographic microscope Nikon MA100 equipped with the ERC5s digital camera and the ZEN 2011 archiving program. Specimens were etched in 2% nital solution.The specimens for transmission electron microscopy (TEM) were machined to 3 mm diameter rods and subsequently were sliced into 0.35 mm thick discs while being kerosene cooled. The discs were subsequently ground down to a thickness of 40-50 μm. These specimens were finally electropolished in a twin-jet at room temperature and at 55-60 V using a 25-pct glycerol, 5 pct perchloric acid and 70 pct ethanol mixture.

The retained austenite content was calculated from the *X*-ray intensities. A peak separation technique was used to separate the $(111)_\gamma$ -reflection from the overlapping $(110)_\alpha$ -reflection. Retained austenite was determined by method of direct comparison (Averbach-Cohen method).

Vickers HV30 hardness measurements were performed according to European Standard EN ISO 6507-1 on the Duramin Z-500 hardness tester.

3 Carbon concentration in austenite

The dilatometry results showed that the relative length change during the formation of bainite increases as the isothermal transformation temperature decreases below the *Bs* temperature, then the amount of bainite formed is dependent on the transformation temperature.

The details for calculation of volume fraction of transformation and determined carbon concentration of the residual austenite are given in Table 3. The carbon concentration in austenite, x_γ, was determined as detailed elsewhere [12] by dilatometry. All details of the determination of carbon content in residual austenite are described elsewhere [12,13]. The obtained results in the area of these issues were discussed in a separate publications [13,14].

Table 3. Parameters for determination of volume fractions of transformation (V_α) and the carbon concentration of the residual austenite x_γ after cessation of bainite reaction in 15HM2VT steel.

Transformation temperature, °C	a_γ nm	a_α nm	$\Delta L/L$ $\times 10^{-3}$ *	V_α **	x_γ mol
460	0.36278	0.28926	3.34	0.74	0.0224
475	0.36290	0.28933	3.03	0.68	0.0185
490	0.36302	0.28939	1.83	0.42	0.0110
508	0.36317	0.28947	1.50	0.35	0.0100
518	0.36325	0.28952	0.92	0.22	0.0085

* Values of the dimentional changes accompanying the complete transformation of austenite to bainitic ferrite at the transformation temperature

** V_α measured data at the transformation temperature

4 Phase diagram

The determined carbon concentrations of the residual austenite at the different temperatures of bainite transformations are compared with the T_0, T_0', A'_3 and A''_3, phase boundaries for investigated steel. It is usually assumed that the point where the bainitic microstructure ceases to change represents full transformation. But in case of bainitic transformation, reaction ceases before the parent phase (austenite) has completely transformed. It means that at any temperature below B_S and in the absence of any interfering secondary reactions only a limited quantity of bainitic ferrite forms before the reaction terminates [15, 16].

The calculated details for phase diagrams of the 15HM2VT steel and the carbon concentrations in austenite, x_γ at selected temperatures are shown in Table 4.

Table 4. Calculated details for phase diagram of the 15HM2VT steel and the carbon concentration in austenite, x_γ at selected temperatures.

Phase boundaries	Transformation temperature, °C							
	700	600	518	508	490	475	460	400
	Carbon concentration on phase boundaries, mol							
A_3'	0.0194	0.0471	0.0762	0.0795	0.0859	0.0917	0.0978	0.1212
A_3''	0.0143	0.0428	0.0727	0.0761	0.0828	0.0887	0.0949	0.1189
$T_0(x_{T_0})$	0.0065	0.0139	0.0215	0.0223	0.0240	0.0254	0.0274	0.0354
$T_0'(x_{T_0'})$		0.0016	0.0106	0.0115	0.0135	0.0152	0.0172	0.0241
x_γ			0.0085	0.0100	0.0110	0.0185	0.0224	

A_3' - $x^{\gamma\alpha}$ - paraequilibrium carbon concentration of austenite in mole fraction,

A_3'' - $x^{\gamma\alpha}$ + 50J - paraequilibrium carbon concentration of austenite in mole fraction but allowing for the 50J/mol stored energy,

x_{T_0} - *T*-zero carbon concentration in mole fraction,

$x_{T'_0}$ - the same but allowing for the 400J/mol stored energy.

In presented diagram (Fig. 1) the reaction is found to stop when the average carbon concentration of the residual austenite is closer to the T_0' curve than the A'_3 boundary. The diagram was calculated as in Ref. [12-13] using a model developed by Bhadeshia [14] based on the McLellan and Dunn quasi-chemical thermodynamic model [15]. The martensite and bainite reaction starts temperatures M_S and B_S were also marked on this diagrams. The paraequilibrium phase boundary is chosen because no substitutional alloying element partitioning occurs during bainite formation. The presented results can be explained when it is assumed that bainitic ferrite grows without diffusion, but any excess of carbon is soon afterwards rejected into the residual austenite by diffusion [16-17].

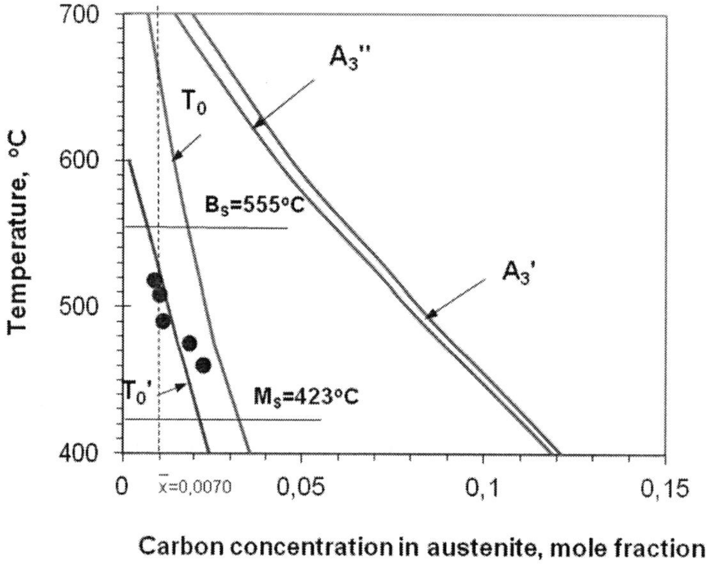

Fig. 1. The calculated phase boundaries A_3', A_3'', T_0 and T_0' for the investigated 15HM2VT steel together with all the experimental data of the measured carbon contents of the untransformed austenite (black circles).

If on the other hand, the ferrite grows with an equilibrium carbon concentration, then the transformation should cease when the austenite carbon concentration reaches the A'_3 curve [18,19]. Thus, it is found experimentally that the transformation to bainite in studied steel does indeed stop close to the T_0 boundary (Fig. 1) and the analysis suggest that bainite grows by displacive mechanism of transformation, but carbon atoms partition into residual austenite shortly after growth is terminated. Similar results have previously obtained by Bhadeshia and Christian [19] and by Ławrynowicz and Barbacki for other alloys [13, 20].

5 Retained austenite and hardness

Volume fraction of retained austenite after normalizing followed by tempering at 500 and 550°C is presented in Fig. 2. The measured hardness change after normalizing and tempering of 15HM2VT steel is shown in Fig. 3 for different tempering temperatures.

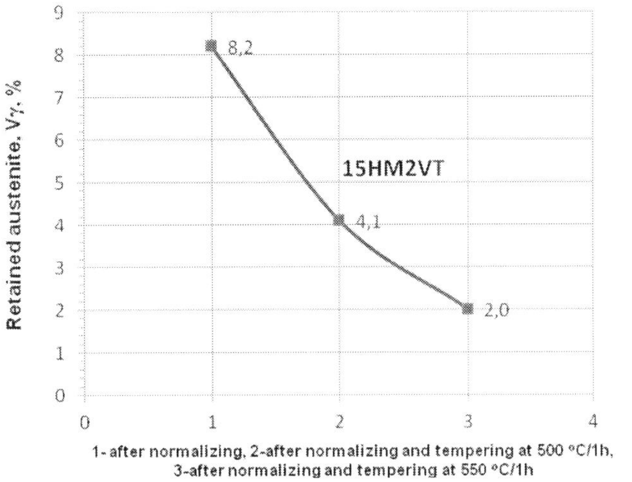

Fig. 2. Volume fraction of retained austenite after normalizing and tempering at 500 and 550°C of 15HM2VT steel.

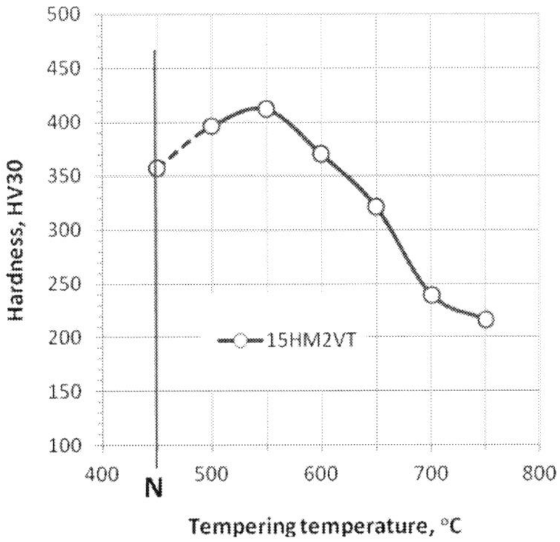

Fig. 3. Hardness change (HV30) after normalizing and tempering of 15HM2VT steel (N-means hardness after normalizing).

6 Microstructural analysis

Light optical microscopy and transmission electron microscopy micrographs of the investigated steel were taken on the normalized and tempered condition. The microstructure of the 15HM2VT steel is shown in Figures 4-7. It was found that for 15HM2VT steel the volume fraction of retained austenite decreased with increasing the tempering temperature (Fig. 2). The volume fraction of retained austenite after tempering above 600°C was beneath the level of resolution of applied X-ray diffractometer.

Microstructure of typical bainite after normalizing of the experimental15HM2VT steel is seen in Fig. 4. The morphology of the bainite is similar to low carbon lath martensite, where dislocated laths are separated by films and blocks of retained austenite. Careful examination of this microstructure shows no evidence of carbides precipitation. This structure, therefore, belongs to upper bainite assuming that upper bainite in this steel is a structure composed of carbide free bainitic laths with interlath retained austenite films [16,19].

Figure 5 shows microstructure obtained by normalizing and tempering at 500°C for 1hour. Transmission electron microscopy revealed that the sheaves were composed of small laths of ferrite and the microstructure contains only bainitic ferrite and carbon enriched residual austenite films. Bainitic ferrite laths of thickness about 0.2÷0.3μm form separately one from another. Retained austenite was present between the ferrite laths as thin discontinuous films. Carbides were difficult to observe in ferrite laths of thickness thinner than 0.3 μm. In this case films of residual austenite can accumulate all the excess carbon from supersaturated ferrite laths of thickness less than 0.3 μm. It is a possibility that, the carbon concentration of untransformed austenite is decreased by the carbide formation and then the reaction can proceed to a larger extent. The hardness of the steels is sensitive to changes in tempering temperature. With tempering at 550°C, the hardness of the steel grows slightly compared to the hardness after normalization (Fig. 2). This shows that the way to strengthen the steels is mainly precipitate strengthening. The precipitation behavior of carbides is certain to affect the mechanical properties. The hardness of the steels shows normal response to the increasing tempering temperature above 550°C, i.e. the hardness decreases with the increase of tempering temperature.

It is known that the formation of precipitates consumes dislocations. The carbide precipitation will decrease much the number of dislocations in the matrix, resulting in weakening of dislocation strengthening. On the other hand, the formation of carbides consumes the dissolved carbon, Cr and Mo which would provide a strong solid solution strengthening. Therefore, although the carbide precipitation could produce precipitation strengthening, it is not enough to compensate for the loss of dislocation strengthening and carbon, Cr and Mo solid solution strengthening.

When the tempering temperature is increased, a large amount of carbides form in the matrix. When the steel is heated at 600 °C, a small amount of precipitates form on the lath of the bainitic ferrite and on the prior austenite grain boundaries and a certain amount of precipitates has formed in the matrix (Fig. 6 and Fig. 7).

The hardness rapidly decreases with increasing tempering temperature, especially when tempered above 600 °C. When the steel is heated to 700 and 750 °C, compared to the steel heated to 500 and 600 °C, the number of precipitates in the microstructure within the bainitic ferrite laths and on grain boundaries increases significantly (Fig. 6 and Fig. 7). The carbide phase grows to a large size in a relatively short tempering time (1hour). The increasing amount of carbide phase could decrease the solid solution strengthening effect due to the consumption of dissolved Cr, Mo, V and Ti atoms.

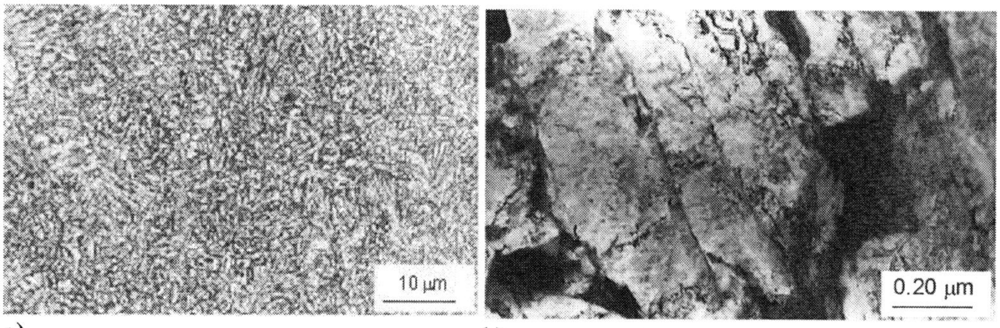

a) b)

Fig. 4. Microstructure of 15HM2VT steel after normalizing, a) light microscopy, etched with 2% nital, b) TEM, thin foil.

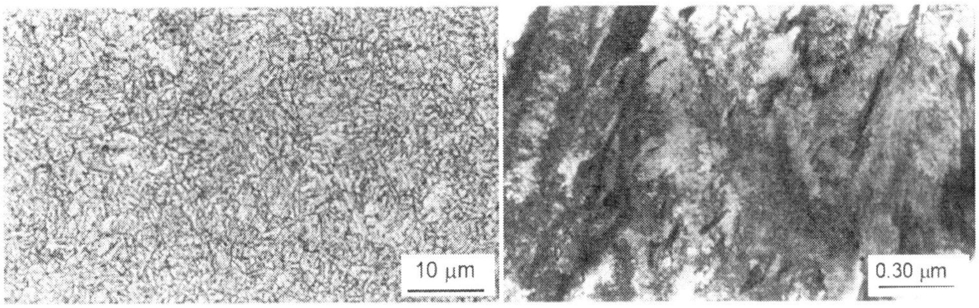

a) b)

Fig. 5. Microstructure of 15HM2VT steel after normalizing and tempering at 500°C for 1h, a) light microscopy, etched with 2% nital, b) TEM, thin foil

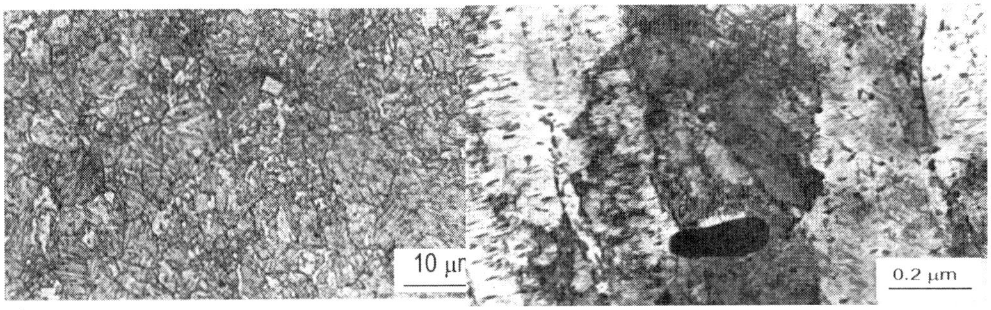

a) b)

Fig. 6. Microstructure of 15HM2VT steel after normalizing and tempering at 600°C for 1h, a) light microscopy, etched with 2% nital, b) TEM, thin foil.

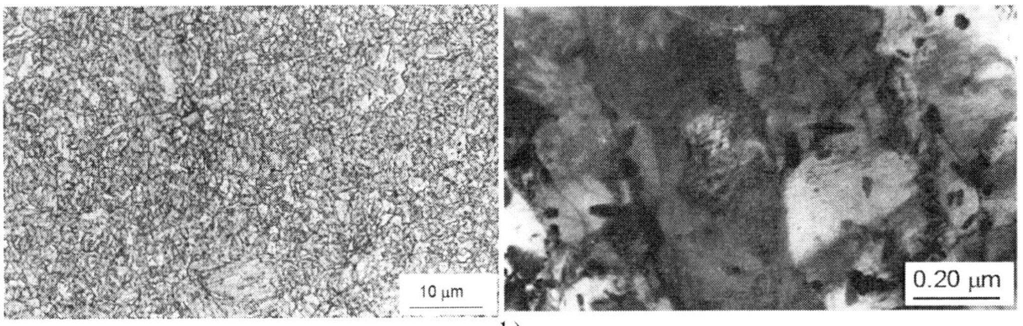

a) b)

Fig. 7. Microstructure of 15HM2VT steel after normalizing and tempering at 700°C for 1h, a) light microscopy, etched with 2% nital, b) TEM, thin foil.

8 Conclusions

The austenite to bainite transformation and microstructure evolution have been studied in experimental low alloyed Cr-Mo-V-Ti steel using high speed dilatometry and transmission electron microscope backed by thermodynamic analysis. The main results obtained are summarized as follows:

1. The bainitic reaction stops prematurely as the carbon content of the residual austenite reaches the T_0' curve well before the paraequilibrium carbon concentration is achieved (given by the A_{e3}' curve). This indicates that the growth of bainitic ferrite is diffusionless.
2. After tempering at elevated temperatures (500-550°C) previously normalized Cr-MoV-Ti steel hardness increases mainly by precipitate strengthening. Tempering at higher temperatures (600-750°C) causes reduction in the dislocation density and polygonization of

bainitic ferrite, previously normalized steel which is reflected in the hardness reduction. Along with the reduction in dislocation density, the $M_{23}C_6$ particles coarsen, allowing the bainitic ferrite laths to transform to more equiaxed subgrains. Along with the coarsening of the $M_{23}C_6$, there is also a coarsening in the M_2X precipitate distribution, although these particles coarsen much more slowly than $M_{23}C_6$.

References

1. R. L. Klueh, OAK RIDGE NATIONAL LABORATORY, Oak Ridge, Tennessee, ORNL/TM, **176**, 1-56 (2004)
2. B.S. Motagi, R. Bhosle, Int. J. of Eng. Res. Dev. **2**, 07-13 (2012)
3. A. Zieliński, G. Golański, M. Sroka, Mater. Sci. & Eng., **A 682,** 664–672 (2017)
4. H. Zheng-Fei, Thermal Power Plants, 1-266 (2012)
5. M. Dziuba-Kałuża, A. Zieliński, J. Dobrzański, M. Sroka, Arch. of Mater. Sci. and Eng. **66,** 21-30 (2014)
6. M. Sroka, A. Zieliński, J. Mikuła J., Arch. Metall. Mater., **61,** 1315–1320 (2016)
7. A. Zieliński, M. Miczka, B. Boryczko, M. Sroka, Arch. of Civ. and Mech. Eng. **16**, 813-824 (2016)
8. J. Shen, H. Yang, Y. Li, S. Kano, Y. Matsukawa, Y. Satoh, H. Abe, J. of Alloys and Comp. **695,** 1946-1955 (2017)
9. R. Kumari, G. Das, J. of Mater. & Metall. Eng. **5**, 7-14 (2015)
10. T. Shrestha, S. F. Alsagabi, I. Charit, G. P. Potirniche and M. V. Glazoff, Metals **5,** 131-149 (2015)
11. T. R. Jacobs, Elevated temperature mechanical properties of line pipe steels, A thesis submitted to the Faculty and the Board of Trustees of the Colorado School of Mines, (2015)
12. Z. Ławrynowicz, Inter. Jour. of Eng. **12**, 81-86 (1999)
13. Z. Ławrynowicz, A. Barbacki, Adv. in Mater. Sci. **2**, 5-32 (2002)
14. H.K.D.H. Bhadeshia, Bainite in steels 2nd ed., London, The Inst. of Mat., (2001)
15. R.B. McLellan, W.W. Dunn, J.Phys.Chem. Solids. **30**, 2631-2637 (1969)
16. Z. Ławrynowicz, Adv. in Mat. Sci. **11**, 13-19 (2011)
17. Z. Ławrynowicz, Mater. Sci. and Technol., **18**, 1322-1324 (2002)
18. F.G. Caballero, M.K. Miller, S.S. Babu, C. Garcia-Mateo, Acta Materialia **55**, 381-390 (2007)
19. H.K.D.H. Bhadeshia, J.W. Christian, **21A,** Metall Trans. A, 767–797 (1990)
20. Z. Ławrynowicz, Mater. Sci. and Technol., **20**, 1447-1454 (2004)

Stress tests on the truck frame for the purpose of the method to determine the endurance and reliability of special vehicles

Paweł Gotowicki[1], *Przemysław* Siminski[1,*]

[1]Military Institute of Armour and Automotive Technology, Poland

Abstract. The frame is the most vulnerable construction component of special purpose trucks. It is vital to provide reliability of that component. The article describes the stress and torsion angle tests, caused by load for the purpose of the method to determine the endurance and reliability.

1 Introduction

The process to implement new military vehicle is quite tedious and time consuming. One of the reasons is the long operation period, approx. 20 to 30 years. One of the basic test types that need to be performed during the initial phase, the prototype phase [1], are the endurance and reliability tests. Such tests include the determination of relevant indicators based on many test track cycles performed. [2] The assumption is to perform certain conduct for different surface types. High variability of conditions and surface profiles results in selective conditions for vehicle assemblies, where weak links are relatively easy to find and thus reducing the vehicle endurance and reliability.

The profile and the conditions of military vehicle operation impose the necessity to seek methods increasing the test performance, especially in terms of their duration. One of the approaches assumes the use of the hypothesis of defect accumulation in material to compare the load interference, described using static characteristics. It allows to compare the fatigue strength under different operating conditions and consider the rate of defect accumulation increase depending on the number of units/cycles. The obtained rates of fatigue strength increase depend on the road profile, which unfortunately rarely remains stable for unsurfaced roads. The main problem is to develop load spectrum for vehicle frame using cycle method. [3]

On the other hand, the vehicle should ensure movement smoothness. [4, 5] The frame, tires and suspension system should limit the interference: surface roughness, movement resistance since they influence the driver psycho-physical state. [6]

The tests for the needs of military vehicles on the influence of dynamic loads on movement smoothness as well as endurance and reliability do not provide strictly precise answer.

* Correspondence address: przemyslaw.siminski@witpis.eu

Therefore, there is the need to frequently monitor the test roads profiles [7, 8] and to determine the influence of fatigue cycles on stresses in selected key construction nodes.

Therefore, during the first stage of works to develop global method, the specialized measurement equipment was acquired and the tests of military vehicle load impact on the stresses in vehicle frame as well as torsion angle under static conditions were performed.

2 Measurement system

The measurement setup for strain monitoring consists of red-out systems (fiber optic interrogator by Deminsys with specification and two fiber Bragg gratings that act as strain sensors. Measured data are transferred to PC via 100Mbps LAN. Due to the similar Bragg wavelength of both FBGs, they were connected separately to two channels of the interrogator.

When the force is applied to the FBG, its Bragg wavelength shift in the reflected/transmitted spectrum is observed due to the grating length (and thus grating period) and refractive index changes.

$$\frac{\Delta\lambda}{\lambda_0} = k * \varepsilon * \alpha_s * \Delta T \qquad (1)$$

Strain values were measured and recorded using digital-storage-oscilloscope Hioki 8847 with Hioki 8969 Strain Unit and a Kyowa DB-120A Bridge Box for Dynamic Strain Measuring Instruments. Self-temperature-compensated (composite and steel) gages Vishay CEA-06-250UN-120 in the three-wire quarter-bridge circuit arrangement were used. The stresses were determined by the strain gages set on the axial center of side members of chassis. The gage length is 6,35 mm and nominal grid resistance 120Ω. The dynamic response of this strain measuring system is 20 kHz. The σ and τ stress values were calculated from the measured strain ε, according to Hooke's Law for the uniaxial stress state. Hioki 8847 Memory HiCorders are high-speed waveform monitoring and recording devices that can simultaneously measure multiple physical phenomena such as voltage, current, temperature, vibration, acceleration and rotation due to complete isolation between channels and from each channel to the ground. VISHAY CEA-06-250UN-120 universal general-purpose strain gages: Strain range ±5%, Temperature range –75° to +175°C, Grid resistance 120Ω±0.3%, Gage length 6.35, Overall length 10.54, Grid width 3.05, Overall width 3.05, Matrix length 13.2, Matrix width 5.6; Gage factor 2,14±0.5%, Transverse sensitivity +0,6±0.2%, Temperature coefficient of gage factor +1,2±0.2%/100°C. KYOWA DB-120A Bridge Box for Dynamic Strain Measuring Instruments: Number of Input Channels 1; Connecting target Strain gage; Strain Method: Quarter bridge 2-wire system, Quarter bridge 3-wire system, Half bridge system, Opposite side 2-active-gage system and Full bridge system; Applicable Strain Gage Resistance 120Ω.

3 Tests

The test object was a truck, STAR 1466 used in the military. Its load capacity is 6 tones and the power transmission system configuration is 6x6. The tests were initiated with the measurement of STAR 1466 curb weight using three different load variants:

- With real curb weight (without driver),
- With additional substitution weight of 5,902 kg, 3,147 kg placed on vehicle open load-carrying body and without load.

The vehicle was fully equipped with full fuel and fluid tanks.

Next, using laser measuring instrument, the axle articulation was measured for different location of height hoist under the wheels of each axis, using three load variants described above. At the same time the vehicle height from surface and bumper height were measured in order to determine the torsion angle in the transverse plane.

Fig. 1. Forcing of frame deformation.

The example measurement results of STAR 1466 axle articulation is presented in Table 1.

Table 1. Results of frame articulation for example load variants.

STAR 1466 axle articulation – measured values				
Item	Measured parameter	Measurement result	Measured parameter	Measurement result
1	2	3	4	5
Vehicle in the basic location – variant 0				
1.	The hydraulic hoist height – left wheel of 1st axis	0 m	The hydraulic hoist height – right wheel of 3rd axis	

STAR 1466 axle articulation – measured values				
Item	Measured parameter	Measurement result	Measured parameter	Measurement result
2.	The 1st axis wheel height – left side	0.590 m	The 1st axis wheel height – right side	0.576 m
3.	The front bumper height – left side	0.926 m	The front bumper height – right side	0.903 m
4.	The frame height in the vicinity of 1st axis – left	0.943 m	The frame height in the vicinity of 1st axis – right side	0.926 m
5.	The frame height between the 2nd and 3rd axis – left	0.932 m	The frame height between the 2nd and 3rd axis – right side	0.922 m
6.	The rear bumper height – left side	0.442 m	The rear bumper height – left side	0.391 m
Vehicle in location – variant 1				
7.	The hydraulic hoist height – left wheel of 1st axis	0.213 m	The hydraulic hoist height – right wheel of 3rd axis	
8.	The 1st axis wheel height – left side		The 1st axis wheel height – right side	0.567 m
9.	The front bumper height – left side	1.111 m	The front bumper height – right side	0.960 m
10.	The frame height in the vicinity of 1st axis – left	1.032 m	The frame height in the vicinity of 1st axis – right side	0.978 m
11.	The frame height between the 2nd and 3rd axis – left	0.952 m	The frame height between the 2nd and 3rd axis – right side	0.925 m
12.	The rear bumper height – left side	0.426 m	The rear bumper height – left side	0.352 m

The object was prepared for stress measurements by taping extensometers to 3 selected points on the vehicle frame (Fig. 4). These points were selected based on the operation experience as potentially exposed to substantial stresses.

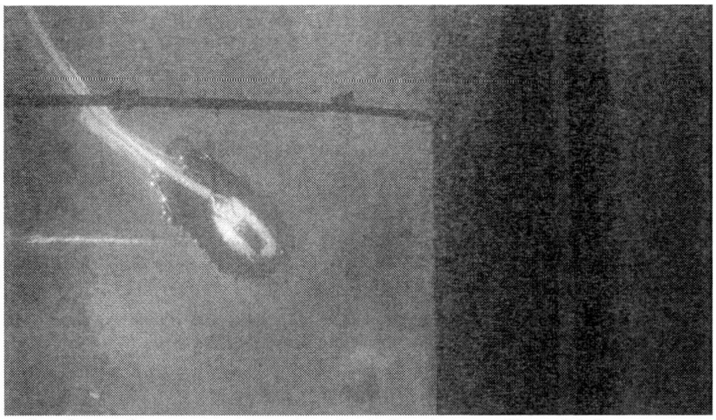

Fig. 4. The example of taped extensometers to selected points on the vehicle.

The stress result measured in the selected frame points for different vehicle loads is presented in Fig. 5. The wheel was lifted with the stroke of every 10 mm within the range of 0 mm to 50 mm and 50 mm to 0 mm.

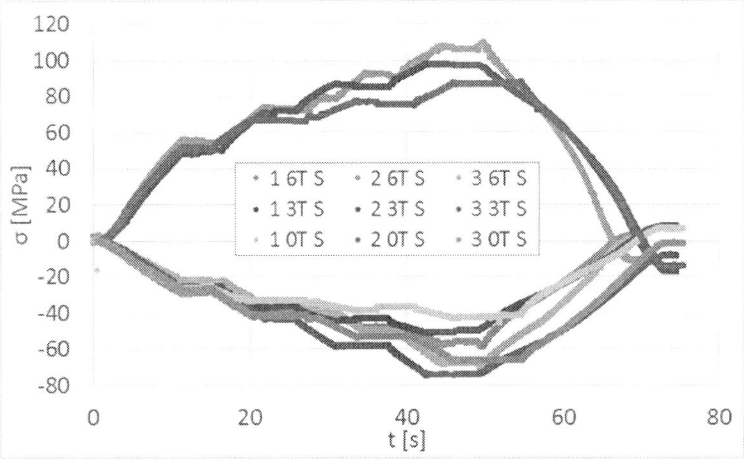

Fig. 5. The stress characteristics (normal stress and shearing stress of beam profile) caused by frame torsion for different vehicle loads: 1 – shearing stress τ in the middle of beam web profile; 2 – normal stress σ on the upper part of beam profile; 3 – normal stress σ on the lower part of beam profile.

As per the expectations, the highest stress values with the value of 110 MPa were obtained for the vehicle load of 6 tones. The highest stress values are present in the normal direction on the upper part of the beam profile. In the tested STAR 1466 model this part is especially prone to damage. The operation practice showed that in points 2 and 3 there are cracks during the off-road drive with maximum load. The tests [9] proved that the stress in the truck standard frame, moving with low speed of 40 km/h on relatively even surface are at the level of 80 MPa. The measured value of 110 MPa for 18G2A unalloyed steel is already a critical value in the static state.

The frame torsion angle was measured to thoroughly observe the impact of open load-carrying body load. These test results are presented in Fig. from 6 through 8.

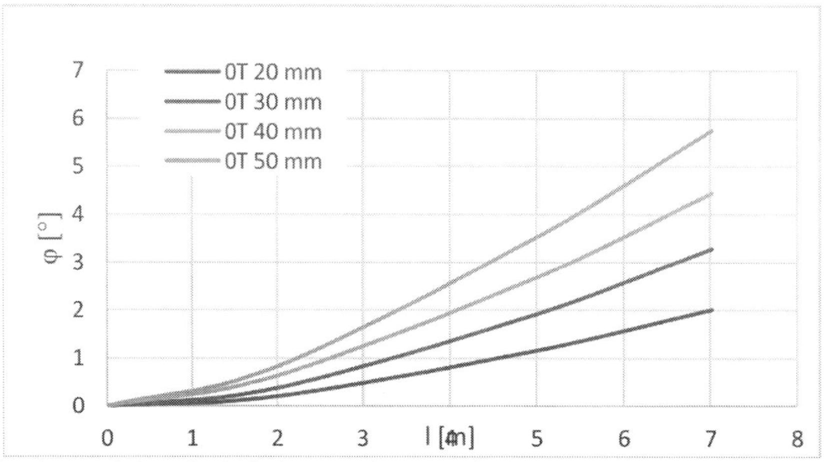

Fig. 6. The characteristics of vehicle frame torsion angle for different forcing without load in open load-carrying body

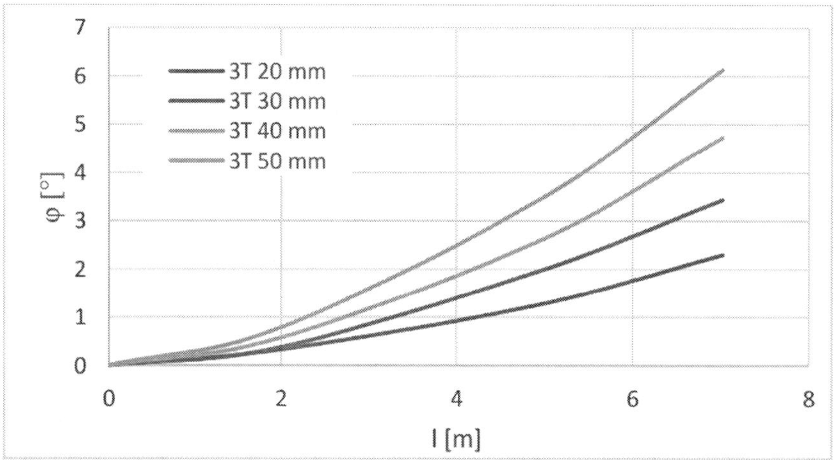

Fig. 7. The characteristics of vehicle frame torsion angle for different forcing with load o 3 tonnes in open load-carrying body

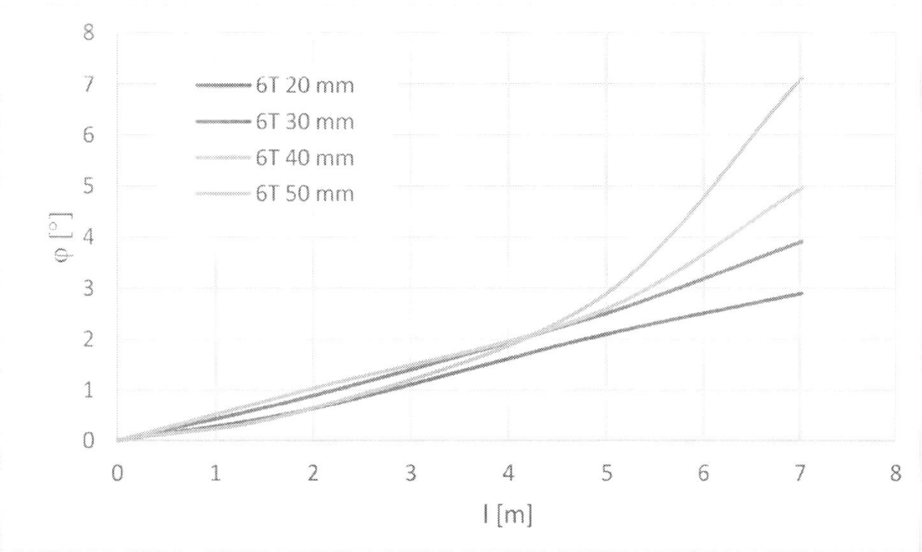

Fig. 8. The characteristics of vehicle frame torsion angle for different forcing with load of 6 tons in open load-carrying body

The frame torsion angle changes in a non-linear manner and the higher dynamics of angle change takes place for the load of 6 tons and 50 mm forcing under the wheel. This is the information that the frame operation on torsion during off-road drive should be analyzed during each new development or fame design.

The presented test results can be used to verify the numeric model of vehicle frame. The designed measurement system can be used to dynamic tests on unsurfaced roads.

The deformation measurement method using type FBG optical sensors indicates features and parameters that allow the performance of tests of composite and metal construction elements and vehicles with static, percussive and operational loads.

The presented methods of experimental analysis of deformations and stresses are used to:

- Determine the characteristics of construction material mechanical properties
- Test the distribution as well as deformation and stress values in the construction elements or in their models, especially when determining the stress concentration.
- Determine the internal forces in construction elements subject to operational load.
- Determine own stresses, e.g. assembly, casting, welding stresses and stresses from heat loads etc.
- Determine the characteristic values for deformation and construction movement state (displacement, speed, acceleration).

References

1. S. Niziński, Eksploatacja obiektów technicznych (Instytut Technologii Eksploatacji, 2002)
2. P. Simiński, Wojskowe Pojazdy Kołowe, (2015)
3. TOP 01-1-010A Vehicle Test Course Severity (Surface Roughness)
4. https://www.matecconferences.org/articles/matecconf/pdf/2015/12/matecconf_icmee2015_02002.pdf
5. https://www.sciencedirect.com/science/article/pii/S002248980600005X
6. W. Luty, W. Kupicz, M. Zwierzyński, *Examination and assessment of special purpose vehicle crews' comfort while operating in off-road conditions* **49** (Zeszyty Naukowe Wyższa Szkoła Oficerska Wojsk, 2017)
7. TOP 02-2 TOP 01-1-010A -506A Wheeled and Tracked Vehicle Endurance Testing
8. TOP 1-1-030 RAM-D and ILS Analysis
9. M. Miros, Wpływ napraw spawalniczych na właściwości eksploatacyjne elementów nośnych pojazdów ciężarowych, PhD thesis, (Politechnika Śląska 2010)

The process of diagnosing an electrostatic precipitator's state during operation in the energy sector

Aleksandra Brzezińska[1,*], and *Bogdan* Żółtowski[1]

[1]University of Science and Technology, Mechanical Engineering Department, Al. Prof. S. Kaliskiego 7, 85-796 Bydgoszcz, Poland

Abstract. The paper proposes procedures for diagnosing the electrostatic precipitator's technical state in terms of its key function as a facility in a complex waste gas dedusting system. The presented state diagnosing procedures concern the assessment, genesis and forecast of the electrostatic precipitator's state during its rational operation.

1 Introduction

The Polish energy sector started developing rapidly in the 1960s when forced industrialisation triggered a sudden demand for electricity. When power plants were constructed, no one realised that power units would have to be temporarily shut down for reasons other than emergencies or repairs [1]. To facilitate the process of planning service activities, modernisations and repairs of any failures of the facility, an appropriate methodology for controlling the electrostatic precipitator's operation and diagnosing its individual components' degradation should be applied. The actions taken will make it possible to improve the electrostatic precipitator's operation and will significantly reduce costs of maintaining its operability during the facility's operation [18].

2 Environmental protection in the energy sector – legislation

Long-term draft amendments to energy policy were issued by the European Commission as a Green Paper including issues related to a European strategy for sustainable, competitive and secure energy [2]. Directive 2010/75/EU of the European Parliament and of the Council of 24 November 2010 on industrial emissions (replacing Directive 96/61/EC (IPPC) and a BAT reference document) brought about key transformations in Polish law, more specifically environmental protection regulations [3]. The most important amendments relate to making regulations on emission standards for industrial lobbies more restrictive. In particular, the IED Directive provides for reducing sulphur dioxide, nitrogen oxides and dust generated by combustion plants. The Polish energy sector is also subject to two Regulations of the Minister of the Environment providing for intensifying pro-ecological activities:

- Regulation of the Minister of the Environment of 30 October 2014 on the requirements for the measurement of emissions and the measurement of the amount of water taken [4].
- Regulation of the Minister of the Environment of 4 November 2014 on the emission standards for certain types of the installations, fuel combustion sources and incineration and co-incineration equipment [5].

Activities to modernise energy facilities enable the energy sector to meet the requirements of EU directives and to achieve a high level of efficiency of individual components of a power unit. However, supervision to ensure the correct operation of power

* Corresponding author: aleksandra.brzezinska91@gmail.com

facilities, including their installations, and appropriate controls to comprehensively diagnose complex facilities, including electrostatic precipitators, are most important.

3 Structure of a thermal power plant

The thermal power plant has a complex structure which, apart from its main equipment (boiler, turbine, generator, transformer), consists of multiple circuits and pieces of auxiliary, control, signalling, security and measurement equipment. Its structure should ensure the lowest possible cost of electricity generation and the greatest possible efficiency of individual transformations [6]. Some part of energy generated in the power plant serves own needs. It is drawn directly from generator terminals and sent via a transformer to specific power plant equipment. The most important pieces of power plant equipment serving own needs include: water and steam circulation equipment (feed pumps, condenser equipment), combustion equipment (fans, dust collectors, including electrostatic precipitators), fuel transportation equipment and machines, coal pulverisers, turbine operating equipment, a generator cooling system, measurement, security, control and signalling equipment, lighting [7].

As certain equipment needs to be in constant operation, it has a back-up power supply system (two or three pieces of equipment are installed to serve a similar purpose, but normally only one of them is active) and some part of electricity in power plants is used to constantly power the most important equipment in the energy conversion process [15, 16].

3.1 Electrostatic precipitator as a power unit component

Apart from fuel combustion installations, power units include fuel dedusting equipment. Development trends in this area focus on making their structure and operation more efficient [8]. The electrostatic precipitator is among complex machines making up the power unit whose structure can be divided into five characteristic groups [9, 10]:

- Structural components: foundations for the ESP's heavy structure to be laid on them later, casing of dumping hoppers for ash and the equipment's frame, i.e. its support structure with pivot bearings and a chamber divided into an upper and lower part.
- Mechanical part: bearings mounted on pillars, rapper shafts with hammers, a diffuser and a confusor.
- Electrical equipment: transformer units and their installations, automation and control systems, auxiliary equipment.
- Thermal insulation group: mineral wool resistant to high temperatures.
- A system of stairs, landings and platforms to access inspection manholes located on the roof, chamber side walls and dumping hopper side walls.

Any damage to one or more of the ESP's indicated components does not have to lead to the facility's complete shutdown, but any undesirable event related to the occurrence of damage makes the operation of equipment less efficient in a certain way. Complex machines can be considered as the so-called "renewable facilities" only if such damage to the facility's components can be repaired and if performed activities lead to making such equipment operable again [17]. The repair of the facility by a qualified team should not normally interfere with key tasks performed on site.

4 Conditions for diagnosing electrostatic precipitators' state

Measurements in the energy industry are made regularly during a warranty period after which they are only made upon order by an investor in accordance with PN-Z-04030-7:1994 "Tests for dust content. Measurement of concentration and mass flow rate of particulate matter in waste gases by gravimetric method" [11]. There is neither a predetermined time interval, which would indicate deadlines for tests to diagnose the electrostatic precipitator's current state, nor a list of parameters to be checked during inspections. The checks to date have covered the values, which are set out in the

Regulations of the Minister of the Environment, relating to combustion equipment, i.e. NO_x, SO_x and dust. Laboratories, which make measurements, test processes, systems and energy equipment, have their own list of offered parameters which can be tested during a check, but it is a person ordering the test who ultimately chooses values to be measured.

If a new emission standard for dust cannot be met, the first idea is usually to modernise the existing electrostatic precipitator power supply system or to invest in a new facility. Enhancements or repairs of electrostatic precipitators are not the cheapest solutions and often make it necessary to shut down a boiler for a few months. Therefore, it is so important to use an appropriate system to diagnose electrostatic precipitators' technical state in order to set deadlines for the machine's repairs, modernisations or inspections, in order to systematically plan works to improve the facility's operation and ultimately to help reduce the electrostatic precipitator's operational costs.

5 Process of diagnosing the electrostatic precipitator's technical state

The process of diagnosing the technical state of the electrostatic precipitator as a complex facility, which is among critical equipment in the power unit, should clearly make it possible to identify certain components, parts, material in the context of recognising their degradation. The procedures for diagnosing the electrostatic precipitator's state and controlling the facility's operability, which are proposed in this paper, list individual steps dedicated to the Maintenance Services which develop maintenance, modernisation or service plans for the facility. Additionally, the specific procedures were supplemented with algorithms to ultimately localise damage and to control the progressive degradation of the existing ESPs' state [12].

5.1 Procedure for assessing the ESP's technical state

To correctly assess the electrostatic precipitator's technical state, it is important to use tools providing a full insight into its current operation. Given its important function (purification of waste gases from harmful dust particles resulting from the fuel combustion process), the electrostatic precipitator must be guaranteed failure-free operation. The proposed scope of the procedure includes an assessment of the machine's (electrostatic precipitator's) technical state performed during operation and control activities to analyse the facility's operation on an ongoing basis, and activities to detect damage: fatigue, aging, random damage, and the electrostatic precipitator's deteriorating structural and mechanical components. As part of the procedure for assessing the electrostatic precipitator's technical state, the following procedure for individual teams supervising the facility's operation should be distinguished:

1. ***Assessment of the technical state based on the visual inspection and detection of failures, damage, defects, changes in the waste gas purification process during the electrostatic precipitator's operation:***
   - diagnosis of failures by telephone based on mA and kV values, and information provision (company performing tests for the energy area);
   - internal and external inspection of electrostatic precipitators to detect the cause of failure (mechanic at the Maintenance Department together with the Head of the Maintenance Department) if damage prevents the facility from further operation;
   - technical supervision of repair works on site (moderniser/repairer responsible for the area of electrostatic precipitator repairs);
   - operational optimisation of mechanical and electrical equipment (moderniser/repairer and the company performing tests for the energy area);
   - implementation of a procedure for preventing electrostatic precipitators' inoperability (NZU) depending on changes in the combustion process (Head of the Maintenance Department);

- operational test – testing and monitoring electrostatic precipitators' operation (moderniser/repairer, the Head of the Maintenance Department and the company performing tests for the energy area);
- technical report together with documentation on the removal of the cause of failure or the repair of malfunction (moderniser/repairer, the Head of the Maintenance Department and the company performing tests for the energy area).

2. ***Assessment of the electrostatic precipitator's technical state based on an algorithm for selecting a test to control the technical state of and localise damage to the facility:***
   - preparation of data for their acquisition;
   - optimisation of a set of values of diagnostic parameters (only if set Y is large, e.g. m>10);
   - organisation of the data set by determining set $\{s_i(\Theta_k), i=1,\dots, 1; k=1, \dots, K\}$;
   - testing the significance of changes in values of diagnostic parameters $\{y_j(\Theta_k)\}$ depending on the state of set $\{s_i = f(\Theta_k); \Theta_k \in (\Theta_1, \Theta_b)\}$, i.e. which diagnostic parameter "best" describes the state of the equipment's inoperability – s_i;
   - specification of a Boolean matrix:
     - 1 – when a change in the state of the equipment's inoperability – s_i causes significant changes in values of diagnostic parameter y_j;
     - 0 – when a change in the state of the equipment's inoperability – s_i causes no significant changes in values of diagnostic parameter y_j.
   - determination of an optimal method for selecting a test to control the state and to localise damage depending on the size of the set of diagnostic parameters and the machine's operation time:
     - development of an optimal method for selecting a test to control the machine's state. The test to control the state (TCS) to be selected based on the Boolean matrix – the TCS as a vector of logical values (0, 1) of diagnostic parameters and its analogical vector of states: $\{(y_1, \dots, y_n, \dots, y_N)\} \Rightarrow S_0 \vee S_1 = \{(s_1, \dots, s_n, \dots, s_N)$ for the state of the equipment's operability S_0 and the state of the equipment's inoperability S_1;
     - testing the impact of the machine's operation time on an optimal diagnostic test;
     - development of an optimal method for selecting a test to localise damage to the machine (test to localise damage (TLD) to be selected based on the Boolean matrix).
   - interpretation of results of tests to control the electrostatic precipitator's (ESP's) state.

5.2 Procedure for the genesis of the ESP's technical state

The procedure covers the genesis of the ESP's technical state performed after assessing the state during the electrostatic precipitator's (ESP's) operation. Symptom genesis is an innovative approach which makes it possible to register values of the ESP's diagnostic parameters and states during operation. Based on collected data and the equipment's visual inspection, it allows for determining what caused and what conditions accompanied the state of the equipment's inoperability. As part of the procedure for the genesis of the electrostatic precipitator's technical state, the following procedure for individual teams supervising the facility's operation should be distinguished:

1. ***Genesis of the technical state based on registered values of the electrostatic precipitator's diagnostic parameters and states during its operation to date:***
   1. constant monitoring of the following: discharge electrode rappers, collection electrode rappers, hopper heating, stand-off or rotating insulator heating, hopper ash level sensors' state, high voltage systems in each zone, dust collection from under hoppers (separate installation);

2. continuous measurement of the ESP's diagnostic parameters: concentration of waste gas components, i.e. SO_2, NO_X, CO, H_2O, O_2 and dust, volumetric flow rate of waste gas, moisture content in waste gas, temperature of waste gas, temperature inside the ESP's chamber, static pressure;
3. when the state of the equipment's operability changes into the state of the equipment's inoperability, it is necessary to take dust samples and determine dust particle size and content of: Cd, Cr, Ni, Pb, V, Cu, Al, As, Zn, H_2SO_4, NH_3, TOC (total organic carbon), TC (total carbon), share of flammable and non-flammable parts;
4. collecting all measurement results, performing necessary calculations and determining the relationship between changes in the genesis value of the diagnostic parameter with a genesis error and the limit value of the diagnostic parameter;
5. determining the relationship between changes in the genesis value of the diagnostic parameter with a genesis error and the limit value of the diagnostic parameter (algorithm for the genesis of the ESP's technical state according to the scheme of estimating values of diagnostic parameters) [13, 14].

The set of parameters to be optimised: by calculating values of criterion functions and weights of diagnostic parameters based on input data for the selected facility (ESP) in the Dedicated Diagnostic System for the ESP (DDSESP). As a criterion for selecting the diagnostic parameter(s), one should assume the maximum weight values wj and select diagnostic parameters according to the criterion adopted. The procedure allows for selecting diagnostic parameters as prefered by a system operator (Head of the Maintenance Department) and updating weigth values for new sets of selected diagnostic parameters [12, 13, 14].

The genesis of the ESP's state to be performed by:

- determining the genesis value of the diagnostic parameter and the genesis error (only for a discrete event notation);
- determining the minimum distance of the genesis value of the diagnostic parameter to its limit value and testing the impact of operational conditions (number of parameters, number of time series) on the state's genesis;
- visualising and recording approximation or interpolation functions for the selected ESP's parameters;
- estimating the cause of damage to the machine based on the minimum distance of the genesis value of the diagnostic parameter to its limit value, values of the machine's process parameters, values of ambient parameters and additional events.

Collecting materials and information on causes of damage to the system of dust collection from under hoppers (taking into account hopper ash level sensors' state) together with taking a series of photographs to illustrate the actual state of individual mechanisms and parts of the dust collection system. It is also necessary to record the date and location of the damage. Information necessary for data generation based on the algorithm for the genesis of the ESP's technical state according to the scheme of estimating values of diagnostic parameters.

5.3 Procedure for forecasting the ESP's technical state

The basic purpose of the procedure for forecasting the ESP's technical state is to present a sequence of tasks necessary for presenting a track record of past events, which occurred during the ESP's operation, so that unfavourable events in the ESP's operation (ESP's damage putting the facility at risk of being shut down) can be predicted in the future based on the data collected. As part of the procedure for forecasting the electrostatic precipitator's technical state, the following procedure for individual teams supervising the facility's operation should be distinguished:

1. ***Forecast of the ESP's technical state based on values of parameters of the ESP's state during operation and the ESP's completed modernisations and repairs:***
   - developing a model process of diagnosing the ESP based on identified symptoms and collected information on: operation of discharge electrode rappers, collection electrode rappers, hopper heating, stand-off or rotating insulator heating, hopper ash level sensors' state, high voltage systems in each zone, dust collection from under hoppers (separate installation).

Having analysed the process of diagnosing the ESP in detail, the following necessary diagnostic parameters for the ESP were determined: concentration of waste gas components, i.e. SO_2, NO_x, CO, H_2O, O_2 and dust, volumetric flow rate of waste gas, moisture content in waste gas, temperature of waste gas, temperature inside the ESP's chamber, static pressure. The model process of forecasting the ESP's state must be developed based on information on the course of modernisation, repair activities and documentation presenting the genesis and assessment of the ESP's technical state.

   - collecting local values of the measure of similarity between instantaneous states and model time series;
   - generating a model of the ESP's forecast state also based on an algorithm for forecasting the ESP's technical state;
   - calculating forecast values of the measure of probability based on collected data on the ESP's operation, modernisations and repairs;
   - coming to conclusions on future states of selected mechanisms and parts of the tested facility (ESP).

2. ***Algorithm for forecasting the ESP's technical state:***
   - forecasting the value of diagnostic parameter y_{j*}:
     - by Brown-Mayer adaptive method, grade 1 (B-M1), with coefficient α=(0.1-0.9) for forecast horizon τ=3ΔQ set for time interval (Q1,Qb) – the method used when a forecast variable is almost constant and minor random fluctuations occur;
     - by Holt adaptive method with coefficient α=(0.05-0.3) and β=(0.4-0.9) for forecast horizon τ=3ΔQ set for time interval (Q1,Qb) – the method used when persistent time trends in a time series are observed;
     - by analytic methods (linear, exponential and power methods of the first, second, and third grade) for forecast horizon τ=3ΔQ set for time interval (Q_1,Q_b) – the method allows for finding a trend function which will meet the set criteria and will optimally adapt to the forecast variable's time series (it is important to select an appropriate analytical form of the trend function);
     - setting a deadline for the next servicing and diagnosing of the machine Θ_d.

6 Summary

The coal power stations have the right number of starts machines (and exceeding permissible parameters that occur in such states), sooner or later there are problems and damages. Cracks appear, followed by failures of lower or higher severity. Paradoxically, older power units are technologically more adapted to load changes, because the steam parameters are not exorbitant. For some power units, further solutions are introduced enabling undervaluation lower of power. Unfortunately, most of the power units are currently probably already largely exploited. The number of work hours the power units goes in hundreds of thousands, slowly exceeding the thresholds allowed, and their current work is based on multiple modernizations. Unfortunately, the history of the power units exploitation is usually not collected, as a result, the actual state of the devices is not known.

Increases over the years, probability of occurrence of more failures and deterioration of the availability coefficients. Of course each of the consortium (manufacturing companies) reports internal reliability parameters and costs incurred for the repair of facilities, but the additional cost alone caused by large load changes and more boiler withdrawals from use, unfortunately it is a problem that is frequently off in the juxtaposition. Therefore, it is necessary to supervise the correct operation of the power facilities, including installations is entering in its composition, and conducting reliable controls aimed at comprehensive diagnosis of complex objects. For financial reasons, the introduction of a methodology to monitor the ESP degradation and control of the current operation of the facility is of key importance.

References

1. M. Kudełko, *Czynniki determinujące rozwój krajowego sektora energetycznego – analiza wrażliwości,* Journal of Management and Finance, **13**, 255-256 (2015).
2. Green Paper, *A European Strategy for Sustainable*, Competitive and Secure Energy, Brussels, 17-18 (2006).
3. EU, *Directive 2010/75EU of the European Parliament and of the Council of 24 November 2010 on industrial emissions (integrated pollution prevention and control),* (Recast), 1-2 (2010).
4. Ministerstwo Środowiska, *Rozporządzenie Ministra Środowiska z dnia 30 października 2014 r. w sprawie wymagań w zakresie prowadzenia pomiarów wielkości emisji oraz pomiarów ilości pobieranej wody,* **2014/1542**, 1-3 (2014).
5. Minister for the Environment in Poland, *Regulation of the Minister for the Environment of 4 November 2014 on the emission standards for certain types of installation, fuel combustion plants and waste incineration or co-incineration facilities,* Dziennik Ustaw, **2014/1546**, 1-4 (2014).
6. B. Zaporowski, *Technologie wytwarzania energii elektrycznej dla polskiej elektroenergetyki*, Energy Policy Journal, **18**, 29-44 (2015).
7. T. Chmielniak, A. Rusin, H. Łukowicz, *Modernizacja i rewitalizacja istniejących bloków węglowych ważnym warunkiem stabilności sektora wytwarzania elektryczności,* Energy Policy Journal, **19**, 49-59 (2016).
8. A. W. Yuen, *Collector Current Density and Dust Collection in Electrostatic Precipitators,* VDM Verlag Dr. Mueller, 1-7 (2008).
9. F. K. Fuliful, A. A. Hamood, *Collection Efficiency of Wire– Duct Electrostatic Precipitator,* LAP Lambert Academic Publishing, 11-20 (2017).
10. H. A. Ziedan, *Modeling of Corona Discharge in Wire-duct Electrostatic Precipitators,* LAP Lambert Academic Publishing, 35-44 (2016).
11. PN-Z-04030-7:1994, *Ochrona czystości powietrza. Badania zawartości pyłu. Pomiar stężenia i strumienia masy pyłu w gazach odlotowych metodą grawimetryczną,* PKN, 1-7 (1994).
12. B. Żółtowski, D. Landowski, B. Przybyliński, *Projektowanie Eksploatacji Maszyn,* Wydawnictwo Naukowe Instytutu Technologii Eksploatacji - Państwowego Instytutu Badawczego, 212-215 (2012).
13. H. Tylicki, B. Żółtowski, *Genezowanie stanu maszyn*, Wydawnictwo Naukowe Instytutu Technologii Eksploatacji - Państwowego Instytutu Badawczego, 29-45 (2012).
14. H. Tylicki, B. Żółtowski, *Rozpoznawanie stanu maszyn,* Instytutu Technologii Eksploatacji - Państwowego Instytutu Badawczego, 11-66 (2010).
15. M. Żółtowski, *Informatyczne systemy zarządzania w inżynierii produkcji.* Wydawnictwo ITE-PIB, Radom 2011.

16. B. Żółtowski, *Metody inżynierii wirtualnej w badaniach stanu, zagrożeń bezpieczeństwa i środowiska eksploatowanych maszyn*. Wyd. UTP, Bydgoszcz 2012.
17. B. Żółtowski, *Badania zagrożeń utraty zdatności środowiskowych systemów technicznych*. WU UTP, Bydgoszcz 2013, s.181.
18. A. Brzezińska, B. Żółtowski, *Dozorowanie procesu technologicznego pracy bloków energetycznych w diagnostycznej strategii eksploatacji*, Opracowanie wewnętrzne, 2017.

Material combination to mitigation of behind armour debris after shaped charge jet attack

Marek Szudrowicz[1,*]

[1]Wojskowy Instytut Techniki Pancernej i Samochodowej, 05-070 Sulejówek, Okuniewska 1, Poland

Abstract. When a shaped charge jet perforates the steel armour plate, the residual jet is accompanied by a much wider cloud of behind armour debris (BAD) generated by stress wave interactions. BAD plays an important role in the evaluation of survivability of vehicle crew and components in a vehicle. The Behind Armor Debris (BAD) cloud caused by shaped charge jet impact was investigated. In order to simulate different configuration of material to mitigation of debris cloud the thickness of ceramics target tiles, glass and aramid laminate liners in contact with the armour plates was varied.

1 Introduction

Armour Piercing Discarding Sabot (APDS) projectiles and High Explosive Anti-Tank (HEAT) warheads are the main weapons used to damage or destroy armoured vehicles. If a such a warhead hits a vehicle, the highest danger for the crew is posed, apart primary jet, by fragments of the primary jet and armour spalls, which may reach a speed o 1000 m/s in case of heavier fragments and about 3000 m/s in case of the lighter ones. Additionally, effects related to changes of temperature and pressure should be considered (Fig. 1) [1].

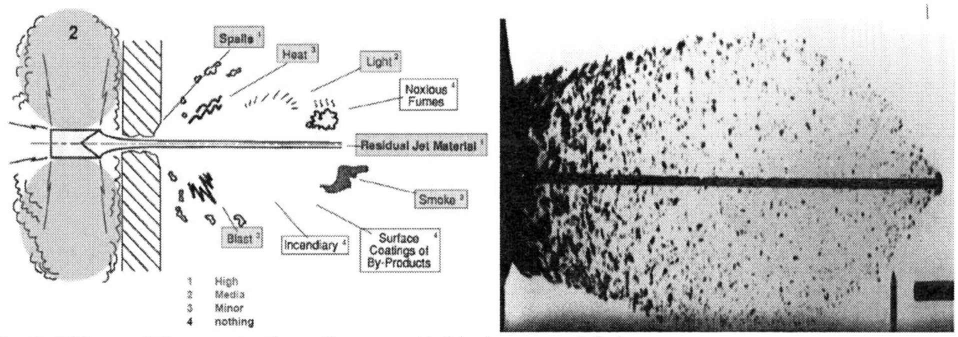

Fig. 1. Effects of the penetration of armour. Behind armour debris.

The following means are employed in design solutions in order to increase the chances of a vehicle crew to survive when vehicle is in danger of being hit by a HEAT warhead:

- detonation of the warhead or otherwise destroying it at some distance from the vehicle,
- disturbing, breaking or fragmenting the shaped charge jet produced after detonation of the warhead,
- reduction of the depth of penetration,
- minimizing of the above-mentioned detrimental effects caused by shaped charge jet penetrating the armour [2,3,4,5,6,7].

Destruction of a warhead, without triggering its detonation by its fuse (without shaped charge jet formation), takes place at some distance from the vehicle using active protection systems, for e.g. that jam the guiding systems or shoot down the projectile, or shortly before

* Corresponding author: marek.szudrowicz@witpis.eu

a warhead hits the vehicle, using active systems (that shoot down the projectile) and passive systems (designed to damage or destroy the warhead) [8,9].

If the warhead detonate and a shaped charge jet is formed, appropriate means should be used to make it happen at such a distance from the armour that chances of penetrating it is low. If a detonation of a warhead occurs at a distance exceeding its 6-8 calibres, the probability of penetrating the armour starts to decrease. This means that for e.g. warhead with a calibre of 75-80 mm (PG-7 warheads fired from the RPG-7 launcher) such a distance should be larger than ca. 500 mm.

This method is implemented by the installation of screens, mainly made of horizontally arranged steel bars or rods, or by using net screens. Such a design of screens makes it also possible to destroy the warhead fuse and consequently prevent the shaped charge jet formation. Two mechanisms are mainly used here:

- short-circuit the fuse circuit,
- deforming the warhead conical liner.

These mechanisms operate only when the warhead fuse does not touch any element of the screens structure, i.e. the warhead flies past bars, wires or metal elements of net. At the same time, the distance between adjacent bars or wires and metal elements of the net must be lower that the warhead calibre [11].

The probability of ensuring protection for crews of vehicles is one of the four main parameters that influence selection of a protection system. The other parameters are the level of protection characterized by the type of threat, weight of the system and its costs. For ballistic resistance, in bulletproof tests, the probability of crew protection is calculated mainly as the share of the armoured surface resistant to perforation in the total vehicle surface area. In case of bar or net screens, the probability of protection depends not only on how much of the vehicle surface is covered with such screens but, due to factors determining operation of the mentioned-above warhead-destroying mechanisms, also on their design, i.e. the percentage share of open space in the protection system installed on the vehicle. The open space is the surface for which the fuse of the flying warhead will not touch any part of the structure and a high probability exists that the warhead will be destroyed without generating a shaped charge jet.

In order to improve resistance to shaped charge jet penetration, vehicles are often fitted with additional external armour that often performs also functions related to protection against kinetic projectiles. This group of solutions includes explosive reactive armour (ERA).

Due to its effectiveness in the protection against HEAT warheads, ERA armour is widely used on heavy armoured vehicles [12,13,14]. However, installation of such armour brings some negative consequences, e.g. explosion of a reactive cassette affects the vehicle components and may cause injuries to soldiers around the vehicle. Development work on other reactive armour types results in designs free of explosives (Non Explosive Reactive Armour - NERA) but employing inert materials that react to the interaction between the shaped charge jet and the armour using deformable plates that disturb the jet. According to the literature, materials used in such sandwich arrangement are e.g. steel plates and rubber [15,16,17,18,19].

Apart from ERA or NERA armours, group of additional external solutions include space armour structures and composite armours based on ceramics and light layers of metallic or non-woven materials.

If all the above-mentioned means fail and the shaped charge jet perforates the main armour (hull of vehicle), the ultimate protection of the vehicle interior and the crew is the spall-liner installed inside the vehicle. It is employed mainly in order to minimize damage and its installation should result in the complete reduction of the angle of debris cloud [20,21].

The article presents results of studies with shaped charge carried out using different material combinations designed to reduce the angle of debris cloud with consideration given to selected model solutions of external armour.

2 Materials and methods

Interactions between the shaped charge jet and model material combinations were recorded using the SCANDIFLASH XR 450 set with two X-ray sources arranged in mutually perpendicular directions. The tested material was placed between the X-ray sources and film cassettes so that the generated shaped charge jet was perpendicular to the cassettes and parallel to the line running between the X-ray sources and the cassettes (Fig. 2).

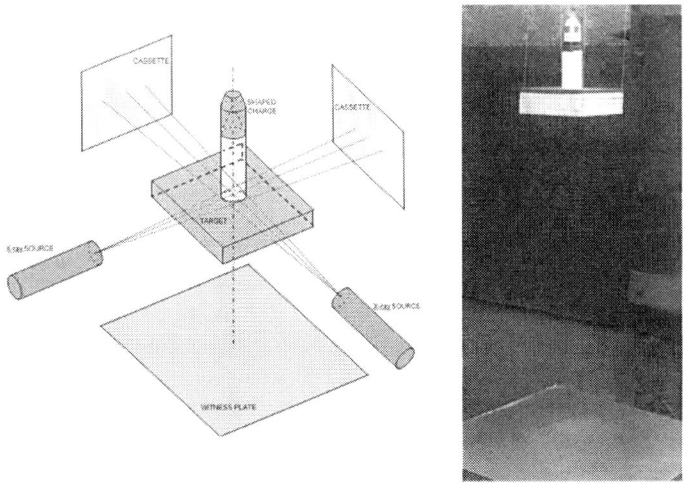

Fig. 2. Schematic view of the test configuration system.

The distance between the upper surface of the tested material combination and the middle of the charge base was adjusted with a PVC tube. During the tests it was 80 mm. Only in the case of testing the influence of distance on jet hitting effects, distance was increase to 160 mm and 240 mm.

In order to recording results of the jet and its fragments perforating the tested material, a steel witness plate with dimensions of 500x500x1.5 mm was installed under the sample. Its distance from the lower surface of the tested material was 660 mm. When the charge was detonated alone, without any material, the distance between the witness plate and the lower edge of the charge was 760 mm.

During each test, photographs were taken at different time intervals from triggering the detonation of the shaped charge. This made it possible to determine the speed of the jet. Additionally, consideration was given to the degree of continuity and shape of the jet, distribution of fragments recorded on X-ray photographs and distribution of penetrations in witness plates.

The tested materials were:

- steel armour plate modelling vehicle main armour with thickness of 7 mm, density 7800 kg/m^3, hardness 500 HB, yield strength 1450 MPa, tensile strength 1700 MPa, elongation 7 % - **HH**
- Al_2O_3 ceramic tile with thickness of 10 mm, density 3870 kg/m^3, hardness 1440 HV, bending strength 379 MPa, modulus 365 GPa - **BC**
- aramid laminates with thickness of 8,5 mm, density 1100 kg/m^3, tensile strength 300 MPa, modulus 7 GPa, elongation 11 % - **LA**
- glass laminates with thickness of 10,4 mm, density 1960 kg/m^3 tensile strength 485 MPa, modulus 25 GPa, elongation 4 % - **LG**

- St3 steel plate with thickness of 1,5 mm, rubber core, St3 steel plate with thickness of 1,5 mm - **NERA**

The study covered the following material combinations:

- steel high hardness armour plate 150x150x7 mm,
- steel high hardness armour plate 150x150x7 mm with Al_2O_3 ceramic tiles with thicknesses of 10, 30, 60 and 90 mm (arranged as stacks of 100x100x10 mm tiles), ceramic tiles installed on the charge side,
- steel high hardness armour plate 150x150x7 mm and aramid laminates 150x150 mm with thicknesses of 8.5, 17 and 25.5 mm installed behind the armour plate in relation to the charge,
- steel high hardness armour plate 150x150x7 mm and glass laminates 150x150 mm with thicknesses of 10.4, 20,8 and 41.6 mm installed behind the armour plate in relation to the charge,
- steel high hardness armour plate 150x150x7 mm and two glass laminates 150x150 mm with a total thickness of 20.8 mm installed behind the armour plate at a distance of 20 mm (air gap),
- steel high hardness armour plate 150x150x7 mm and two aramid laminates 150x150 mm with a total thickness of 17 mm installed behind the armour plate at a distance of 20 mm (air gap),
- sandwich 100x200 mm combinations: 1.5 mm thick steel sheet, 3 and 6 mm thick rubber core, 1.5 mm steel sheet. Such material combinations were arranged at an angle of 60° in relation the materials below them, i.e. 20 mm thick ceramic tiles of Al_2O_3, steel high hardness armour plate and laminates: 10.4 mm thick glass-type and 8.5 mm thick aramid type.

The study was carried out with shaped charges H-Fe-39-150. The charges were fitted with a conical liner of copper with an angle of 42° and weight of 37 g. The explosive was RDX. Its weight was 32 g. The charges were prepared for triggering the explosion centrally with a primer. The maximum outer diameter of the liner was 37 mm and the diameter of the steel casing was 39 mm.

Ability of shaped charge jet to penetration armour plates, confirmed during the preliminary tests on the stack of armour plates is 80 mm.

3 Results and discussion

3.1 Preliminary tests

In the preliminary test, speed of the shaped charge jet was established on the basis of photographs taken at different time intervals from triggering the shaped charge, as well as there were determined effects made directly on the witness plate by the jet.

The estimated average speed of the shaped charge jet in the air is 7450 m/s. The witness plate is hit by a dispersed jet. The average angle of flight of fragments is equal to 10.8 degrees. Additionally, a large hole is made as a result of hitting the witness plate by so-called cluster created together with the shaped charge jet (it is part of the charge liner which was not use to create the narrow, long, jet).

Furthermore, studies were carried out on effects accompanying the penetration of a single high hardness armour plate modelling the main armour (vehicle hull), which was used as the reference for other tested materials.

Photographs after the jet penetration in single armour plates show the flight of the plate spalls. Apart from the jet and its fragments, also spalls from the armour plate hit the witness plate (Fig. 3). Large holes in armour plates result from the penetration of the plate with clusters created together with the jet. The average angle of flight of jet fragments and plate spalls is equal to 24.9 degrees.

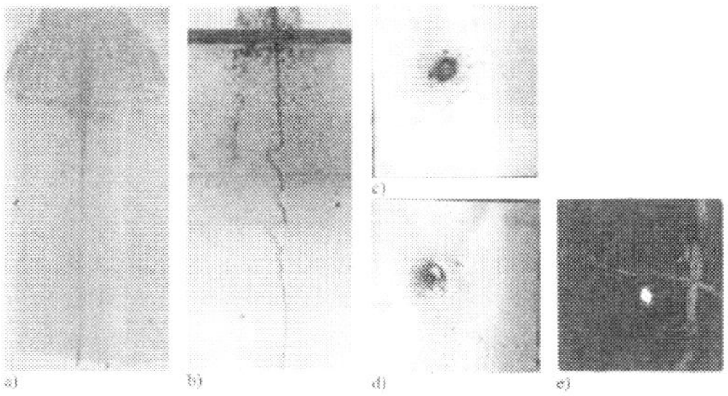

Fig. 3. a) view of a shaped charge jet in air and b) after the armour plate perforation for t = 60 μs, c) witness plate (500x500 mm) after hitting by jet without armour plate and d) after hitting by a jet the armour plate, e) view of perforated armour plate.

3.2 Tests of materials combinations

3.2.1 Studies on penetration of ceramic tiles and the armour plate

Ceramic tiles were installed on the armour plate. They were used in order to erode the jet and disturb its movement before it reaches the armour plate.

When 10 mm and 30 mm thick ceramic tiles were used, the jets shown in the picture (Fig. 4) are disturbed to a degree similar to that observed in the pictures of a jet penetrating an armour plate without ceramic tiles.

The average angles of flight of fragments are equal to 24.7 and 24.5 degrees (Fig. 5).

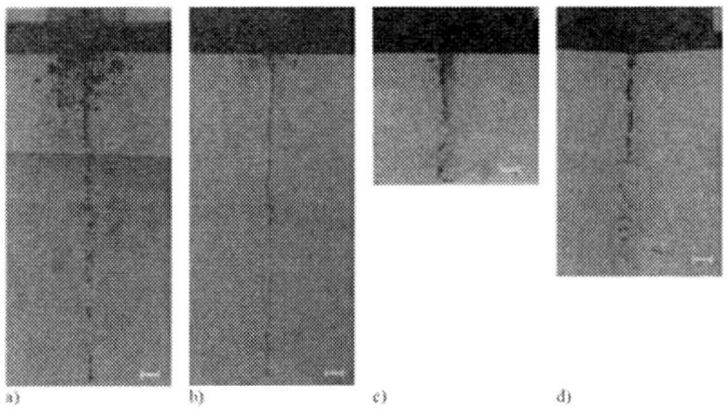

Fig. 4. View of the jet after perforation ceramic - armour plate specimens, for ceramic tiles: a) 10 mm, t=60 μs; b) 30 mm, t=60 μs; c) 60 mm, t=90 μs; d) 90 mm, t=120 μs.

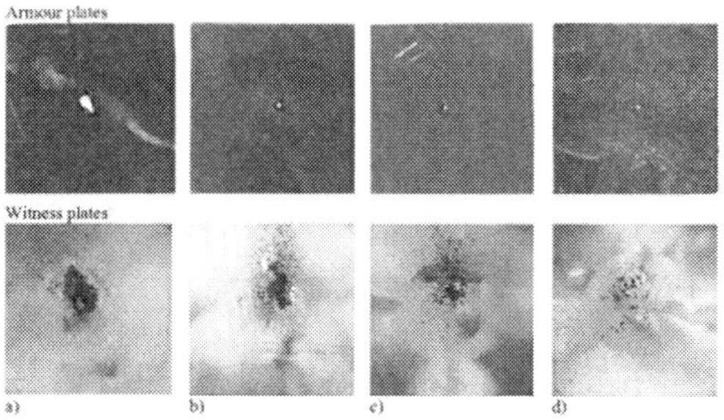

Fig. 5. Comparison of the armour plates (150x150 mm) and view of the witness plates (500x500 mm) after perforating by jet the ceramic - armour plate specimens, for ceramic tiles: a) 10 mm; b) 30 mm; c) 60 mm; d) 90 mm.

The estimated jet tip speeds after perforating the ceramic-armour plate specimen are shown in Table 1.

Table 1. The jet speeds after penetrating the ceramic-armour plate specimens.

Thickness of ceramic [mm]	Jet speed [m/s]
10	6,700
30	6,550
60	2,770
90	2,300

The hole after the cluster hitting is only seen for 10 mm thick ceramic tiles. In case of 60 mm and 90 mm thick ceramic tiles, a cloud consisting of jet fragments and plate spalls can be seen in the pictures. The speed of the fragments is still high and the witness plate is perforated. The average angles of flight of jet fragments are equal to 21.8 and 20.6 degrees. The disturbed jet is not so effective anymore in perforating the plate. As can be seen, the jet speed and diameter of the hole resulting from the perforation of the armour plate with a shaped charge jet are significantly smaller.

A study was made on how the distance between the shaped charge and the armour plate affects the process of penetration. For the combination of 20 mm of ceramics with an armour plate, tests were made using a charge located at distances of 160 and 240 mm, which is equivalent to about 5 and 7.5 calibres of the shaped charge. In the course of penetration through this combination of materials, the jet produced at a distance of 160 mm is still continuous. In case of the charge moved to a distance of 240 mm, after the penetration of the armour plate, the jet is fragmented (Fig. 6). At the same time, the increased distance resulted in the fragmentation of the jet even before it reached the ceramic tiles. In this case holes made in the witness plates are significantly smaller. In the armour plates, not one but several holes were made.

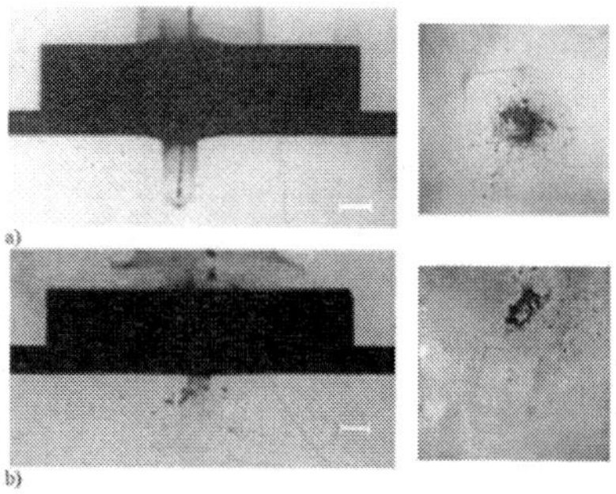

Fig. 6. Photograph of the combination of 20 mm thick ceramics and an armour plate after it was hit by a jet from a distance of: a) 160 mm b) 240 mm

3.2.2 Studies on the penetration of the armour plate with aramid laminate

The laminate layers were applied mainly in order to stop, as much as possible, spalls from the armour plate. The photographs show that at the initial phase of the penetration, such spalls are stopped by the double or triple layer of aramid laminate. However, for longer time intervals (t = 90 μs), the protection offered by the laminate layer is ineffective and the photographs show a cloud of spalls under the laminate. This fact is confirmed by the photographs of the witness plates and dimensions of holes in the aramid laminate (Fig. 7).

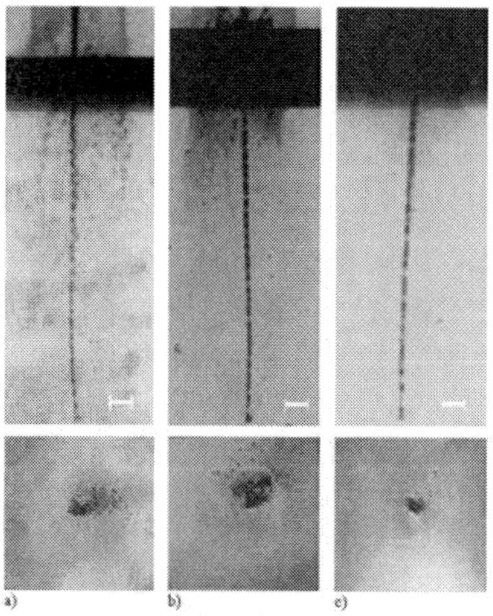

Fig. 7. View of the jet and the witness plates (500x500 mm) after hitting the armour plate – aramid laminate specimens with: a) single laminate, b) double laminate, c) triple laminate.

Additionally, it can be shown that aramid laminate layers do not significantly affect the jet head speed as compared to the speeds estimated for the bare armour plate (table 2). However, they reduce the angle of flight of spalls, which has values of 23.9, 10.4, and 8.2 degrees, respectively.

Table 2. The jet speed for the combination of the armour plate with aramid laminates.

Thickness of aramid laminate [mm]	Jet speed [m/s]
8.5 (1 layer)	6,740
17 (2 layers)	6,250
25.5 (3 layers)	7,120

3.2.3 Studies on penetration of the armour plate combined with glass laminate

A similar effect was achieved with a glass laminate layer installed under the armour plate (table 3). Additionally it was found out that holes in the witness plates were slightly more scattered than for the same number of aramid laminate layers (Fig. 8). The angle of flight of spalls was equal to 21.4, 14.3, and 11.2 degrees, respectively.

Table 3. The jet speed for the combination of the armour plate with glass laminates.

Thickness of glass laminate [mm]	Jet speed [m/s]
10.4 (1 layer)	7,165
20.8 (2 layers)	7,240
31.2 (3 layers)	6,200
41.6 (4 layers)	6,550

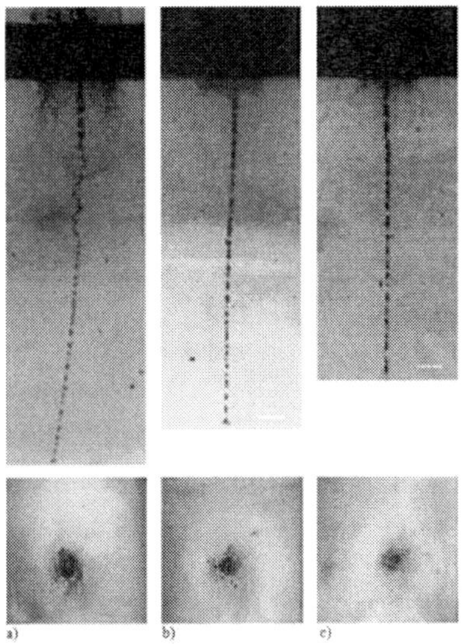

Fig. 8. View of the jet and the witness plates (500x500 mm) after hitting the armour plate – glass laminate specimens with: a) single laminate, b) double laminate, c) four laminate layers.

Tests were made to find out if an increased distance between the laminate and the armour plate increases the effectiveness of the spall catching effect. Two layers of aramid laminate or two layers of glass laminate were installed at a distance of 20 mm under the armour plate (air gap). The laminate layers can stop only some part of spalls. The number and arrangement of holes in the witness plate show that in this configuration both laminates offer similar effectiveness in catching fragments (Fig. 9).

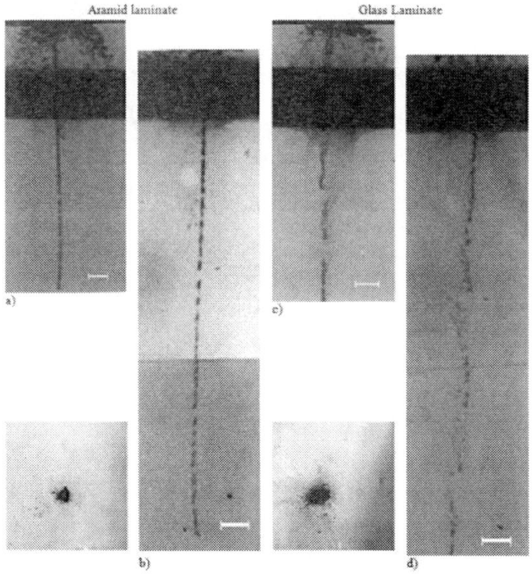

Fig. 9. View of a jet and witness plates (500x500 mm) after perforation of armour plates - 20 mm air gap - aramid laminates: a) t = 40 μs, b) t = 60 μs after perforation of armour plates - 20 mm air gap - glass laminates c) t = 40 μs, d) t = 60 μs

3.2.4 Studies on penetration of material combinations

The expectation was that laminates would catch spalls from the armour plate and jet fragments more effectively when the shaped charge jet was disturbed by a ceramic layer before reaching the armour plate. Consequently, the following combination of materials was used: 20 mm of ceramics - armour plate - 20 mm air gap - 2 laminates. The analysis of the X-rays shows that the jet maintains its direction (jet fragments continue the flight in the same direction) and the quantity of spalls from the armour plate below the laminate layer is low. The tested configuration is relatively effective in catching the jet fragments and spalls, which is confirmed by the small diameter of the main hole in the witness plate and the low scatter of small spalls (Fig. 10). What draws attention, is the smaller hole in the armour plates. As already established before, this effect is caused by the ceramic tiles.

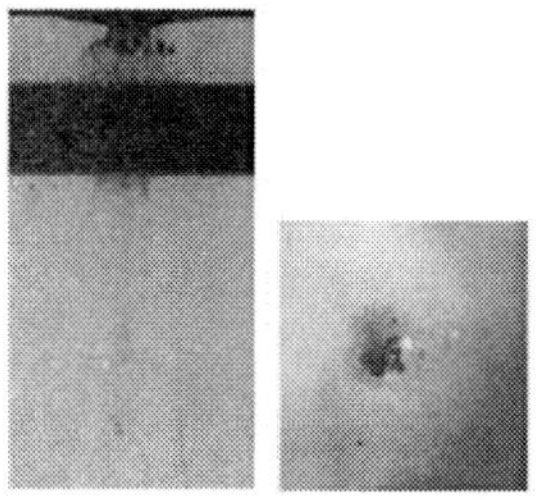

Fig. 10. View of a jet (t = 60 μs) and witness plates (500x500 mm) after perforation of ceramic tiles - armour plate - 20 mm air gap - glass laminates.

Additionally, studies were made on the reaction of the sandwich type combination steel plate - rubber - steel plate to the shaped charge jet and on how its presence in front of the ceramic tiles affects further penetration. The following combination of materials was chosen: 20 mm of ceramics - armour plate - 1 glass laminate - 1 aramid laminate (Fig. 11).

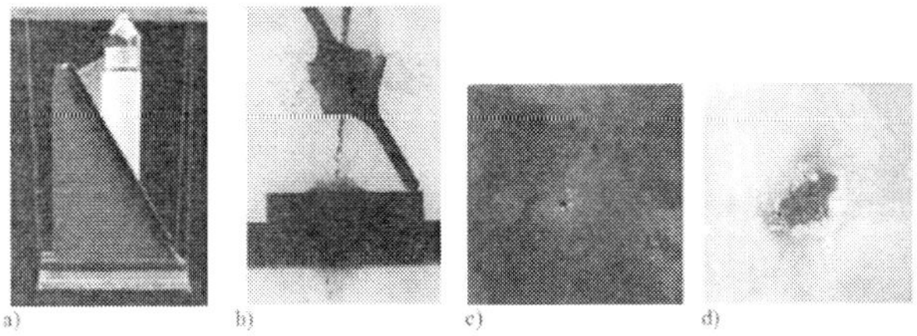

Fig. 11. a) sandwich configuration with a rubber core and ceramics of 20 mm - armour plate - glass laminate - aramid laminate; b) view of the jet perforating sandwiched materials, c) armour plate (150x150 mm) and d) witness plate (500x500 mm) after secimen perforation.

The analysis of the test results shows that the hit of jet head causes violent, local deformation of external sheet metal, which movement disturbs the jet. In one instance of the tests made on thus configured materials, the witness plate was not penetrated. The average angle of flight of jet fragments and spalls is equal to 13, 9.1 and 6.1 degrees.

Graphical summary of the results is shown in figures 12 and 13.

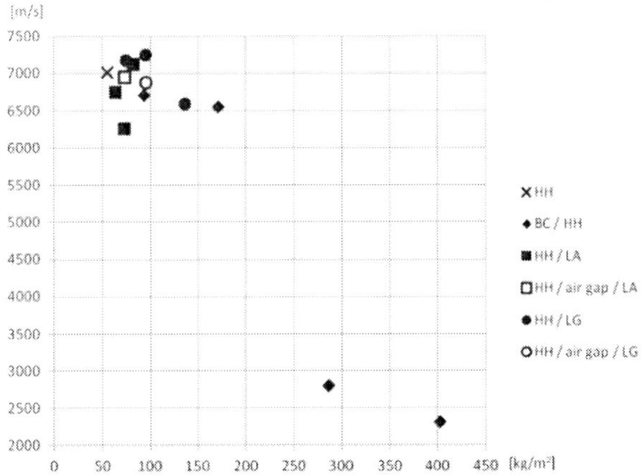

Fig. 12. Influence of material combinations aerial density on the jet speed, after material perforation.

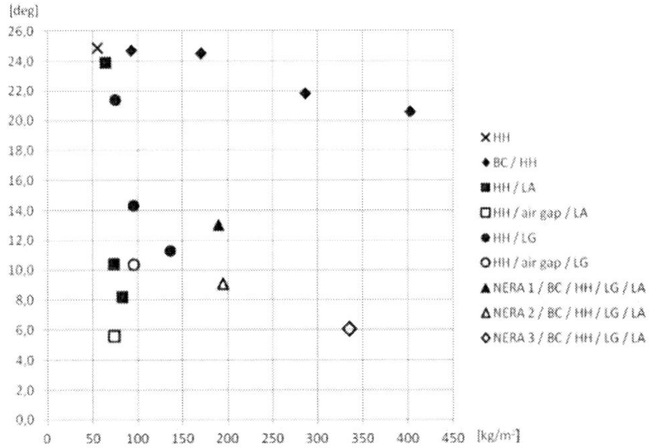

Fig. 13. Influence of material combinations aerial density on the angle of flight of spalls.

4 Conclusions

The use of different materials does not cause significant changes in the jet speed after their perforation, in addition to the two cases for thick layers of tiles from in the front of plate armour. The performed studies show that if armour is designed with the principles specified in this paper, it is possible to achieve a solution that minimizes both damage made inside the vehicle and probability of sustaining injuries by its crew. According to these principles:

- a HEAT warhead should be forced to detonate at some distance from the main armour, the element that triggers detonation should also disturb the continuity of the jet,
- employed layers that erode the jet and reduce its diameter before it reaches the main armour,
- a spall liner should be installed at some distance behind the main armour, the studies showed that the most effective were aramid laminates

The usefulness of these solutions has been confirmed in the performed studies where, in one case, no perforation of the witness plate was recorded. For this solution, the mass efficiency was calculated.

The mass efficiency was calculated as the ratio of the aerial density of tested material combination to the aerial density of armour plates which protect against perforation over the shaped charge jet. Areal density of materials combination is 335 kg/m^2. In experimental studies it was concluded that the shaped charge jet perforate 80 mm of armour plates and its areal density is 620 kg/m^2. Therefore, the mass efficiency is equal to the 1,9.

5 References

1. D.R. Kennedy, *The History of the Shaped Charge Effect, the First 100 Years*, (DTIC, 1983)
2. O.V. Svirsky, N.P. Kovalev, B.A. Klopov, V.V. Ashurov, V.A. Krutyakov, Int. J. Impact Eng., **26**, 735 (2001)
3. G.A. Shvetsov, *Ballistics '99* (18th International Symposium on Ballistics, San Antonio, Texas, 1999)
4. Z.X. Yin, C.M. Ma, S.X. Li, G.Q. Cheng, Mater. Sci. Eng. A **379**, 443 (2004)
5. M. Mayseless, R. Genussov, Int. J. Impact Eng., **23**, 585 (1999)
6. V. Madhu, T. Balakrishna Bhat, Defence Sci. J., **61**, 394 (2011)
7. A. Sirois, Y. Baillargeon, A. Vallée, G. Bergeron, *Technical Report TR 2012-063* (Defence R&D Canada, Valcartier, 2012)
8. R. Panowicz, M. Konarzewski, J. KONES, **22**, 193 (2015)
9. J.D. Powell, D.L. Littlefield, *Technical Report ERL-TR-3108* (Ballistic Research Laboratory, Aberdeen Proving Ground, 1990)
10. W.P. Walters, J.A. Zukas, *Fundamentals of Shaped Charges* (John Wiley & Sons, New York, 1989)
11. M. Szudrowicz , J. KONES, **18**, 639 (2011)
12. H.S. Yadav, B.M. Bohra, G.D. Joshi, S.G. Sundaram and P.V. Kamat, Defence Sci. J. **45**, 207 (1995)
13. M. Held, Propellants Explos. Pyrotech, **29**, 244 (2004)
14. M. Held, Propellants Explos. Pyrotech, **26**, 97 (2001)
15. M. Brand, M Mayseless, R. Genussov, *Proceedings - 19th International Symposium on Ballistics* (19th International Symposium on Ballistics, Interlaken, 2001)
16. K. Thoma, D. Vinkier, J. Kiermeier, U. Deisenroth, W. Fucke, Propellants Explos. Pyrotech, **18**, 275 (1993)
17. E. Lidén, A. Helte, A. Tjernberg, *Proceedings - 23th International Symposium on Ballistics* (23rd International Symposium On Ballistics, Tarragona, 2007)
18. M. Held, Propellants Explos. Pyrotech, **26**, 191 (2001)
19. Z. Rosenberg, E. Deckel, Int. J. Impact Eng., **21**, 297 (1998)
20. I.Horsfall, *Proceedings - 22nd International Symposium on Ballistics* (22nd International Symposium on Ballistics, Vancouver, 2005)
21. *Report nr FOI-R-1751-SE* (FOI Defense Research Agency Weapons and Protection, 2005)

Usability Techniques Possible to Use in Diagnostic Tools Interface Improvement

Ryszard Bielski[1,*] , Karol Kufel[1], Marcin Bielski[2], Hendra Jantanata[3], Cezary Graul[4]

[1] Gdańsk University of Technology, Faculty of Management and Economics, Poland
[2] Shiprepair Yard Nauta SA, Newbuildings Division Gdańsk, Poland
[3] LGT Indonesia, Batam, Riau Island, Indonesia
[4] UTP University of Science and Technology in Bydgoszcz, Faculty of Management, Poland

Abstract. The development of diagnostic tools has caused a rapid increase in the amount of information generated and presented, necessitating the use of computer displays as an interface to data presentation and the use of modern diagnostic tools. The paper presents methods for testing usability that can be used in designing and improving the quality of interfaces of modern diagnostic tools that use communication through graphical user interfaces.

1 BACKGROUND

Currently, industrial devices are becoming more and more complex, require more and more complex procedures and at the same time expected to be more and more reliable in their operation.

Fig. 1. Power Plant Control Room view.

A natural consequence of meeting the requirements of reliability is the installation of an increasing number of sensors, which, however, generate significant amounts of

* Corresponding author: rysbiels@pg.edu.pl

measurement data. Their acquisition, processing and interpretation would not be possible without the use of modern technologies, especially computers. Examples of control panels of a nuclear power plant and a modern passenger plane are presented in Fig. 1 and Fig. 2.

Fig. 2. Airbus A 380 Cockpit view.

At the same time, the amount of diagnostic data generated compels us to deliberate their presentation on computer screens. That is why the quality of the presentation of diagnostic devices on displays is such a critical factor. Their similarity to the screens of computer applications allows the adaptation of techniques known from usability in computers for the analysis, evaluation and evaluation of the quality of diagnostic device interfaces.

2 USABILITY TECHNIQUES POSSIBLE TO USE IN DIAGNOSTIC TOOLS INTERFACE IMPROVEMENT

Techniques and usability testing (Usability) are used for testing the usable quality of web services and applications. Many of their methods have been developed. Their knowledge allows to improve IT products and interfaces, identify problems and create recommendations for design. Some research show, that there are elements that may disturb in comfortable use of graphical interfaces. They may be for instance too many moving elements, overspecification or wrong, not expected, misleading answers form automated system [6, 17, 24]. This may be overcame by using guides in development phase, especially in case of older users, who need simplified interfaces [16, 23]. Also emotion-centered techniques, like Kansei engineering may be helpful in delivering high quality, intuitive solutions [7]. Such approach is necessary in research on barriers in the use of all digital products, including diagnostic tools. This publication relays on report [14] and is a review of techniques in this area. Main techniques are shortly presented in following points.

2.1 Analysis of the context of use (Context of use analysis)

Analysis of the context of use is a method of obtaining information about the product and its use. During the session with the participation of key persons of the group is being created description of the product itself, information about users, tasks performed by users and the environment in which it is performed is obtained. Sessions should be performed as soon as possible, usually during the product planning phase [8]. However, the results of this analysis can be helpful as a reference point also in later stages of product preparation and creation, especially if they are updated on a regular basis in the meantime.

2.2 Task analysis

Task analysis determines what the user should do (in the sense of actions or processes) to achieve the assumed goal. A careful analysis of tasks can help to understand the system and the flow of information inside. Analysis of tasks allows for proper development and allocation of tasks within the system, which allows for proper assignment of functionalities in later stages. This is important for correct construction and subsequent system integrity. In particular, it may be a threat that insufficiently accurate diagnosis of tasks performed in the tested system usually increases the risk of expensive problems (resulting from lack of knowledge about the tasks necessary to perform) in the later stages of the project [12].

2.3 Focus groups

A focus group is a meeting of all parties with interests (stakeholders) in a given process. Views on the essential elements are cited by the parties. Meetings can be recorded for later analysis. This method is useful at an early stage of product specification or as a way of obtaining comments after a certain period of use. It allows to obtain a multilateral perspective in terms of requirements and identification of elements to which attention should be paid [13].

2.4 Observation

Observation methods require the researcher to see users while they are working. During the process, the researcher makes notes on activities that can be observed. The observation may be direct (physically next to the observed person, which may affect the behavior of the observed person) or indirect (using technical means – which is more neutral), including by means of video recording for later analysis [20].

2.5 Request for context (Contextual inquiry)

The Contextual inquiry is one of the best methods if you only need to understand the environment in which users perform tasks. Basically, it is a structured interview method, based on a few basic principles that distinguish it from reporter interviews [8]. It is the opposite of the evaluation process and seeks to discover. It is usually used in the early stages of work in order to get feelings about the work being done, how to do it, what is the flow of information and so on.

2.6 Scenarios

Scenarios are characteristics of users and their tasks in a specific context. They allow to visualize the user working with the product in order to achieve a specific goal [15]. The goal of using scenarios in the early stages of design is to improve the quality and user-friendliness of the user's requirements presented to the project team. In later stages scenarios can be used during design and evaluation.

2.7 Card sorting

Card sorting is a technique that allows designers to see how users link elements of a website or application interface into groups. Each element is saved on a small card (in a typical card sorting session about 30 to 80 cards are allocated). Users, working alone or in pairs, sort the received cards into groups, and then give the groups names. The designer can use this information to better organize the site structure. There is a variation of this method, called Reverse Card Sorting, which provides users with main category names along with cards showing information and features available on the site. The examined persons, instead of combining functions into groups, assign them to predefined main categories [21, 26].

2.8 Interview techniques

In this method, users (experienced and / or new) are asked by in-depth questions to get subjective opinions about the feelings of using the product. Interviews may take the form of predefined lists (structured) and / or may allow users to freely formulate opinions (unstructured) [1]. The type, detail and objectivity of the collected data depend on both the type of interview and the researcher's experience.

2.9. Questionnaires (Survey)

Questionnaires include sets of saved questions asked to large numbers of users. They can be useful in acquiring information, including on how to work and settings for processes and interfaces. First of all, we distinguish between closed questions, in which the respondent chooses from predefined answers and open answers, where he can give any answer. The cantaterers are then processed using statistical methods [22].

2.10 Parallel design

Several different project groups prepare parallel versions of the interface concept. The purpose of this is to prepare and evaluate various interfaces before choosing the one that will be directed for further work. Project groups must work independently because the goal is to get as much diversity as possible. Designers are not allowed to discuss projects until they are presented. One of the concepts or a combination of the best features among many presented concepts may be chosen [11]. Although parallel design may seem expensive, because many concepts are created without their implementation, it is in fact an effective way of discovering a wide spectrum of possible project concepts and selecting the optimal design.

2.11 Paper prototyping

The method of paper prototypes uses the simplest materials to perform interface simulation based on elements drawn (and / or cut, attached) on paper. Its purpose is to study the requirements (or feelings) of users. After the paper prototype is prepared, a member of the research team sits down in front of the user and plays the role of a computer (namely changes in the interface - content of the monitor) by moving the interface elements in response to the actions taken by the user. The user makes selections and launches interface elements using fingers to indicate (and voice for possible data entry). Users are instructed about required tasks, and during their performance are encouraged to share thoughts, insights and feelings. Members of the research team note observations and obtained comments [4, 18]. This method is useful when it is easy to simulate the operation of the system and it is not necessary to very accurately reproduce the screen elements.

2.12 Cognitive walk through

The step-by-step process (although in Poland too the English name - cognitive walk through) of the tested product or system is used along with collecting the observed reactions of key team members or typical users. One or two team members usually conduct the transition process (subsequent steps in using a product or system) while others comment on their feelings and observations [2].

2.13 Heuristic evaluation

Heuristic evaluation is an expert study method that identifies general usability problems that can be expected when using a product. It usually requires at least three system experts in accordance with established guides, rules and lists. The result is noted observations often ordered in accordance with their significance. It is a quick and effective method [3].

2.14 Participatory evaluation

It is a cost-effective technique for identifying utility problems in prototypes. It consists in encouraging the cooperation of end users and members of project teams in order to identify problems and find solutions. The result is a qualitative information about the difficulties in use when trying to perform tasks and other elements of the interface that cause problems [10].

2.15 User-based testing of design / interaction

This method offers a relatively fast and cheap way to evaluate an existing product or prototype by the user.

When testing the system, it focuses on completing the task and getting feedback, in particular on where users are unable to complete tasks or need help in completing them. Particular emphasis is put on testers representing typical users. Detailed test records are not needed. Researchers make records during the interaction of testers performing the imposed tasks and identify the most important problems [25].

In the case of interaction testing, it is suitable for assessing both the high quality of interactive prototypes and Web sites. The surroundings and the tested product should be as close as possible to reality. Researchers in this case also take notes, they can also use audio and / or video recordings. Then, the observations are analyzed and relevant metrics used in the field of usability.

2.16 Eyetracking

The eye-tracking method allows the analysis of the perception of websites from the user's perspective (in opposition to, among others, process perspective or service perspective). The history of this research method dates back to the nineteenth century, when the first eye-tracking devices were used by psychologists and physiologists. In the thirties of the last century, research devices appeared that can be considered conceptually similar to contemporary ones [5]. Currently, eye-tracking studies are carried out using computer hardware. The advantage of eye-tracking research is the 'natural' way for the respondent to acquire data, as well as the limited conscious control of the respondent over the course of the study itself [9]. In the field of marketing, these studies are used in many areas, including assessment of advertisements, packaging or websites [19].

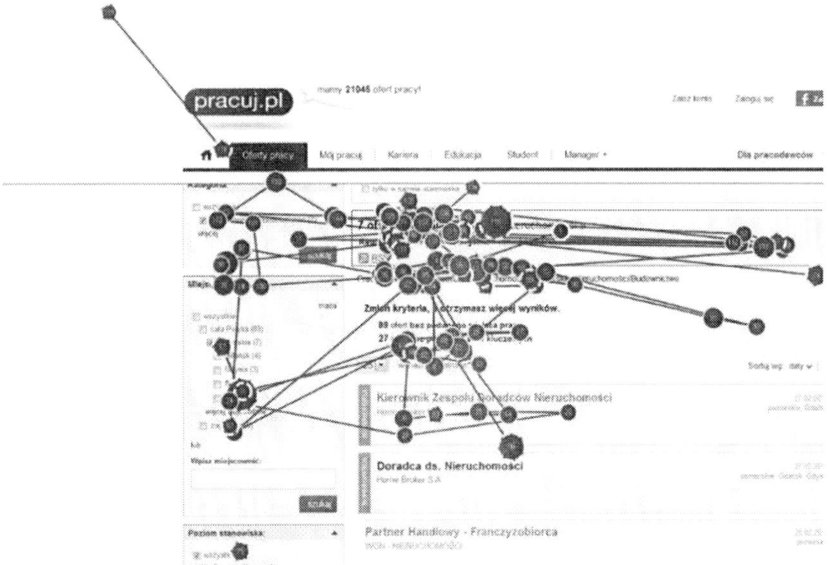

Fig. 3. Sample Gaze-plot of site pracuj.pl.

Sample test result using eyetracking gaze-plot method is presented on Fig. 3 [19].

Each of the described methods has its own characteristics. This implies that during whole process of preparing, developing and use of interfaces of diagnostic tools, different methods may be used. In particular, one may point three consecutive phases of the preparation process of the diagnostic tool interface design. These are: specification of requirements (which takes place mainly before whole design process starts), concept design and testing and evaluation (usually after tool is designed). The potential use of methods in individual design phases is presented in Tab. 1.

Table 1. Usability techniques possible to use in the diagnostic tools interface design process.

	Requirements specification	Concept design	Testing and evaluation
1. Analysis of the context of use	v		
2. Task analysis	v		
3. Focus groups	v		
4. Obsservation	v		
5. Request for context (Contextual inquiry	v		
6. Scenarios	v	v	
7. Card sorting	v	v	
8. Interview techniques	v	v	
9. Questionnaires (Survey)	v		v
10. Parallel design		v	
11. Paper prototyping		v	
12. Cognitive walk through		v	v
13. Heuristic evaluation		v	v
14. Participatory evaluation		v	v
15. User-based testing of design / interaction			v
16. Eyetracking			v

3 CONCLUSIONS

The development of diagnostic tools has caused a rapid increase in the amount of information generated and presented, necessitating the use of computer displays as an interface to data presentation and the use of modern diagnostic tools. At the same time, the increased complexity of modern devices has caused an avalanche increase in the demand for the use of diagnostic tools and the necessity of designing their interfaces so that they are simple and user-friendly. Usability techniques presented in this paper may be used for improving quality of modern diagnostic tools in order to make them more user-friendly.

References

1. J. Anderson, F. Fleek, K. Garrity, F. Drake, Integrating Usability Techniques into Software Development, January/February 2001, IEEE SOFTWARE (2001)
2. M. H. Blackmon, P. G. Polson, M. Kitajima, C. Lewis, Cognitive walkthrough for the web, CHI '02 Proceedings of the SIGCHI Conference on Human Factors in Computing Systems, ACM (2002)
3. J. St. B. T. Evans,The heuristic-analytic theory of reasoning: Extension and evaluation, Psychonomic Bulletin & Review, June 2006, Vol. 13, Iss. 3 (2006)
4. Eileen G. Abelse, Marilyn Domas White, Karla Hahn, A user-based design process for Web sites, Internet Research, Vol. 8 Iss. 1 (1998)
5. Gofman A., Moskowitz H.R., Fyrbjork J., Moskowitz D., Mets M. (2009), Extending Rule Developing Experimentation to Perception of Food Packages with Eye Tracking, Open Food Science Journal; Vol. 3 (2009)
6. Kossecki, P.; Wachowicz, J. Świerczyńska-Kaczor, U.;. Possible Stress Factors for Elderly Users in Online Human Resources Services, in: Psychology&Health vol. 28, Taylor&Francis (2012)
7. Ludwiszewski, B.; Redlarski, K.; Wachowicz J.. Możliwości zastosowania metody Kansei Engineering w projektowaniu usług on-line dla osób zagrożonych wykluczeniem cyfrowym, in : Interfejs użytkownika. Kansei w praktyce, eds. Sikorski M., Marasek K., pp. 88-96, Warszawa, PJWSTK (2011)
8. M. Maguire, Context of Use Within Usability Activities, Int. J. Human-Computer Studies (2001)
9. Meißner M., Decker R. (2010), Eye-Tracking Information Processing In Choice-Based Conjoint Analysis, International Journal of Market Research; Vol. 52 Issue 5 (2010)
10. M. J. Muller, L. Matheson, C. Page, R. Gallup, Methods & tools: participatory heuristic evaluation, interactions, Interactions, Vol. 5 Iss. 5, Sept./Oct. 1998 ACM (1998)
11. J. Nielsen ; J.M. Faber, Improving system usability through parallel design, Computer, Vol. 29 Iss. 2 • Feb. 1996, IEEE (1996)
12. Task Analysis, usability.gov, https://www.usability.gov/how-to-and-tools/methods/task-analysis.html, accessed 2018-05-03
13. J. Nielsen, The use and misuse of focus groups, IEEE Software, Vol.: 14 Iss.: 1 , IEEE
14. R3UEMs: Review, Report and Refine Usability Evaluation Methods: The 3rd COST294-MAUSE International Workshop, 5th March 2007, Athens, Greece, eds. Scapin D., Law. E (2007)
15. T. Rafla, P. N. Robillard, M. Desmarais, Investigating the impact of usability on software architecture through scenarios: a case study on web systems, http://trafla.tripod.com/TR-Article_Final.pdf, accessed 2018-05-12
16. Redlarski, K.;, Kossecki, P.; Ludwiszewski, B.; Świerczyńska-Kaczor, U.; Wachowicz, J.. Czynniki wpływające na efektywność korzystania z multimedialnych serwisów internetowych przez osoby starsze, in: Edukacja dla bezpieczeństwa, pp. 59-70 (2011)
17. Redlarski, K.; Wachowicz, J.;, Ludwiszewski B.. Online services for the elderly towards a better quality of life: What encourages, what discourages? In: Psychology & Health 28, (2012)
18. S. Reinhard, M. Tscheligi, V. Giller. "Paper prototyping - what is it good for?: a comparison of paper- and computer-based low-fidelity prototyping." CHI Extended Abstracts (2003)..

19. Świerczyńska-Kaczor, U.; Wachowicz, J.. Ocena dostosowania internetowych serwisów rekrutacyjnych do potrzeb osób starszych - metoda eye-tracking, in : Rozwój potencjału społecznego organizacji - wyzwania w XXI wieku, eds. Klincewicz K., Grzywacz W., pp. 215-230, Warsaw University (2012)
20. S. M. Thompson, Remote observation strategies for usability testing, Information Technology and Libraries; Chicago Vol. 22, Iss. 1, (2003):
21. Wachowicz J. Card Sorting,, in: R3UEMs: Review, Report and Refine Usability Evaluation Methods: The 3rd COST294-MAUSE International Workshop, 5th March 2007, Athens, Greece, eds. Scapin D., Law. E., (2007)
22. Wachowicz J. Data Gathering Methods: Review - Questionnaire, in: R3UEMs: Review, Report and Refine Usability Evaluation Methods: The 3rd COST294-MAUSE International Workshop, 5th March 2007, Athens, Greece, eds. Scapin D., Law. E., (2007)
23. Wachowicz, J.. Information Society in Poland – Similarities and Differences in the Perception of ICT Between Generations, in: Information Systems: Development, Research, Applications, Education. 9th SISGSAND/PLAIS EuroSymposium 2016 Gdańsk, Poland, ed.: Wrycza S., LNBIP 264, Lecture Notes in Business Information Processing, Springer (2016)
24. Wachowicz, J.; Ludwiszewski, B.; Redlarski, K. (2012). Factors Disturbing Internet Usage in Opinion of Elderly Users, in: Managerial Challenges of the Contemporary Society, Vol. 4, 2012, Risoprint, Cluj-Napoca, Romania, (2012)
25. L. White, H. Almezen, N. Alzeidi, User-based testing of GUI sequences and their interactions,. ISSRE 2001. Proceedings.of 12th International Symposium on Software Reliability Engineering, 2001, IEEE xPlore (2001)
26. J. R. Wood, L. E. Wood, Card Sorting: Current Practices and Beyond, Journal of Usability Studies, Vol. 4, Issue 1, November 2008, (2008)

Defects of Smartphones reported by users in social media and questionnaires

Robert Kostek[1,*], and *Katarzyna* Samek[1]

[1]University of Science and Technology, Al. Prof. S. Kaliskiego 7, 85-796 Bydgoszcz, Poland

Abstract. The main aim of this article is to identify the most frequent damage of Smartphones and problems reported by users in a questionnaire and on social media. Social media are a relatively new issue; they are used by companies to get information about offered products and advertise them. On the other hand, they are used by users to exchange opinion about products and write recommendations; thus these opinions are studied. At the same time planned obsolescence is applied, which makes shorter the replacement cycle. Some of damage and problems observed by users are the results of planned obsolescence. The obtained data were processed statistically.

1 Introduction

In the United States of America almost three-quarters of internet users use social media [1]. A similar situation is observed in China, which is the largest market considering the number of internet users [2]. Usually social media (SM) are understood as social networking websites. The most important feature of SM is the publication of information available to many internet users. Users publish texts, pictures, sounds and films. Moreover, people build relationships with other people; these relationships depend on activities, areas of interest, backgrounds, accepted convention, country and internet-based applications. There is a diversity of SM web pages and activities. For more information please see [3,4].

Social media provides a large amount of information directly from users, which can be used by companies to: quickly react on opinions, change offered products and increase sales [5-7]. This information is free of charge, but is large and not organised, thus data processing is difficult (large amounts of data). On the other hand, information about defects and problems with products can be quickly spread, which can lead to an image crisis; because information published by Facebook friends is more trustworthy than television advertisements. Thus information about defects of Smartphones published in SM was studied in this article.

On the other hand, planned obsolescence is a policy of designing products with an artificially limited lifetime. This results in increase in sales [8]. The lifetime can be limited for example by, non-replaceable batteries, thin layers of varnish and coatings, brittle screens, brittle or soft materials, operating-system updates, applications running in the background and slowing the processor down [9,10]. New applications drain the phone's battery, take CPU time and thus frozen screens are observed. After two years a Smartphone is no longer functional [11]. These problems should be reported in SM; moreover, a questionnaire survey was conducted. The obtained results are presented below.

2 Investigation of customers' opinions

Questionnaire surveys were conducted in UTP (University of Science and Technology in Bydgoszcz) to obtain information about smartphones. In the UTP studies, 7,700 students,

* Corresponding author: robertkostek@o2.pl

220 of them fulfilled the questionnaires, but 212 of the questionnaires were analysed. Questionnaires should be short, whereas questions should be clear, otherwise responders make mistakes. First market shares were studied. The obtained results are presented in Fig. 1, 90% confidence intervals were calculated for the UTP students. Moreover, market shares for fourth quarter of 2017 year are presented [12]. Generally, similar results are obtained, but some trade marks are more popular among young people, which is positive sign for them.

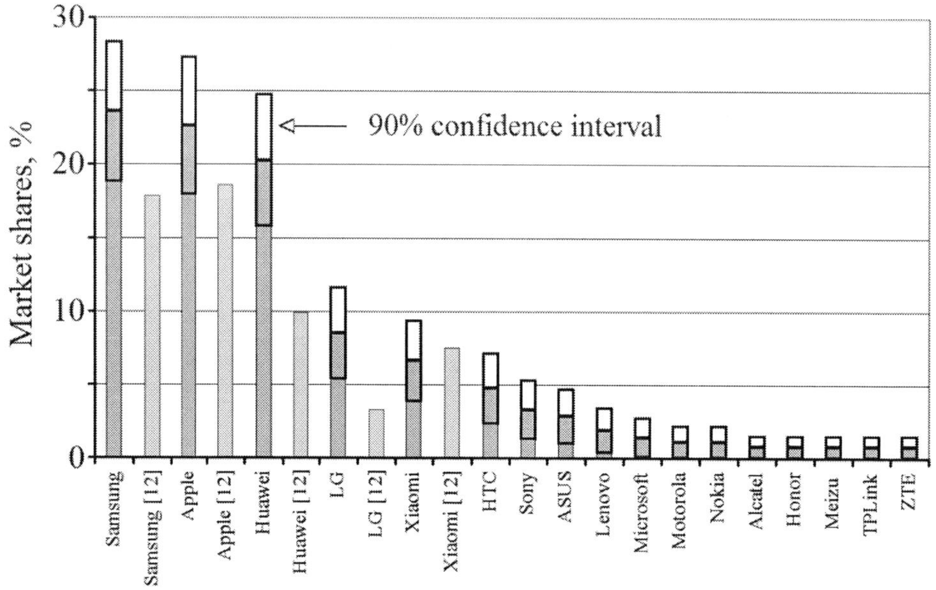

Fig. 1. Market shares reported by companies [12] and estimated for the UTP students.

Next, customer satisfaction with quality of smartphones was studied. Percentages of three groups of customers (satisfied, no-opinion, dissatisfied) were calculated based on independent research and 14,918 opinions published on the internet [13]. Generally, students are more satisfied with smartphones, than internet users (Fig.2). Internet users publish 2.5 times more negative opinions and 8.1 times more neutral opinions; which suggests that problems with products are more powerful motivation to write an opinion than proper functioning. Moreover, satisfaction with batteries was studied. One third of the students are dissatisfied with the quality of batteries; the same problem is reported in one third of internet opinions (Fig. 4); the reasons for this dissatisfaction are presented below. Around 70% of students recharge batteries at least once a day and they are difficult to replace. 69% of batteries are "non-removable". Moreover, only 4.7% of batteries were replaced. This shows that durability of battery is a key issue and limits the durability of Smartphones. A number of charge cycle is near 600, which gives roughly 600 days (almost two years). Again, 16% of respondents recharge their battery twice a day, 3.7% thrice a day, and 1.4% four times a day. This shows that recharging the battery twice a day is hardly acceptable and leads to a quick end of any battery. This problem can be solved with power bank and installation of an extra large battery (Fig.3). Around 46% of students use power banks, which provide an opportunity to recharge any battery anywhere.

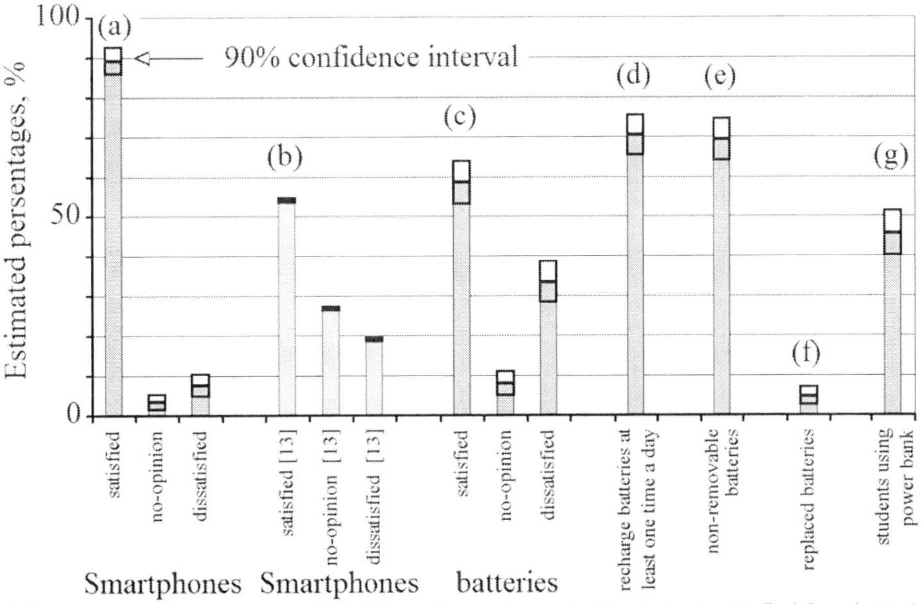

Fig. 2. Percentages of: customers satisfied from Smartphones (a, b), students satisfied from batteries (c), batteries recharged at least one time a day (d), non-removable batteries (e), replaced batteries (f), and students using power bank (g).

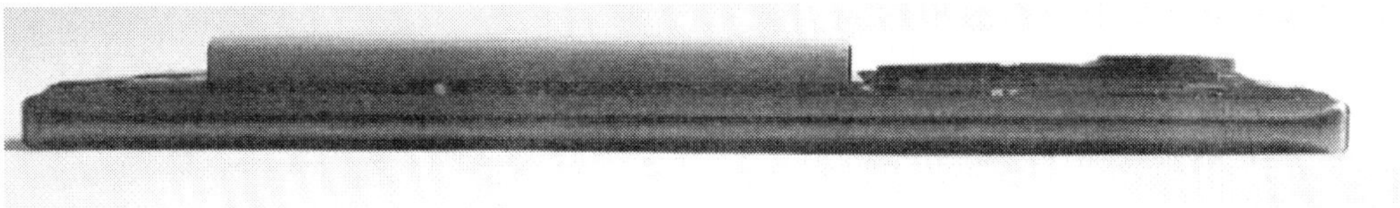

Fig. 3. Extra large battery installed in a Smartphone.

Apart from any problems with a battery, a few more defects were reported by users in social media and questionnaire surveys. A total of 289 customer opinions published on internet were analysed (Fig. 4) [13]. Freezing of a Smartphone was reported by 55% of students and 29% internet users. Freezing is a sign of problems with software and hardware. The main reasons can be a slow processor, overheating, applications working in the background, installation of actualisation and many other reasons. A scratched back cover was reported by 48% of students and 11% internet users (Fig. 5cd). The main reason of this problem is application of soft and brittle materials. If the back is painted, which reduces production cost for a variety of offered colours, then varnish is partially removed; which finally is unaesthetic. The solution is an application of a colour polycarbonate and proper surface texture. The screen is a critical part of a Smartphone, thus any defect of this element causes problems (Fig. 5ab). A screen can be scratched if is covered by foil or can be cracked. A cracked screen does not react properly on touch, whereas a scratched screen can reduce video quality. Problems with touch screen were reported by 37% of students and 10% internet users, whereas cracked screens were reported by 23% of students and 8% internet users. More problems are reported in the questionnaires, because they were listed, whereas an internet user can forget about some problems. Nevertheless, the order of problems is the same; week battery, freezing, scratched back cover, not working touch screen and cracked screen.

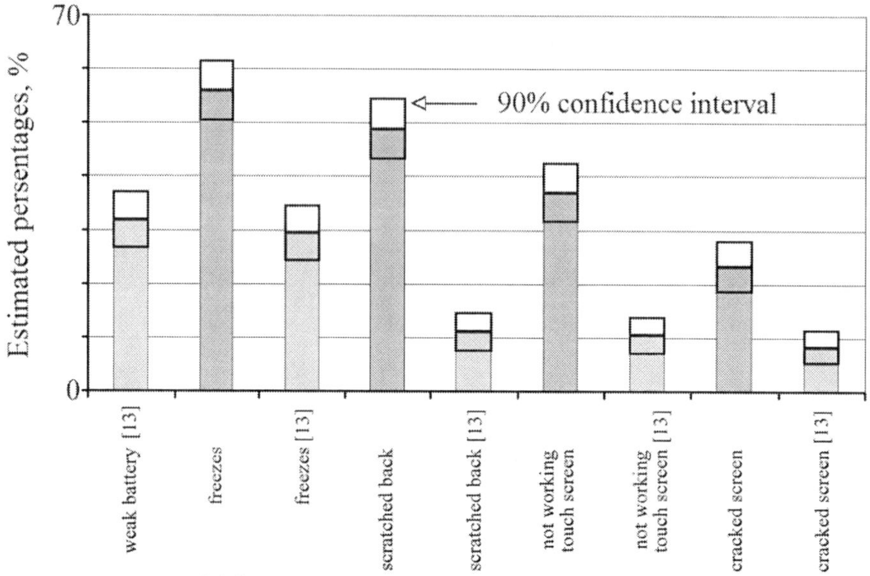

Fig. 4. Percentages of defects reported in questionnaires and internet opinions [13].

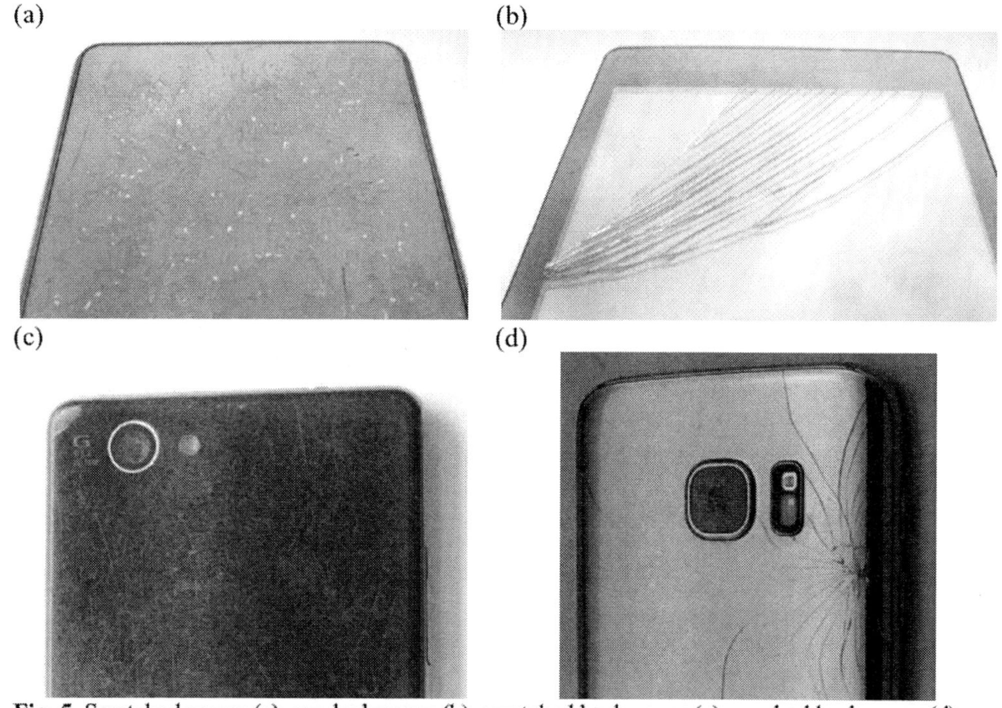

Fig. 5. Scratched screen (a), cracked screen (b), scratched back cover (c), cracked back cover (d).

The described problems can be solved by producers, but they do not ask universities about any solutions [14]. Moreover, reported defects are coupled with planned obsolescence, and are thus are accepted by producers. The average age of a Smartphone is 16.6 months which gives a 90% confidence interval in (15.4; 17.8) months. This leads to a conclusion, that the average lifetime is 33.2 months, which gives 90% confidence interval in (30.7; 35.6) months. On the other hand, the average lifetime declared by students is 29.2 months, which gives a 90% confidence interval in (28.0; 30.4) months. This shows good agreement between these two results. Some of the Smartphones are eliminated by screen defects, next by weak battery and finally by the processor slowing down. Only 34% of students declared that the observed problems are the result of intentional politics. To better illustrate this problem empirical cumulative distribution function is presented (Fig. 6). Very few Smartphones are older than 36 month, which shows that the lifetime is clearly limited. Moreover, Smartphones are removed from the market between the 24th and 36th month of their lives, which is clear from this graph. This issue will be studied in the next article because it is very interesting.

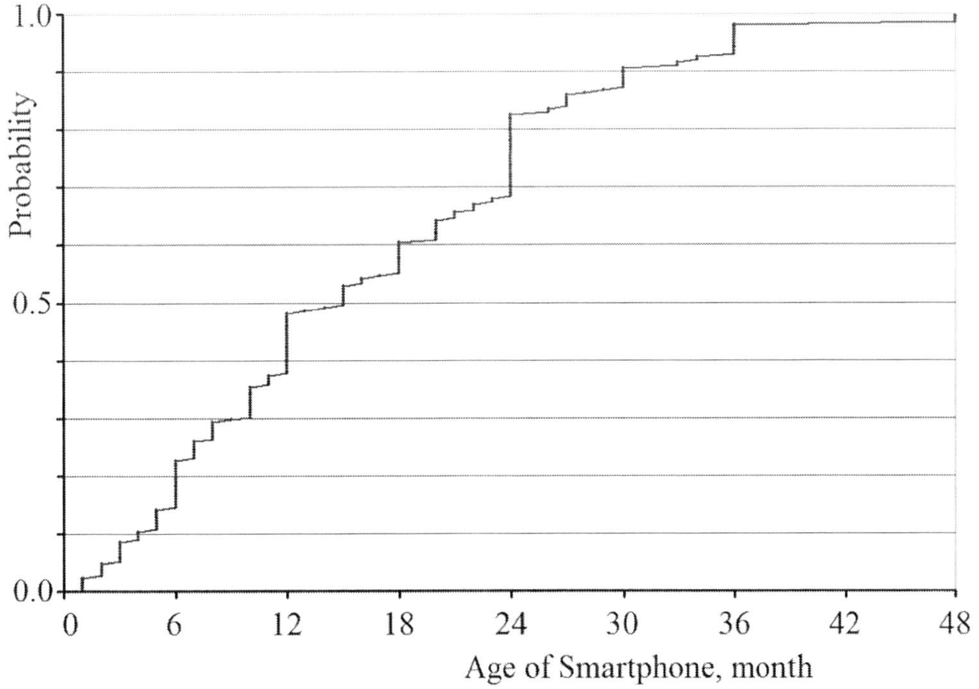

Fig. 6. Empirical cumulative distribution function of age of Smartphones.

3 Conclusions

Defects of Smartphones reported by students and internet users were studied in this article. Internet opinions and surveys provide a similar picture of this issue. Similar hierarchy of problems is observed, but percentages are varied. This is a result of the application of various research methods. Internet opinions provide valid information, which can be used by companies to improve theirs products, but they apply planned obsolescence. Given internet opinions are various, from upright to unremarkable. The obtained results provide an opportunity to estimate the durability of Smartphones in the range from 30.7 to 35.6 months with a probability of 90%. Which shows that smartphones are carefully designed.

References

1. http://www.pewinternet.org/2015/10/08/social-networking-usage-2005-2015/ (1.06.2018)
2. https://www.internetworldstats.com/top20.htm (1.06.2018)
3. S. Edosomwan, S.K. Prakasan, D. Kouame, J. Watson, T. Seymour, J. Appl. Manag. Entrep. **16**, 3 (2011)
4. A. Kaplan, M. Haenlein, Bus. Horiz. **53**, 1 (2010)
5. F. Parveen, N.I. Jaafar, S. Ainin, Manage. Decis. **54**, 9 (2016)
6. A.S. Abrahams, J. Jiao, G. A. Wanga, W. Fanc, Decis. Support. Syst. **54**, 1 (2012)

7. D. Kepren, *Likable Social Media. How to Delight You Customers, Create an Irresistible Brand, and Be Generally Amazing on Facebook (and Other Social Networks)*, McGrawHill, New York. pp 9-10 (2011)
8. A. Paiano, G. Lagioia, A., McGrawHill, Resour. Conserv. Recycl. **73**, pp 162-171 (2013)
9. http://www.komputerswiat.pl/artykuly/redakcyjne/2016/06/najczestsze-usterki-w-smartfonach-czytelnikow-komputer-swiata-raport.aspx (1.06.2018)
10. http://www.dailymail.co.uk/sciencetech/article-2709502/Does-Apple-deliberately-slow-old-models-new-release-Searches-iPhone-slow-spike-ahead-launches.html (1.06.2018)
11. E.M. Huang, K. Yatani, K.N. Truong, J.A. Kientz, S.N. Patel, IEEE Pervasive Comput. **8**, 1 (2008)
12. https://komorkomania.pl/35592,najwieksi-producenci-smartfonow-pomiedzy-samsunga-a-apple-a-wskoczyl-gracz-ktorego-wiekszosc-moze-nie-znac (1.06.2018)
13. mgsm.pl (1.06.2018)
14. K. Smiatacz, J. Niezurawska, L. Niezurawski, K. Samek, *INTED Proceedings*, pp. 8022-8031, 2016

The research of traffic offences by pedestrians on the unregulated crossings with humps

Antonina Korzhova[1], and *Denis* Kapski[2,*]

[1] Scientific-research Center of road traffic, Belarusian National Technical University, Belarus
[2] Automotive and Tractor Faculty, Belarusian National Technical University, Belarus

Abstract. The article presents the results of the study of traffic offenses by pedestrians on the unregulated crossings equipped with humps. The analysis of behavioral features of pedestrians is made taking into account traffic conditions and intensity of transport and pedestrian streams.

1 Introduction

The traffic management is a management of people who the most part of time are uncontrollable and react according to the knowledge and beliefs in traffic [1]. In this regard statement, the social relations in traffic are defining.

The social relations can divide into several groups conditionally: the relations between the state and traffic participants (drivers, pedestrians, passengers) and the relations between traffic participants (the driver – the driver, the driver – the passenger, the driver – the pedestrian, the driver – the cyclist, the pedestrian - the cyclist). In city conditions, about 80% of all accidents in the country are made. And pedestrians are exposed to the greatest danger (50% of accidents with victims) [2].

2 The research of pedestrians behavior in crosswalks zones

The communication between drivers and pedestrians and also the state and pedestrians is defined the Traffic regulations generally. The chapter 4 regulates basic pedestrian's rights and duties [3]. The pedestrian has advantage before drivers when crossing a carriageway if he was convinced that the cross the carriageway is safe (points 16.1 and 17.2).

The definition of safety of a pedestrian exit to the carriageway is subjective. It has no accurate formulation and does not give a fair form of the transition moment on the road from safe to dangerous. According to the Traffic regulations, traffic safety is the condition of traffic providing the minimum probability of emergence of danger in process and the road accident. The danger to the traffic is the change of traffic conditions or technical condition of the vehicle menacing to traffic participants safety and forcing the driver to reduce the speed or to stop.

First of all when carrying out a research social relationship between pedestrians and drivers and behavior of pedestrians upon crossing of the carriageway on crosswalks with "guard", with a hump (as a physical measure of the speed reduction of traffic flow) was determined.

Objects of a research were chosen according to transport and pedestrian loading characteristics and an location of a hump in a crosswalks zone (fig. 1), also the category of the street and quantity of strips for traffic in a research zone (table 1) was taken into account. Basic data have been processed in the program complex "RTF-Road traffic flows" [4].

The using of humps is most effective on sites of accidents concentration (from the point of view of total losses - emergency, economic, ecological) which main reason is exceeding the traffic speed. By the legislation of the Republic of Belarus, the traffic speed in such places is limited up to 20 km/h, like pedestrian and residential zones. As the exception the humps

* Corresponding author: d.kapsky@gmail.com

application is allowed on sites of the accidents concentration on streets and roads of settlements with the general speed restriction of 60 km/h if all other measures for decrease in accident rate either are impossible, or are inefficient. At the same time the daily traffic intensity shouldn't exceed the following values, depending on building type:

- ✓ multystoried, dense, bilateral – 500 buses/days;
- ✓ multystoried, dense, unilateral or multystoried, free, bilateral – 750 buses/days;
- ✓ multystoried, free, bilateral or one-storey, bilateral – 1000 buses/days;
- ✓ one-storey, unilateral – 1250 buses/days.
- ✓ vacant (in a radius up to 50 m) the territory of the settlement – 1500 buses/days.

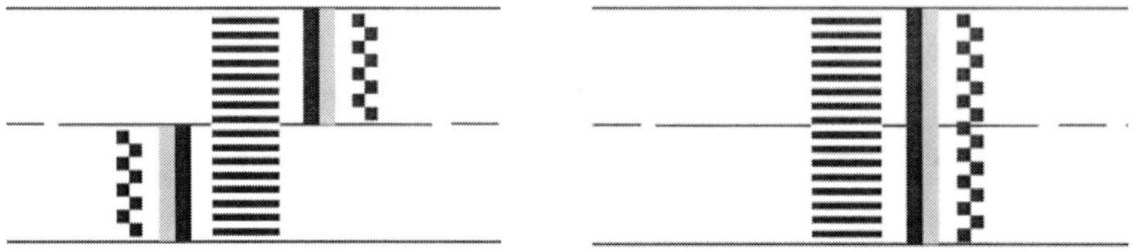

a) b)

Fig. 1. The standard variants of humps allocation in a crosswalk zone (a) – variant 1; – variant 2).

Table 1. The characteristics of the searched pedestrians crossings (Minsk, Belarus).

Наименование объекта	Кол-во полос, шт.	Расположение ИН	ИД транспорта, авт. /час	ИД пешеходов, пеш. /ч	$К_{пн}$
ул. Н. Орды, 11	2/2	вариант 1	912	180	1,08
ул. П. Бровки, 12/1 (односторонняя)	3/0	вариант 2	886	168	1,01
ул. Руссиянова, 48	3/2	вариант 1	392	172	1,16
ул. Одинцова, 65	2/2	вариант 1	1048	112	1,05
ул. Я.Коласа, 12	2/2	вариант 1	1384	220	1,05
ул. Карбышева, 13/2	2/2	вариант 1	404	156	1,12
ул. Кульман, 22	1/1	вариант 2	480	210	1,00
пр. Пушкина, 19 (вдоль школы)	1/1	ИН2	480	128	1,00
ул. Богдановича, 258	1/1	ИН2	900	72	1,16
ул. Одоевского, 107	1/1	вариант 1	2268	210	1,1
ул. Я.Коласа, 18	2/2	вариант 1	1264	296	1,1
ул. Хмельницкого, 47/1	1/1	вариант 1	420	49	1,02

The using of humps on streets with traffic intensity with more than 200th vehicles/hour leads to increase of total losses in traffic due to growth of economic and ecological losses because of braking before a hump and the subsequent dispersal. It should be noted that the using of humps leads to economic and ecological losses round the clock and does not depend on presence of the traffic participants (pedestrians, cyclists, etc.) for whose protection they also are intended [5 - 12].

At a research of behavioral pedestrians reactions upon crossing of the carriageway the following movements were traced and fixed:

- ✓ whether the pedestrian was convinced of danger lack when approaching to crossing;

- ✓ whether the first action is duplicated;
- ✓ other behavioral reactions at stay on the carriageway.

The analysis of the received results has shown that 28% of pedestrians don't comply with paragraph 17.2 of the Traffic regulations, that is aren't convinced of safety of an exit to the carriageway, and 53% don't look around even during crossing (fig. 2).

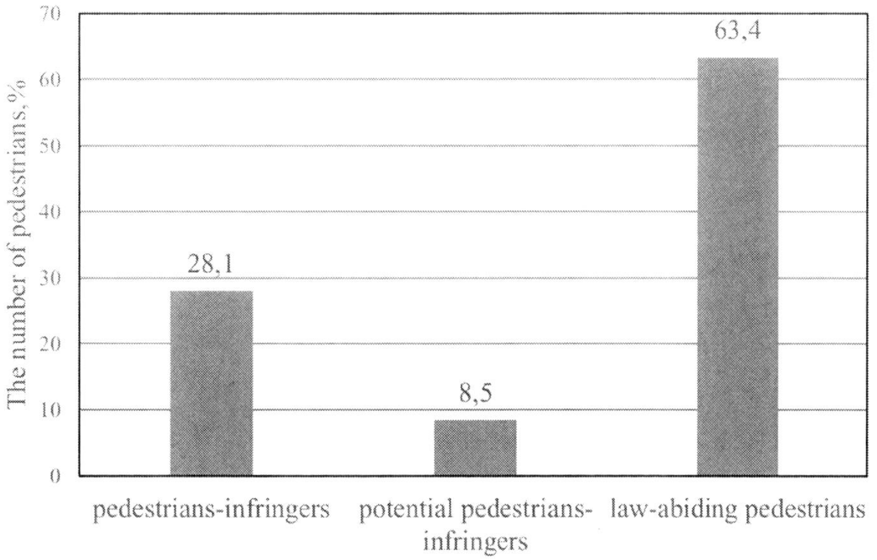

Fig. 2. The distribution on behavioral reactions when crossing the carriageway on crosswalks with humps.

It is necessary to allocate the group of potential pedestrians-infringers who actually can create danger at an exit to the carriageway. Pedestrians of this group look to the right at the right-hand traffic in our country on the sidewalk before the crossing, and to the left look already on the carriageway or do not look absolutely.

The way of crossing of the carriageway by pedestrians was fixed when carrying out a research also (fig. 3).

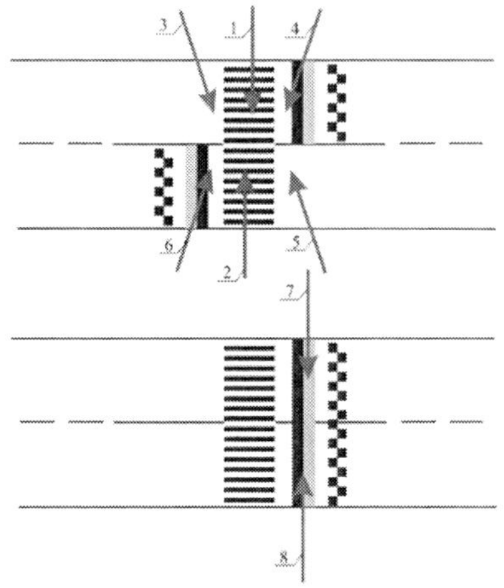

Fig. 3. Ways of crossing by pedestrians of the carriageway in zones of crosswalks with humps.

As a result, we have received that 8% of pedestrians cross the carriageway on trajectories 3 – 6 and 92% - passed, according to the Traffic regulations.

It should be noted that in places with accurate allocation of a approach zone to the crossing (the walking path is separated from the carriageway by a lawn, zones of approach are executed by a tile or are asphalted) all pedestrians passed the carriageway according to the Traffic regulations. In places where the approach zone to the crossing is a part of the sidewalk and it is not designated, for example, by a pedestrian protection, some pedestrians passed the carriageway near the designated crossing.

The calculation of a infringers share has been executed on the basis of the obtained data:

$$\Delta n = (n_o + n_i) / n \quad (1)$$

n_o – the number of the pedestrians who have ignored danger;
n_i – the number of the pedestrians crossing the carriageway not on the crosswalk;
n – the number of measurements.

The annual number of violations on each the research object was determined by a formula:

$$P_o = Q_n * \Delta n * F_y \quad (2)$$

Q_n – the total pedestrians intensity on the crossing, people/h;
Δn – a pedestrians-infringers share;
F_y – annual fund of time, h/year (table 2).

Table 2. The pedestrians-infringers share and annual number of violations of the rules of crossing of the carriageway by pedestrians on research objects (Minsk, Belarus).

The name of an object	The share of the pedestrians-infringers	Annual number of violators, violations/year
st. N. Ordi, 11	0,14	90 720
st. P. Brovki, 12/1	0,49	296 352
st. Russiyanova, 48	0,34	210 528
st. Odincsova, 65	0,23	92 736
st. Y. Kolaca, 12	0,80	633 600
st. Karbisheva, 13/2	0,20	112 320
st. Kulmana, 22	0,06	45 360
av. Pushkina, 19 (near school)	0,31	99200
st. Bogdanovicha, 258	0,46	82800
st. Odoevskogo, 107	0,11	46200
st. Y.Kolasa, 18	0,29	214600
st. Hmelnicskogo, 47/1	0,68	83300

3 Summary

It is necessary to pay attention to the social relations in the organization and management of traffic on the basis of the received results and also tendencies. Mistakes in a regulation of such relations can lead to considerable costs as their influence not only on social losses, but also on emergency, economic and, indirectly, on ecological, is obvious [13].

Within the executed research, it is possible to recommend the following:

- ✓ to accurately regulate installation sites of humps on streets categories and transport and pedestrian loadings;
- ✓ to fulfill precepts of law of the drivers and pedestrians communication. The accurate formulation of safety of a pedestrian exit to the carriageway is necessary. It is possible with constructive proposals on a street road network;
- ✓ for decrease in number of violations (crossing not on the crosswalk) needs to be allocated accurately zones of an exit to the carriageway for pedestrians using structural elements or technical means of the traffic organization.

References

1. Y. Vrubel. *Dangers in traffic* (2013).
2. http://gaiminsk.by/statistika [date of access - 02.08.2016]
3. *The traffic regulations of Republic of Belarus* (2015).
4. D. Kapski, D. Mozalevski, M. Miroshnik, A. Korzhova, V. Kuzmenko, A.Polkhovskaya, E. Kostyukovich. *The certificate No. 222 of 17.09.10 on registration of computer programs in the National center of intellectual property*.
5. Y. Vrubel, D. Kapski, E. Kot. *The definition of losses in the traffic* (2006).
6. D. Kapski, S. Azemsha. *The traffic security audit* (2015).
7. D. Kapski, A. Korzhova, S. Skirkovsky. *The transport in planning of the cities* (2015).
8. P. Preston. *A surge of the humps. Sleeping policemen sprout with the daffodils. But it's more to do with budgets than braking* (2006).
9. http://www.ador.ru/data/files/static/audit_07.pdf. [date of access - 12.07.2016].
10. *Vision Zero Initiative Traffic Safety by Sweden* [electronic resource]. http://www.visionzeroinitiative. com/ru/concept [date of access - 19.03.2016].
11. *The Project // Vision Zero Oregon Neighbors Working for Safer Streets* [electronic resource]. http://www.visionzerooregon.org/the-project/ [date of access - 19.03.2016].
12. J. Whitelegg. *Vision Zero: Adopting a Target of Zero for Road Traffic Fatalities and Serious Injuries* (2006).
13. Y. Vrubel. *Losses in traffic* (2003).

Possibilities of the electrochemical drilling process's (ECM) vibroacoustic analysis

Adam Troszyński[1,*] , and *Michał* Liss[1]

[1]University of Technology and Life Sciences, Faculty of Mechanical Engineering, Al. Prof. S. Kaliskiego 7, 85-796 Bydgoszcz, Poland

Summary. In the course of metals treatment, there occurs a direct interaction of a tool with the processed material. It is connected with occurrence of different physio-mechanical processes significantly influencing the material's properties, among the others such as the surface's roughness. An attempt to estimate the shaped material's surface in the course of the ECM process based on the measurements of the high-potential plate's oscillations, is presented in the article.

1 Introduction

Coming in of many constructional materials of special properties among which we may single out tool steels, self-bonded carbides, composite materials and titanium alloys, is the result of dynamic development of the material engineering. Newly developed materials are more and more often used on responsible constructional elements, as they are characterized by good mechanical properties, among the others by big hardness, high resistance to abrasion and corrosion. A common feature characterizing the above listed materials is their exceptionally difficult workability [1].

Electrochemical machining ECM is one of the processing methods used for shaping elements made of constructional materials irrespective of their mechanical properties. Thanks to these properties, it is widely used among the others in the engine industry, aeronautical industry, as well as in biology, chemistry, electronics and medicine [2,3,4].

The quality of the conducted electrochemical machining (ECM) is most often assessed based on the shaped sample's surface, verification of which can be realized only following completion of the whole process, conducting the roughness ' measurement. No possibilities to monitor of the machining ECM process in the course of its performance, results in generation of losses in the situation, when the machining process brings no desired effect, in the form of for ex.: expected roughness of the shaped good's surface.

This article presents the attempt to assess the quality of the electrochemical drilling process ECM, one of the more important parameter of which is the obtained shaped surface's roughness . So, an attempt was made to connect the characteristic parameters of vibrations' signals obtained from a high-potential electrode with roughness of a shaped sample's surface.

2 Electrocheical machining ECM

Contactless electrochemical machining also called profiling, is one of many popular methods of a surface's shaping. Removal of the surface material's layer requires connecting of a positive electric current source's pole to a machined object (PO), while to a high-potential electrode (ER) of a negative electric current's pole. The interelectrode space is powered with electrolyte, thanks to what there take place processes of an electric charge's exchange, the consequence of what is dissolution of a machined object giving it a shape of a defined precision. In the interelectrode slot we deal with the mixture of electrolyte, hydrogen particles

* Corresponding author: adatro000@utp.edu.pl

and products formed as a result of electrochemical dissolution, which significantly decide on the ECM machining precision [5,6].

The fact, that the precision of the machined surface is the higher the smaller is the distance between the high-potential plate (ER) and the machined object (PO) is a particular feature in the electrochemical machining ECM. In case of precise machining, applying of the interelectrode slots' thicknesses below 0,1 – 0,05 mm is necessary.

Such a small a slot's thickness presents a problem with running a stable drilling ECM process. It is caused by many factors, among the others by liberation of gas phase in electrolyte, heating up of electrolyte by the flow of electric current through a center, and low electrolyte's flow speed. The above listed factors make the intensive removal of dissolution products and heat from the interelectrode space impossible to a considerable degree. It increases the risk of occurrence of the critical ECM state and electrical discharges damaging the high-potential plate and a machined object [7].

Fig. 1. Electromechanical machining ECM stand.

The electrochemical drilling machining was conducted at the following input parameters:

- interelectrode voltage U – 14 V,
- original slot h – 0,1 mm,
- feed motion speed PO – 0,1 mm · mm^{-1},
- electrolyte temperature – 293 K,
- electrolyte type – $NaNO_3$ 15%,
- machining time – 300 s,
- stream of the electrolyte Q volume – 0,004 $m^3 \cdot mm^{-1}$,
- frequency of transverse vibrations – 10 Hz,
- transverse vibrations' amplitude – 0,2 mm.

Application of a complex movement of a high-potential plate in the feed plane is one of many methods enabling operation at very small interelectrode slots. Periodically changing the thickness of the interelectrode slot makes it possible to correct the machining precision at a minimum slot, while in case of maximum slot – good electrolyte's exchange. Appropriately selected kinematics of a high-potential plate's movement as well as frequency of its vibrations, influences the machine surface's precision [8].

The studies were conducted on 10 samples having the diameter 35 mm made of tool steel NC6 in an improved condition having the hardness 55 HRC. In figure 2-a there is presented a sample prior to drilling, while in figure 2-b a sample following its electrochemical drilling.

a) b)

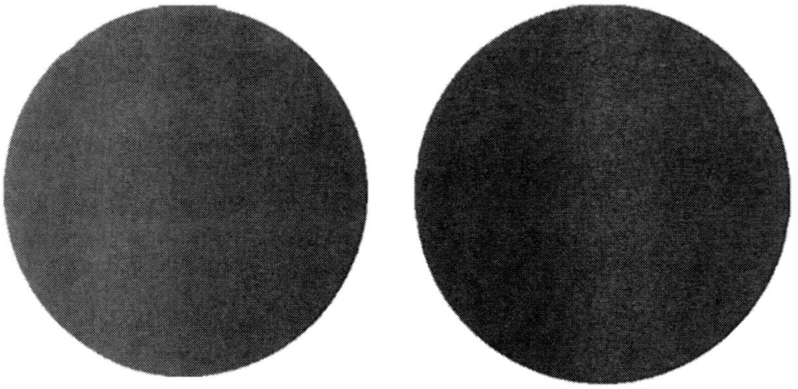

Fig. 2. A sample used in ECM tests: a) a sample prior to its ECM machining ECM, b) sample after its ECM machining.

3 Vibroacoustic analysis in the course of the ECM machining

Apart from application of the impulse current, introduction of a high-potential plate's vibration is one of the methods increasing the machining's precision. The periodically changing thickness of an interelectrode slot, makes it possible to use smaller thicknesses of an interelectrode slot as compared to machining without vibrations, as the periodically occurring bigger thicknesses of the interelectrode slot create good conditions of an electrolyte's exchange and carrying the dissolution products off. From the analysis of the literature it results, that introduction of a high-potential plate's vibrations increases the machining precision and its reliability [9,10,11] and in case of high vibrations' frequencies, the dynamic effects occurring in the interelectrode slot become important. So, monitoring of vibroacoustic parameters of high-potential plate occurring during material's processing process is necessary.

In the conducted studies, the forced vibrations of a high-potential plate were registered in one axis with the use of piezoelectric sensor of PCB Piezoelectronics. The observed frequencies' band was determined as 100 [Hz]. According to the Shanon theory, there was assumed the sampling frequency amounting to 200 [Hz]. Modulations of the registered vibration' signal most often result from operation of servomotors causing the electrode's movement and the flow of the electrolytic agent, what has been determined by conducting appropriate analysis of the amplitude-frequency characteristics. However, an appropriately selected sensor and the place of its assembly contributed to the fact, that these phenomena did not significantly influence the form of the determined characteristics FFT.

Analysis of the electrochemical drilling process' vibration spectra came down to finding the frequency of the high-potential plate's forcing together with subsequent harmonious ones and observing the values of the vibrations' amplitude (in the form of acceleration [m/s^2]). The selected amplitude-frequency characteristics is presented in figure 3.

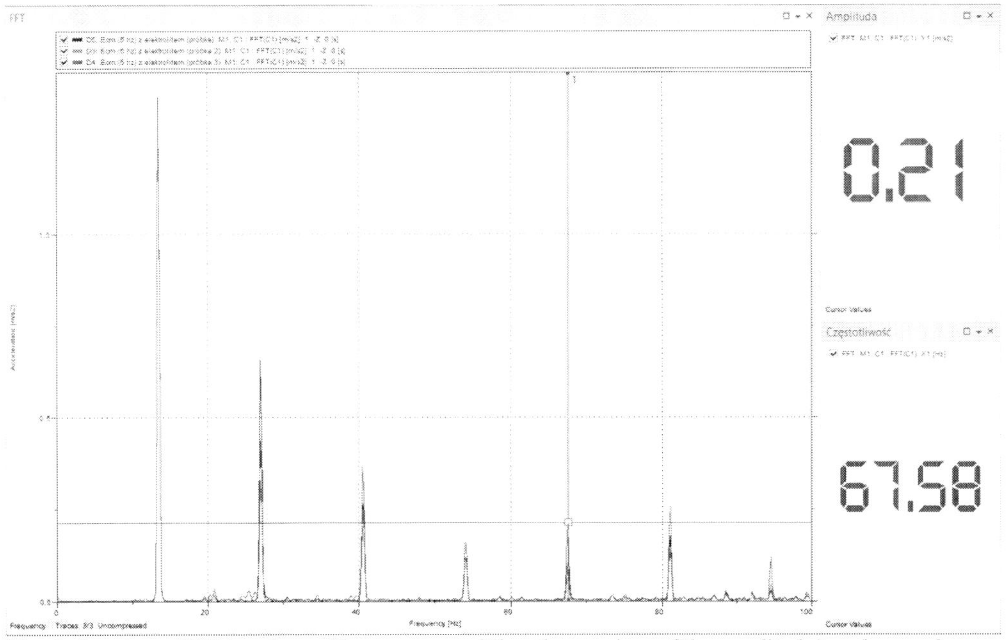

Fig. 3. Spectrum FFT together with a cursor enabling the preview of the amplitude's values and vibrations' frequency.

Dynamic identification of the ECM process was conducted also based on the analysis of the registered time courses. The selected time courses are presented in figure 4.

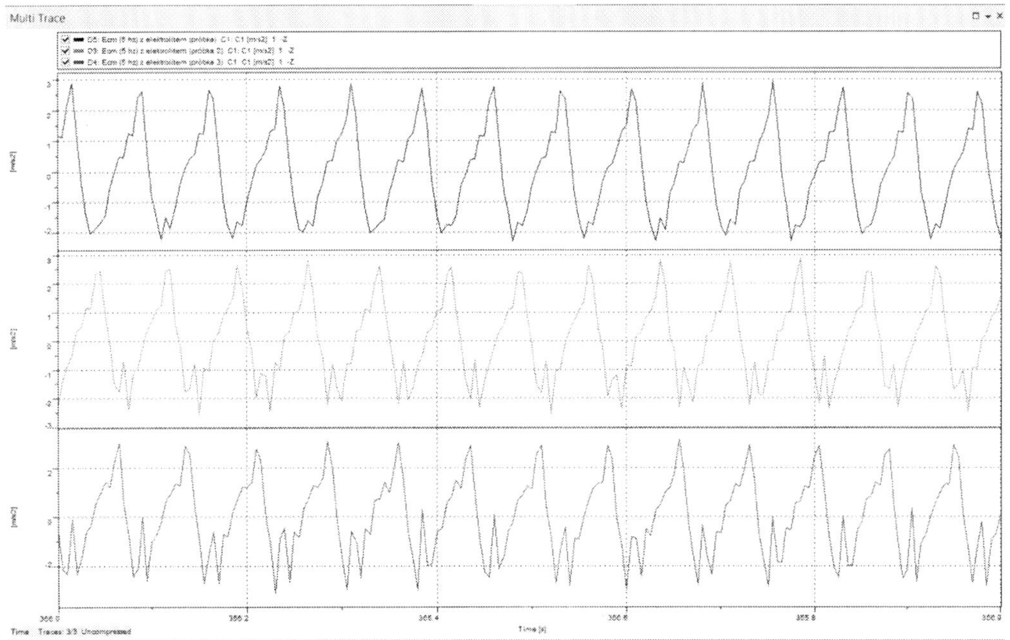

Fig. 4. Selected section of time courses registered during the electrochemical drilling process ECM.

4 Assessment of the surface after the ECM process

The surface obtained after the electrochemical drilling, was inspected with respect to the surface roughness. The Ra parameters were used to assess the tested samples, which are the mean arithmetic deviation of the roughness' profile, with the use of which we obtain information on quality of the machined surface. Measurement of the tested surface was performed on the surface roughness meter make Mahr, in accordance with the direction presented in the figure (fig. 5).

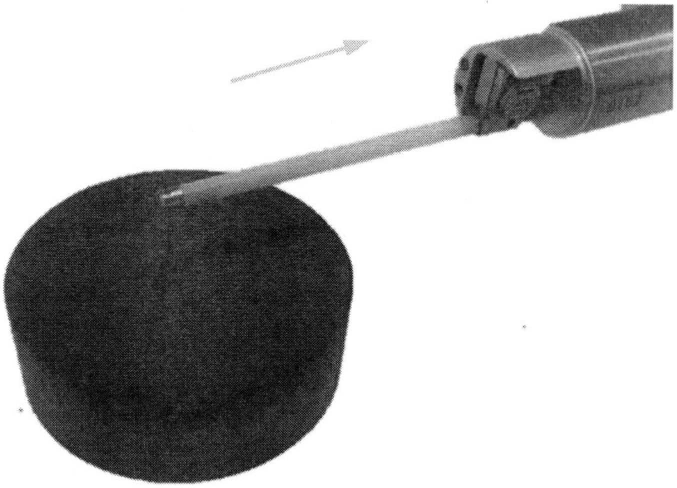

Fig. 5. Measurement of roughness following the ECM machining.

Roughness of the studied samples was measured in three places in order to check the roughness' distribution along the whole sample's surface. Figure 6 presents the places in which the above- mentioned measurement was conducted.

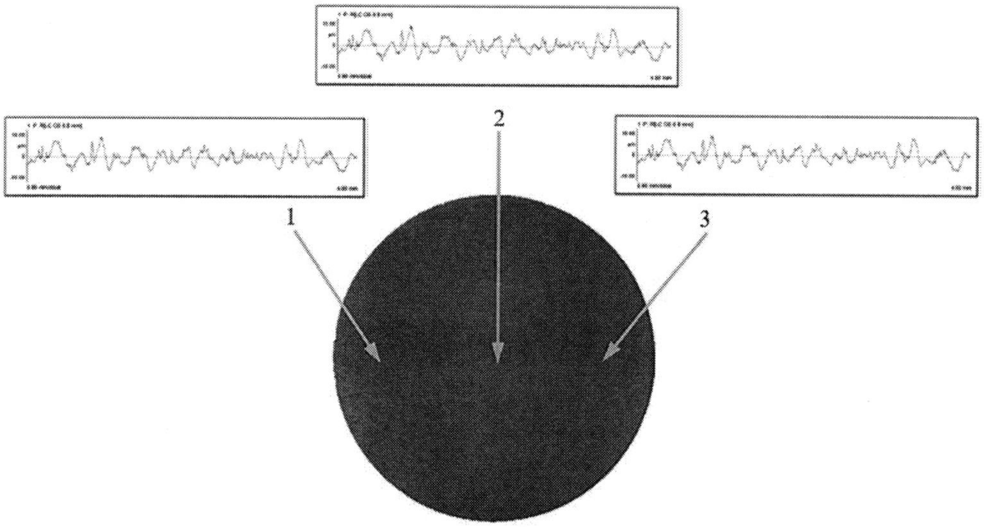

Fig. 6. Places of the coarseness measurement on the studied sample.

5 Results and their analysis

The obtained results from the roughness measurements of the studied samples' surfaces are presented in table 1. In this table there are also presented the results of measurement from individual places of the examined sample and the mean value for the considered samples has been determined. Based on the determined mean roughness values Ra it has been defines, that the highest quality of the surface's performance with the electrochemical drilling method concerns the samples: 1,6 and 8. In the remaining samples lower quality of the surface's performance was observed.

Table 1. List of the mean values of the measured roughness parameter Ra.

	Place of measurement 1	Place of measurement 2	Place of measurement 3	Mean value Ra [μm]
Sample 1	2,38	2,76	2,09	2,41
Sample 2	3,09	2,07	3,09	2,75
Sample 3	2,51	2,52	3,19	2,74
Sample 4	2,78	2,63	2,89	2,77
Sample 5	2,91	2,45	3,10	2,82
Sample 6	2,45	2,15	2,65	2,42
Sample 7	2,65	2,55	2,79	2,66
Sample 8	2,54	2,31	2,65	2,50
Sample 9	2,75	2,49	2,91	2,72
Sample 10	3,02	2,61	2,81	2,81

Having determined the quality of the shaped surfaces in ten samples, subsequently there were analyzed the registered time courses. Analysis of the time courses showed distinct changes in the course of the process of the electrochemical drilling's duration between samples of the highest surface's quality and the samples in the case of which the surface's quality was lower.

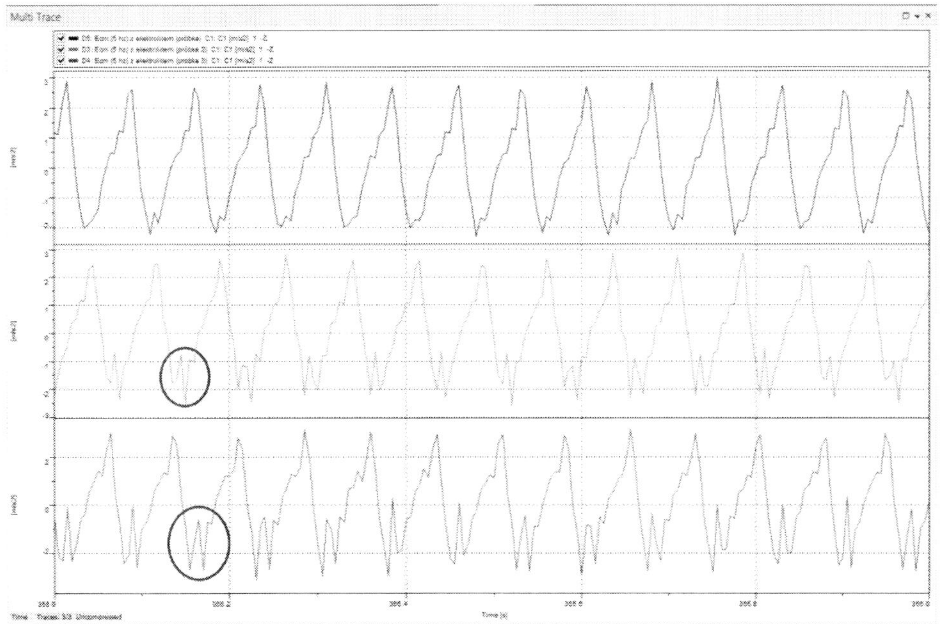

Fig. 7. Presentation of the time courses from the electrochemical drilling process for the sample 1 (black colour), sample 2 (orange colour) and the sample 5 (green colour).

Analysis of the obtained time courses from the samples representing three different surface's qualities point at a similar impulse character of a high-potential plate in the course of material's machining. However, additionally it was also observed, that in case of the samples in the case of which there occurred a lower quality of the shaped surface, in the time courses there occurred changes of the vibrations' signals in the form marked in figure 7. These changes occur only in case of samples of low quality of the performed surface, what causes, that this tendency may be identified just with that phenomenon.

The observed effect was tried to be confirmed by making analysis of individual amplitude-frequency courses. The registered forcing frequencies' values, their harmonious and vibrations' amplitudes, are listed in the table 2.

Table 2. List of frequency amplitudes for individual samples subject to electrochemical drilling machining ECM

	Amplitude [m/s2]						
Frequency [Hz]	13,48	26,95	40,43	53,91	67,58	81,05	94,53
Sample 1	1,37	0,66	0,26	0,12	0,11	0,18	0,03
Sample 2	1,34	0,33	0,20	0,11	0,07	0,14	0,04
Sample 3	1,37	0,43	0,37	0,16	0,21	0,26	0,12
Sample 4	1,32	0,56	0,23	0,10	0,08	0,18	0,02
Sample 5	1,28	0,47	0,28	0,13	0,07	0,17	0,11
Sample 6	1,38	0,62	0,28	0,14	0,09	0,22	0,04
Sample 7	1,35	0,65	0,31	0,13	0,10	0,17	0,01
Sample 8	1,36	0,60	0,27	0,18	0,15	0,18	0,02
Sample 9	1,33	0,57	0,39	0,15	0,20	0,23	0,13
Sample 10	1,31	0,54	0,33	0,12	0,03	0,22	0,13

The values presented in table 2 are also presented in a graphical form in figure 8.

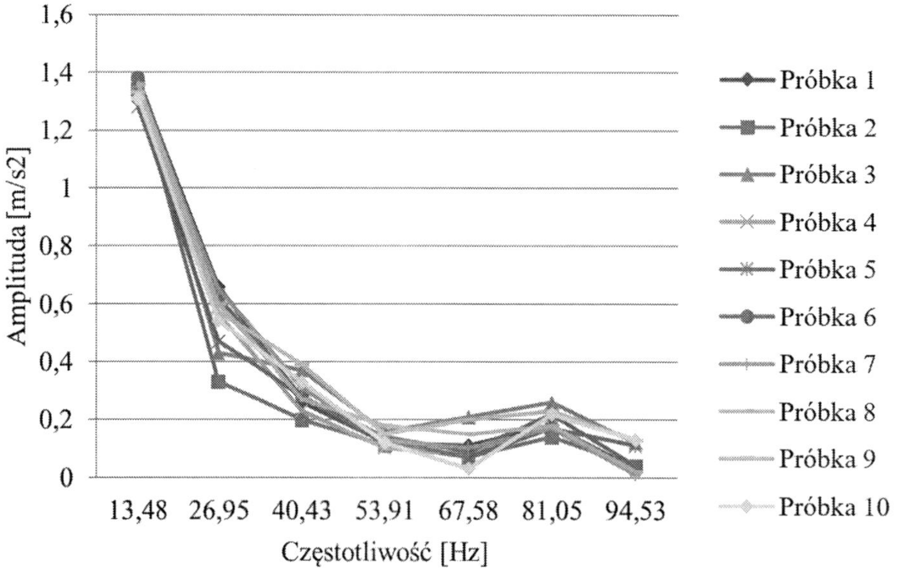

Fig. 8. Diagram of the oscillation amplitudes' values depending on the frequency of the high potential plate's input functions.

Unfortunately, the analysis of the FFT spectra ambiguously shows, that the quality of the shaped sample's surface may be confirmed based on the analysis of only the values of oscillations' amplitude assigned to forcing frequencies.

6 Summary

Dynamic identification of electromechanical machining (ECM) was based on the analysis of the time and frequency courses. The presented results clearly show, that monitoring of such a machining process is better based on the analysis of individual time courses registering the high-potential plate's movement. The changes taking place at the time of the electromechanical drilling process may be observed much better on time courses, what is pointed out in this publication. An individual dependence between the machined surface's quality and the changes in the time characteristics has been observed among the others in these courses' analysis. Additionally, the vibroacoustic analysis has confirmed the assumed high-potential plate vibrations' amplitude obtained with the use of a cam powered with belt transmission. One should have in mind the fact, that the vibroacoustic parameters' values often depend on many sources' operation, of different frequencies and amplitudes. The obtained results of a stand's vibration and the presented time courses' and amplitude-frequency diagrams', are the preliminary results and require further studies.

LITERATURE

1. S. Zaborowski, Elektrochemiczna intensyfikacja obróbki ściernej materiałów trudno obrabialnych, Prace Instytutu Technologii Maszyn i Automatyzacji Politechniki Wrocławskiej, Monografia, Oficyna Wydawnicza Politechniki Wrocławskiej, Wrocław (2000)
2. P. Frankowski, W. Mielnicki, Kształtowanie elektrochemiczne łopatki sekwencją procesów ECM – PECM, Inżynieria Maszyn, R. 16, (2011)
3. P. G. Pott, Modern electrochemical machining in practice, Proc. ISEM-9, s. 146-150, Nagoya (1989)
4. A. Ruszaj, Application of ECM machining in the Polish industry. Mat. SECM-88, s. 247-283, Lubniewice (1988)
5. L. Dąbrowski, Podstawy komputerowej symulacji kształtowania elektrochemicznego, Pr. Nauk., Mechanika 154, Wyd. Politechniki Warszawskiej, Warszawa (1992)
6. J. Kozak, Kształtowanie powierzchni obróbką elektrochemiczną bezstykową, Pr. Nauk., Mechanika 41, Wyd. Politechniki Warszawskiej, Warszawa (1976)
7. K. Łubkowski, Stany krytyczne w obróbce elektrochemicznej, Pr. Nauk., Mechanika 163, Oficyna Wydawnicza Politechniki Warszawskiej, Warszawa (1996)
8. T. Paczkowski, Symulacja komputerowa obróbki elektrochemicznej powierzchni krzywoliniowych elektrodą roboczą o złożonym ruchu translacyjnym, Wydawnictwo Uczelniane UTP, Bydgoszcz (2012)
9. B. Bhattacharyya, M. Malapati, J. Munda, A. Sarkar, Influence of tool vibration on machining performance in electrochemical micromachining of copper, International J. of Machine Tools & Manufacture 47, 335-342, Kolkata (2007)
10. M. S. Hewidy, S. J. Ebeid, T. A. El-Taweel, A. H. Youssef, Modelling the performance of ECM assisted by low frequency vibrations, J. of Materials Processing Technology 189, 466-472, Egypt (2007)
11. J. Kozak, Analiza procesu obróbki elektrochemicznej elektrodą drgającą, Mat. konf. EM'97, 204-212, Bydgoszcz – Golub – Dobrzyń (1997)

Vibration diagnostics of concrete block

Mariusz Żółtowski[1,*1], *Michał* Liss[2], *Jozef* Melcer[3]

[1] UTP University of Science and Technology in Bydgoszcz, Faculty of Management, Poland
[2] UTP University of Science and Technology in Bydgoszcz, Faculty of Mechanical Engineering, Poland
[3] University of Zilina, Faculty of Civil Engineering, Slovakia

Abstract. Presented study indicates the possibility of using complex estimators of vibrational energy flow used in quality testing of concrete blocks, as part of a dedicated research methodology and detailed methodologies developed in many previous studies by the author. Theoretical analysis and practical verification of the sensitivity testing measures of complex vibration processes indicate the wide possibilities of their applications. Recognizing the need to improve the testing methods of building structures for assessing their degradation status - this paper presents significant results of the research procedure in the field of effectiveness verification of proposed mutual measures in wall element bench tests.

1 Introduction

Identification in building constructions becomes a method of assessing changes in the structure operational state, and the modal analysis methods with related vibration estimators well reflects the wall elements destruction. Searching for mapping the state of building material with the parameters of their vibration models, in bench tests and on real objects, made it possible to assess the similarity of models, the accuracy of decisions and the effectiveness of the used test method. Assessment of the building structures dynamic state with the help of vibration energy propagation measures requires the association of the structure features to assessed object with the set of measures and assessments of the initial processes.

Dedicated research procedure algorithm during vibrational energy flow analysis in the study of machines and building structures includes [9]:
- formulating a research problem;
- acquisition and processing of vibrational energy measures;
- qualitative optimization of vibration symptoms;
- models of binding measures with the degradation state;
- inference.

In general, the methodology for degradation state testing includes detailed procedures for the development of a data acquisition system, their processing and statistical inference - repeatable in many different issues of vibration signals testing.

Measurements of vibrations along with their physical interpretation (the proposed model of the vibration signal passing through the material) constitute the basis for acquiring information in the study of selected masonry components degradation. Basic research in this work conducted to confirm the assumptions and conditions of specific destructive tests verify the validity of the adopted models and based on the recorded vibration signals characterize the dynamic properties of the tested materials. Elements of modelling theory and simulation research (necessary even in the properties recognition) allow to determine whether any research object (building material) in the simplified modelled with the

* Corresponding author: mazolto@utp.edu.pl

parameters m, k, c - a system with one degree of freedom - can be treated as a rigid system linear [7].

Considerations of this work indicate that the properties of such modelled object elements can be analysed by describing them with differential equations (solving the own problem) or the same information can be obtained by measuring vibrations (a, v, x - parameters of the vibration process) and on the basis procedure of determination vibration mutual measures [1,2,9].

The detailed discussion of the vibration energy flow analysis and modal analysis methods indicates the development of software and measurement stations for the implementation of practically difficult measurement techniques in the non-dismantling assessment of entire building structures destruction status, or only in the research of structure elements, e.g. wall elements or segments.

Described in a definition and from the practical application of the vibration process measure (due to the process of averaging during their determination called estimators) indicate a large range of activities necessary during the acquisition and processing of information, activities programmed during the computer development and presentation of measurement data.

2 Vibrations in degradation tests

When researching real systems (constructions, structures, machines, devices), the basic problem is to determine the value of vibroacoustic energy stored, dispersed and transferred by individual elements of these systems. Knowledge of these quantities serves to assess material effort, fatigue, diagnostic tests and noise prediction, and facilitates the design of system components (eg vibroinsulation).

The development of measurement methods, especially measurements of energy quantities, significantly expanded the possibilities of sound radiation testing by structures and allowed to calculate the acoustic power radiated to the far field based on measurements in the near field. Methods of quantitative and qualitative research of vibroacoustic energy propagation in space with complex boundary areas have been developed. It related to the quantitative assessment of vibroacoustic energy accumulated in the construction elements and the assessment of energy radiated by these elements, as well as energy carried by various routes.

Fig. 1 presents basic (exemplary) waveforms of measured amplitude in time for subsequent measurements, which are the basis for further program studies for distinguishing information about state changes.

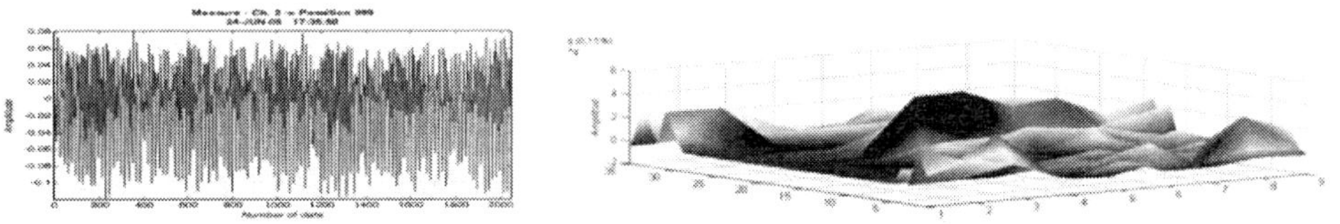

Fig. 1. Measurements of amplitude changes in time for subsequent vibration measurements.

The basic result of the time course measurement and its spectral analysis are shown in Fig.2. These are the main types of signal transformation used later in this work, because all other vibration signal characteristics are based on the above courses.

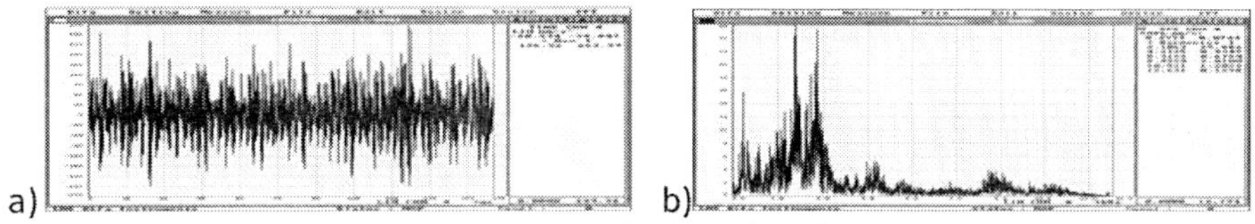

Fig. 2. Time course a). and the spectrum of the vibration signal b).

Vibration processes are mostly accidental processes, that's why they are further processed in the processor according to different algorithms to obtain different types of signal estimators. These estimators, as opposed to signals, are already determined and depending on their type they reflect the various characteristics and information properties of the signals.

Estimators, as numerical measures of the vibration process, are defined in three categories of signal description: time, frequencies and amplitudes [1,9,12] and as determinate quantities reflecting the state of the evaluated material can be compared with the state standards of the tested materials.

The available literature [1,4,7,8,9,11] presents the definitions and properties of various vibration estimators, distinguishing personal estimates - concerning one signal and mutual estimators - regarding two signals.

New tools in this area of research concern the possibilities of using the methodology to study the vibrational energy flow, as well as the modern acquisition and processing of vibration measures - to assess the quality of masonry elements, which form the basis for consideration in this work. In practical applications, they allow for a better understanding of the behavior of complex structures, optimization in the design process and assessment of dangerous states. In the latter area the content of the subject is placed, looking for measures to assess the state of destruction of new, long-term wall elements, and often with an unknown state of destruction and the wall's safety factor of many buildings.

The methodology for testing the vibrational energy flow used in the considerations of this work has been adapted from the dynamic field of mechanical structures. Its usefulness is greater for homogeneous structures of steel structures than for building materials such as masonry elements, where each test result is related to the structure, shape and physical state of the sample. Therefore, it is feared that the heterogeneity and coarse grain structure of concrete and ceramics may be an obstacle to the effectiveness of very precise measuring techniques. The task of this work is to assess the usefulness of measures of vibrational energy flow in the diagnosis of homogeneous masonry elements (bricks, blocks, blocks) and to develop recommendations for testing more heterogeneous masonry fragments.

The tasks of this work cover in detail the following issues: modeling and description of the process of destruction of masonry elements, the use of vibrations in the description of changes in their dynamic state; implementation of destructive testing of wall elements and segments, application of a software system for acquisition and processing of measurement data and statistical inference (SYMPTOMS GENERATE, VIOMA, OPTIMUM, SVD programs). The method of solving the main job includes methods, measures and procedures for: archiving the results of measurements of building materials programmed destructions, their processing aimed at selecting the best diagnostic measures and determining the relationship between the state of load (destruction) of the material tested and the values of measures.

The obtained results of the work carried out indicate the vibration measures characterizing the destruction of the tested wall elements, determined by a bench test method, based on measurements and processing of the vibration signal. They also assess wall elements destruction model and creates the basis for the construction of simple diagnostic procedures for the needs of the construction products control system placed on the market and for the assessment of the safety class of masonry walls.

These measures include [3,5,8,10]:

- transmittance: the dynamics of the system in the frequency domain can be described by means of the equation

$$X(j\omega) = H(j\omega)F(j\omega) \qquad (1)$$

where: $X(j\omega)$ is the vector of the system response spectra, $F(j\omega)$ is the vector of forcing forces spectra, $H(j\omega)$ is the frequency response matrix;

- FRF (spectral transition function): this is a frequency function that describes the relationship between the force of force F(t) and the acceleration of vibrations a(t) as the response signal. The frequency response function (FRF) can be described as the quotient of the Fourier transform value of the force that forces the oscillation $F(\omega)$ by the Fourier transform of the response signal $A(\omega)$:

$$FRF = \frac{\sum F(\omega)}{\sum A(\omega)} \tag{2}$$

The FRF function is a complex frequency function, the module of which presents the amplitude-frequency characteristic, whereas the argument is the angle of the phase shift between the forcing force and the response;

- coherence function: is a measure of the consistency of two oscillation processes $x(t)$ and $y(t)$) and is defined as follows [1,4,7]:

$$\gamma_{xy}^2(f) \equiv \frac{|G_{xy}(f)|^2}{G_{xx}(f)G_{yy}(f)} \tag{3}$$

Regardless the transfer paths character of the transition paths, the coherence function always assumes a value of 1, if only the signals x(t) and y(t) come from the same source. For more than one source, the non-negative coherence function is always less than one (from 0 to 1). The coherence function as a measure of the coherence of two signals has good diagnostic properties, because the appearance of a new fault signal violates the coherence of the previous ones, so that the coherence function decreases.

The new research tools developed in this work mainly concern the assessment of the applicability of the proposed vibration measures, modern vibration processing and statistical cause and effect reasoning in the study of selected building materials.

3 Research methodology

3.1 Research object

Identification tests as well as basic tests for randomly selected 6 concrete blocks were carried out on laboratory stands, deliberately adapted and suitably equipped. Providing a limited number of degrees of freedom is ensured by suspending the tested blocks on a thin line. Signal reception sensors were mounted on wax, and the directions of extortion were taken as shown in Fig. 3.

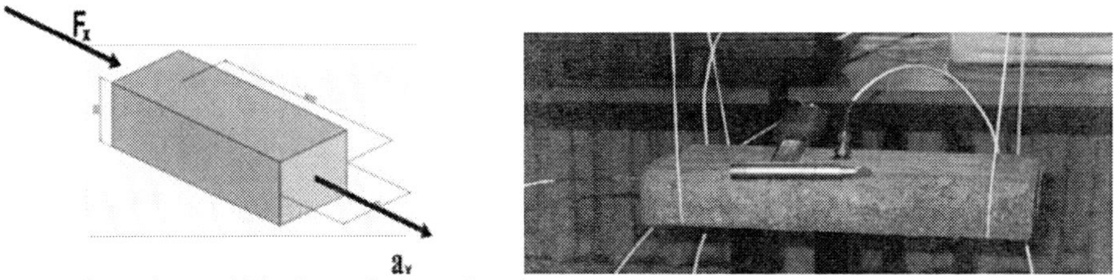

Fig. 3. Mounting of tested blocks and piezoelectric sensors.

The LMS TEST.XPRESS state-of-the-art measuring equipment was used to measure the time of force and response of the system as well as the determination of the H(f), FRF and coherence functions. This software enables to analyze the vibrational energy flow in elements and any other building constructions.

The LMS TestXpress software allows for the recording of a forced signal and responses in the form of time courses and can generate mutual functions and their numerical values (area under the curve) used in the study of this work.

3.2 Research position

An interesting case of the conducted research of this work were the results obtained for a broken concrete block, with a distinct crack in the tested element, as shown in Fig.4.

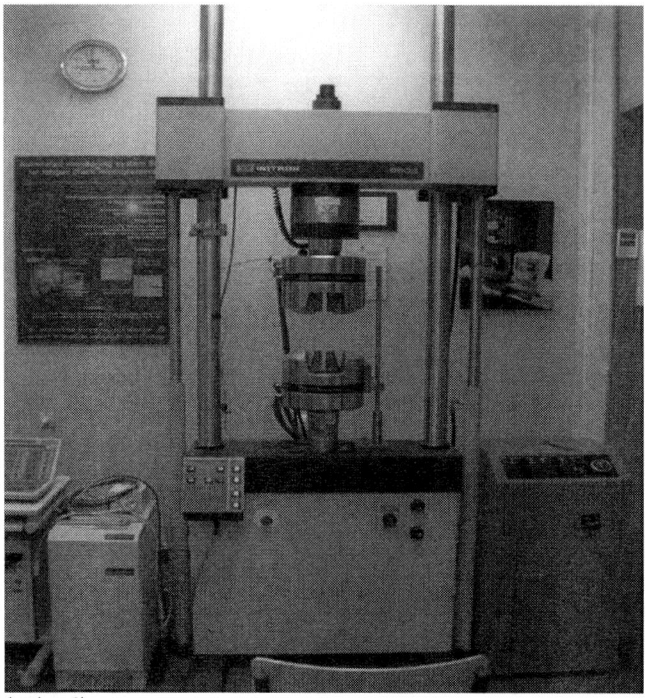

Fig. 4. INSTRON hydraulic press.

In the tests on the INSTRON 8502 testing machine, the range of the limit loads for the tested concrete blocks was determined for destruction, which allowed to determine the test loads for detailed tests. The destructive tests were carried out on 6 concrete blocks and after averaging the results, the values of loads for basic research were proposed.

3.3 Findings

Time passes of the force of force (its stability is ensured by the APB-200 system) and the corresponding vibration acceleration courses for the group of 6 concrete blocks tested are shown, for example, in the coordinate system: acceleration - time in Fig. 5 for elements without a destructive load. Similar analyzes were recorded for further processing for vibration-inducing forces, as well as for the loads highlighted on the Instron machine.

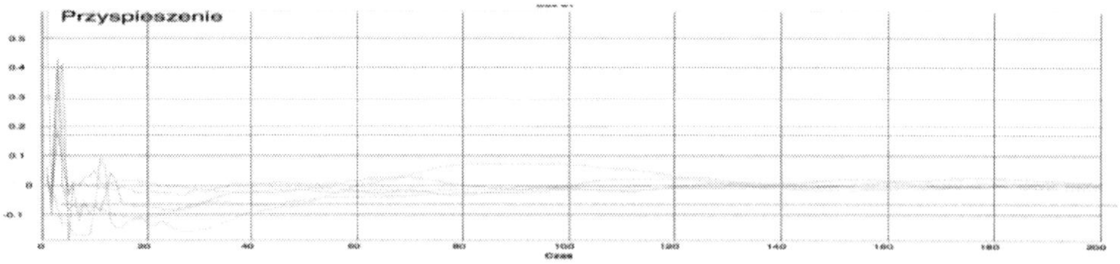

Fig. 5. Amplitudes of vibration acceleration of the tested block for various F0x loads.

The numerical values of the vibration signal measures for the tested concrete block, with various strength loads recommended in the tests of this work, are presented in Table 1.

Table 1. Measurement data of pre-selected measures.

Wartość siły	FRFr	FRFu	H(f)r	H(f)u	coher	delta	Urms	w1	w2
0	1824,87	40,32	20,8	1,57	179,48	0,916	0,011	4	40
50	1115,24	-1502,55	8,56	22,56	107,01	0,992	0,011	2	40
100	326,82	-340,92	-3,01	24,57	92,48	0,969	0,011	2	30
150	-1084,96	-397,68	-14,48	21,46	88,45	0,959	0,011	2	50

In the following drawings, the modes of forced force and the corresponding amplitudes of vibration accelerations for the tested wall elements are presented with the loads tested F1, F2, F3, respectively, specified in table 1. It is worth noting that the tested concrete blocks were loaded during the force tests: 50, 100, 150kN.

Considering all the conducted sensitivity analyzes, using different statistical methods used in this work, it is possible to further propose as the main following measures of concrete block destruction state:

FRF_r, FRF_u, $H(f)_r$, $H(f)_u$, U_{RMS}, coher, ω_2.

Because of the tests carried out for the indicated loads along the Z axis, the results presented in Fig. 6 were obtained.

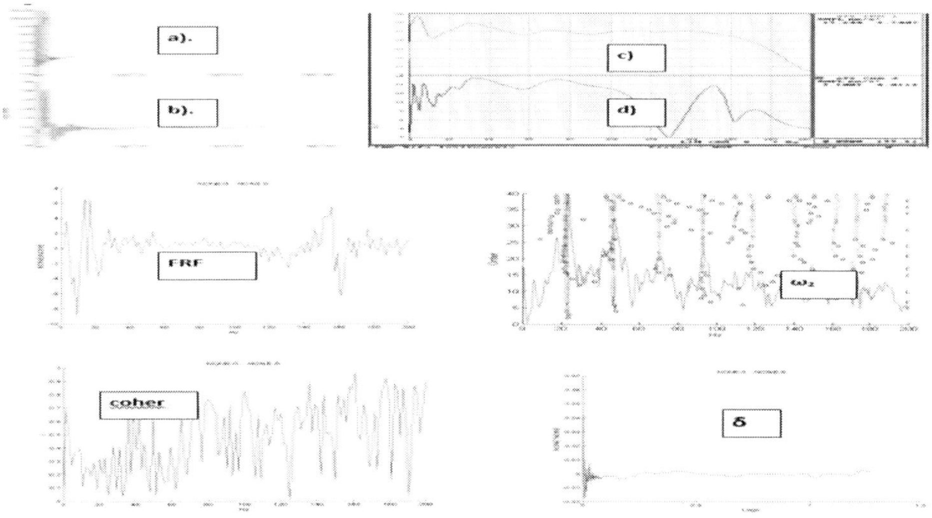

Fig. 6. X-axis test results (no load) - cracked block.

Fig. 6 shows the courses of force (a), its vibration acceleration amplitude (b), vibration spectrum of these signals (c, d) and exemplary measures of destruction: FRF function, values of natural frequency from the stabilization diagram with the characteristic frequency ω_2, coherence function (coher) and decay indicator of the mutual correlation function δ. The differences in the obtained measures for the broken block in relation to the undamaged one are clearly visible, which is mainly manifested by the appearance of the second natural frequency around 150 Hz - (see Fig. 6.d), FRF and ω2.

All functional waveforms were further developed (discretized) numerically and used during qualitative inference using the previously proposed statistical methods [7, 9, 10].

Differentiation of vibration state measures for a fit and damaged pulley in the study of this work is presented in the collective drawing in the form of stabilization diagrams and field values under the curves of the FRF and Coherence functions

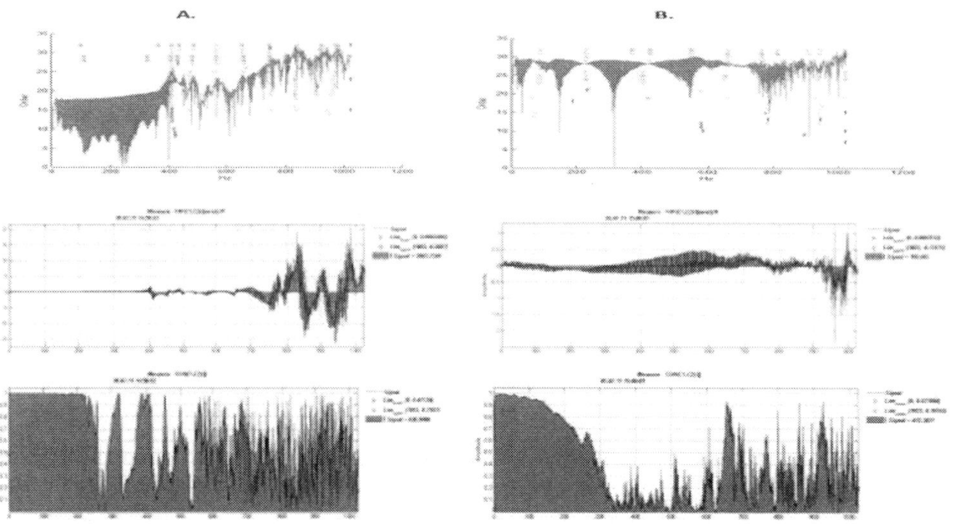

Fig. 7. Stabilization diagrams, FRF and Coherence functions for a suitable (A.) and damaged concrete block (B.).

The use of all data, usually collected in excess in the experiment and obtained during processing, to decide about the state of destruction of a masonry element is often neither possible nor necessary. It is natural, therefore, to strive to reduce the initial data, including the data sets of the examined objects and information describing them, which should be replaced by relatively small sets of data respectively aggregated and ordered. In general, such data reduction may consist of:

- information selection, i.e. a reduction in the number of measures describing the materials tested;
- replacing the continuous range of variability of the measure with discrete and sometimes binary changes;
- reduction in the number of materials tested representing individual status classes.

The reduction methods are of great importance for the problem of information selection, which:

and). they minimize the risk or average probability of classification error of the object;

b). maximize the assumed objective function;

c). they maximize (with limitation) information indicators of features.

The optimal reduction methods (a) minimize the average risk or average probability of error in the classification of the object's condition. Practically, the importance of these methods can be significantly reduced due to ignorance of probability distributions. In such a situation, at least some examples of observations from individual classes must be known a priori. Types of data reduction b) and c) fall within the scope of methods used in mathematical statistics and methods of sampling and quantizing signals aiming at adapting the signal form to the requirements related to information processing.

In terms of the selection of information, the procedures of OPTIMUM, SVD, and SIBI recommended in many other studies [13]. The use of functional measures in the processing has been replaced by numerical measures, proposing determining the values of the functional waveform with the numerical value of the field under the curve, as shown in Fig.7.

The implementation of algorithms for the indicated methods illustrated in Fig. 8, allows distinguishing information-related vibration measures used to build a cause-and-effect model that forms the basis of the proposed assessment procedure for the degradation of the tested concrete block.

The next steps of the calculation procedure in the OPTIMUM method are presented in the table 2 [6,9,13].

Table 2. Vibration measures processed in the OPTIMUM [9] procedure.

	FRFr	FRFu	H(f)r	H(f)u	coher	delta	Urms	w1	w2
f1	2,287	1,207	5,108	0,611	0,364	0,033	0,000	0,400	0,204
max (f1)	5,108								
f1*	0,448	0,236	1,000	0,120	0,071	0,006	0,000	0,078	0,040
f1**	1,000								
f2	0,985	0,030	1,000	0,743	0,874	0,430	0,001	0,775	0,316
max (f2)	1,000								
f2*	0,985	0,030	1,000	0,743	0,874	0,430	0,001	0,775	0,316
f2**	1,000								
Li	0,553	1,235	0,001	0,917	0,937	1,145	1,414	0,949	1,179
1/Li	1,810	0,810	1000	1,090	1,067	0,873	0,707	1,054	0,848
wi	0,002	0,001	0,992	0,001	0,001	0,001	0,001	0,001	0,001

The algorithm of processing the measures in the graphic interpretation for the OPTIMUM method is shown in Fig.8A, and the result of the statistical elaboration in the SVD procedure in Fig.8B.

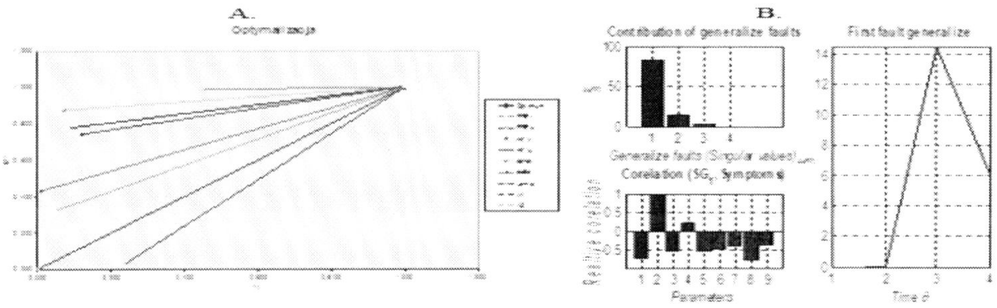

Fig. 8. Graphical presentation of sample results from OPTIMUM and SVD.

A further procedure in accordance with the methodology proposed for the study of vibration energy distribution proposed in the author's work was to determine good measures of the state of destruction of the concrete blocks tested and to establish cause and effect relationships between the state of block destruction and its measures. This was obtained by examining the functional relationship separately for each variable, i.e. for each measure individually characterizing the state of destruction and when testing the function of several variables, for all measures simultaneously. The study of these mathematical relations was carried out using simple regression and multiple regression, and the corresponding equations are shown directly on the regression curve drawings.

For the distinguished measures of degradation status, regression partial models were determined, for example shown in Fig.9. To approximate the form of the function between the state of destruction and the measures of this state, the most preferred algebraic polynomials in the statistics, with the especially recommended second-degree polynomial, were adopted. The obtained relations indicate how many changes in the load of the tested wall elements affect changes in the values of the vibration signal selected in the examination procedure (individual or multiple).

Measurements of approximation errors of the examined functions were determined by the multidimensional correlation coefficient R2, also called the coefficient of determination, taking values from the interval [0,1]. Its values are given in regression drawings, the higher its value, the better the correlation - better adjustment of the regression curve to the measurement results.

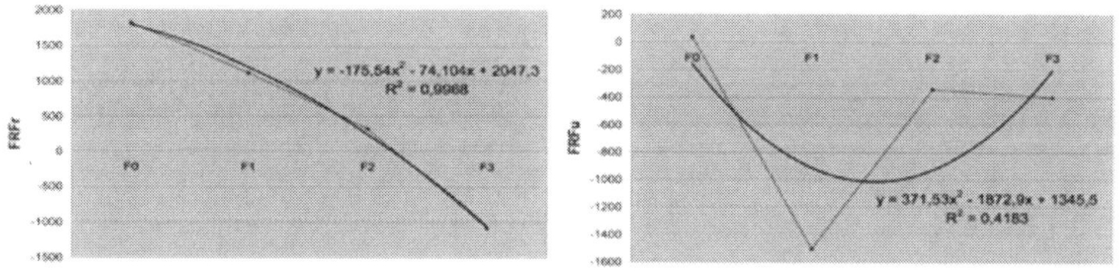

Fig. 9. Partial models for the indicated vibration measures

The expected effect of data reduction for the adopted quality of state classification is to simplify the structure of the data set, simplify the classification algorithm, shorten the calculation time, increase the speed of the system operation, reduce technical implementation costs.

Finally, the state degradation measures for a concrete block are: FRFr, H(f)r, coher. used in accordance with the model shown in Fig.10.

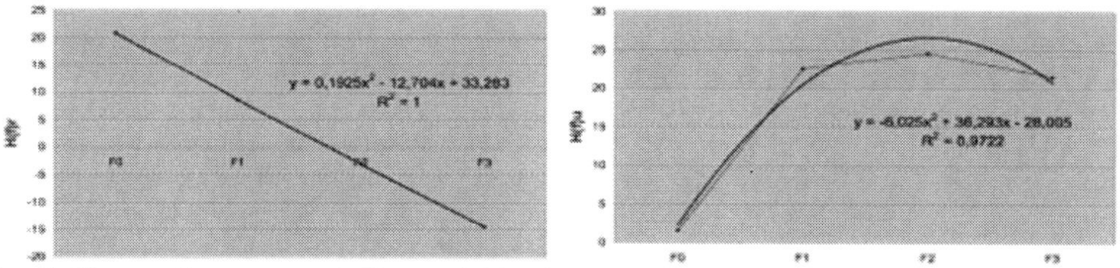

Fig. 10. The resulting cause and effect model for the tested concrete blocks.

The developed model for investigations of concrete block degradation that captures measures well differentiating the state of destruction is the basis for building a test method useful in industrial production.

4 Summary

Vibrations measurements along with their physical interpretation form the basis of the information obtained in the study of the destruction of selected building structures or their masonry components.

The exemplification of the method on a group of concrete blocks carried out in this study confirmed the compliance and effectiveness of the proposed dedicated methodology developed for the study of degradation of materials and structures in construction. The assumptions of the developed research methodology were positively verified and the effectiveness of complex vibration measures in the application for the assessment of degradation of masonry materials was confirmed, which promises well for the needs of real-life research on long-term objects.

The considerations of this study confirm the usefulness of the proposed algorithms and programs supporting the implementation of the research methodology, that is: the acquisition and processing of information, the reduction of excess information, statistical inference and the presentation of the study results.

The bursting of the block can be located by means of the coherence function (which already has an indication in the mathematical record for energy evaluation between points) and mutual measures: transmittance and its inversion - FRF.

Examining the structure state or its elements thanks to vibration measurements (acceleration a, velocity v, displacement x), can predict and prevent the possible failure time and prevented it by proper repair. What is more, by tracking the amplitudes of individual spectral components or other measures of the vibration process that constitute the destruction of elements, their condition, so important for the correct exploitation of long-term building structures, can be assessed without any problems.

References

1. B. Żółtowski, L. Castaneda, *Structural diagnosis of rail vehicles and method for redesign.* Diagnostyka Vol.15 (2014).
2. B. Żółtowski, M. Łukasiewicz, *Diagnostyka drganiowa maszyn.* (ITE-PIB 2012).
3. B. Żółtowski, M. Łukasiewicz, T. Kałaczyński, *Techniki informatyczne w badaniach stanu maszyn* (UTP Bydgoszcz 2012).
4. B. Żółtowski, M. Żółtowski, *Vibration signals in mechanical engineering and construction* (ITE-PIB 2015).
5. B. Żółtowski, C. Cempel, *Engineering of diagnostics machines* (PTDT, ITE – PIB 2004).
6. M. Żółtowski, *Operacyjna analiza modalna w badaniach konstrukcji budowlanych* (WU UTP Bydgoszcz 2012).
7. M. Żółtowski, *Investigations of harbour brick structures by using operational modal analysis.* Polish Maritime Research No 1, (2014).
8. M. Żółtowski, B. Żółtowski, *Vibrations signal to the description of structural damage of dynamic the technical systems* (XIII International Technical Systems Degradation Conference 2015).
9. M. Żółtowski, *Investigations of harbour brick structures by using operational modal analysis.* Polish Maritime Research No. 1/(81) vol.21 (2014).
10. M. Żółtowski, M. Liss, *The use of modal analysis in the evaluation of welded steel structures.* Studies and Proceedings of Polish Association for Knowledge Management Vol.79 (2016).
11. M. Żółtowski, M. Liss, *Zastosowanie eksperymentalnej analizy modalnej w ocenie zmian sztywności prostego elementu konstrukcyjnego.* Studies and Proceedings of Polish Association for Knowledge Management Vol. 80 (2016).
12. M. Żółtowski, R.M. Martinod, *Technical Condition Assessment of Masonry Structural Components using Frequency Response Function (FRF).* Masonry International Journal of the International Masonry Society Vol.29,No.1 (2016).
13. M. Żółtowski, R.M. Martinod, *Quality identification methodology applied to wall-elements based on modal analysis.* Civil Engineering the Athens Institute for Education and Research, Emerald, Athens (2015).

THEORY AND PRACTICE OF VIBRATION ANALYSIS IN ELECTRONIC PACKAGES

Igor Kovtun[1,*], *Juliy* Boiko[2], *Svitlana* Petrashchuk[1], and *Tomasz* Kałaczyński [3]

[1]Technologies and Design Faculty in Khmelnytsky National University, Department of Art and Technical Graphics, 29016, 11 Institutska street, Khmelnytsky, Ukraine
[2]Programming and Computer, Telecommunication Systems Faculty in Khmelnytsky National University, Department of Telecommunication and Radio Engineering, 29016, 11 Institutska street, Khmelnytsky, Ukraine
[3]Faculty of Mechanical Engineering UTP in Bydgoszcz, Department of Vehicle Engineering, Al. Prof. S.Kaliskiego 7, 85-796 Bydgoszcz, Poland

Abstract. The paper represents theoretical and experimental vibration analysis of electronic packages. The research objectives are bearing parts, such as printed circuit boards, walls of the cases and electronic components, in enclosed electronic packages subjected to vibration during their operation. The theoretical model of the oscillation system representing electronic package installed on the support subjected to vibration is described. Vibration transmission from the shaker to the tested electronic package is explained as the complex problem to solve. The experimental measurement verified significant difference in amplitudes of vibrations transmitted inside the package and incident to. The elements of plate bending theory is used for main boards, case walls and other planar rectangular structures considered as equivalent homogeneous plates, in order to estimate their natural frequencies and forms of free cross oscillations and hence to identify their possible resonances with purpose to avoid them in operation. The significant attention was paid to the experimental research of dynamic characteristics, representing results in form of Chladni figures and amplitude frequency responses.

1 Introduction

Practice of vibration tests and also usage of electronic packages installed on the supports sustaining vibration during their operation testified of damages in some components or failures of electronic packages although the vibrations remained under acceptable limits [1]. Vibration measured at locations where such damages occurred indicated that at some excitation frequencies the damaged components and parts sustained vibration amplitude, velocity or acceleration (vibration overloads) 10 times higher than correspondent parameters simultaneously measured on the shaker table or on the support of electronic package that receives and transmits vibration while operation.

In case when printed circuit boards (PCBs) and other bearing parts of electronic packages sustain resonance oscillations the dynamic loads multiply and may cause breakages of components and units or malfunctions because of deviations of their electronic parameters [2].

Therefore, special attention is paid to finding natural oscillation frequencies of bearing parts in electronics by using calculative and experimental methods in order to identify their possible resonances and hence to avoid them and correspondent dangerous operational modes.

* Corresponding author: dr.igorkovtun@gmail.com

Complexity of this problem arises due to the lack of input data needed for calculations. As the matter of fact electronic packages represent complex structures comprising many components whose vibration characteristics such as mass and stiffness are different [2] and the knowledge of their real values is insufficient. Moreover, characteristics of components mounted on PCB and in this way connected are no longer individual and may vary as being influenced by other nearly situated components during vibration.

In calculations the main boards, case walls and other planar rectangular structures are normally considered as equivalent homogeneous plates [3] whose theory is quite complicated, long, full of simplifications and assumptions, and incomplete.

Therefore, the current study is focused on using elements of these theories and basic formulas in order to calculate and estimate natural frequencies of free cross oscillations of the rectangular plates and assess their efficiency and applicability to the planar construction elements of the electronic packages in particular to PCBs. The significant attention was paid to the experimental research conducted by vibration tests, representing results in form of Chladni figures and amplitude frequency response, which allowed assessing vibration for the real structures subjected to the real operational conditions.

2 Theoretical model of oscillation system

The dynamic characteristics of any object including electronic packages are normally measured while vibration tests performed on the shakers. The shaker with the tested object installed on it and the clamps fastening the object to the shaker are the elements of one oscillation system. In the common case such oscillation system can be represented by the simplified mass-spring-damper model shown in fig. 1. The tested object and the clamps are assumed as the total mass *m*.

This installation is not considered as absolutely rigid therefore the model (fig. 1) is represented as the total mass *m* supported by the shaker by the spring with stiffness *k* and the damper with damping coefficient *c*.

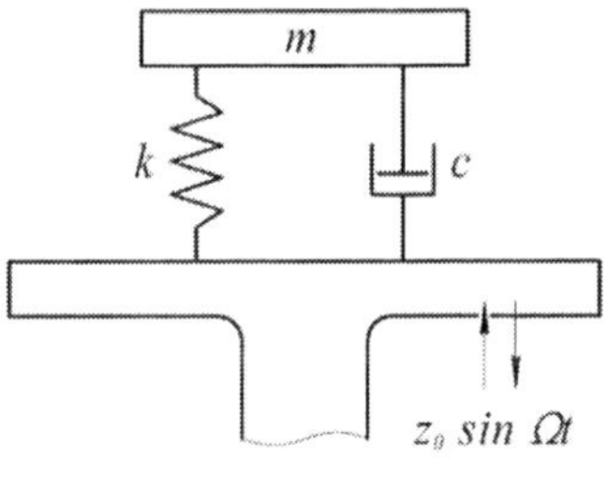

Fig. 1. Oscillation model.

The movable shaker table (shaker) generates harmonic oscillations described by the function $z_0 sin\Omega t$, where z_0 – amplitude, Ω – angular frequency, t – time of oscillations.

Downward movement of the shaker by the value $z_0 sin\Omega t$ causes spring extension that is expressed by formula $z - z_0 sin\Omega t$, where z stands for mass vertical downward displacement. The elastic force of the spring is expressed as $-k(z - z_0 sin\Omega t)$ and the damping force proportional to velocity is $-c(z' - z_0\Omega \cos\Omega t)$.

On the base of second Newton's law the following oscillation equation of the system is obtained:

$$mz'' = -k(z - z_0 sin\Omega t) - c(z' - z_0\Omega \cos\Omega t),$$
$$mz'' + k(z - z_0 sin\Omega t) + c(z' - z_0\Omega \cos\Omega t) = 0, \quad (1)$$
$$mz'' + cz' + kz = kz_0 sin\Omega t + cz_0\Omega \cos\Omega t).$$

Mass m movement with respect to vibrating shaker is found by spring extension derivatives taken with assumption that mass m displaced at the initial moment together with the spring by $z_1 = z - z_0 sin\Omega t$ then $z = z_1 + z_0 sin\Omega t$.

Derivatives are:

$$z' = z_1' + z_0\Omega \cos \Omega t,\ z'' = z_1'' - z_0\Omega^2 sin\Omega t. \tag{2}$$

Substituting these expressions into equation 1 gives the expression:

$$mz_1'' - mz_0\Omega^2 sin\Omega t + cz_1' + cz_0\Omega \cos \Omega + kz_1 + kz_0 sin\Omega t = \\ = kz_0 sin\Omega t + cz_0\Omega \cos \Omega t \tag{3}$$

The final simplified expression is:

$$mz_1'' + cz_1' + kz_1 = mz_0\Omega^2 sin\Omega t, \tag{4}$$

where mz_1'' – mass inertia; cz_1' – damping force; kz_1 – spring elastic force; $mz_0\Omega^2$ – inertial exciting force.

Such excitation of oscillation is called inertial [4].

Thus, the statements is made that the mass m (which includes both tested object and clamping appliance) moves relative to vibrating base (shaker table) in the same way as the mass m moves on motionless base (shaker), when mass is exposed to harmonic excitation force by amplitude $mz_0\Omega^2$. Forced oscillations of such single-mass system have been well studied and are not in objective for the present research.

Noteworthy is that amplitude and phase of such oscillations depend on proximity of excitation force frequency Ω to cycling frequency ω_c (5) of natural oscillations for mass m.

$$\omega_c = \sqrt{\frac{k}{m}}. \tag{5}$$

The formula for oscillation of the system with mass m, stiffness k, and damping $c = 0$ is:

$$z = \frac{z_0 \sin \Omega t}{1 - \left(\frac{\Omega}{\omega_c}\right)^2} \tag{6}$$

Thus, the issue is made that mass oscillation becomes close to shaker oscillation only when ω_c is considerably higher than Ω.

As Ω approaches to ω_c , the mass oscillation grows and goes out of phase.

When $\Omega = \omega_c$ the system undergoes resonance so that mass oscillation tends to maximum (∞) and phase difference tends to 90°.

After passing the resonance and further more after, mass oscillation decreases approaching to zero and phase delay increases so the phase difference ultimately reaches 180°. Thus, mass oscillations (in the tested object) will be directed contra-phase to the shaker oscillation and will likely quench them. Actually, since $\Omega = 10\omega$ the oscillations of the shaker will not be transmitted to the mass m.

Theory of two-mass systems assumes possibility to quench oscillations of one mass by the contra-phase oscillations of another one connected by the spring. This is the foundation of oscillation dynamic quench principle. However, this issue is not for consideration in the present paper. The important issue, which is theoretically proven, is so that if mass installation on the shaker is not absolutely rigid then shaker oscillation does not have to coincide with that of the mass.

Incidentally, the shaker oscillations are desirable to be transmitted from the shaker to the tested object without distortion at 1:1 scale for all excitation frequencies. But the reality appears to be even more complicated.

3 Vibration tests of electronic blocks

The vibration tests were performed for electronic blocks representing electronic packages in the enclosure case. The type of blocks chosen for the test had damages detected during typical production tests. Vibration graphing was applied during the test. Five vibration sensors were installed in places of potential damages on the case surface (fig. 2).

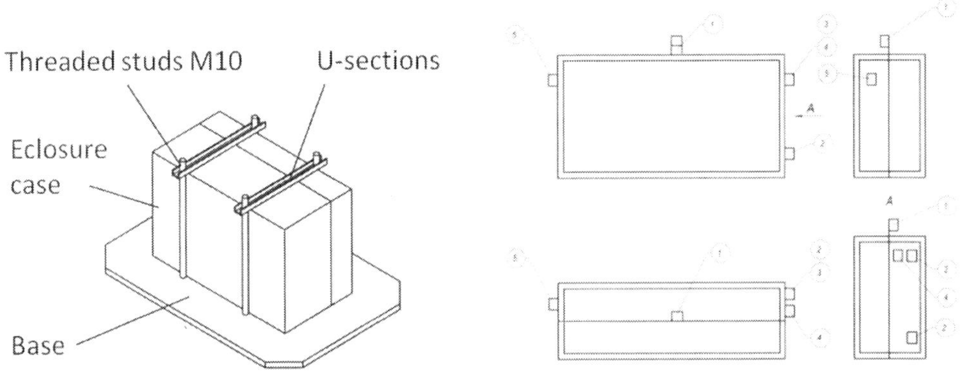

a) b)

Fig. 2. Vibration test installation: a) clamping appliance; b) vibration sensors installation.

The vibration sensor 6 was set on the shaker platform (not shown in the fig. 2). Then in accordance to production testing technology the tested block was fastened on the base by four threaded studs M10 and two U-sections; the base was tightened to the shaker platform by two bolts M10. It should be mentioned that production technology did not provide for the bolts fastening control.

The experiment was conducted in the following order. In accordance to the technical specifications the blocks were subjected to vibration tests by the fixed frequency method. The shaker was set to the specific (frequency and amplitude) oscillation mode.

Vibration sensors were connected to vibration measuring instruments. The given oscillation frequency was controlled by the sound generator.

To identify resonance of the individual parts or the whole block in given frequency range of vibration the sweep frequency method was applied.

The measurement results were obtained at these resonance frequencies at specified and constant amplitude of the shaker oscillation (A = 0.1 mm).

Analysis of collected data testified that oscillation amplitudes, recorded from vibration sensors on the tested block, 8-25 times exceeded the generated ones. For an instance, the real oscillation amplitude reported by vibration sensors reached from 1.35 mm (sensor 2) to 1.85 mm (sensor 1) and 2.7 mm (sensor 5) at oscillation frequency 80 Hz whilst the shaker amplitude was 0.1 mm. During the tests conducted at frequencies lower than 70 Hz this difference was negligible.

Another bearing block was subjected to vibration test. To identify overloads in PCBs inside the block two piezosensors ABC 032 were installed. The generated overloads were controlled by another two piezosensors set in the places of block attachment onto the shaker table.

The experiment was conducted as follows. The tested block with sensors attached was installed on the platform of the shaker in two positions: horizontally when vibration was directed along the PCBs maximal stiffness; and then at the side position when direction of vibration coincided with direction of PCBs minimal stiffness (excitation force was directed perpendicularly to the PCB plane).

Then the block was exposed to vibration acceleration of about 4-14 g read from vibration sensors 1 and 2, which were set in places of block attachment to the shaker, and vibration overloads of PCB were read from sensors 3 and 4. This procedure was conducted in both horizontal and side positions of the tested block.

The vibration overloads of PCB measured during the tests at the resonance frequencies with vibration acceleration driven to the places of block attachment demonstrate that vibration sensors set in different places of the shaker table measured different vibrations, what testifies that vibration unevenly distributed over the shaker table, and also that PCB vibration at resonance excitation may differ from vibration driven to places of block attachment by dozens of times, what confirms theoretical results obtained before.

4 Elements of plate vibration theory

Printed circuit boards, case walls and other planar rectangular structures used in electronic packages are represented as homogeneous rectangular plates to perform their vibration analysis and calculations. The plate is assumed an object of geometry, in which height H is at least 5 times less than its length and width.

The coordinate origin is oriented in the plate so that axes X and Y are constraint to its middle plane to which axis Z is perpendicular. The cross uniform inertial load bends the plate and its mid-plane becomes the middle elastic surface, which divides the plate into stretched and contracted areas. The same assertion refers to neutral fibers and neutral section axis in beams theory [4]. As is the case with beams vertical displacements of points in mid-surface are called plate deflections z, which are assumed small in comparison with plate thickness H. As well as for the case of beams the assumption can be made within second order accuracy that when the plate is deflected then the fibers of its mid-surface do not change their length, what means that analogically to neutral axis and neutral fibers the mid-surface is the neutral one and all its points receive displacement along axis Z while deformation.

Analogically to the planar section hypothesis in beam bending theory [4] the kinematic hypothesis or straight normal hypothesis is used in plate theory, in accordance to which the planes, having been normal to middle plane of the plate before deformation, remain normal thereafter. The deformation image of the plate reminds that of the beam with an exception that it deforms in two planes XZ and YZ simultaneously.

Points in the mid-surface are assumed with the second order accuracy to receive only vertical displacements, which are deflections z, and points vertically remote from mid-surface sustain more than just vertical displacements but also horizontal ones, which are derived from displacements along axes X and Y.

In accordance to straight normal hypothesis each normal to mid-surface as being deformed displaces as absolutely rigid piece equal to plate thickness H, so all points of the deformed plate can be found by drawing perpendiculars of length $H/2$ from correspondent mid-surface point. Thus, finding mid-surface equation $z = z(x,y,t)$ grants finding positions of all points in the plate and hereby using Hook's law finding stress and solving strength problems by using strength theories [4].

The general expression for the plate deflection, which is function $z(x,y,t)$ obtained by using mentioned theories is represented by the following formula:

$$\frac{\partial^4 z}{\partial x^4} + 2\frac{\partial^4 z}{\partial x^2 \partial y^2} + \frac{\partial^4 z}{\partial y^4} + \frac{\rho H}{gD}\frac{\partial^2 z}{\partial t^2} = \frac{q}{D}, \qquad (7)$$

where ρ – plate material density kg/cm^3; g – standard gravity, m/s^2; D – plate cylindrical stiffness, kgf·cm; q – intensity of load distributed over the plate surface, kg/cm^2; t – time, s.

Cylindrical stiffness D is expressed as:

$$D = \frac{EH^2}{12(1-\mu^2)}, \qquad (8)$$

where E – elastic modulus of the plate material, N/m^2; μ – Poisson's ratio.

If necessary, the presentation of expression (1) can be shortened by implementing harmonic operator [4] named as Laplace operator:

$$\Delta = \frac{\partial^2}{\partial x^2} + \frac{\partial^2}{\partial y^2}, \qquad (9)$$

and biharmonic operator:

$$\Delta\Delta=\left(\frac{\partial^2}{\partial x^2}+\frac{\partial^2}{\partial y^2}\right)\left(\frac{\partial^2}{\partial x^2}+\frac{\partial^2}{\partial y^2}\right)=\frac{\partial^4}{\partial x^4}+2\frac{\partial^4}{\partial x^2\partial y^2}+\frac{\partial^4}{\partial y^4}. \quad (10)$$

Then the shortened equation for the plate becomes written as:

$$\Delta\Delta(z)+\frac{\gamma H}{gD}\frac{\partial^2 z}{\partial t^2}=\frac{q}{D}. \quad (11)$$

The case can be considered more particularly. When q is function of time t, then equation (7) describes forced cross oscillation of plate driven by time variant distributed load q; when q is constant and time invariant, then plate sustains free or natural cross oscillation; when q = 0, then the following equation is obtained:

$$\Delta\Delta(z)+\frac{\gamma H}{gD}\frac{\partial^2 z}{\partial t^2}=0, \quad (12)$$

which describes free oscillations of unloaded plate.

A problem that describes free oscillation assumes the inertial force as the load:

$$m\cdot a=\frac{\gamma H}{g}\frac{\partial^2 z}{\partial t^2}=m\frac{\partial^2 z}{\partial t^2}, \quad (13)$$

where a – acceleration, m/s^2; $\frac{\gamma H}{g}=m$ – plate mass, kg.

Plugging this expression into (6) gives:

$$D\left(\frac{\partial^4 z}{\partial x^4}+2\frac{\partial^4 z}{\partial x^2\partial y^2}+\frac{\partial^4 z}{\partial y^4}\right)+m\frac{\partial^2 z}{\partial t^2}=0. \quad (14)$$

5 Natural frequency and form of oscillation

For the case when all edges of the plate freely rest on supports solution of equation (14) is:

$$z(x,y,t)=w(x,y)e^{j\omega_0 t}. \quad (15)$$

The amplitude function $w(x,y)$, which is called natural oscillation form of the plate is defined as:

$$w(x,y)=A_{u,f}\sin(u\pi x/a)\cdot\sin(f\pi y/b), \quad (16)$$

where a, b – length and width of the plate correspondently; u, f – number of sinusoid half-waves in direction of axes X and Y correspondently.

Dependence (15) is demonstrated in fig. 3. Any line parallel to axis X converts into sinusoid, which consists of u half-waves in interval (0, a), and the line parallel to axis Y has f half-waves. As shown in fig. 3 u = 2, f = 3.

Grid lines drawn from the nodes of half-waves divide plate into six sections, signs in which indicate direction of their deflection in specified moment of time: positive sign indicates upward deflection; negative sign – downward deflection, therewith signs shown in the left indicate deflection in ZY plane and those in the right – in ZX plane.

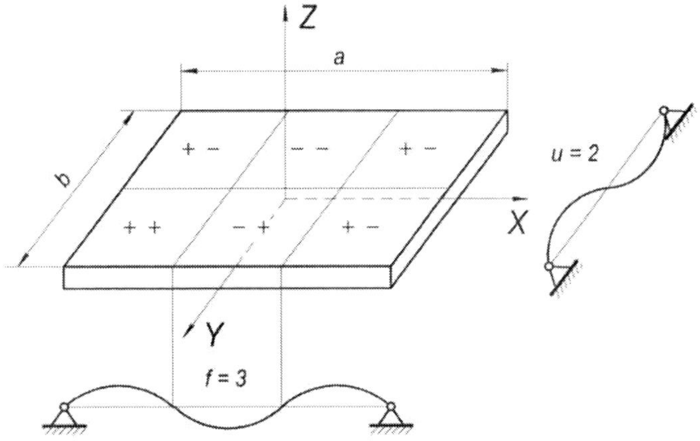

Fig. 3. Plate deformation and oscillation forms.

This solution complies with boundary conditions in outer edges:

- $z = 0$ and $\frac{\partial^2 z}{\partial x^2} + \mu \frac{\partial^2 z}{\partial y^2} = 0$ when $x = 0$ and $x = a$;
- $z = 0$ and $\frac{\partial^2 z}{\partial y^2} + \mu \frac{\partial^2 z}{\partial x^2} = 0$ when $y = 0$ and $y = b$ correspondently,

which indicate deflections and bending moments equal to zero in the edges (that rest on supports) of the plate.

Substitution of solutions $z(x,y,t)$ (15) and (16) into (12) gives:

$$\left(\frac{u\pi}{a}\right)^4 + 2\left(\frac{u\pi}{a}\right)^2 \cdot \left(\frac{f\pi}{b}\right)^2 + \left(\frac{f\pi}{b}\right)^4 = \omega_0 \frac{\rho H}{D}. \quad (17)$$

Wherefrom natural frequency of oscillation is:

$$\omega_0 = \pi^2[(u/a)^2 + (f/b)^2]\sqrt{D/\rho H}\,. \quad (18)$$

The represented method for finding natural frequency of oscillation is also applicable for the case when rectangular plate is freely supported on two parallel edges, what is normally the major method for PCBs installation in multilevel cases.

The fact is that real natural frequencies often differ from values estimated by demonstrated formula (18) because of the whole range of factors, such as: PCB population with discrete components and their non-uniform distribution over its area; conductive tracks, pads and other features etched from copper, which change stiffness of the PCB substrate; complexity to estimate fastening force in PCB fixtures to the cases of electronic packages and others. Noteworthy is that oscillations of PCB substrate and installed components are mechanically connected and in resonance frequencies the near components may involve each other into the mutual oscillations and in this way change parameters of the whole oscillation system: mass, stiffness, damping etc. Hence the conjecture is made that natural frequency estimation is just approximate and the real values can only be obtained experimentally.

To verify this conjecture the natural frequencies were calculated by formulas (18), and then measured experimentally. The research was conducted for the batch of PCBs with fiberglass epoxy substrate populated with integrated circuits uniformly distributed over its surface. The PCBs parameters were: $H = 1.22$ mm; $a = 200$ mm; $a / b = 1$; $Q_c / Q_b = 1$. The natural frequency calculated by the formula (18) equaled to 75 Hz. Meanwhile the average value of natural frequencies measured experimentally was 150 Hz.

Therefore, the represented estimation method is only considered as rough and approximate but accurate evaluation of natural oscillation frequency in the whole system of PCB substrate with installed components requires application of more complex models or experimental methods of the research.

6 Chladni figures and frequency response

The forms of oscillations for PCBs or electronic components were also offered to define by application of Hook's and Chladni method [5]. The idea of Hook's and Chladni method was applied to research forms of oscillations for PCBs using designed experimental installation (fig. 4).

The surface of the tested PCBs was wetted by kerosene and poured by fine-disperse quartz sand. Varying excitation frequency of vibration driven to the tested plate its oscillation amplitude rises at the resonance. This effect testifies of excitation and natural frequencies coincidence, what results in resonance. Therewith sand is dropped from the oscillation antinodes down to places, where board is not vibrating, which are considered to be oscillation nodes. In the nodes sand remains still. Besides, in these places the highest stresses appear. This method was successfully applied to find forms and natural frequencies of oscillation for main boards in four different variants of boards installation and also for bearing case walls in three orthogonal planes.

In the electronic packages PCBs are installed as in horizontal so in vertical positions what was provide by experimental cross appliance shown in fig. 4. The appliance provides simultaneous testing of 12 PCBs as the maximum.

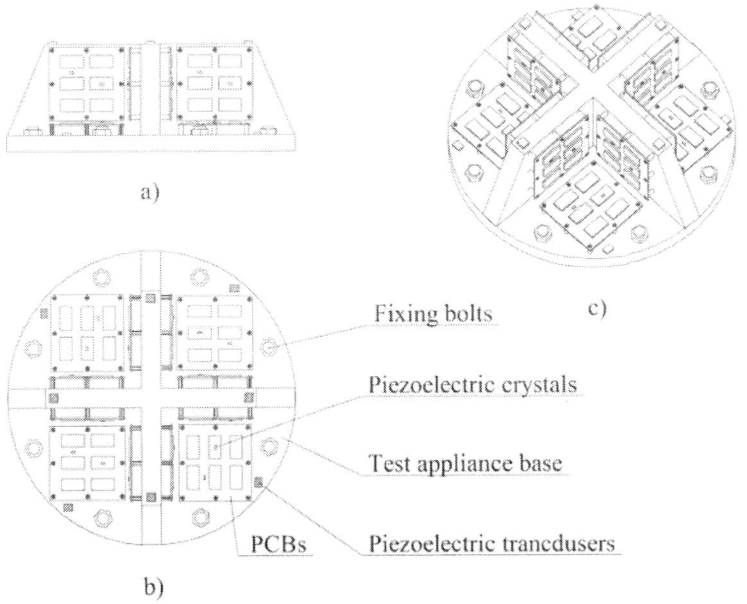

Fig. 4. Experimental vibration testing cross appliance presentation: a) front; b) top; c) pictorial projections.

Tests were conducted for PCBs populated with three types of electronic modules, which differed by their elastic and inertial characteristics. Masses of tested electronic modules M1, M2, M3 were 60, 80, 100 g correspondently.

The natural frequencies, forms and amplitudes of oscillations of PCBs were searched in frequency diapason of 2000-5000 Hz, which corresponds to operational conditions of electronic packages flying onboard. The test results are represented in form of Chladni figures (fig. 5, 6) and also by amplitude frequency responses, which were defined by averaged amplitude ratios A_{mdl} / A_{brd} (fig. 7), where A_{mdl} and A_{brd} – oscillation amplitudes of electronic modules and the main boards correspondently. The amplitudes of oscillations in electronic modules were measured with consideration of irregular distribution of vibration acceleration over the shaker platform.

The Chladni figures demonstrate that researched objects have complex forms of oscillations, what testifies of complex strain and hence stress conditions for both boards and electronic modules [6].

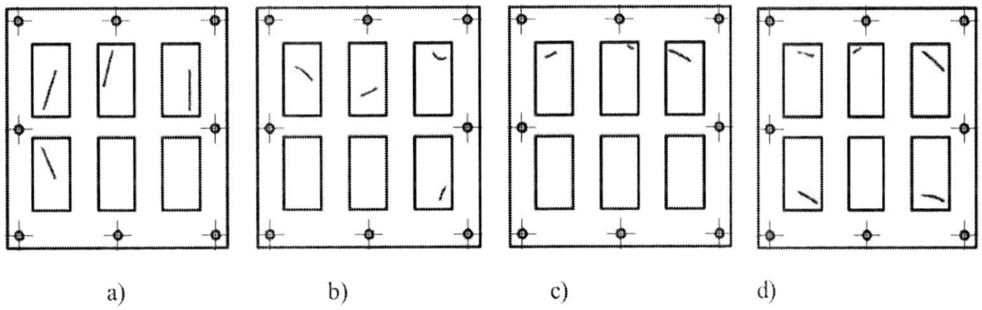

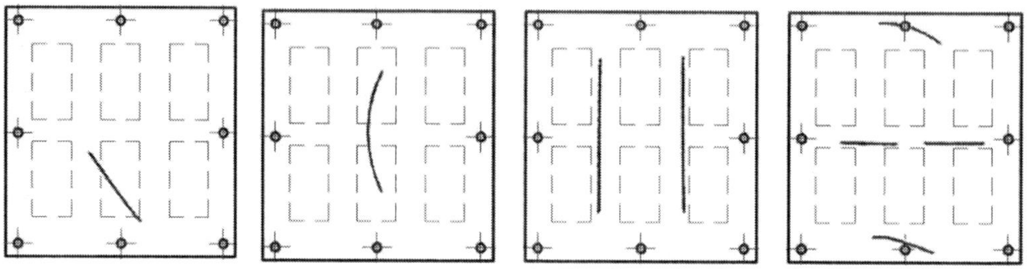

e) f) g) h)

Fig. 5. Chladni figures for PCBs installed in horizontal position and tested by vibration at their natural frequencies (Hz): a) 2580; b) 3373; c) 3501; d) 3666; e) 4880; f) 2061; g) 2227; h) 2799.

Analysis of amplitude frequency responses (fig. 6) testifies of higher resonant response for boards installed in vertical position than in horizontal one. This fact is explained by the difference of elastic characteristics of tested objects which vary depending on whether vertical or horizontal installations. Besides, obtained results indicate of having dense spectrum of natural frequencies, which depends on design of electronic modules installed on the main boards.

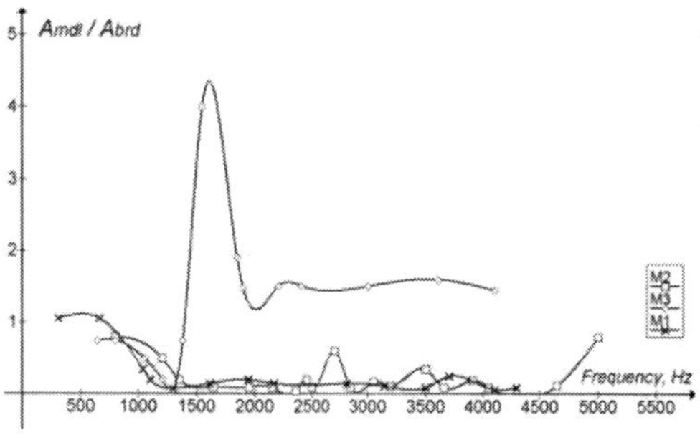

a)

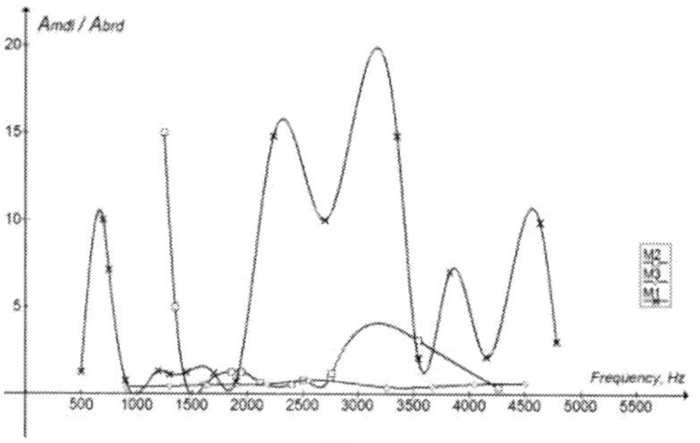

b)

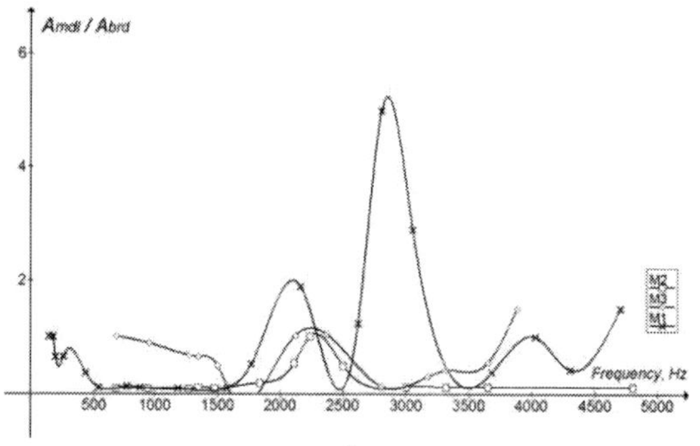

c)

Fig. 6. Amplitude frequency response for main boards installed in experimental vibration test cross appliance: a) vertically in X plane; b) vertically in Y plane; c) horizontally.

Thus, obtained experimental results testify that measuring dynamic characteristics of PCBs populated with electronic components or modules requires consideration of PCB installation position in the electronic package and electronic component distribution over the PCB area.

7 Conclusions

The theoretical model has been developed for oscillation system that represents electronic packages installed on supports subjected to vibration. The theoretically proven issue is made that vibration parameters of supports do not characterize vibration inside the electronic packages unless their connection is absolutely rigid.

Printed circuit boards, case walls and other planar rectangular structures used in electronic packages are represented as homogeneous rectangular plates in the offered plate bending model to perform their vibration analysis, however such estimation is approximate in comparison with the real values obtained experimentally.

Considering complexity, which becomes disadvantages of the traditional techniques for vibration tests performed by shakers, the experimental setup was created to define natural frequency and form of oscillation of PCBs and electronic components by using reverse and direct piezoelectric effects.

The Chladni figures, applied to vibration analysis of printed circuit boards and installed electronic components, demonstrated complex forms of oscillations, what testifies of complex strain and hence stress conditions for the research objects.

Analysis of amplitude frequency responses of printed circuit boards populated with electronic components testify that measuring their dynamic characteristics requires consideration of PCB installation position in the electronic package and electronic component distribution over the PCB area.

References

1. G.S. Aglietti, C. Schwingshackl. Analysis of Enclosures and Anti Vibration Devices for Electronic Equipment for Space Applications. School of Engineering Sciences, Aeronautics and Astronautics, University of Southampton, UK (2010)
2. T. Hamano, Y. Ueki, T. Nakasuji, K. Fujimoto. Destruction mechanisms resulting from vibration load in PCB-mounted electronics. ESPEC Technology Report No.17, 9th Symposium on Microjoining and assembly technology in electronics (2003)
3. R.M. Jones. Buckling of bars, plates and shells. Virginia, USA: Blacksburg, Bull Ridge Publishing (2006)
4. G.S. Pisarenko, V.A. Agarev. Strength of materials. (Kiev: Technika, 1967)
5. T. D. Rossing. Chladni's law for vibrating plates. American Journal of Physics, vol. 50, issue 3, pp. 271–274, 1982, doi.org/10.1119/1.12866
6. I. Kovtun, J. Boiko, S. Petrashchuk, G. Bauriene, K. Pilkauskas. Effects of the strain transmission from the main board to the installed electronic components. MECHANIKA, vol. 22(6), pp. 489-494, (2016), DOI: http://dx.doi.org/10.5755/j01.mech.22.6.16891

The decision making process of a state technical facilities based on rough set and vibroacoustic estimates

Rafał Boniecki[1,1] and *Mirosław* Miciak[1]

[1]University of Technology and Life Sciences in Bydgoszcz, Institute of Telecommunications and Computer Science, Al. prof. S. Kaliskiego 7, 85-796 Bydgoszcz, Poland

Abstract. The article presents idea and develops of an algorithm to support decision-making process for assessment a technical object state, based on vibroacoustic signals and rough set theory. The article discusses how to implement the main elements of the proposed system using Java technology, with Swing graphics library and JFreeChart component. FirebirdSQL engine was used to be a data base for the objects of diagnosis. JDBC technology will be used for generating queries and communication between the application and database.

1 Introduction

Assessment of a technical object state poses the biggest problems connected with diagnostics of technical devices. Methodology of diagnostic tests involves performing classification tests, which include the following diagnostic steps [3]:

- diagnosis - a process involving assessment of a machine state in time $_b$,
- genesis – a process involving reconstruction of the machine previous states, eg. through defining the cause of the machine detected state, in the time of its damage,
- prognosis – a process involving assessment of the machine future state, eg. through scheduling the date of the machine next servicing Θ_d.

The main problems that occur during identification of a technical object state are:

- formulation of the machine diagnosis, prognosis and genesis purpose,
- change of the machine state during its operation,
- description of the machine state by means of its characteristics and dependence on diagnostic parameters,
- solution of the machine state diagnosing,
- solution of the machine state prognosis,
- solution of the machine state genesis.

An analysis of a technical object state is complicated due to operational - measurement data which are often characterized by large dimensionality. In defining the notion of high dimensionality, we mean that data sets (eg. measurement data corresponding to operational states) have a big number of attributes (variables). In practice, it often happens that many of these attributes are strongly correlated with each other and it is enough to consider only a small subset to provide a full picture of the described machine state or identify some regularities (or irregularities) in the object. Although, there are several methods for analysis of technical object operational states, it is usually impossible or hardly possible to identify the object usability only by analysing all the symptoms. Thus, among many tasks, which are involved in machine operation and maintenance, those which enable to reduce dimensionality of the set of symptoms are very important.

---

1 Corresponding author: raboni@utp.edu.pl

The major problems connected with assessment of a machine state include: formulation of the state diagnosing, prognosis and genesis goal, change in a machine state during its operation, description of the state by means of its characteristics and parameters of the examined state signal and the dependence between the state characteristics and diagnostic parameters. Thus, the main problems that emerge while solving so formulated problems are:

- choice of 'the best' diagnostic parameters describing the machine current state and its changes during operation,
- determination of the diagnostic parameter boundary value,
- determination of the diagnostic parameter generated value and estimation of the technical object state, value performed by it in the past.

The main goal of the study was to design and create an application which will be useful in the process of making decisions on the technical object state assessment. Additionally, the application to be developed should enable communication between computers, regardless of available equipment and the system platform. The software will be written in Java language with the use of graphic library Swing and JFreeChart. FirebirdSQL engine was chosen to be a database storing the objects, whereas, communication and generation of queries, between the application and the database, will be carried out by JDBC technology.

2 Algorithm supporting the process of making decisions on technical object state

Identification of a technical object state, on the basis of available qualitative information, indicates the class of assignment. In this study, a method for assessment of a technical object state is proposed, based on statistic information obtained from vibroacoustic signals. Figure 1 shows an algorithm supporting the process of making decision on the technical object state.

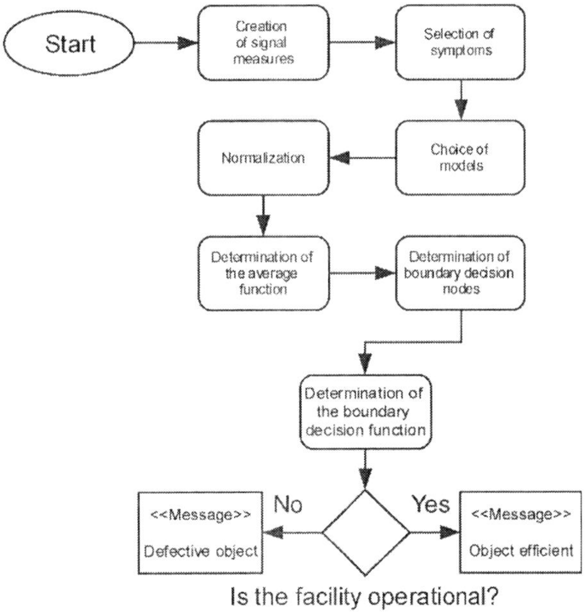

Fig. 1. The proposed algorithm supporting the process of making decision on the state of a technical object, on the basis of vibroacoustic signals and rough set.

2.1 Acquisition of measurement data and determination of signal measures

According to the developed methodology [4], the first stage of the presented algorithm (fig.1.) implementation is acquisition of measurement data. Measurement data is stored in text files UNV, and for the needs of the created application, the data must be processed to provide the form of a unified dynamic structure. Measurement data, due to a random number of channels, is stored in a two-dimensional dynamic structure whose dimensions depend on the number of the signal samples and on the number of channels.

In the second stage, measures of vibroacoustic signals are created to form selected estimates of the vibration. In the second stage, measures of vibroacoustic signals are created to form selected estimates of the vibration. These measures are most frequently determined from signals in the domain of time, from their transforms in the domain of frequency or time and frequency.

Table 1 shows a list of denotations for selected estimators (symptoms), used in the developed application.

Table 1. List of denotations of selected estimators (symptoms).

	Denotation of estimator	**Description of signal**
1.	Media	Mean value of signal
2.	RMS(t)	Global value of effective value calculated for a signal in the domain of time
3.	RMS(f)	Global value of effective value calculated for a signal in the domain of freq.
4.	RMSp	Global value of spectral power top value
5.	Max	signal top value
6.	Min	Minimal value of signal
7.	Kurtosis	Kurtosis of data series
8.	Sesgo	Measure of data series asymetry
9.	D. Estdandar	Standard deviation
10.	Factor Forma	Coefficient of shape
11.	Factor Cresta	Top coefficient
12.	Factor Impulso	Coefficient of impulse
13.	Factor Holgura	Coefficient of clearance
14.	Fu (v)	Rice's frequency for vibration speed
15.	F u(d)	Rice's frequency for vibration displacement
16.	H(v)	Coefficient of vibration speed harmoniousness

2.2 Standardization of symptoms

A set of attributes undergoes (in the form of symptom columns) initial data processing to enable their comparison and further analysis. Due to the type of processed data, standardization was used, in result of which, the variable mean expected value assumes zero and the mean variance is equal to one. The most commonly used standardization method is, the so called, standardization z, which can be expressed by the below given formula:

$$z = \frac{x - \mu}{\sigma} \tag{1}$$

where: x – value of symptom, μ - mean value of symptom, σ – standard deviation of symptom.

2.3 Selection of technical object symptoms

There can be many symptoms from one or more signals (measurement channels) but most of them are significantly interfered or independent of the monitored phenomenon. Thus, only those which are correlated with the phenomena should be accepted. Thus, only those which are correlated with this phenomenon, should be selected. The number of such signal measures must be big enough to compensate unavoidable random interferences in each of them. The proposed solution involves choosing symptoms with the use of rough set theory. This technique provides methods for determination of the most important attributes of a computer system without losing classification ability [9]. Rough set theory was developed by Zdzisław Pawlak [1,2,6,7], is based on the concept of upper and lower approximation of the set, approximating space and models of sets. Data is presented in the form of a table, whose columns correspond to the attributes (symptoms) and rows corresponds to particular objects (states of technical objects). Such a record is called an information system [9,10]. The basis of rough set theory is a reduct, which is the main part of the information system. Thanks to this, it is possible to distinguish all the distinguishable objects from the original set of attributes, hence, reducts are minimal subsets of attributes maintaining characteristics of the whole set of attributes. Literature provides many methods for determination of sets of information system reducts. For the needs of the application, the authors have taken into consideration the methods of: Johnson, global, local [5,8]. Significance of the symptoms is described based on the determined set of reducts, where the attributes are ordered according to the number of occurrence in the set. An exemplary set of reducts is presented in table 2.

Table 2. Set of reducts.

	Number	**REDUCTS (no. of symptom)**
1	9	{ 1, 3, 6, 7, 10, 12, 29, 31, 35 }
2	9	{ 1, 3, 6, 7, 10, 22, 29, 31, 35 }
3	8	{ 1, 3, 6, 10, 11, 12, 15, 31 }
4	9	{ 1, 3, 6, 10, 11, 12, 29, 31, 35 }
5	8	{ 1, 3, 6, 10, 11, 15, 22, 31 }
6	8	{ 1, 3, 6, 10, 11, 22, 29, 31 }
7	8	{ 1, 3, 6, 10, 11, 22, 31, 35 }

Symptoms ordered according to significance are presented in table 3.

Table 3. Symptoms ordered according to significance.

Symptom	**1**	**3**	**6**	**10**	**31**	**11**	**12**	**29**	**35**	**22**	**15**
Number	7	7	7	7	7	5	4	4	4	3	1

2.3 Determination of boundary decision function on the basis of *t* symptoms with the same weights

Undertaking a decision on the basis of a *t*-dimensional model can be formulated in the following way:

- In interval <*a;b*> of data there are $(t-1)(n+1)$ of different points $x_0, x_1, \ldots x_t(n+1)$ which can be called decision nodes, and the values of function $y(x)$ in these points are, respectively:

$$f(x_0)=y_0, f(x_1)=y_1, \ldots, f(x_t(n+1))=y_t(n+1)$$
$$S'_n=\frac{w_1 y_1(x_n)+\ldots+w_{t-1} y_{t-1}(x_n)}{w_1+\ldots+w_{t-1}} \tag{2}$$

- Points determined by the symptom which is the first to achieve boundary value do not belong to decision nodes, for differentiation, these points are called extreme decision nodes and the function value in this point is:

$$y = f_s(x) \tag{3}$$

- When decision nodes are available, one can determine mean decision nodes in *<a;b>* interval, being a weighted mean for *t-1* symptoms measures:

$$\acute{S}_n = \frac{w_1 y_1(x_n) + \ldots + w_{t-1} y_{t-1}(x_n)}{w_1 + \ldots + w_{t-1}} \tag{4}$$

where $w_0, \ldots w_{t-1}$ are weights of particular symptoms.

$$\acute{S}_n = \frac{w_1 y_1(x_n) + \ldots + w_{t-1} y_{t-1}(x_n)}{w_1 + \ldots + w_{t-1}}$$

- When mean decision nodes are available, *n*-degree polynomial is used for approximation to achieve a continuous function, further referred to as the mean function, for *t-1* symptoms, whereas the function value in this point is:

$$y = f_{t-1}(x) \tag{5}$$

- In order to determine a boundary decision function, the notion of prophylactic or preventive distance is introduced into *<a;b>* interval. Prophylactic distance dx_0 is a non-negative number, unequivocally assigned to a segment, meeting the following conditions:

$$|AB| = 0 \, then \wedge only \, then, when \, A = B \tag{6}$$

$$|AB| = |BA| \tag{7}$$

$$|AC| \leq |AB| + |BC| \tag{8}$$

where, *A,B* and *C* are random points of a plane.
Number $|AB|$ is a prophylactic distance of point *A* to point *B* which is expressed by:

$$|AB| = dx_0 = \sqrt{\left(\frac{max}{\sin(t-1)} f_n(x_0) - f_{n-1}(x_0)\right)^2} \tag{9}$$

$$d_{x0} = \sqrt{¿\left(\max_{0 \leq n \leq t-1} f_n(x_0) - f_{t-1}(x_0)\right)^2} \text{ when}$$

$$QUOTE\, gdy\, A = \left(x_0, \max_{0 \leq n \leq t-1} f_n(x_0)\right) i\, B = (x_0, f_{t-1}(x_0))\, A = \left(x_0, \max_{0 \leq n \leq t-1} f\right. \tag{10}$$

and

$$QUOTE\, gdy\, A = \left(x_0, \max_{0 \leq n \leq t-1} f_n(x_0)\right) i\, B = (x_0, f_{t-1}(x_0))\, B = (x_0, f_{t-1}(x_0)) \tag{11}$$

- Knowing prophylactic distance dx_0, in point x_0, boundary decision nodes are determined in point x_0 which belongs to interval *<a;b>*. Boundary decision node in point x_0 is defined as:

$$QUOTE\, gdy\, A = \left(x_0, \max_{0 \leq n \leq t-1} f_n(x_0)\right) i\, B = (x_0, f_{t-1}(x_0))\, Bw_0 = \left(x_0, dx_{¿} x_0\right. \tag{12}$$

- Having available decision nodes in interval <*a,b*> we approximate them with *n*-th degree polynominal to obtain a continuous function, further known as boundary decision function for t symptoms, and the value of the function in this point is:

$$QUOTE\, gdy\; A=\left(x_0,\ \max_{0\le n\le t-1} f_n(x_0)\right) i\, B=\left(x_0, f_{t-1}(x_0)\right) y=f_D(x) \tag{13}$$

- Using boundary decision function, decision on the technical object state can be made. with the use of symptoms.

2.5 Selection of models describing damage evolution

After implementation of the process of selection of symptoms, best describing a given object, it is possible to start building a model describing damage evolution of a given object. It enables comprehensive identification of a machine state using deck or stationary diagnostic systems, operating on the basis of the described algorithm. Power and expotential models of symptoms Asr, Xsz, Asz, Ask, were determined for the selected symptoms, by means of the proposed algorithm (section 2.4) for object Sil24_d2 (values of parameters of particular models are presented in table 4).

Table 4. Power and expotential models of selected symptoms for Sil24_d2 object.

Symptom	Power model	Expotential model	Boundary values
Asr_Sil24_d2	4.89491*x.^0.146727168	18.0803*exp(1.9551414E-6.*x)	50
Xsz_Sil24_d2	0.14341*x.^0.071569542	0.26399*exp(1.7536301E-6.*x)	5
Asz_Sil24_d2	24.05980*x.^0.132450920	77.03300*exp(1.8367322E-6.*x)	200
Ask_Sil24_d2	7.02050*x.^0.147649869	25.7803*exp(2.0315695E-6.*x)	69

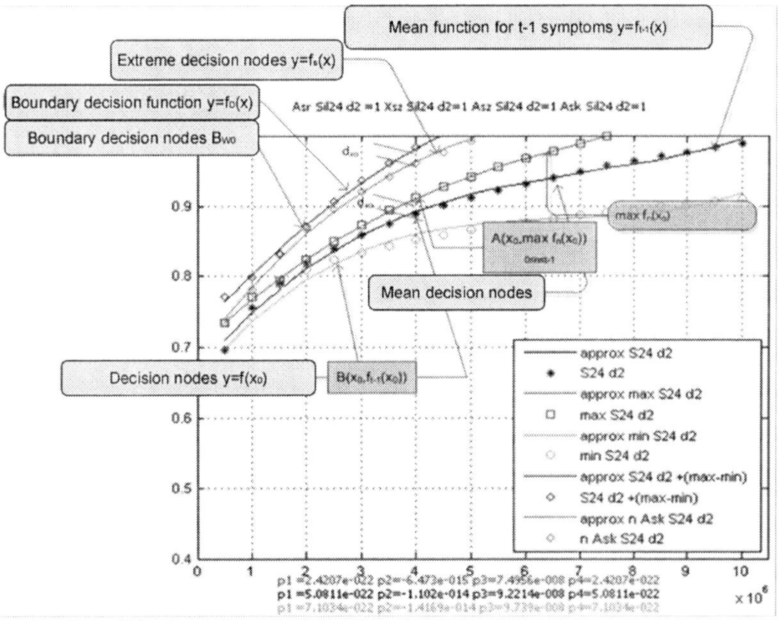

Fig. 2. Determination of the boundary decision function.

On the basis of the determined model Sil24_d2, it is possible to define a place where the boundary value is achieved.

3 Conclusions

The article presents implementation of an algorithm to support making decision on the state of a technical object, based on vibroacoustic data. Implementation of the application key elements has been presented. Application of Java technology provides the application with mobility (independence of the operational system). Utilization of open-source technology significantly decreases the implementation costs. The application uses interface JDBC (Java DataBase Connectivity) for connection with SQL Firebird base. Base sql includes 12 tables, 108 procedures using UDF libraries (User Defined Function).

References

1. J. Bazan, M. Szczuka, A. Wojna, M. Wojnarski, *On Evolution of Rough Set Exploration System,* Lecture Notesin Artificial Intelligence 3366, 592–601,Springer-Verlag (2004)
2. J. Bazan, M. Szczuka, J. Wroblewski, *A New Version of Rough Set Exploration System,* Lecture Notes in Artificial Intelligence 2475, 397–404, Springer-Verlag (2002)
3. L. Bedkowski, *Elementy diagnostyki technicznej,* WAT, Warszawa (1991)
4. R. Boniecki, M. Miciak, R. Wiatr, *Modele nienadzorowanego grupowania obiektów technicznych z wykorzystaniem metod inteligencji obliczeniowej* , Studia i materiały Polskiego Stowarzyszenia Zarządzania Wiedza, 49, 35-48. Bydgoszcz (2011)
5. A. Ohrn, *Discernibility and Rough Sets in Medicine*, Tools and Applications. PhD Thesis, Dep. of Comp. and Inf. Science, Norwegian Univ. of Science and Technology. (1999)
6. Z. Pawlak, *Information Systems - theoretical foundations*. WNT, Warszawa (1983)
7. Z. Pawlak, *Rough sets*, Int. Journal of Parallel Programming 11. 341–356. (1982)
8. M. Salamo, E. Golobardes, *Global, Local and Mixed Rough Sets Case Base Maintenance Techniques,* Recent Advances in Artificial Intelligence Research and Development. 217-234, (2004)
9. E. Szydlowska, *Algorytmy selekcji atrybutów w zadaniach eksploracji danych*, XIII Konferencja użytkowników i deweloperów ORACLE, 262-267. Zakopane (2007)
10. B. Walczak, D.L. Massart, *Rough sets theory*, Gemoemetrics and Intelligent Laboratory Systems 47, 1-16 (1999)

Triboacoustic diagnostic fixed joints of machines

Aaleksandr Dykha[1,*], and *Sergij* Matyukh[1]

[1]Khmelnitskyi National University, 29016, Instytutska 11, Ukraine

Abstract. The experimental results of the tribo-acoustic control in real time the integrity of nominally stationary friction joints under conditions of fretting are presented. The disturbance in the integrity of the tribosuppression as a result of wear has been reliably detected both by the change in the shape of the acoustic histograms of the operating noise and by the change in the noise spectrum due to the appearance of strong mid-frequency modes of the contact gap. The appearance of two oppositely directed wave energy cascades in the regime of fretting has been observed: a direct cascade from mod the pump in the direction of high frequencies and the opposite – side to the low infrasonic frequencies. The former has been associated with microplastic deformations in the zones of actual contact, while the latter has been associated with the dynamic generation of slow angular motions in the friction system.

1 Introduction

Modern technological conditions of production activity present increased requirements to the accuracy and efficiency of measurements, to the quality of control and diagnostics of various aggregates at the stage of their design, manufacturing, testing, operation and repair. Traditional methods of non-destructive testing are widely used at the manufacturing and repair stage. However, when used in a working environment, there are significant difficulties. Acoustic methods, being one of the most informative, provide, on the one hand, the identification of various defects of the object of control, and, on the other hand, the ability to carry out operational measurements of the technological parameters of various processes. Modern development of analog and digital technology allows to significantly increase the sensitivity and accuracy of acoustic measurements, to expand the possibilities of their application for diagnostics and measurement of parameters of dynamic objects and processes.

In tribotechnics, acoustic methods have been widely used for analysis and to control friction [1–0]. The acoustoemission of the ultrasonic frequency range using special contact-type sensors has become the most widely employed [1, 3, 4, 8]. Contactless acoustic control of sound and near infrasonic frequency bands based on standard or special software and hardware is much easier in technical implementation. However, it requires an additional analysis to isolate the useful signal against background noise. In the past, triboacoustics has primarily been used for mobile functional compounds [2, 5, 6]. It should seem that fretting in nominally motionless joints is hard to hear in the overall operating noise of an installation, mechanism, or machine. However, due to the mutual dependence of the constituents of dynamic processes proceeding in a complex technical system, a disturbance in the integrity of any such joints will inevitably affect the amplitude level and the spectral distribution of the operating noise. Due to the relatively low slip rates in fretting modes, the characteristic frequencies of acoustic emission fall precisely into the acoustic and near

* Corresponding author: tribosenator@gmail.com

infrasonic range of the spectrum that enables using conventional sound recording equipment and standard computer software.

The aim of this work is to perform the contactless triboacoustic control of the integrity of a nominally motionless friction joint under fretting conditions, as well as to determine the nature and direction of the major wave processes of energy and momentum transfer in real time.

2 Experimental

The experimental setup used in our study for fretting tests has been described in detail elsewhere [10]. It does not contain rotating elements that would create an increased level of contact noise (Figs. 1, 2). According to the plane–ball scheme, fretting is studied with contact between unhardened steel 30 HGSA (the leading upper sample 1 in Fig. 1) with a ball of the LH-15 steel bearing with a diameter of 12 mm (Fig. 1, lower driven counterbody 14). The sample was preliminarily ground and subsequently polished to a surface roughness of the 9th class (Ra = 0.2–0.3 µm). The reciprocating motion of the leading sample was produced using an electromagnet and a spring system. The amplitude of its vibrodisplacement Hs was 20 µm. The driven counterbody (ball) was clamped in the frame of the tensor bar (Fig. 2). Two-thirds of the normal contact load in the joint (~50 N) was due to the weight of the drive and one-third was due to the torque wrench force (Fig. 1) that gradually attenuated during the fretting. The main working frequency of vibrations was 100 Hz.

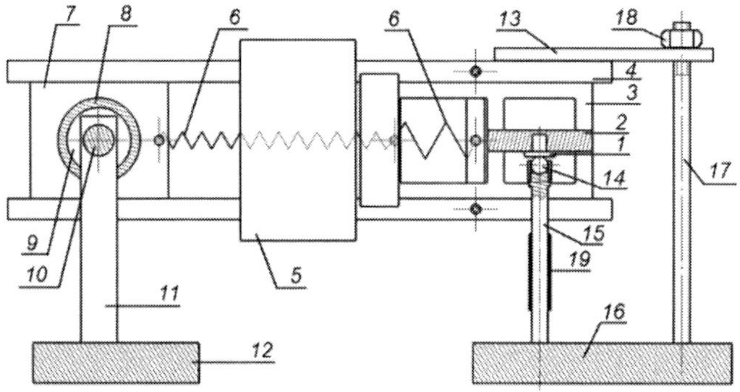

Fig. 1. Below the figure. Schematic of the experimental setup: (1) leading sample of a pair of friction; (2) sample holder; (3) carriage; (4) guides; (5) electromagnet; (6) a system of springs; (7) plates; (8) the bushing; (9) radial-thrust bearings; (10) shaft; (11) shaft supports; (12) base; (13) clamping plate; (14) driven counterbody (ball); (15) stretch bar; (16) base; (17) hairpin; (18) nut; (19) strain gages.

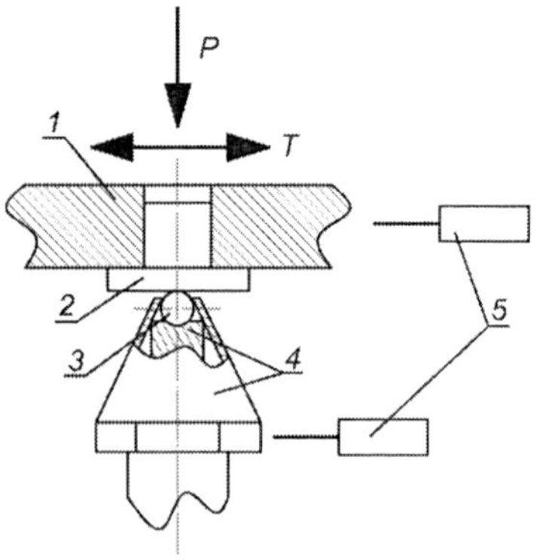

Fig. 2 Contact diagram: (*1*) sample holder; (*2*) sample; (*3*) counterbody (ball); (*4*) locking system counterbody; (*5*) inductive proximity sensors; *P* is clamping force, *T* is tangential force.

To record the acoustic signal, a portable microphone Media-Tech SFX MT 383 microphone installed at a distance of several decimeters from the friction unit was used. Data were processed using the Audacity 2.0.4 program. Online plotting of the contact hysteresis loop in coordinates of the displacement of the driven counterbody versus the tangential force was used as an additional informative channel. The displacement and tangential force were measured using an inductive microdisplacement sensor and a bridge of strain gages on the beam, respectively. The acoustograms were prerecorded in the idle ode of the setup operation, i.e., with an almost vertical position of the drive guides. Then, in their horizontal position of the drive guides, the joint was initially lamped while tests were performed in the fretting mode, i.e., from the initial run-in process to the disturbance in the joint integrity resulted in a qualitative hanged in both the level of the amplitude and the spectral composition of the acoustic signal.

3 Results and Discussion

Figure 3 shows the acoustograms recorded for 1 min each as the testing develops. Acoustograms are useful for studying fretting in general.

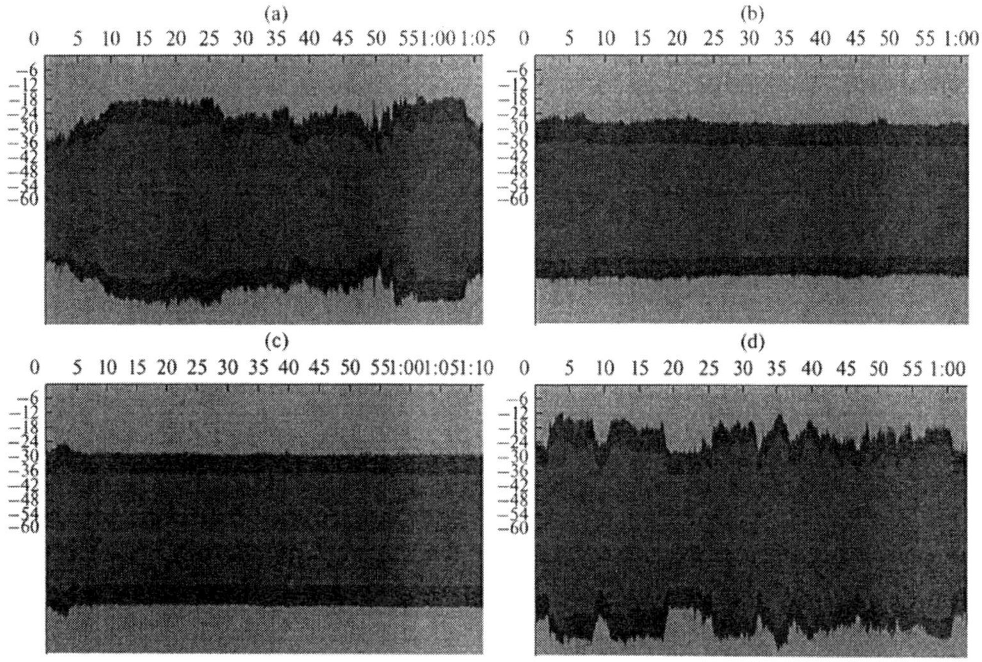

Fig. 3. Acoustograms of working noise recorded at (a) 1st, (b) 15th, (c) 55th, and (d) 88th min. Horizontal axis represents time in minutes and seconds and the vertical axis represents the signal level in dB.

The low-frequency instability that was specific to the initial stage of run in (Fig. 3a) decreased significantly in about 1 × 105 cycles of oscillations (Fig. 3b), leading to a longer period of optimal joint operation (~5 × 105 cycles) (Fig. 3c). A notable disturbance in the integrity of the joint occurred after approximately 88 min of testing (~5.3 × 105 cycles) (Fig. 3d).

The corresponding hysteresis loops were more similar, but less informative (Fig. 4). Nevertheless, at the final stage (Fig. 4d), the increase in amplitudes of high-frequency signal components was noticeable and followed by a decrease in the tangential force on the driven counterweight and a decrease in the amplitude of its oscillations due to slippage in the joint contact.

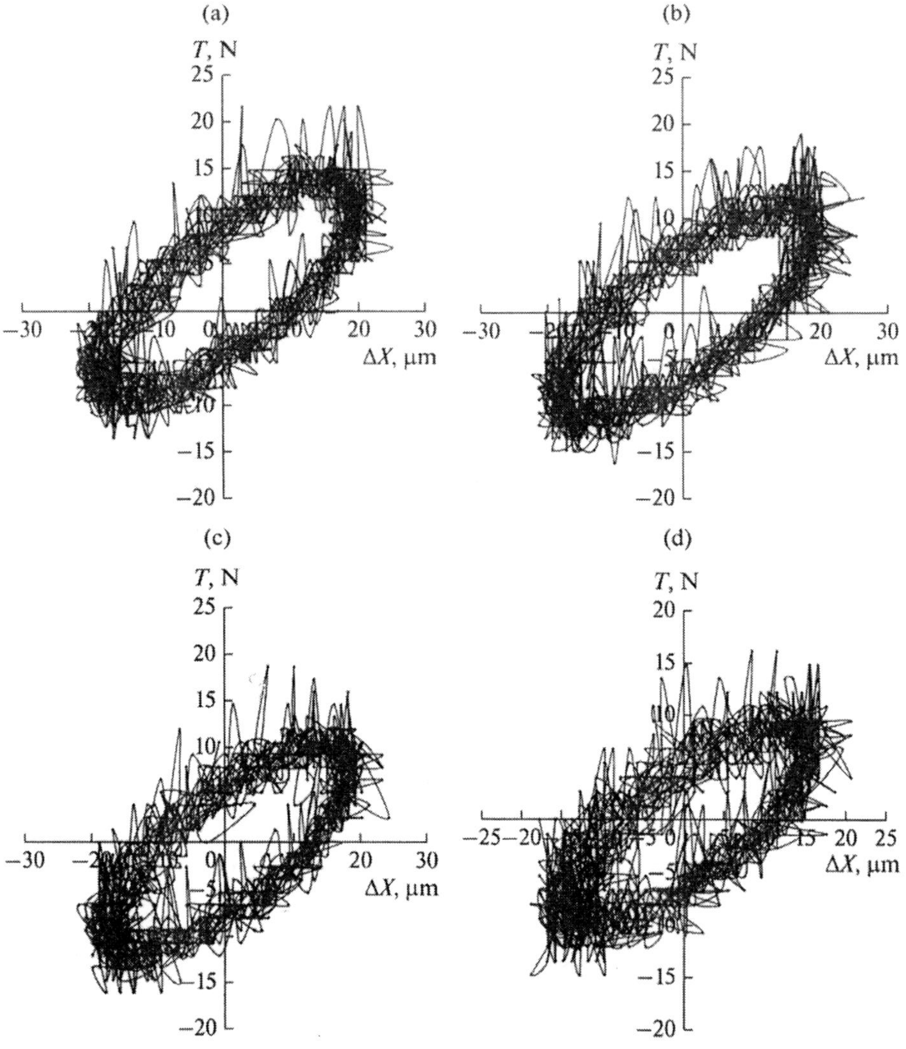

Fig. 4. Contact hysteresis loops plotted for (a) 1st, (b) 15th, (c) 55th, and (d) 88th min.

Of greatest interest in this case, the Fourier spectra of the above ecostorm – Fig. 5. They are dominated by two main pump frequency disturbances, 50 Hz and 100 Hz and their harmonics arising from the nonlinearity in a dynamic system and a corresponding disharmony of vibrations. These harmonics at the initial stage of running-in (Fig. 5a) form a well-defined direct energy cascade towards high frequencies with the maximum at frequencies of 1-5 kHz.

The occurrence of this maximum is due to quasi-normal contact vibrations induced by the processes of microplastic deformations at the stage of running-in. The other – the reverse energy cascade is directed from the pumping modes in the low frequency near-infrasound region of the spectrum (Fig. 5a). Its origin is connected with guided slow angular movements of the guides of the actuator, which has a direct impact on the rapid vibration of the sample leading to friction and, consequently, on the processes of contact interaction in the zone of fretting.

This kind of low-frequency self-modulation of the friction process often is at the expense of cyclical damage accumulation and separation of wear particles [6]. In this case, such cyclicity is a consequence, not the cause of automodulation arising from the activation of the available degrees of freedom and feedback in the overall dynamic system of friction. Thus, the friction acts as a global (system-wide) and not just local (mesophysical) process.

In addition, we analyzed the Fourier spectra of the above-mentioned acoustograms and confirmed the qualitative change in the spectral composition of the acoustic emission of friction, which was associated with a disturbance in the joint integrity.

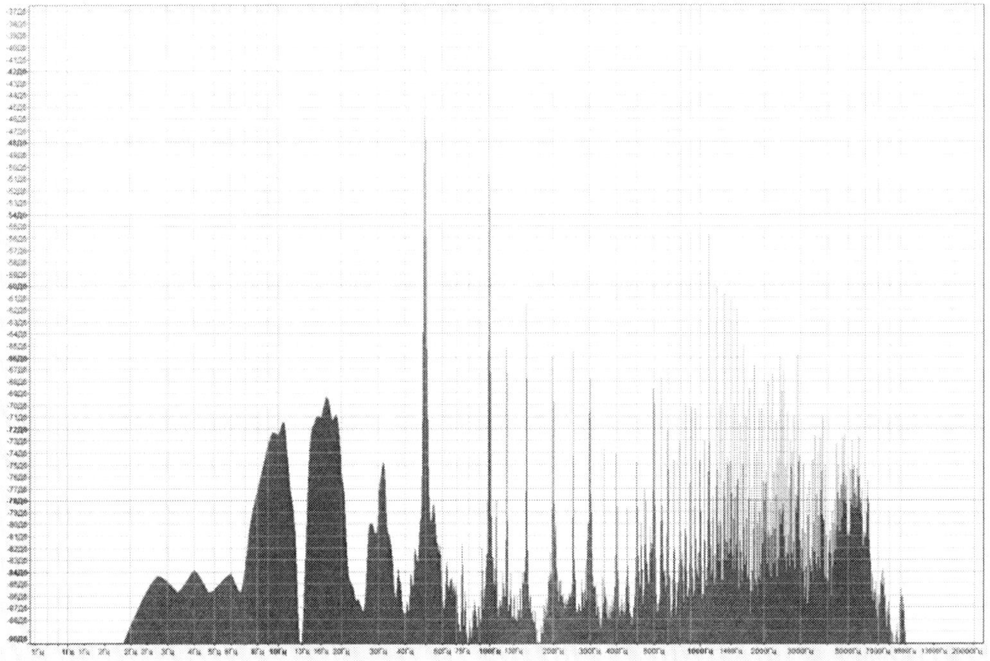

a

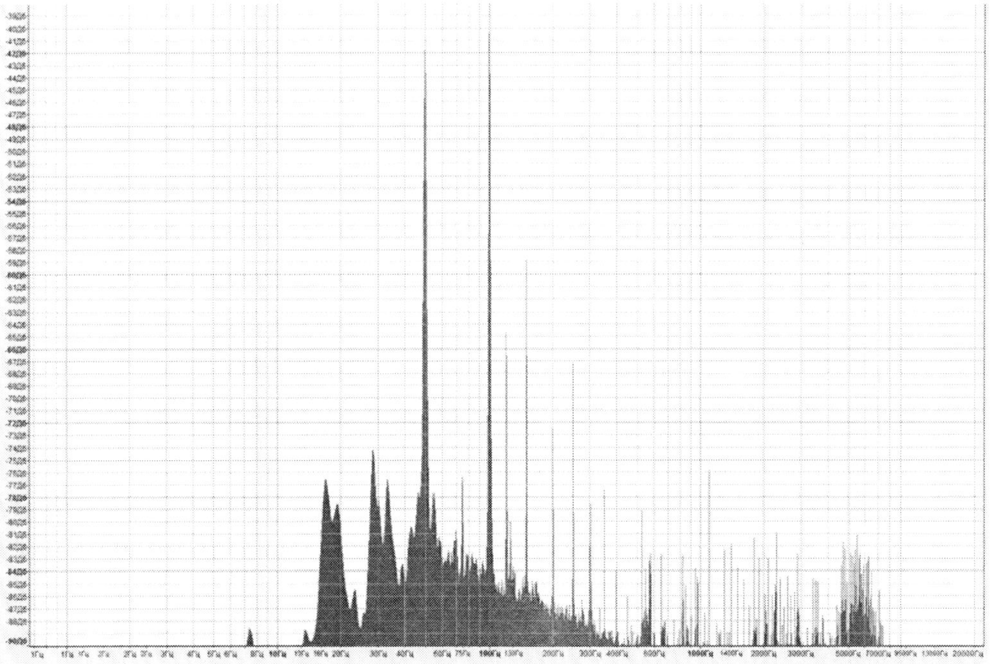

b

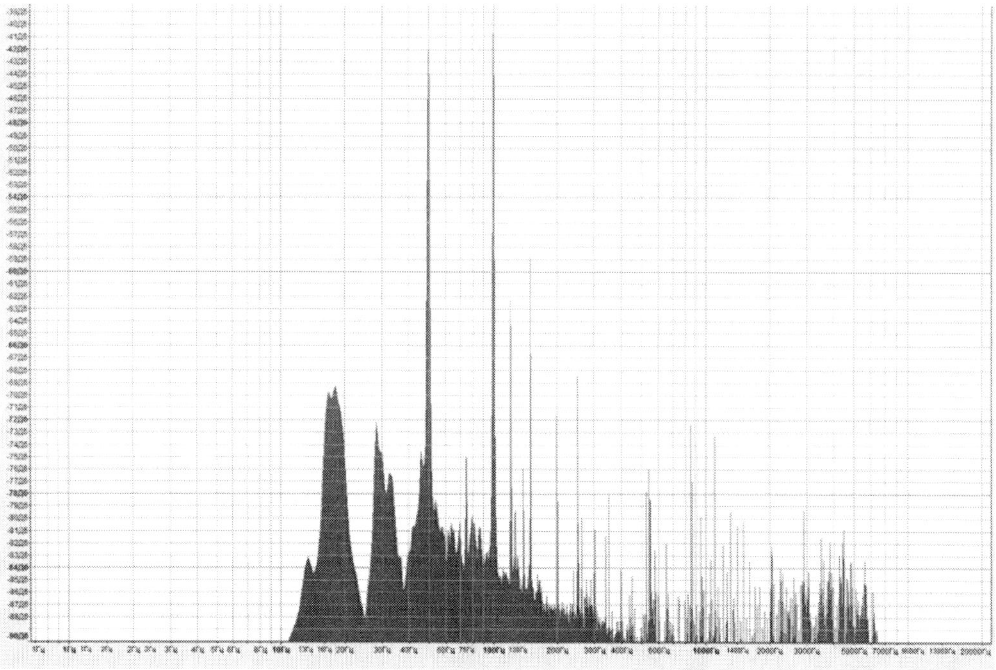

c

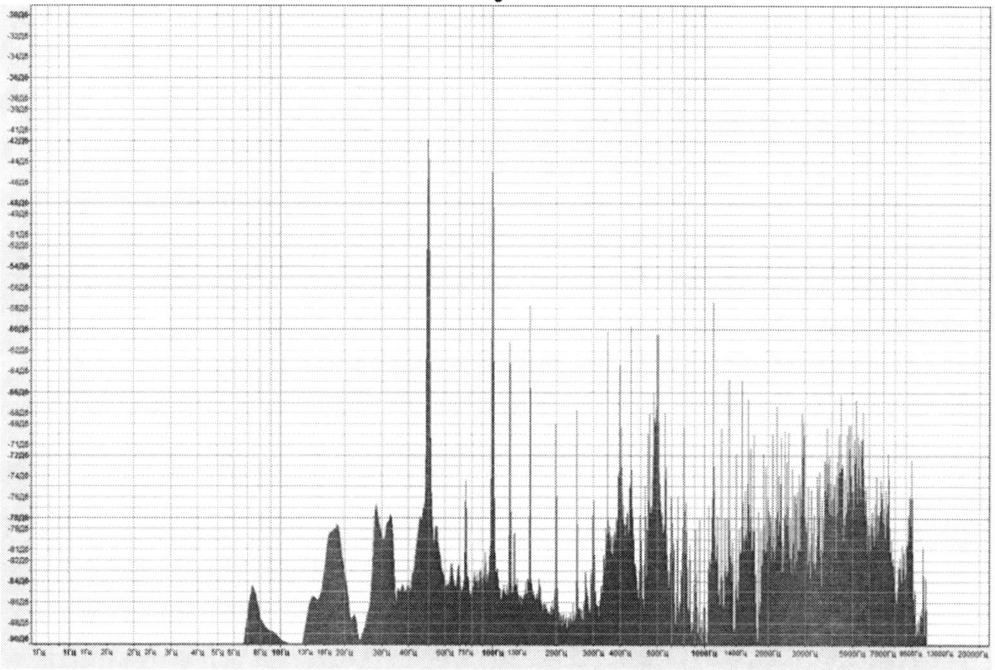

d

Fig. 5. Fourier spectra of acoustic emission signals recorded on the 1st (a), 15 (b), 55 (c) and 88 minutes (d) minutes of testing.

In the run mode of the connection (Fig. 5 b, c) is characterized by a substantial weakening of the high frequency and midrange components of the acoustic spectrum, show a decrease in the initial rate of microplastic deformation. General view of the spectra in this case is close to the mean spectra of the idle mode of operation.

A qualitative change in the spectral composition of the acoustic emission of friction associated with the violation of the integrity of the connection is clearly seen in the spectrum of Fig. 5 d. Here again amplified high frequency components, and in addition to this there are two strong mid-frequency components of the ≈400 Hz and about 600 Hz. The latter is necessary to characterize the modes of the contact gap, because their appearance is directly related to the wear of units, with a corresponding decrease in the contact stiffness and slippage in the contact.

Thus, use a public domain sound recording equipment and standard computer software allows, at least with respect to low-noise experimental installations, to carry out contactless acoustic control of the integrity of nominally stationary friction joints under conditions of fretting. Violation of the integrity of such joints is recorded as a change in the overall shape of ecostorm operating noise, and (more reliably) – characteristic changes in the spectrum of sound with a strong mid-frequency components of the spectrum arising from the contact gaps.

Low frequency self-modulation of the friction process, which is manifested in the spectrum of the acoustic emission is a consequence of the dynamic targeting slow angular motion in the system friction. A-modulation is a mistake to ascribe directly of cyclical damage accumulation and separation of wear particles.

4 Conclusions

The use of triboacoustic monitoring the fretting is practical for the real-time detection of a disturbance in the integrity of nominally motionless friction joints. This disturbance was determined by the change in the spectrum of the acoustograms of working noise, as well as by the characteristic changes in the sound spectrum due to the appearance of strong mid-frequency components caused by the occurrence of contact gaps. Low-frequency friction self-modulation results from the dynamic generation of slow angular motions in the friction system during the reverse energy cascade and is directed from the pumping modes toward low frequencies, which manifests in acoustograms and the spectrum of acoustic emission. In parallel with this, the direct energy cascade (toward high frequencies) is due to plastic deformations in the zones of the actual joint contact.

References

1. A. Sviridenok, N. Myshkin, T. Kalmykova and O. Kholodilov. *Acoustic and Electrical Methods in Tribotechnics,* Minsk, Nauka i Tekhnika (1987)
2. A. Akay, J. Acoust. Soc. Am., V. **111**, 4 (2002)
3. V. Vlasov, N. Mel'nichenko and E. Reizer, Trenie Iznos, 1989, V. **10**, 2 (1989)
4. Yu. Fadin, A. Leksovskii, B. Ginzburg and V. Bulatov, Pis'ma Zh. Tekh. Fiz., V. **19**, 5 (1993)
5. B. Gritsenko, Trenie Iznos, V. **26**, no. 5, (2005)
6. A. Kolubaev, E. Kolubaev, I. Vagin and O. Sizova, Tech. Phys. Lett., V. **31**, 10, (2005)
7. V. Rubtsov, E. Kolubaev, A. Kolubaev and V. Popov, Tech. Phys. Lett., V. **39**, 2, (2013)
8. V. Zaporozhets and V. Stadnichenko, J. Frict. Wear, V. **36**, 3, (2015)
9. A. Dykha, and A. Kuz'menko, J. Frict. Wear, V. **37**, 4, (2016)
10. V. Kurskoi, V. Slashchuk and A. Slashchuk, UA Patent 94006, Byull. Izobret. 20, (2014)

Measures of the vibration function pass in the wall element diagnostic test

Daniel Lubos[1], and *Mariusz* Żółtowski[2,*]

[1] University of Zilina, Faculty of Civil Engineering, Slovakia
[2] UTP University of Science and Technology in Bydgoszcz, Faculty of Management, Poland

Abstract. Building constructions as well as their constituent structural elements must meet the strength requirements, which will not jeopardize the safety of their use. As part of the research experiment, the focus was on the issue related to strength testing properties of masonry elements using a non-invasive test method involving the measurement of vibrations. 40 samples of bricks. Half of them were deliberately damaged so that it would be possible to assess the suitability of the measurement method used and its variability resulting from damage to the masonry element.

1 Introduction

Modal analysis is widely used in removing defects caused by infrastructure vibrations, structure modification, updating of the analytical model, or state control, and is also used to monitor the vibrations of structures in the aviation industry and land engineering mechanics [2,5,3].

The vibration test of building structures is not only to the measurement of the vibrations we are interested in. In order for researchers to be able to obtain interesting measurement values (such as the FRF function, stabilization diagrams and accompanying vibration estimators), it is necessary to carry out a whole sequence of operations, which results in the possibility of obtaining interesting data and results.

As part of this study, the methodology of investigations of masonry elements and course of conduct in order to obtain reliable results of conducted vibration tests were described. It was used to describe this process an example of the use of experimental modal analysis in the study of brick elements.

Traditional experimental modal analysis (EAM) uses input (excitation) and output (response) and measures it to estimate modal parameters, consisting of modal frequencies, attenuation and vibrations. However, traditional EAM has some limitations, such as [1,6,8]:

- in traditional EAM, artificial excitation is normally carried out to measure the vibration frequency;
- traditional EAM is usually carried out in a laboratory environment, but in many cases the true state of degradation may differ significantly from those tested in a laboratory environment.

This article presents the results of testing a brick masonry element in a dry environment using experimental modal analysis and appropriate software (partly authoring) used to carry out and visualize the results of such research.

2 Building structure dynamics

One of the basic criteria used in the design of modern building structures are the dynamic properties of the structure. They have a direct impact on system vibrations, noise emitted,

* Corresponding author: mazolto@utp.edu.pl

fatigue strength and structural stability. Analyzes of dynamic properties in most cases encountered in practice are made based on analysis of the behavior of the conglomerate model [4,7,10].

In most cases structural models (MES) are used to describe the dynamics of the structure, which consists in discretization of the system with a continuous distribution of parameters with the adoption of some simplifying assumptions, eg. related to the deflection line of the modeled element (transfer method). However, the models built in this way, for the purposes of dynamics analysis, give approximate results, the use of which is very limited. They require tuning based on the knowledge of properties measured on the actual object.

The use of vibrations in the study of the degradation (quality) of building structures results from the following reasons:

- vibration processes reflect physical phenomena occurring in constructions (deformations, stresses, cracks), on which the degree of destruction (fitness) and proper functioning depend, which results from the nature of the vibration process spreading;
- easy measurement of vibration processes in conditions of normal operation of the facility, without having to turn it off and special preparation, it enables a non-removal assessment of the state of destruction;
- vibration processes are characterized by high speed of information transfer in a time unit, defined by the Shanon formula:

$$C = F\,\mathrm{lg}_2(1+\frac{N_S}{N_Z}) \tag{1}$$

- depending on the spectral width of the process F and the ratio of the power of the useful signal NS to the power of interference noise NZ;
- vibration processes are characterized by a complex time, amplitude and frequency structure, which ensures proper evaluation of the state of the entire structure as well as its individual elements.

During the operation of the structure, due to the existence of a number of external factors (extortion of the environment, extortion from other structures) and internal (aging, wear, cooperation of elements) in the structure there are disturbances of equilibrium states which propagate in the elastic medium - the material from which the structure is built. The disorder is dynamic and maintains the conditions of balance between the state of inertia, elasticity, damping and extortion. This results in the dissipation of wave energy, their deflections, reflections and mutual overlap. The existence of sources and the spread of disturbances causes the occurrence of vibrations of the construction elements and the surrounding environment.

When separating the input processes, structure and output processes in the analysis of the dynamic state of the structure, one should remember about their random character.

The internal input, treated as a set of forcing quantities - defining the structure (shape, quality, clearance, etc.) and the method of elements cooperation is shaped in random conditions during production, which is revealed by random properties during operation.

External input, defining the conditions of cooperation of structural elements with other elements of the system (load changes, speed, environmental impact) is in practice also random.

The richness of the possibility of randomness and the existence of disturbances is the reason for additional assumptions about the inputs and the transformations of the structure destruction states. They refer to assumptions about linearity, stationarity and ergodicity of objects models and processes [6,8,9].

3 Measuring software

The latest measurement equipment purchased from SignalLab cooperation named Sigview was used to measure the time of force and response of the system as well as to determine the FRF function. This software allows you to easily perform modal analysis of brick elements, as well as any other building constructions - fig.1.

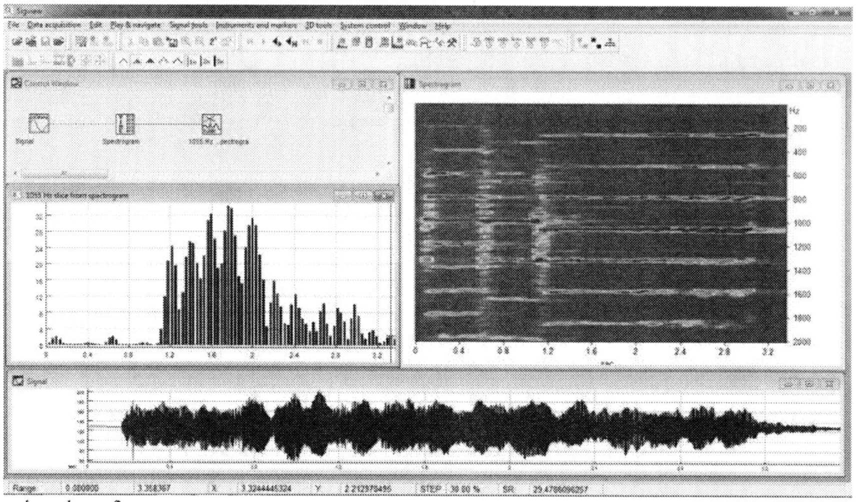

Fig. 1. Sigview interface.

In the measurement system, all the data needed to be calibrate, and the measuring path have been defined. For the needs of research carried out in this stage, it was started by defining the number of active measuring channels. Their number is limited only by the number of inputs on the measurement card, which is different for different models of measurement segments.

Two measurement channels were defined for the measurement using experimental modal analysis. According to the theoretical assumptions of the experimental modal analysis, the first sensor was reserved for a modal hammer (vibration excitation), and in place of 2 piezoelectric sensor was connected to measure the element's response to extortion - Fig.2.

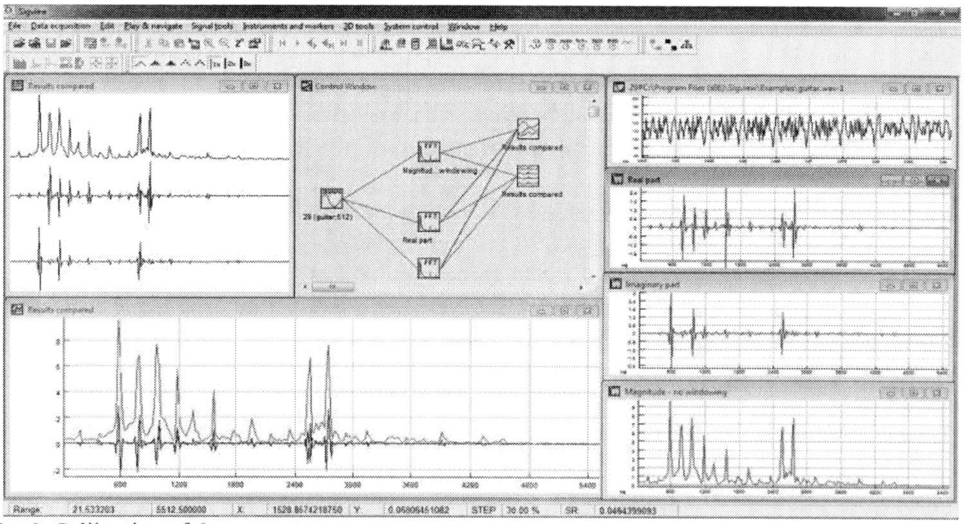

Fig. 2. Calibration of the sensors.

As part of the examinations that served to describe the course of investigation during the research, brick wall elements were examined - Fig.3. In the implementation of this kind of experiment, experimental modal analysis was used for measurements. For this purpose, 40 bricks were tested, out of which 20 samples were damaged intentionally, to show differences in the obtained test results, which demonstrates the suitability of a given test method for assessing masonry component degradation.

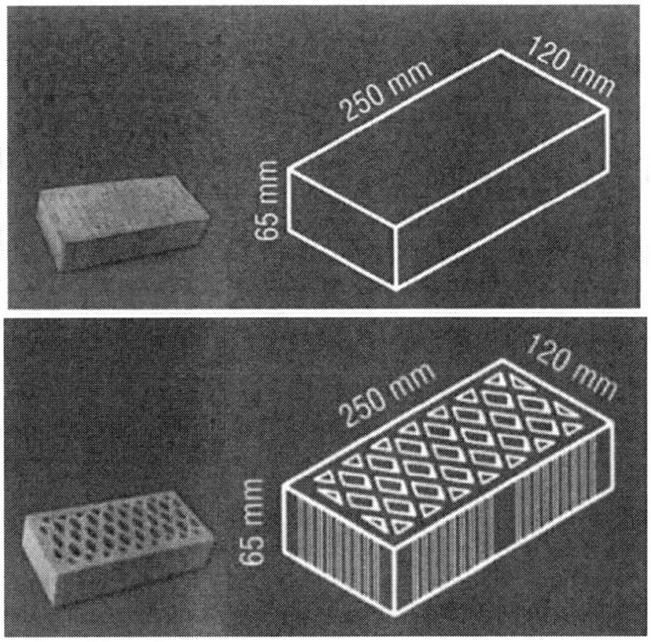

Fig. 3. View of samples subjected to measurements.

After calibrating the measurement system, it was possible to start measurements. According to the assumptions of the modal analysis, the samples were suspended on an un-stretchable line, which allowed to release all ties. Tests carried out only in the Z axis, because since they were brick masonry elements, it is interesting to pass the vibration signal in the direction of the compressive forces acting on the wall. During the measurement, the vibrations were forced by a modal hammer in the -Z direction, while the reception took place in the + Z axis due to the sticky sensor at the bottom of the element. As a result of the conducted tests, the time of force of forced force (modal hammer) and time courses of response (piezoelectric sensor) were obtained, and their visualizations are shown below in Fig.4.

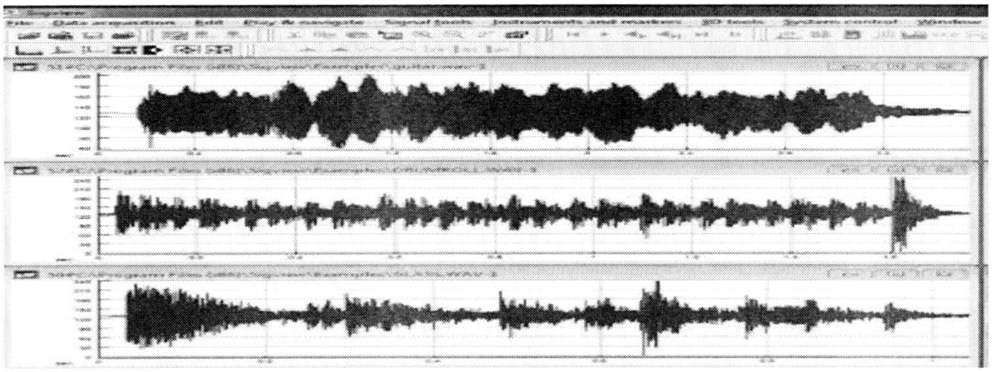

Fig. 4. View during measurement registration in Sigview software.

4 Methodology of data acqusition in the form of frf functions generation and calculation of it's surface area

After recording the time courses of force and response, these results should be subjected to further transformation. The aim of these operations is to obtain the FRF transition function and later to create stabilization diagrams on its basis, thanks to which it is possible to generate the natural frequency of the material being tested. The course of action is described in detail below to obtain the results that are interesting for the researcher. Two new programs operating in the MATLAB environment have been developed for further data processing.

During the vibration tests, the possibilities of acquiring new cognitive values, which may be useful to assess the destruction of the tested wall elements, were analyzed. Such information, apart from time and force transitions and stabilization diagrams together with the generated natural frequencies of the tested elements, may carry the numerical value of the surface area of the obtained signal transition function by the element being tested. To be able to calculate the area under the function curves, a new proprietary software named FUNCTIONS ANALYSIS was created. The program interface view is shown in the figure below.

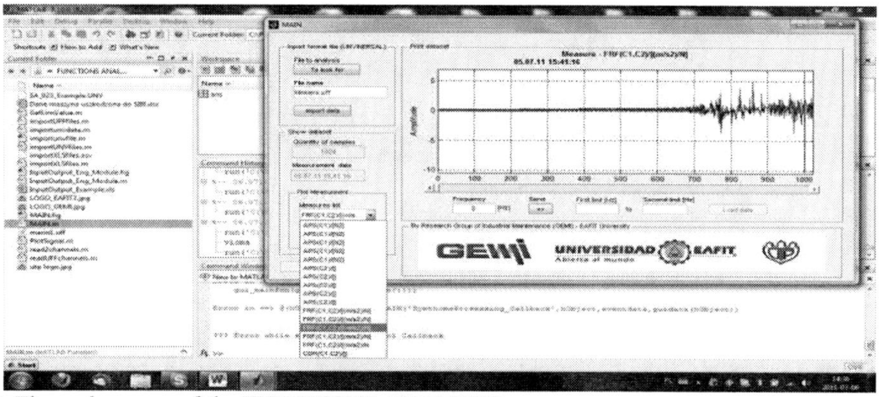

Fig. 5. The main screen of the FUNCTIONS ANALYSIS program.

The range of functions that we mark in the program for counting the surface area we are interested in is graphically marked by red. In the example drawing below, all the functions have been marked and the area of functions in the full frequency range was calculated - Fig. 6.

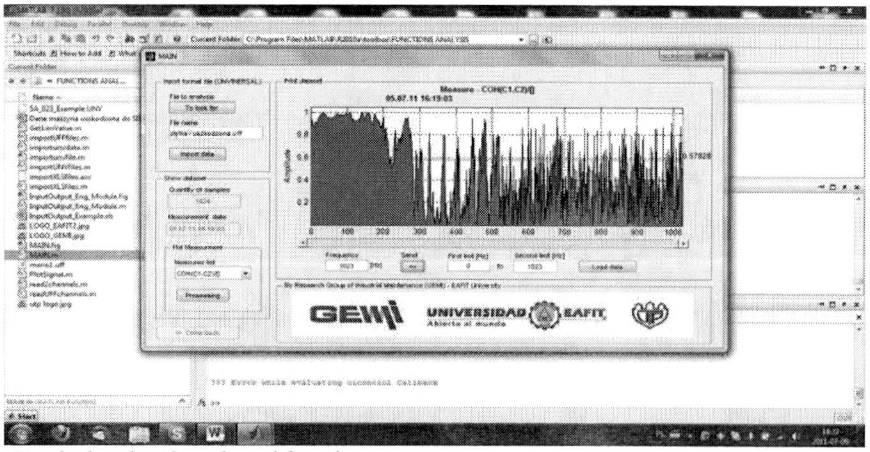

Fig. 6. Graph showing the selected function areas.

5 Results

During the tests, the vibration signal transition function was generated by the structure (FRF function) and stabilization diagrams with the applied vibration frequencies for each element.

The following are the averaged results of the tests in the form of stabilization diagrams, which were prepared for 10 measurements (10 extortions and 10 responses) for the good components and damaged wall elements.

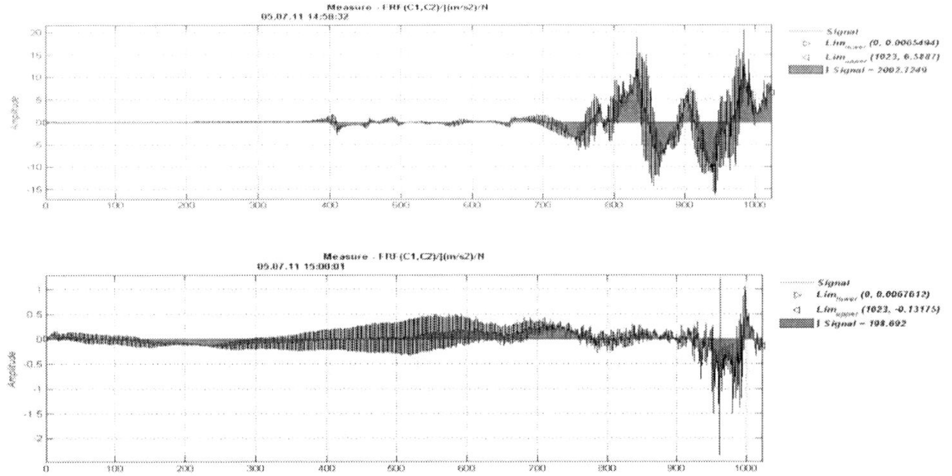

Fig.7. Area of the FRF function for solid and damaged brick in the X axis.

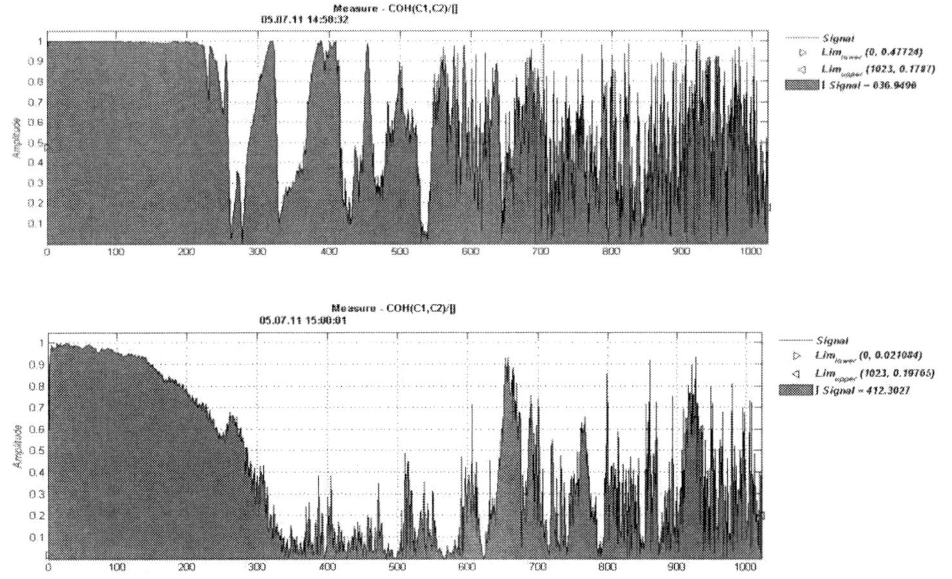

Fig.8. Coherence function surface fields for full and damaged brick in the X axis

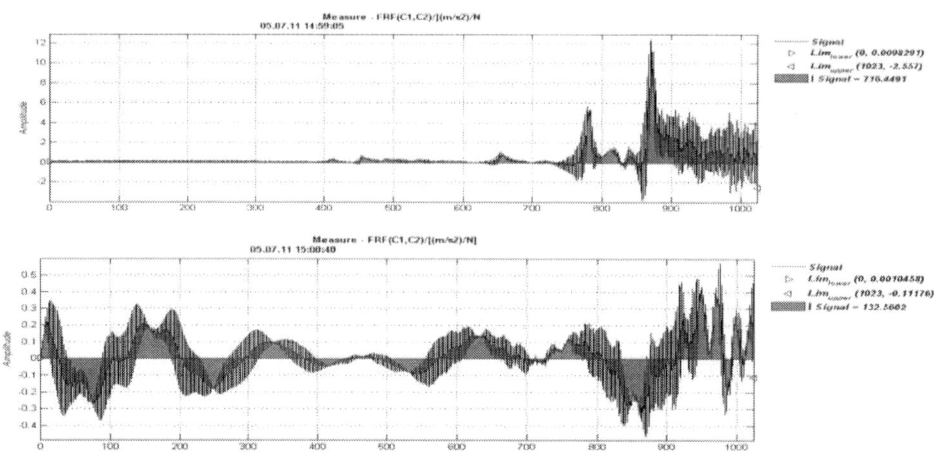

Fig. 9. Area of the FRF function for solid and damaged brick in the Y axis.

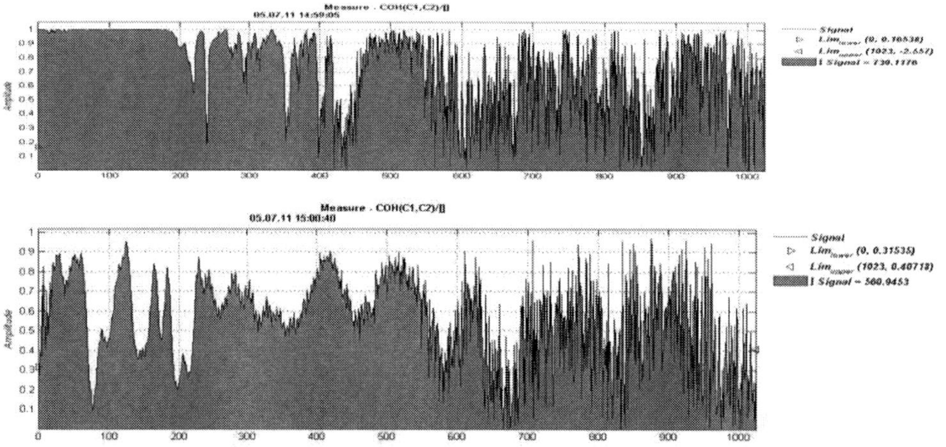

Fig. 10. Coherence function surface fields for full and damaged brick in the Y axis.

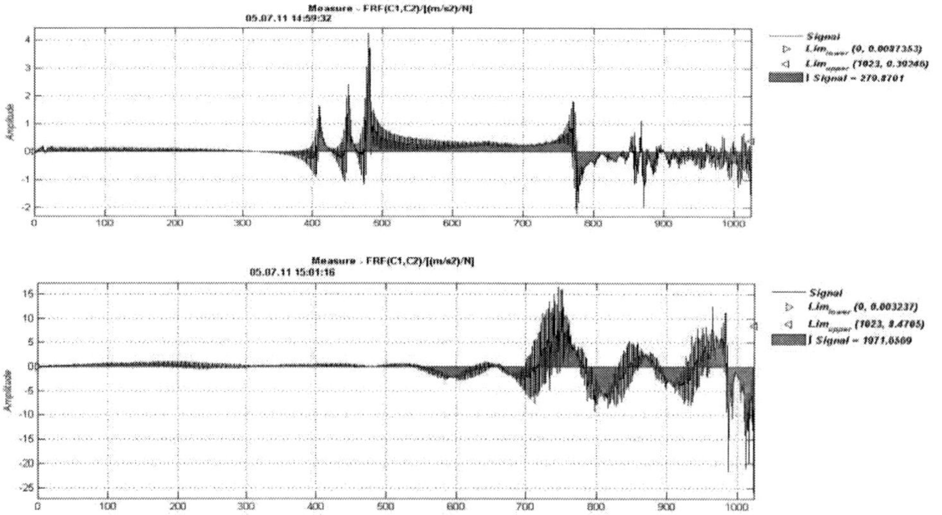

Fig. 11. Area of the FRF function for solid and damaged brick in the Z axis.

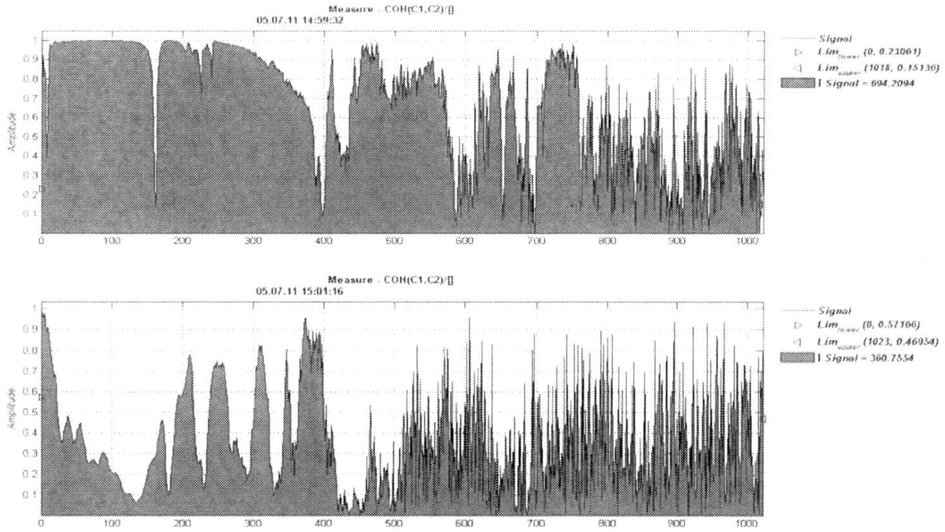

Fig.12. Coherence function surface fields for full and damaged brick in the Z axis.

6 Summary

The publication describes a thoroughly complex process of obtaining measurement data, which can be considered as a unified system of exploitation of test results and their validation and generation of significant values of natural frequency

The presented research results indicate that it is possible to distinguish between material properties, which affects the ability to distinguish their strength properties. The tests also confirmed the suitability of measuring equipment for tests using exploitative modal analysis performed on real structural elements.

References

1. B. Żółtowski, M. Żółtowski, *Vibration signals in mechanical engineering and construction* (ITE-PIB 2015)
2. B. Żółtowski, C. Cempel, *Engineering of diagnostics machines* (PTDT, ITE – PIB 2004)
3. M. Żółtowski, *Investigations of harbour brick structures by using operational modal analysis.* Polish Maritime Research No 1, (2014)
4. M. Żółtowski, B. Żółtowski, *Vibrations signal to the description of structural damage of dynamic the technical systems* (XIII International Technical Systems Degradation Conference 2015)
5. M. Żółtowski, *Opis drganiowy konstrukcji budowlanych.* Logistyka nr.6/2014 (2014).
6. M. Żółtowski, *Investigations of harbour brick structures by using operational modal analysis.* Polish Maritime Research No. 1/(81) vol.21 (2014)
7. M. Żółtowski, M. Liss, *The use of modal analysis in the evaluation of welded steel structures.* Studies and Proceedings of Polish Association for Knowledge Management Vol.79 (2016)
8. M. Żółtowski, M. Liss, *Zastosowanie eksperymentalnej analizy modalnej w ocenie zmian sztywności prostego elementu konstrukcyjnego.* Studies and Proceedings of Polish Association for Knowledge Management Vol. 80 (2016)
9. M. Żółtowski, R.M. Martinod, *Technical Condition Assessment of Masonry Structural Components using Frequency Response Function (FRF).* Masonry International Journal of the International Masonry Society Vol.29,No.1 (2016)

10. M. Żółtowski, R.M. Martinod, *Quality identification methodology applied to wall-elements based on modal analysis*. Civil Engineering the Athens Institute for Education and Research, Emerald, Athens (2015)

Application of surface frequency functions to assess the state of the kinematic pair elements

Maciej Matuszewski[1*], *Małgorzata* Słomion[1], *Janusz* Musiał and *Andres* Bustilo[2]

[1]University of Science and Technology in Bydgoszcz, Faculty of Mechanical Engineering, 85-796 Bydgoszcz, Poland
[2] University of Burgos, Department of Civil Engineering, Avda Cantabria s/n, Burgos 09006, Spain

Abstract. In this paper the possibility of using frequency functions of the kinematic pair surface layer elements was analyzed for evaluation of the surface layer condition. The function of spectral power density and autocorrelation which were accepted for analysis was characterized. Evaluation of the elements condition was made on the quantify way as changes in degree of isotropy and in the qualitative way based on the graphs of the analyzed frequency functions. After tribological tests the surface structures were adopted to assess the changes. Spectral graphs for structures without cooperation (directly after-machining) and for the different friction distance, were analyzed. Conditions during tribological tests were taken as permanent. The determined degree of isotropy basis on the frequency functions and received graphs of spectra power density function and autocorrelation become useful in evaluation of the surface layer changes during its transformation.

1 Introduction

Each technical object is characterized by a set of functional qualities. In case of machines their capabilities, e.g. durability and reliability are result of the features represented by the elements that create the structural combination of the object. These combinations are a set of kinematic pair so it can be concluded that the aforementioned set of functional qualities is determined by the features of individual kinematic pairs.

As the numerous research shows [1-8] it was found that the functional characteristics depend to the greatest extent of the surface layer condition, which is formed throughout the entire production process. Condition of the surface layer is mainly determined by the surface stereometry, i.e. the outer part of the surface layer. The stereometric shape of the surface is defined as the surface geometrical structure. The stereometric surface structure is a set of surface micro roughness which are traces of machining operation or effects of the wear process. Roughness, waviness, degree of surface isotropy, traces directionality of machining, shape deviations and surface defects are assumed as the basic quantities describing the geometric surface structure [9, 10]. Depending on the distribution of the characteristic elements of the geometric surface structure, the surface may be anisotropic or isotropic.

The reason for functionality loss of the elements forming friction nodes is limited wear resistance their surface layer on the damaging effects of external forces. Influence of these factors (mechanical, thermal, chemical, electrical) over the cooperation of kinematic pair, causes changes in the surface layer. These changes are generally the reason of loss kinematic pair ability to fulfill the design-related functions in a different grade.

The evaluation of the surface layer cooperating elements changes is described by quantitative and qualitative changes. Various roughness parameters, weight loss, changes in

*Corresponding author: matus@utp.edu.pl

linear dimensions are the quantities accepted for the description of the quantitative changes [11-14]. Whereas, the evaluation of qualitative changes is made among others on the basis of photographs of the examined surfaces or spectral analyzes of these surfaces [10, 15]. On the basis of qualitative analysis we get information about general features of the geometric surface structure (roughness and directionality), effects of wear (traces of wear) and type of possible damage.

In this paper frequency characteristics were described for the quantitative and qualitative analysis of the geometric surface structure condition, which describe the frequency structure of the tested surfaces.

2 Frequency surface functions

Space-frequency functions are commonly used and these are mainly spectral power density and autocorrelation. Frequency analysis is a source of information of the manufacturing process. Articles [16-19] present examples of processing monitoring with the use of frequency functions and conclusion on these grounds about various process factors, e.g. machine cutting forces and their components, whereas in papers [20, 21] these functions were used as an information about cutting edge condition of the cutting tool and machine tool. On the basis of these functions, it is possible to determine isotropy degree of the geometrical surface structure.

Functions of spectral power density and autocorrelation describe the frequency surface structure. Functions of spectral power density and autocorrelation during the manufacture and shaping the geometric surface structure, as well as wear process are useful in an analysis the surface topography.

The surface autocorrelation function is a symmetrical real-valued function. Its value at zero dislocation is the maximum value equal to the variance of the irregular crests heights. It is a measure of the dependence of the data value in one position from their value in the second position and reacts very sensitively to any surface structure interferences [22, 23]. In research, depending on their purpose, one of the above functions is generally used. It results from mutual relations, namely the power spectral density function is a Fourier transform of the autocorrelation function. The power spectral density function illustrates how the deviation of surface irregularities decomposes with the surface frequency.

Generally, on the basis of the spectral analysis of the surface structure, we can draw the following conclusions [22, 23]:

- low frequency components are more significant in the spectral analysis, the role of high frequency components is negligible,
- on the periodic surfaces in one or two perpendicular directions, the energy is concentrated in the space of proper frequency along the same directions,
- on the anisotropic surfaces energy is concentrated along the perpendicular direction to the structure directionality (symmetry),
- in case of mixed anisotropic surfaces with the participation of a short-wave random component, decisive are long-term components defining surface anisotropy,
- in anisotropic random surfaces, characteristic features of random noise appear the background of clearly dominant components with low frequency, while the surface structure retains anisotropy and directionality,
- random noise dominates in random isotropic structures and the features of anisotropy are negligible.

3 Analysis of exemplary applications of frequency functions

Selection of the various surface roughness parameters to determine condition of the kinematic pair surface layer elements is carried out on the basis of scientific research, statistical correlation between them and availability of equipment. Selected parameters which describe the surface structure with reference to its functional qualities may not be always clearly classified. Because of this the choice is always more or less subjective. Therefore, use of the geometric surface structure description based on its mathematical functions, especially on the surface autocorrelation or spectral power density seems to be a procedure without a subjective model parameters selection error. Frequency analysis makes it possible to assess the geometric surface structure independently from machining type.

It can be confirmed by an comparative analysis of the geometric surface structure presented in this paper [23]. The analyzed surfaces (Fig. 1) were obtained after the following processing:

- throwing at high cutting speed,
- longitudinal grinding,
- deep grinding,
- superfinish.

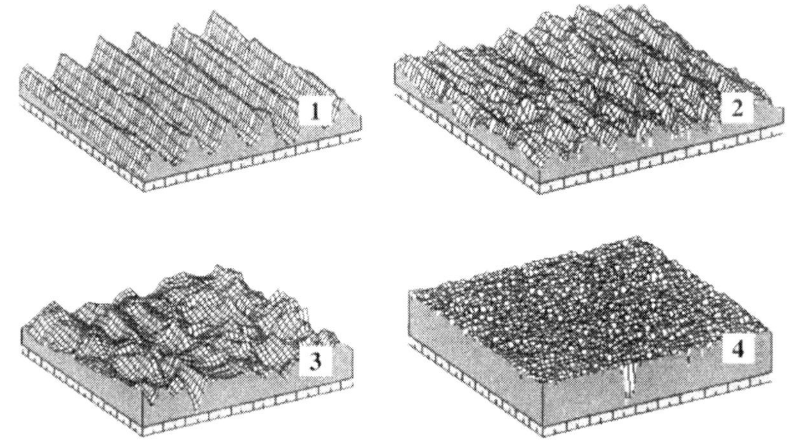

Fig. 1. Obtained surfaces by: 1 – throwing, 2 – longitudinal grinding, 3 – deep grinding, 4 – superfinish [23]

The surface after throwing and longitudinal grinding is clearly anisotropic, the surface after superfinish is isotropic, while the surface after deep grinding is mainly anisotropic.

Figure 2 presents the graphs of the autocorrelation function of these surfaces and figure 3- the degree of isotropy in form of rose of directions.

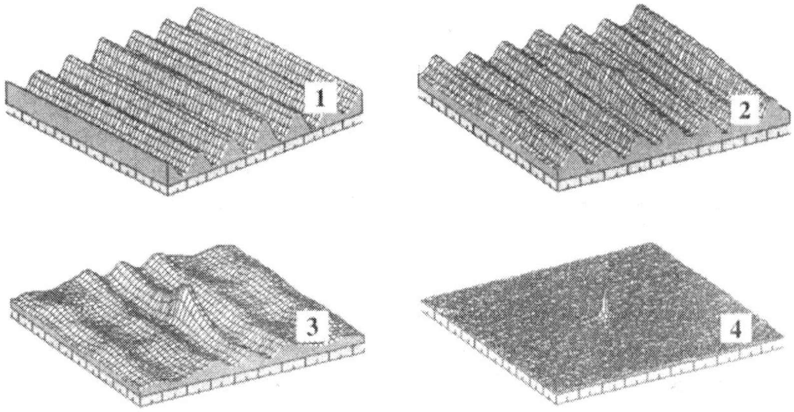

Fig. 2. Graph of surface autocorrelation function (surface designation as in Fig. 1) [23]

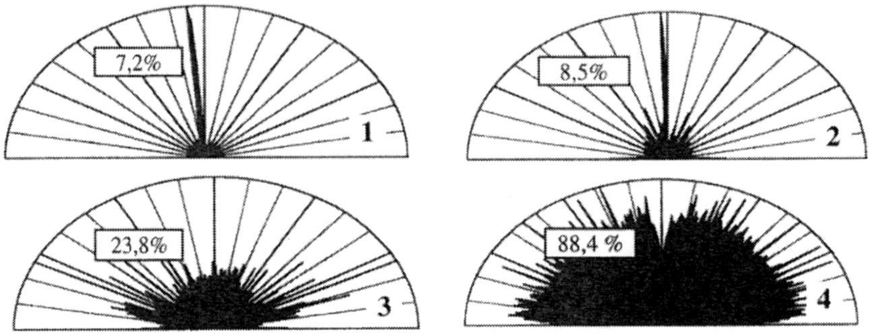

Fig. 3. Degree of surface isotropy (surface designation as in Fig. 1) [23]

At the presented figures the autocorrelation function graphs change can be notice and the isotropy degree can be specified precisely on the basis of this function depending on the analyzed surface topography. Account the surface isotropy determined from the frequency function can be a factor contributing to better evaluation the condition of analyzed surface, especially of its potential functional qualities and tribological characteristics.

In order to verify the usefulness of spectral analyzes evaluation for quantitative and qualitative changes occurring in the surface layer cooperating elements experimental studies were carried out. Specimens was subjected to tribological tests and the degree of isotropy was accepted for quantitative evaluation of changes taking place but for the qualitative assessment graphs of spectral power density was accepted as well as autocorrelation function of the tested structures on a different friction distance. Conditions during tribological tests were mosty taken as permanent. Specimens and counter-specimen were cooperating in the lubricating medium which was machine oil (L-AN 68) and the operating parameters were as follows: relative motion speed: 2,9 m/min (0,05 m/s), loads: 300, 450 and 600 N in the quantitative analysis and 600 N in the qualitative analysis. For this loads the theoretical pressure in contact zone was respectively: 1; 1,5 i 2 MPa. Cooperation angle between characteristic directionality of specimens and counter-specimen was 0°. Specimens were made of 102Cr6 steel, whereas counter-specimen was made of X210Cr12 steel. Hardness of counter-specimen was 50% higher in order to enable changes on the specimens surface. Hardness values were respectively 60 HRC and 40 HRC.

Figure 4 shows changes of the geometrical surface structure isotropy degree in funtion of friction distance.

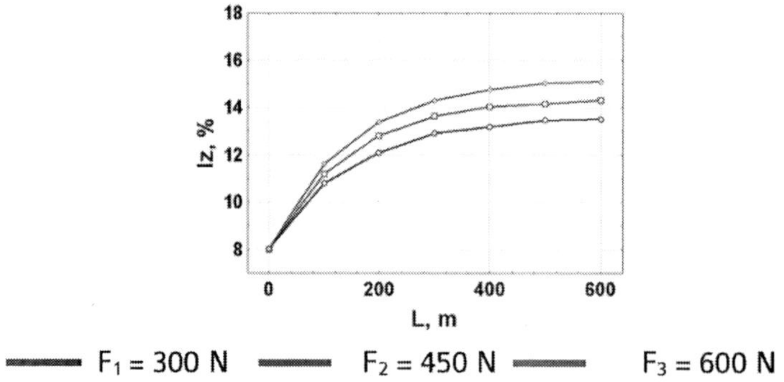

Fig. 4. Change of isotropy degree *Iz* in function of friction distance *L* for different loads

On the basis of graphs it is possible to draw conclusion that the isotropy degree changes with the increase in friction distance. The changes also depend on realized loads and they are the result of surface layer structure change during operation. The value of isotropy level is getting higher which may have the following interpretation: the formation of the stereometric structure is "flattened" and the symmetry of this configuration increases. Single pix and microroughness crest undergo partial or total abrasion. Traces of machining decided about directions of shaping and together with increase of isotropy degree they get distorted. General assessment of description of the surface layer changes on the basis of isotropy degree confirmed the usefulness of these parameter use in the analysis of changes taking place in kinematic pair surface layer elements during operation.

In figure 5 and 6 the quality analysis on the basis of the graphs of the power spectral density function and autocorrelation for selected friction distances (100, 500 and 2000 m) was presented.

a)

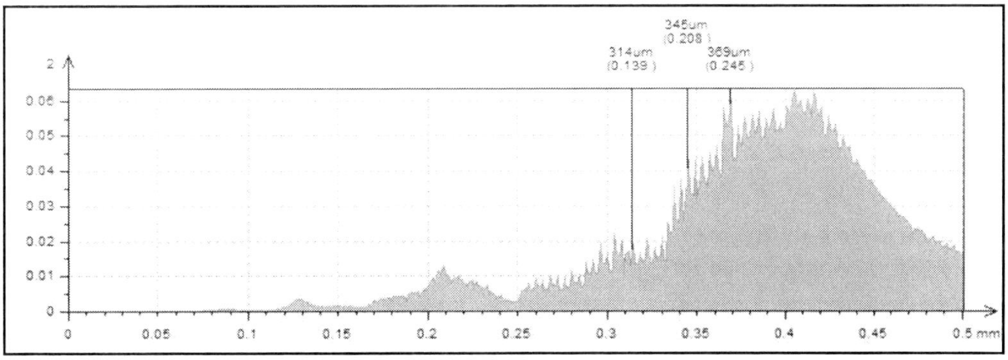

b)

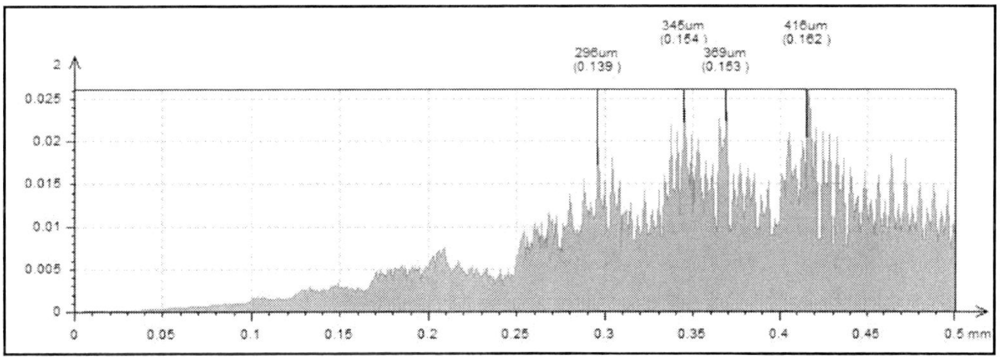

c)

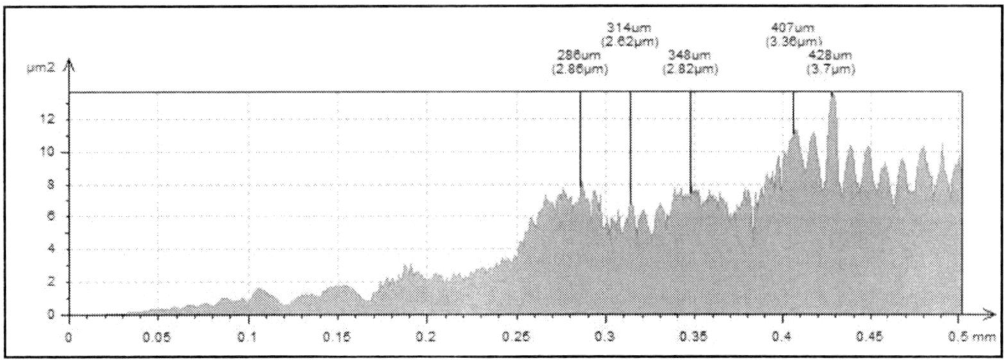

d)

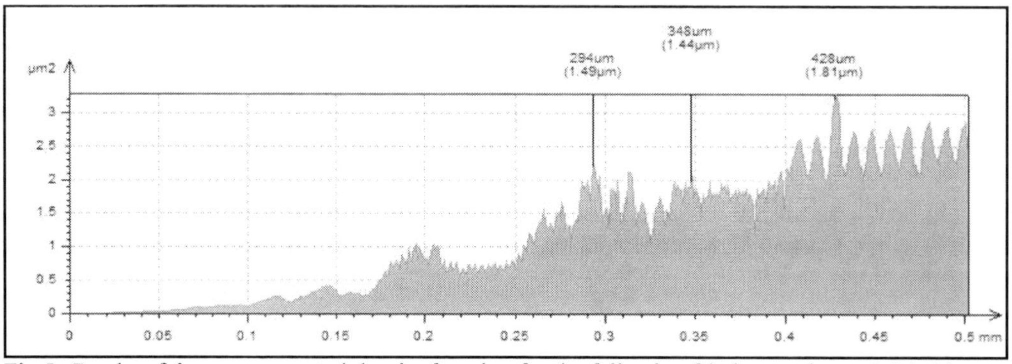

Fig 5. Graphs of the power spectral density function for the following friction sections: a) 0 (without the cooperation), b) 100, c) 500, d) 2000 meters

The above graphs shows how the spectral power density function changes at the particular cooperation stages. At the beginning (Fig. 1a), when the specimen did not cooperate one dominant may be noticed in the graph which follows from the surface character (defined directionality of geometric surface layer) and hence the determined way of machining and accepted parameters. In the remaining graphs there is no longer one dominant and this is due to changes arising as a result of surface cooperation. Traces of treatment disappear and the direction of these traces is not so clearly privileged.

a)

b)

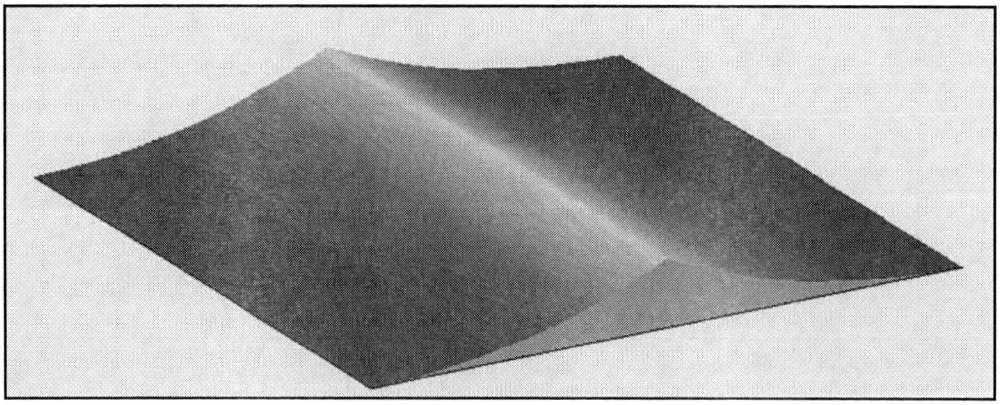

c)

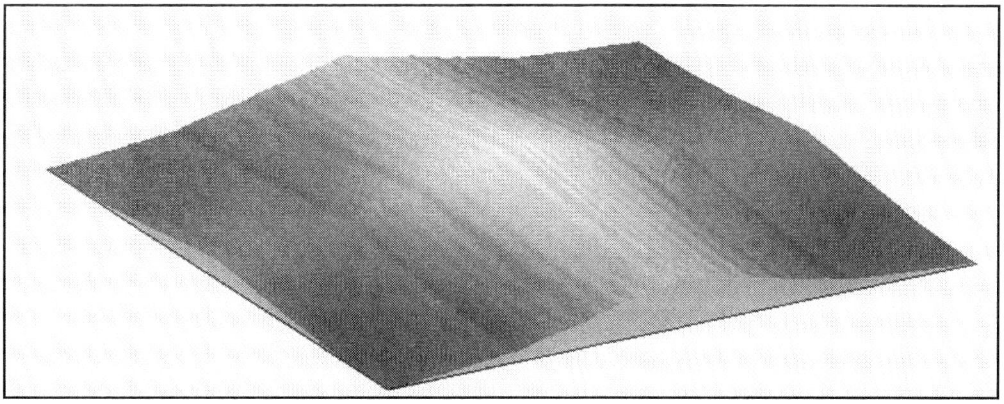

d)

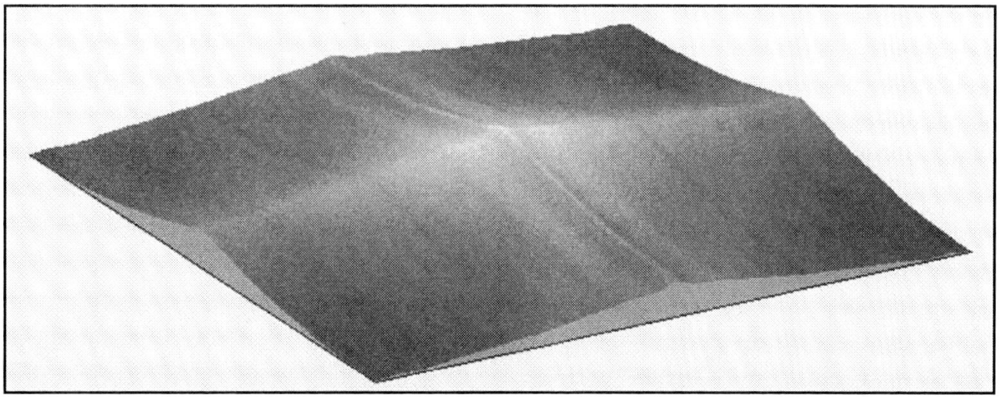

Fig. 6. Graphical form of the autocorrelation function for the following friction segments: a) 0 (without cooperation), b) 100, c) 500, d) 2000 meters

From the presented drawings one can observe how the geometric structure of the anisotropic surface (Fig. 2a) changes into a mixed structure without a clearly privileged direction, but with the still existing directivity of the geometric surface structure (Fig. 2d). However, no full change in the isotropic structure is observed, with complete disappearance directionality of the geometric surface structure.

4 Summary

On the basis of the analyzes carried out in this paper, the usefulness of surface frequency functions to assess the condition of the surface layer of cooperating elements of kinematic pairs was found. Due to the subjective selection to assess the state of the surface structure of parameters that characterize it, e.g. roughness parameters, the use of surface frequency functions for this assessment seems to be a more favorable solution, not subject to the error of subjective model selection of parameters. The obtained graphs of the power spectral density and autocorrelation proved to be useful in qualitative assessment of surface structure. It is possible to analyze the occurring changes of the surface layer during its transformation.

In order to provide a more complete quality assessment of the surface layer of cooperating elements, it would also be necessary to analyze the tested surfaces based on their 2D or 3D images. It is then possible to additionally conclude about the details of surface stereometry.

References

1. J. H. Horng, J. F. Lin, K. Y. Li, *Effect of surface roughness on stell roller scuffing.* Wear, **184**, 203-212 (1995)
2. T. Kałaczyński, M. Łukasiewicz, J. Musiał, R. Polasik, M. Szczutkowski, N. Dluhunovych, J. Wilczarska, T. Kasprowicz: *Analysis of the diagnostic potential research thermovision in the technical state of combustion engine injectors assessment*, 24th International Conference Engineering Mechanics 2018, Svratka, Czech Republic, Book of full texts, Institute of Theoretical and Appiled Mechanics of the Czech Academy of Sciences, Prague, 357-360 (2018)
3. J. Krawczyk, *Tribological properties of selected structural steels (Właściwości tribologiczne wybranych stali konstrukcyjnych).* Tribologia, **4**, 223-233 (2010)
4. E. Mańka, M. Słomion, M. Matuszewski, *Constructional features of ropes in functional units of mining shaft hoist.* Acta Mechanica et Automatica, **12**, 1, 66-71 (2018)
5. M. Matuszewski, T. Mikołajczyk, D. Yu. Pimenov, M. Styp-Rekowski, *Influence of structure isotropy of machined surface on the wear process.* International Journal of Advanced Manufacturing Technology, **88**, 9, 2477-2483 (2017)
6. M. Matuszewski, M. Styp-Rekowski, *Influence of texture direction of kinematic pair elements surfaces on service operated layer transformation.* Tribologia **4**, 87-97 (2006)
7. Ł. Muślewski, *Evaluation Method of Transport Systems Operation Quality.* Polish Journal of Environmental Studies, **18**, 2A, Hard Olsztyn, Olsztyn (2009)
8. Ł. Muślewski, M. Pająk, B. Landowski, B. Żółtowski, *A method for determining the usability potential of ship steam boilers.* Polish Maritime Research, **4** (92), 23, 105-111 (2016)
9. T. Burakowski, T. Wierzchoń, *Metal surface engineering* (*Inżynieria powierzchni metali*), (Publ. WNT, Warszawa, 1995)
10. R. Górecka, Z. Polański, *Metrology of the surface layer* (*Metrologia warstwy wierzchniej*), (Publ. WNT, Warszawa 1983)
11. A. Bustilo, D. Yu. Pimenov, M. Matuszewski, T. Mikołajczyk, *Using artificial intelligence models for the prediction of surface wear based on surface isotropy levels.* Robotics and Computer Integrated Manufacturing, **53**, 5, 215-227 (2018)
12. M. Łukasiewicz, T. Kałaczyński, J. Musiał, J. Shalapko, *Diagnostics of buggy vehicle transmission gearbox technical state based on modal vibrations.* Journal of Vibroengineering, **16**, 6, 3137-3145 (2014)
13. M. Styp-Rekowski, *Geometrical constructional features of special rolling bearings against their exploitational properties*, Proceedings of IVth Symposium INTERTRIBO '90, **C**, 93-96 (1990)
14. M. Styp-Rekowski, E. Mańka, M. Matuszewski, M. Madej, D. Ozimina, *Tribological problems in shaft hoist ropes wear process.* Industrial Lubrication and Tribology, **67**, 1, 47-51 (2015)
15. P. Pawlus, *Surface topography: measurement, analysis, impact* (*Topografia powierzchni: pomiar, analiza, oddziaływanie*), (Publ. Rzeszow University of Technology, Rzeszów, 2006)
16. W. Lubimow, K. E. Oczoś, *Selected issues of shaping surface irregularities in machining processes* (*Wybrane zagadnienia kształtowania nierówności powierzchni w procesach obróbkowych*). Mechanik, **3**, 81-84 (1997)
17. B. Nowicki B., *Geometric structure: surface roughness and waviness* (*Struktura geometryczna: chropowatość i falistość powierzchni*), (Publ. WNT, Warszawa, 1991)
18. B. G. Rosen, R. Ohlsson, J. Westberg, 1995. *Interactive surface modelling, an implementation of an expert system for specification of surface roughness and*

topography. International Journal of Machine Tools and Manufacture, **35**, 317-324 (1995)

19. D. J. Whitehouse, *Process and quality control using surface finish*. Proceedings 6th Symposium IMEKO Metrology for quality control in production, Vienna, 699-710 (1998)
20. J. Raja, *Field testing of machine tool diagnostic techniques using surface metrology*. Annals of CIRP, **32**, 503-506 (1983)
21. Y. S. Wong, A. Y. C. Nee, X. Q. Nee, X. Q. Li, C. Reisdorf, 1997. Tool condition monitoring using laser scattering. Journal of Material Processing Technology, **63**, 205-210 (1997)
22. K. E. Oczoś, W. Lubimow, *Determinism and randomness of surface geometric structures (Determinowość i losowość struktur geometrycznych powierzchni (SGP))*. Pomiary Automatyka Kontrolna, **10**, 2002, 4-6 (2002)
23. K. E. Oczoś, W. Lubimow, *Surface geometric structure (Struktura geometryczna powierzchni)*, (Publ. Rzeszow University of Technology, Rzeszów, 2003)

The comparison of vibro-acoustic impact of chainsaws with electric and combustion drives

Wojciech Rukat[1,*], *Roman* Barczewski[1], *Bartosz* Jakubek[1], and *Mateusz* Wróbel[1]

[1] Poznan University of Technology, *Faculty of Mechanical Engineering and Management, Piotrowo 3 St, 60-965 Poznan, Poland*

Abstract. Chainsaws generate high level vibrations and noise during work. No matter what type of the drive (electric, combustion) they exceed the permissible levels defined for a standard eight-hour working day. The intensity of a vibro-acoustic impact on chainsaw operators depends on many factors. One of them is the type of the drive. This is due to significant differences in saw design (inter alia characteristics of the drive, rotational speeds – quasi-constant for an electric drive and variable for a combustion one). For chainsaws of the same or similar power and different types of the drive the levels of vibrations and noise may vary. The paper compares the noise levels and frequency weighted vibration accelerations for different types of the drive. The tests were performed with the same interchangeable cutting set (guide bar, cutting chain). The repeatability of the cutting process was ensured. The test results may be helpful when choosing the tool.

1 Introduction

There exist various mechanized hand-held tools used for timber production and processing. Chainsaws fall within this product group. Most of tools of this type are driven by means of a combustion engine or an electric engine powered by a battery. This is because the work has to be performed under field conditions requiring mobility or at heights. The mains powered electric chainsaws constitute a separate group of tools. Such tools are mainly used for household and construction works. Because of the character of work a chainsaw operator is particularly exposed to various harmful factors, such as vibrations, noise, chemical substances, and dust. Moreover, moving parts of the cutting system [1-3] cause the risk of various limb and body injuries.

The petrol end electric chainsaws available on the market, and especially the models for non-professional use may be characterized by lower quality of used materials and workmanship. Designers of such models pay less attention to ergonomic features. As a result, this leads to an increase in operators' exposure to the above mentioned factors. An analysis of catalogue data for 57 tools of 10 brands has shown that hand-arm vibration (frequency weighted vector sums of r.m.s. values of vibration accelerations – a_{hv}) declared by manufacturers, recorded at the handles of electric chainsaws (determined according to the EN ISO 60745-2-13 standard [4]) do not exceed in any case the value of 11.2 m/s^2. This is the maximum value of a_{hv} with exposure time of up to 30 minutes. However, the declared a_{hv} values of vibrations exceed the permissible value defined for an eight-hour working day. Their range is between 2.9 and 6 m/s^2.

In case of chainsaws with combustion engines the catalogue data analyzed for 157 tools of 12 different brands show that the a_{hv} values determined according to ISO 7505:1986 [5] or ISO 22867:2011 [6] range between 4.4 and 7.1 m/s^2. These values also do not exceed the maximum value for a short-time exposure (11.2 m/s^2), simultaneously exceeding the permissible values for an eight-hour working day. To minimize negative impacts of vibrations on

* Corresponding author: wojciech.rukat@put.poznan.pl

operators it is justified for them to use personal protective equipment, such as anti-vibration gloves [7].

User's manuals contain also data concerning noise emission. The analysis these data shows that the A-weighted sound pressure level, measured according to [4] for electric chainsaws ranges between 83.7 and 96 dB. In case of petrol chainsaws the A-weighted sound pressure level, declared by manufacturers and determined according to ISO 22868:2011 [8] ranges between 100.1 and 105.3 dB. That is why it is necessary to protect operator's hearing by means of personal protective equipment, such as, for example, earmuffs [9].

Test results presented in some papers indicate, that the vibration level emitted by chainsaws depends on:

- the tool itself [10-15],
- processed material [16-18], and
- the operator [19, 20].

The vibration levels of chainsaws, declared by their manufacturers are purely indicative. This stems from the simplifying assumption, where the same time shares are assumed for different load conditions of the drive in the overall working time of the device. For petrol chainsaws they equal to 1/3 for work at Idle Speed (ID), maximum speed without load (Racing – RA) and maximum load (Full Load – FL) each. Equation (1) allows to determine the value of vibration accelerations for a petrol chainsaw according to [5, 6].

$$a_{\mathrm{hv,eq}} = \sqrt{\frac{1}{3}\left(a_{\mathrm{hv,ID}}^2 + a_{\mathrm{hv,FL}}^2 + a_{\mathrm{hv,RA}}^2\right)} \quad (1)$$

In case of electric chainsaws the shares equal to 1/2 for work at maximum speed without load (Racing – RA) and at maximum load (Full Load – FL) each [4]. The design of an electric chainsaw makes it impossible to work at idle speed. The value of vibration accelerations is determined according to [4] by means of Equation (2).

$$a_{\mathrm{hv,eq}} = \sqrt{\frac{1}{2}\left(a_{\mathrm{hv,FL}}^2 + a_{\mathrm{hv,RA}}^2\right)} \quad (2)$$

The figure below presents percentage time shares for the respective working conditions for petrol and electrical chainsaws according to the standard.

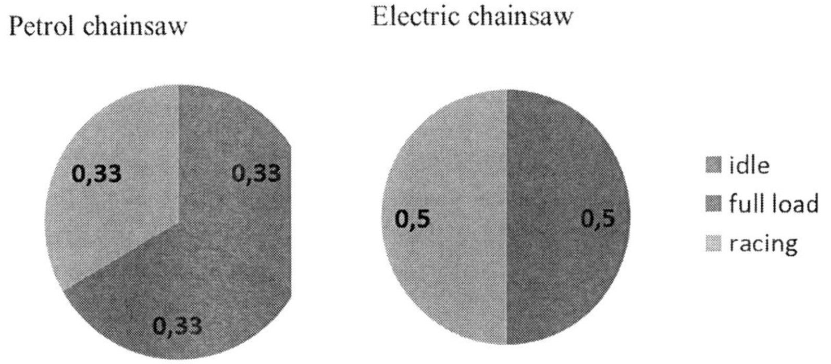

Fig. 1. Time shares of working conditions in the overall working time of a chainsaw depending on the type of its drive

Another important factor is the way vibrations are transmitted to the device handles. The petrol chainsaws are equipped with a vibro-isolation system separating their driving units from the handles. In the older models there were rubber shock absorbers mounted, which have currently been replaced by a system of coil springs or sometimes by combined elements. The electric chainsaws usually do not have any vibration reduction system, and their handles are permanently joined with the driving unit making a whole manufactured as a monolith.

As it was shown in the paper by K. Wójcik [21] a precise determination of time shares for the individual operating modes (ID, RA, FL) in the total operating time depending on the performed processing operation (felling, pruning, cross-cutting) is of key importance at the assessment of vibration and noise exposure.

Determination of a_{hv} according to Equations (1) and (2) is only justified in case of estimation of the impact of local vibrations on a chainsaw operator for a particular type of the device. A comparison of vibration impacts of electric and petrol chainsaws having similar parameters based on these equations would not be completely unbiased. This is due to the assumption of different time shares for the respective operating modes of the devices. Having this in mind the comparison of vibrations and noise was performed based only on measurement results obtained for the operation at full load (FL). The vibrations and noise were recorded during cross-cutting of pine beams with the same cross-sectional dimensions and parameters (grade, moisture content).

The authors of this paper focused on the comparison of vibro-acoustic impacts of chainsaws for non-professional use depending on the type of a driving unit – a combustion engine or an electric one.

2 Experimental

The methodology of chainsaw handle vibration acceleration measurement was based on the guidelines of the ISO 7505:1986 and ISO 33867:2011 standards [5, 6]. The A-weighted sound pressure level measurements were performed according to the z ISO 22868:2011 standard [8]. Two three-axial piezoelectric transducers DYTRAN 3023M2 were used to record the vibration accelerations. The sound pressure was recorded by means of a free field microphone ROGA RG-50 (ICP). Calibration of recording channels was performed using a reference source of vibrations – K10 type RPT97-146 (a_{cal} = 10 m/s^2), and a reference sound source KA-10 ($L_{p,cal}$ = 94 dB at a frequency of 1000Hz). The calibration procedure was performed both before and after the measurements.

The signals from the vibration transducers and the microphone were recorded synchronously using an eight-channel recording unit TEAC LX–10. The recording unit was used as a front-end and it was connected by LAN with a workstation. The A/D conversion was performed at a sampling frequency f_s = 48 kHz. This ensured the noise measurement and analyses to be performed in the whole audible range (from 20 Hz to 20 kHz). The spectral analyses and the determination of measures parametrizing noise - L_A, L_{Lin} and vibrations a_{WX} , a_{WY}, a_{WZ}, a_{hv} were done by means of digital signal processing in the DasyLAB* environment.

The way and place of fixing the transducers to the chainsaw handles are shown in Fig. 2a. Fig. 2b shows the position of the microphone toward the operator's head. The vibration acceleration transducers were connected to the chainsaw handles by means of a SA 50** adapter. To obtain a durable and repeatable coupling between the SA 50 adapters and the handles screw-type band clips were used additionally.

To obtain repeatable parameters of a cutting technique and constancy of ontogenetic traits, all of the tests were performed by one operator (a man; weight 85 kg, height 175 cm). The way the tool was held during the tests and the measurement stand is shown in Fig. 3. Fig. 3a show additionally the directions of reference axes of the coordinate system - its orientation conforms to the device and standards [5, 6].

The cutting was performed on pine beams with a cross-section of 70×90 mm, being in the air-dried state. The wood moisture content was measured by means of POWERFIX Profi + Moisture Meter HG03064C. The moisture content of the wood was equal to 11±2 %. The tests involved an uninterrupted sequence of eight cuttings of a beam across fibers in a material

* Data Acquisition System Laboratory®
** dedicated adapter for accelerometer fixing

without defects. The test conditions were selected in a way that full load of the driving unit was obtained.

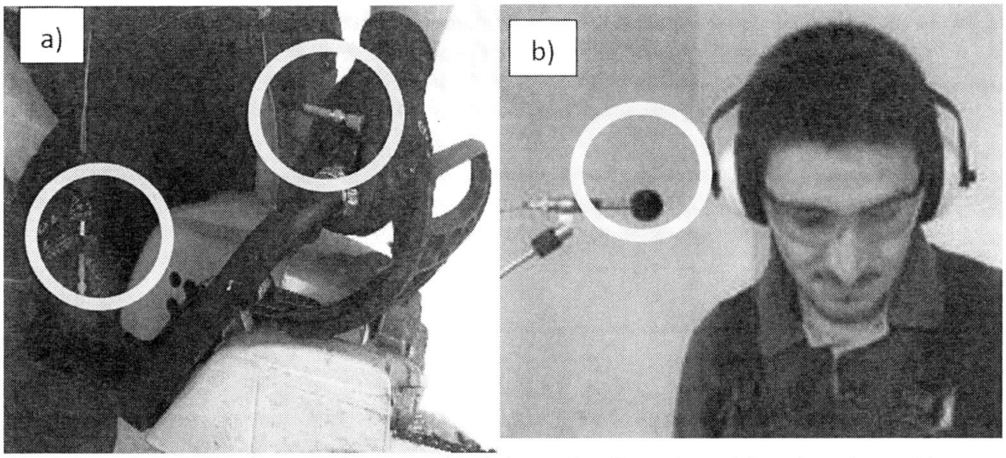

Fig. 2. Location of vibration transducers on the chainsaw handles - a), and the microphone - b)

Electric and petrol chainsaws of a leading brand on the market were subjected to the tests – models for non-professional use. The tests on chainsaws were carried out with the same cutting set (cutting chain, guide bar). This was possible due to the standardization of cutting sets by the chainsaw manufacturer. After installation of the cutting set the same tension of the chain was fixed. Such an approach enables to reduce to a large extent the influence of the cutting set and its settings on measurement results. Technical parameters of the tested chainsaws are shown in Table 1.

Table 1. Technical parameters of the chainsaws being tested

Type of drive		Combustion	Electric
Power [kW]		1.5 (at 10000 RPM)	1.4 (at 7900 RPM)
Maximum linear speed of cutting chain [m/s]		24.8	14.6
Cutting chain	pitch [in]	3/8 P	
	manufacturer's designation	61 PMM3	
	drive link gauge [mm]	1,1	
	type of chain	semi-chisel	
Guide bar length [mm]		350	

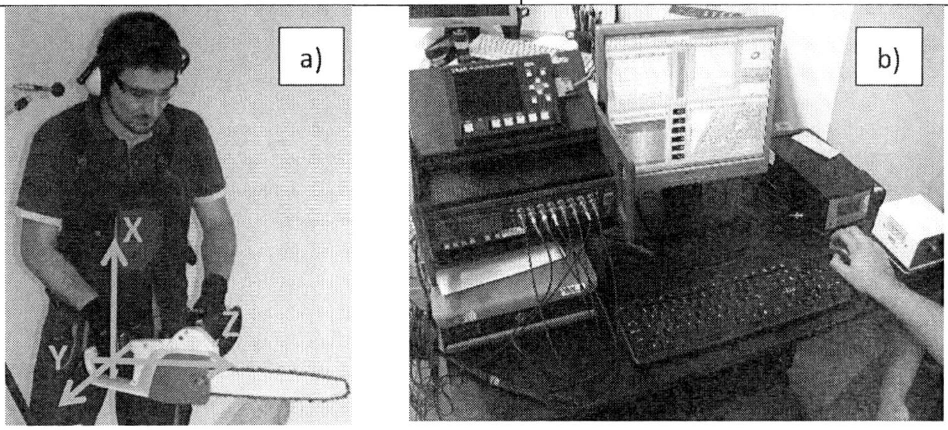

Fig. 3. Operator's body position – a) and the measurement equipment – b).

3 Results

Fig. 4 shows the results of measurement of vibrations of chainsaw handles (taking frequency weighting W_h into account). The averaged values of vector sums a_{hv} (singled out in the figure) and values measured in X, Y and Z directions – a_{hwi} are presented in the form of a bar chart. To determine the average values three signal sequences (lasting approximately 20 seconds each) involving eight cuttings of beam were used. It should be emphasized that the sequences involved only the process of wood cutting – i.e. the FL mode. If parametrization of all operating modes of the driving unit (ID, RA, FL) are taken into account, the determined values of a_{hv} will be slightly lower than those presented in the figure.

In case of the petrol chainsaw the highest values of a_{hwi} were obtained for the X direction, for the rear handle. The highest value of a_{hwi} at the front handle was recorded for the Z direction. The lowest values at the front and rear handles were recorded in the X and Z directions respectively. In this way the earlier test results obtained by the authors [22] have been confirmed. Generally, the values of a_{hwi} measured for the electric chainsaw are higher than those obtained for the petrol one. The only exception is the Y direction at the rear handle. For both electric chainsaw handles the values of vector sums a_{hv} are higher than those for the petrol chainsaw ones. The differences equal 1.81 m/s^2 for the front handle and 1.51 m/s^2 for the rear handle.

Fig. 5 presents a comparison of octave spectra of noise for the tested chainsaw and the A-weighted sound pressure level. The nature of generated noise is similar for both devices. The differences in sound pressure levels for individual octave bands of the tested chainsaws range between 1.9 and 19.8 dB. The biggest difference was noted for the octave band with the center frequency of 125 Hz. High values of the sound pressure levels were recorded in the octave bands with center frequencies of 1 kHz, 2 kHz and 4 kHz. This corresponds to the range of the highest sensitivity of human hearing. That is why the noise generated by petrol chainsaws is particularly burdensome for their operators. A higher sound pressure level for the petrol chainsaw in the octave band with the center frequency of 125 Hz may stem from the phenomena accompanying the driving unit operation at the maximum speed and full load. The highest values of the sound pressure level were recorded in the octave band with the center frequency of 2 kHz. They equal to 95.3 dB and 99.6 dB for the electric and petrol chainsaw respectively. The spectral composition of the noise emitted by the chainsaws with the dominance of components in frequency bands between 500 Hz and 8 kHz causes that the total linear sound pressure level and the A-weighted sound pressure level differ only slightly. The values of the A-weighted sound pressure level equal to 100.3 dB and 105 dB for the electric and petrol chainsaws respectively.

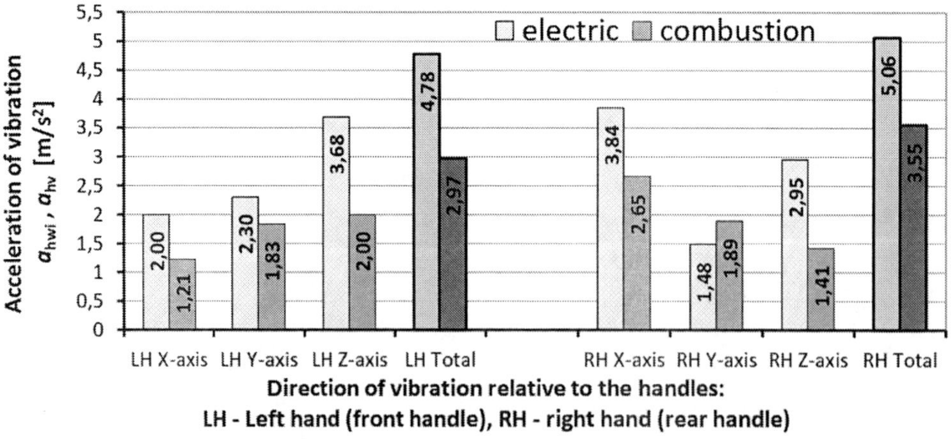

Fig. 4. Frequency weighted chainsaw handle vibration accelerations for the process of cutting.

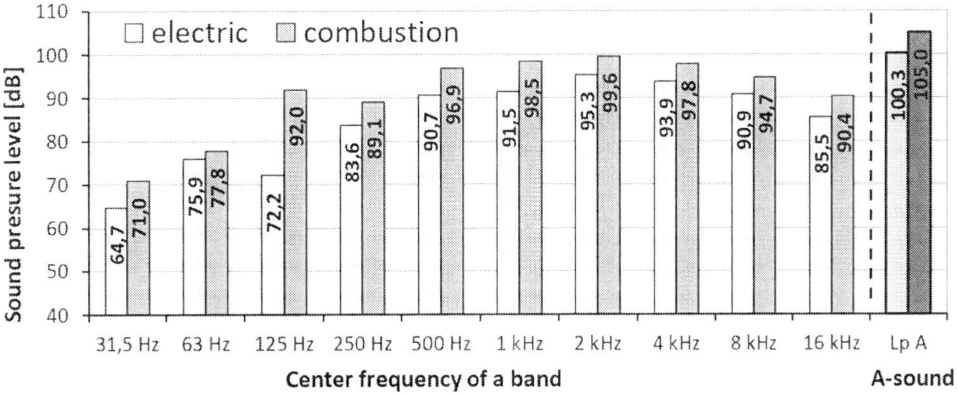

Fig. 5. Comparison of the octave spectra of the noise emitted by the electric and petrol chainsaws during cutting and L_{pA}.

4 Summary

The measures parametrizing the handle vibration accelerations (a_{hwi}, a_{hv}) are higher for the electric chainsaw than those for the petrol one (about 60% for front handle and 42% for rear handle). For both types of drives higher vibration acceleration values are recorded at the rear handle (right hand). Generally, electric chainsaw vibrations for the individual directions are higher than those for the chainsaw with a combustion engine. The only exception is direction Y for the rear handle. The dominant direction of front handle vibrations is the Z direction, and for the rear handle vibrations – the X direction. Based on the obtained results it can be said that it is necessary for an operator to use individual protective equipment, such as anti-vibration gloves. This is particularly important for work with an electric chainsaw.

Higher values (about 5 dB) of the sound pressure level accompany the petrol chainsaw operation. For the tested tools the highest values of the sound pressure level were obtained for the octave band with the center frequency of 2 kHz. Taking the A-weighting into account causes that the dominant components of noise emitted by the chainsaws correspond to the bands of the highest sensitivity of human hearing. Hence, noise will be particularly burdensome for users. The nature of the noise emitted by both types of devices is similar, excluding the octave band with the center frequency of 125 Hz, where the sound pressure level is by 19.8 dB higher for the chainsaw with a combustion engine. The A-weighted sound pressure level is higher in case of work with the petrol chainsaw and it equals to 105 dB, and differs from the value obtained for the electric chainsaw by 4.7 dB. This indicates the necessity of use of personal protective equipment, such as, for example, earmuffs.

Acknowledgement

The results presented are results of the research that was funded with grant 02/21/DSPB/3515/2018 for education allocated by the Ministry of Science and Higher Education of the Republic of Poland.

References

[1] J. Malinowska-Borowska, V. Socholik, B. Harazin, *The Health condition of forest workers exposed to noise and vibration produced by chainsaw*, Medycyna Pracy, **63** (2012); p. 19 – 29

[2] W. Ł. Nowacka, T. Moskalik, *The negative effects of working in forestry with special focus on timber harvesting*, FORRESTRY LETTERS, **105** (2013)

[3] K. Wójcik; *Ocena zagrożenia hałasem operatora pilarki spalinowej podczas okrzesywania i przerzynki*; Zeszyty Problemowe Postępów Nauk Rolniczych; **543** (2009); p. 309 – 318

[4] EN ISO 60745-2-13 Hand-held motor-operated electric tools. Particular requirements for chain saw

[5] ISO 7505:1986 Forestry machinery – Chain saw – Measurement of handtransmitted vibration

[6] ISO 22867:2011 Forestry and gardening machinery – Vibration test code for portable hand-held machines with internal combustion engine – Vibration at the handle

[7] S. Rakheja, R.Dong, D. Welcome, A. W. Schopper, *Estimation of toll-specific isolation performance of antivibration gloves*, International Journal of Industrial Ergonomics, **vol. 30** (2002); p. 71 – 78

[8] ISO 22868:2011 Forestry and gardening machinery – Noise test code for portable hand-held machines with internal combustion engine – Engineering method

[9] E. Kotarbińska, *Zasady użytkowania ochronników słuchu na hałaśliwych stanowiskach stanowiskach pracy, CIOP, 2001*

[10] K. Wójcik, *Influence of the size of chainsaw and lenght of the guide bar on vibrations emmited during delimbing,* Nauka Przyroda Technologie, **24** (2012); p. 1 – 12

[11] K. Wójcik; *Wpływ konstrukcji piły łańcuchowej na drgania emitowane przez pilarki*; Zastosowania Ergonomii nr 2, **3/99**; p. 127 – 134

[12] A. Maciak, *The impact of initial tension on rapidity of dulling of saw cutting chains during cross-cutting of pine wood*, Annals of Warsaw University of Life Sciences – SGGW, Agriculture, **66** (2015); p. 89 – 98

[13] W. Stempski, K. Jabłoński, J. Wegner, *Effect of the edge geometry in a cutting chain on the saw vibration level*, Acta Scientiarum Polonorum, **9** (2010); p. 25 – 33

[14] W. Stempski, K. Jabłoński, J. Wegner, *Relations between top-plate filling angle values of cutting chains and chainsaw vibration levels*, Acta Scientiarum Polonorum, **9** (2010); p. 31 – 39

[15] A. Maciak, A. Gendek, *Effect of cutting with the chainsaw with two pairs of cutting link per section*, Annals of Warsaw University of Life Sciences – SGGW, Agriculture, **50** (2015); p. 59 – 63

[16] J. Skarżyński, K. Wójcik, *Timber hardness impact on the level of vibrations I.C.E.-powered sawing machine grips during cutting*, Inżynieria Rolnicza, **99** (2008); p. 345 – 351

[17] J. Skarżyński, *Influence of wood diameter on the level of vibrations on chainsaw handles during wood cutting*, Technika Rolnicza Ogrodnicza Leśna, **5** (2007)

[18] K. Wójcik, J. Skarżyński, M. Szałek; *Wpływ średnicy drewna sosnowego na drgania elektrycznej pilarki z piłą łańcuchową Stihl E180 przy przerzynce*; Zeszyty Problemowe Postępów Nauk Rolniczych; **543** (2009); p. 395-408

[19] J. Skarżyński, K. Wójcik, *The impact of work technique during timber cutting on vibration and forces on I.C.E.-powered sawing machine grips*, Inżynieria Rolnicza, **99** (2008); p. 425 – 431

[20] K. Wójcik; *Analiza drgań pilarek elektrycznych przy wykonywaniu przerzynki różnymi technikami*; Prace z Zakresu Nauk Rolniczych i Leśnych, tom 101 (2007); p. 265 – 277

[21] K. Wójcik, *Analisys of processing operation time and his percent share in timber harvesting with the chainsaw,* Annals of Warsaw University of Life Sciences – SGGW, Agriculture, **50** (2007); p 71 – 77

[22] W. Rukat, B. Jakubek; *The influence of the cutting tooth design and wear of a saw chain on the vibration level of a chainsaw*; Vibration in Physical Systems, vol. 20 (2017)

[23] PN-83 N-01353 Dopuszczalne wartości przyspieszenia drgań o miejscowym oddziaływaniu na organizm człowieka i metody oceny narażenia

[24] P. Kowalski*; Pomiar i ocena drgań mechanicznych w środowisku pracy według nowych przepisów prawnych*; Bezpieczeństwo pracy, vol.9 (2006); p. 24 – 26

Vibration measures in the diagnostics of technical systems

Bogdan Żółtowski[1,*], *Petra* Dančová[2], and *Jan* Novosád[2]

[1] UTP University of Science and Technology in Bydgoszcz, Poland
[2]Technical University of Liberec, Faculty of Mechanical Engineering, Department of Power Engineering Equipment, Technical University of Liberec, Studentská 2, 461 17 Liberec, Czech Republic

Abstract. In the article author identifies a possible bone of mutual vibration energy measuring application used in the study of standard rail, as part of a dedicated research methodology and detailed methodologies developed in many previous author studies. Theoretical analysis and practical verification of susceptibility testing measurement information complex vibration processes indicates a possibility for their use. Recognizing the need to improve methods of machines and structures testing to assess their degradation state - this work presents important results of the research in the verification field of the proposed measures effectiveness of mutual research bench.

1 Introduction

Ensure the safe and trouble-free operation of machines and structures requires a search or improvement of tools, methods and techniques for non-invasive assessment of the emergency state. A common trend in recent years is the scheme and to organize various business areas targeted for specific purposes. Ensuring repeatable perform actions leading to the intended purpose it is possible, using a variety of situations dedicated research developed methodological procedures allowing for action in various areas.

The problem description and study of changes in the components destruction, materials, design and construction are carried out using the measurement of simple and complex vibration diagnostics methods based on modal analysis [14,15]. An important issue here is the description and propagation study of useful vibration energy is already widely used in diagnostic vibration and a little less used in modal analysis methods [12,16,18].

In this paper the suitability test measures in the monitoring complex process of the rail connections state with the base concrete, which is associated with many problems specific railway construction. The use of composite measurements mapping the interactions measured vibration process is simpler and less subject to error than the complicated procedures simulation based on numerical methods [9,10,20].

2 Mutual measures of vibration process

Research mutual similarities processes vibration through mutual estimators allows to specify: interdependencies between processes, correlation between sources, compliance within one process, changes in the road propagation processes in the machine and structure.

Measurement processes mutual indicate a change of source or variation in the transfer passageways signal extortion and answers. For these measures may include [1,7,13]: cross-

* Corresponding author: bogzol@utp.edu.pl

correlation function, coherence function, transfer function, FRF function *(Frequency Response Function)*.

The process of measurement of the vibration are determined from the measured numerical procedures base signal, which is the value of the measured process or the two measured signals: force and the system response to put force. The concept of using vibration energy propagation of measurement uses the value of the mean square amplitude of dynamic directly proportional to the energy of the signal and written in the form [7]:

$$U = \int_0^T x(t)^2 dt \tag{1}$$

A measure of the propagation efficiency of vibration energy ***H*** binding force signal ***x (t)*** of forcing ***F (t)*** takes the form:

$$H = \frac{\int_0^T x(t)^2 dt}{\int_0^T F(t)^2 dt} \tag{2}$$

Due to the large range of amplitudes of vibration signals is a deliberate to use logarithmic scale for comparing the numerical values of these measures:

$$H_{[\log]} = \log(H) \tag{3}$$

Large potential research opportunities create the analysis results described in the form of a matrix using measures built at different intervals integration or pre-processed waveforms.

The concept measures the propagation of vibration energy refers to the Parseval's theory [7], which indicates the identity of the signal energy presented in the time and frequency domains:

$$\int_{-\infty}^{\infty} x^2(t)dt = \int_{-\infty}^{\infty} \left|X(f)^2\right| df \tag{4}$$

It is thus possible to define the measure of vibration energy propagation also in the frequency domain, and then the equation (2) becomes:

$$H(f) = \frac{\int_{-\infty}^{\infty} \left|X(f)\right|^2 df}{\int_{-\infty}^{\infty} \left|F(f)\right|^2 df} \tag{5}$$

In this relationship ***X (f)*** is the power spectral density of the signal force, ***F (f)*** - the power spectral density of the signal force, and ***f*** is ***the*** frequency in Hertz.

The essence of the measurement of input and response system for e designate of measures of mutual shown in Figure 1.

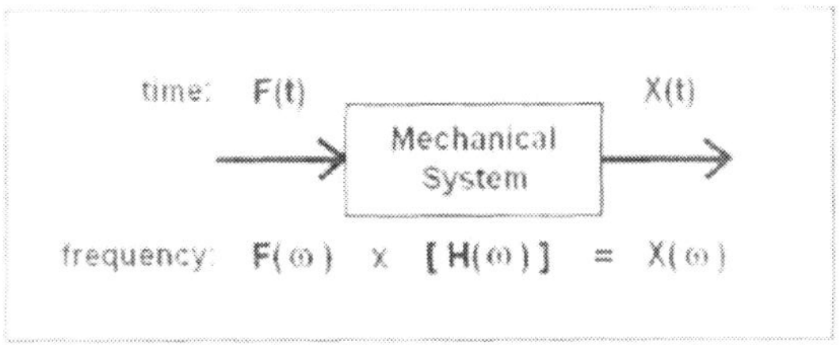

Fig.1. Model components present measurement metrics mutual.

The opportunities created by the presented approach allows to solve many technical problems during testing of real objects, which measure defined below each other are widely use [3,5,6,11].

FRF – FREQUENCE RESPONCE FUNCTION

It is the function describing the frequency relation between the strength force ***F (t)*** and the vibration acceleration ***a (t)*** as ***the*** response signal measured at two points in the system under study. Spectral function of transition *(FRF - Frequency Response Function)* can be described as a quotient of the Fourier transform vibration exciting force *F (ω),* the Fourier transform of the response signal *A (ω)*

$$FRF = \frac{\sum F(\omega)}{\sum A(\omega)} \quad (6)$$

The *FRF* as the main measure of experimental modal analysis is an inversion defined previously transmittance *H (f)* and is referred to as real and imaginary parts. The method of its determination on the actual objects shown in Fig. 2.

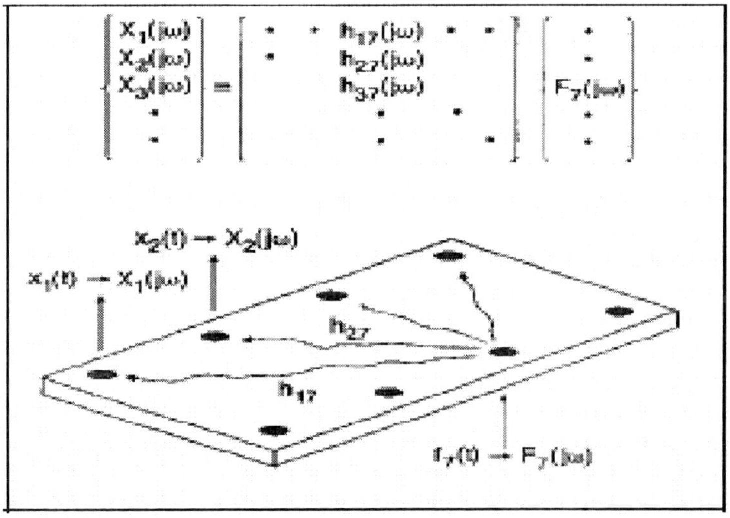

Fig.2. Research *FRF* facility.

The *FRF* is a complex function of the frequency, which the module shows the characteristics of amplitude - the frequency, and the argument is the angle of the phase shift between the power of forcing a response [7,13,17]. The FRF is an inversion of the transmittance, which is an important observation resulting from the analysis and testing of these measures global propagation of energy [19].

Due to the possibility of the use of force or strength kinematic introduced different characteristics depending on the nature of the physical processes the input and output, highlighting: the vulnerability of dynamic mobility, acceleration, dynamic stiffness, impedance and dynamic mass [4.8,10].

3 Research methodology

Verification of the indicated measures of mutual propagation of vibration energy was carried out at a laboratory [1]. Being investigated size vibration energy flow between the points of attachment rails to the sleepers of the track has been switched on Fig. 3.

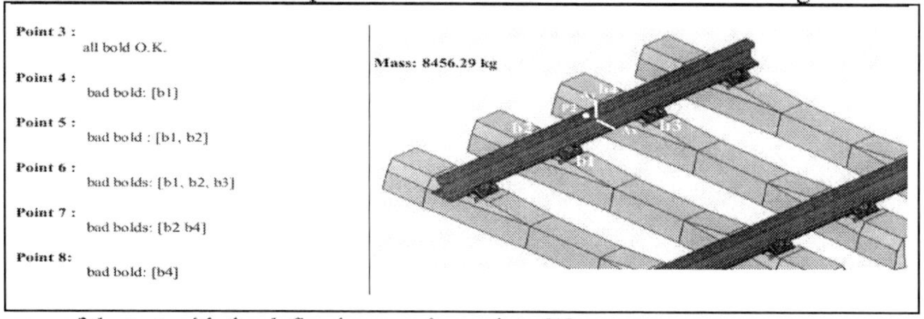

Fig. 3. Diagram of the test with the defined measuring points [1].

Using the studies measured vibration signals extortion and response software was set appropriate measure of each other, where the sample waveforms are shown in Fig. 4-6.

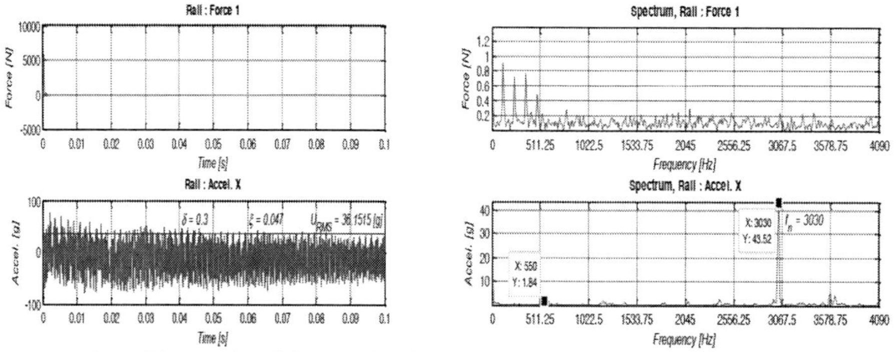

Fig. 4. Examples of the results of the exciting force and response in the time domain and frequency.

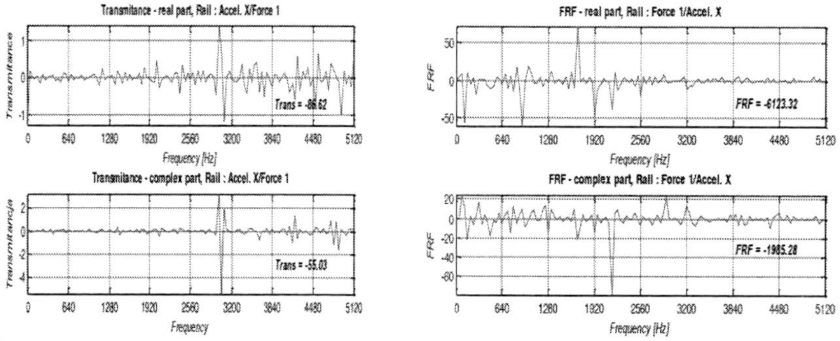

Fig. 5. The response signal and determined its coherence function

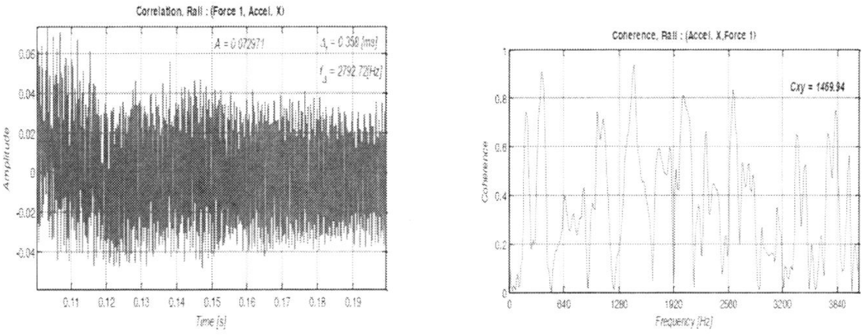

Fig. 6. Examples of waveforms transmittance and FRF.

The completed experiment studied a la the various states of attachment (pressure) between the indicated attachment points give a lot of diff s values (Table 1).

Table 1. Results of bench studies indicated measures of mutual.

Reception points	FRFr	FRFu	H(f)r	H(f)u	Cohere
1 - 2	288.71	-15.05	137.47	4.91	197.03
2 - 1	327.21	248.69	5.67	4.32	112.65
3 - 4	163.73	352.91	-3.42	-0.54	89.41
4 - 3	380.32	-29.54	108,18	11.01	195.27
1 - 3	83,97	892.43	-9.88	-6.97	109.06
1 - 4	587.91	-15.35	55.93	1.92	192.37
2 - 3	456.41	506.42	7.73	18.02	139.29
4 - 2	639.01	-118.72	58.17	10.72	192.04
2 - 4	1,213.83	15.03	17,91	17.99	125.36
3 - 1	679.94	157.66	16,34	11.73	139.08
3 - 2	985.87	-64.41	39.13	6.75	183.6
4 - 1	799.05	38.18	4.55	24.81	106.61

Correct inference requires redundancy of information and statistical analysis to obtain a model of bonding between the assumed technical condition and the value of mutual measures. To implement this project proposed in the work of the author's methodology dedicated research facilities in the "construction" [1,2,9] uses redundancy procedures, OPTIMUM, SVD and modeling cause – effect [7,20].

Using the information contained in Table 1 processed according to the proposed methodology can confirm the good properties of measurement destruction of the cases examined the destruction of joints in order: ***FRFr, FRFu, H(f y, H(f)u, cohere.***

Considering all conducted a sensitivity analysis of the proposed measures of mutual propagation of vibration energy, using different statistical methods used in this work can continue to propose the following as the main measure of the state of destruction:

FRF_y., $H\ (f)_R$, cohere.

For the indicated reduction measures of destruction was determined regression model binding shown in Fig. 7.

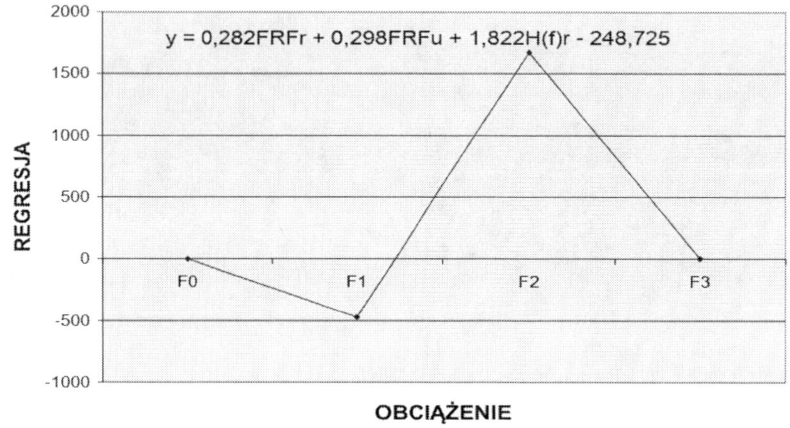

Fig.7. Model binding measures of mutual technical condition of the connection.

The presented methodology of the study allows to simplify the painstaking research conducted in the field of traditional airworthy calls on the raceways, replacing the work of many service employees to maintain fitness for work programmed utilizing available information technology proposed in this study (Fig.8).

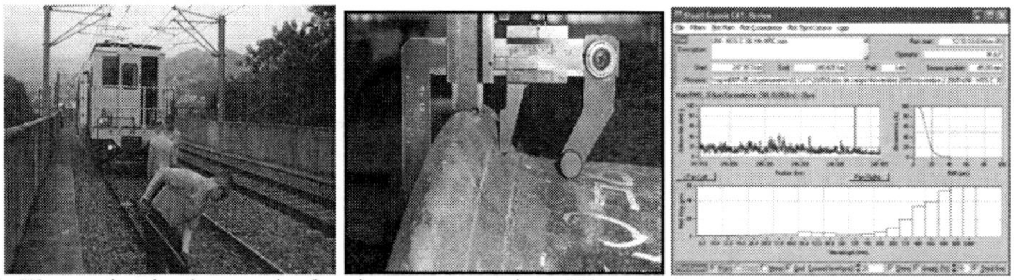

Fig.8. The modernization work of maintenance services [5]

4 Summary

The substance of the development includes the results shows usefulness of mutual propagation measures of vibration energy than the various forms of applications in civil engineering. The main assumption was realized by exploratory research to present the possibilities of practical use of complex estimators vibration - as a tool to solve many fundamental problems with the condition of degradation and building construction elements dynamics.

These issues, it studies exemplify dedicated to the degradation state of the construction methodology procedure for determining the physical aspects and practical applications of propagation of vibration energy measures. The proposed methodology proceedings indicate that the important role and manner of acquisition and use of vibration composite measurements in the construction industry, revealing at the same time the need to take this issue in research the destruction of existing structures. Served briefly elements of vibration theory focused on the physical aspects of the description and use of vibration estimators in the construction industry bring the issues of modeling the dynamic state of different structures and rules for their use.

Changes identification in the building structures degradation state becomes a method of assessing the structure operational status and modal analysis methods and related vibration estimates reflects well the elements degradation.

New ways of obtaining and processing information contained in the processes of vibration often leads to improvement of various operating systems constructions, often leads to creative change existing practical solutions.

References

1. L. Castaneda, B. Zoltowski, *Bases del diagnostico tecnico de maquinas*. EAFIT University, Colombia (2010).
2. C. Cempel, *Vibroacoustical Condition Monitoring*, Ellis Ltd., New York, (1991).
3. H. Tylicki, B. Żółtowski B, *Rozpoznawanie stanu maszyn*. ITE - PIB, Radom (2010).
4. T. Uhl, *Computer-aided identification of mechanical structure models*. WNT Scientific Technical Publishers, Warszawa (1997).
5. B. Żółtowski B, L. Castaneda, *Structural diagnosis of rail vehicles and method for redesign*. Diagnostyka Vol.15 (3) (2014).
6. B. Żółtowski, M. Łukasiewicz, *Diagnostyka drganiowa maszyn*. ITE-PIB, Radom (2012).
7. B. Żółtowski, M. Łukasiewicz, T. Kałaczyński T, *Techniki informatyczne w badaniach stanu maszyn*. UTP Bydgoszcz (2012).
8. B. Żółtowski, M. Żółtowski M, *Vibration signals in mechanical engineering and construction*. ITE-PIB, Radom (2015).
9. B. Żółtowski, C. Cempel, *Engineering of diagnostics machines*. ITE – PIB, Radom (2004).
10. M. Żółtowski, *Operacyjna analiza modalna w badaniach konstrukcji budowlanych*. UTP Bydgoszcz (2012).
11. M. Żółtowski, *Informatyczne systemy zarządzania w inżynierii produkcji*. ITE-PIB, Radom (2011).
12. M. Żółtowski M, *Investigations of harbour brick structures by using operational modal analysis*. Polish Maritime Research **No 1/2014**, (2014).
13. M. Żółtowski, B. Żółtowski B, *Vibrations signal to the description of structural damage of dynamic the technical systems*. XIII International Technical Systems Degradation, Conference, Liptowski Mikulesz (2015).
14. M. Żółtowski, *Opis drganiowy konstrukcji budowlanych*. Logistyka nr.6/2014, (2014).
15. M. Żółtowski, *Investigations of harbour brick structures by using operational modal analysis*. Polish Maritime Research, **No. 1/(81),** vol.21, (2014).
16. M. Żółtowski, M. Liss M, *The use of modal analysis in the evaluation of welded steel structures*. Studies and Proceedings of Polish Association for Knowledge Management, Tom **79**, Bydgoszcz (2016).
17. M. Żółtowski, M. Liss, *Zastosowanie eksperymentalnej analizy modalnej w ocenie zmian sztywności prostego elementu konstrukcyjnego*. Studies and Proceedings of Polish Association for Knowledge Management, T80, Bydgoszcz (2016).
18. M. Żółtowski, R. Martinod, *Technical Condition Assessment of Masonry Structural Components using Frequency Response Function (FRF)*. Masonry International Journal of the International Masonry Society, Vol.29, No**1** (2016).
19. M. Żółtowski, R. Martinod, *Quality identification methodology applied to wall-elements based on modal analysis*. Civil Engineering the Athens Institute for Education and Research, Emerald, Athens (2015).
20. B. Żółtowski, M. Żółtowski, *Vibrations in the Assessment of Construction State*. Applied Mechanics and Materials Vol. 617 (2014).

Verification of the method of equivalent amplitude determination based on two - parameter fatigue characteristic

Karolina Karolewska [1,*], *Bogdan* Ligaj [1]

[1]UTP University of Science and Technology in Bydgoszcz, Faculty of Mechanical Engineering, al. Prof. S. Kaliskiego 7, 85-796 Bydgoszcz, Poland

Abstract. In the most causes the loads which are affected on structural components are various over time and their character changes is stochastic. The stochastic character of operational loads of construction elements in various machine types is depended on many factors, included : work forces variability, environmental conditions, physical properties of components etc. Fatigue life calculation for this type of loads are conducted on the basis of determined sinusoidal cycles set through to use of the cycles counting method. The cycles which are contained to the sinusoidal cycles set are characterized by extensive range of amplitude S_{ai} variation and mean values S_{mi}. Application of S_a-N curve in fatigue life calculations caused disregard of the cycle mean value. This may affect the accuracy of calculations. Taking into account the cycle mean value S_m in the calculations may be realized by determining a substitute cycle with an average value S_m=0 and a substitute amplitude $S_{az} \neq S_a$.

Abbreviations

N	-	number of cycles (fatigue life),
N_0	-	base number of cycles requires fatigue limit ($N_0 = 10^6$),
R	-	stress ratio (expressed by the formula $R = S_{min}/S_{max}$),
S_a	-	stress amplitude of the sinusoidal cycle in MPa,
S_{az}	-	equivalent stress amplitude of the sinusoidal cycle with S_m i S_a parameters in MPa,
$S^{(T)}_{a(R)}$	-	stress amplitude of the sinusoidal cycle with given value of a stress ratio R determining a line of constant fatigue life (N = const.) in MPa,
$S_{f(-1)}$	-	fatigue limit of the sinusoidal cycles with stress ratio R=-1 for number of cycles N_0 w MPa,
S_i	-	value of local stressat the i-th load level in MPa,
S_m	-	average value of sinusoidal cycle stress in MPa,
$S^{(T)}_{m(R)}$	-	the mean value of the sinusoidal cycle stress with given value of a stress ratio R determining a line of constant fatigue life (N = const.) in MPa,
S_{max}	-	maximum value of stress in a sinusoidal cycle in MPa,
S_u	-	material tensile strength in MPa,
S_y	-	yield point of material w MPa,
k	-	exponent in the equation $\psi_N = N^{-k}$,
$m_{(-1)}$	-	exponent of in the formula describing fatigue (Wöhler) graph for the stress ratio R=-1,
ψ_N	-	the sensitivity coefficient of the material for the asymmetry of the cycle for $N \neq N_0$,

* Corresponding author: karolina.karolewska@utp.edu.pl

λ - number of repetitions of a program to fatigue fracture.

1 Introduction

Fatigue life assessment is one of the main criteria during the design process of devices elements. On the basis of the fatigue life calculations under random loads conditions the design features selection of machine elements are made. Fatigue life determination for operational loads requires the use of characteristic describing the material properties, cumulative damage model and load spectrum developed for indicated loads.

Operational loads can be divided into two groups: narrow or broad spectrum. Due to fatigue life between the mentioned loads significant differences refer to cycles participation from a given range of the stress ratio R values and relate to occurrence of cycles with large amplitude values, where the maximum amplitude $S_a \approx S_{max}$.

Calculations of fatigue life for load chracterized by narrow spectrum are carried out by use of block load spectra (the spectra 1D) developed on the basis of determined cycles set by using one of the cycles counting method. The realization of fatigue life calculations in cause of board spectrum loads is based on loads specta describing variability of cycles parameters S_m and S_{ai} (the spectra 2D). The use of this type of spectrum is connected with cycles stress ratio values $-\infty < R < +\infty$ occuarncy in cycles loads and their impact on fatigue life [1].

Replacement of load two-parameter spectrum (the spectra 2D) by one-parametre spectrum (the specta 1D) is based on determination substitite amplitude method S_{az} with the use of two-parameter fatigue charakteristic [2]. For the purposes of the paper two-parameter characteristics denoted as : model IM (Fig. 2a), and model II (Fig.2b) were estabilished.

The aim of the paper is to compare determining fatigue life methods with the use of load spectrum. The load spectra were obtained using one of the cycles counting method, which were full cycles counting method, and also by using two-parameter characteristic: the IM model and II model.

The scope of the paper contains the presentation of determination substitute amplitude methods, comparison of fatigue life calculation results with the experimental test results for indicated operational loads.

2 The method of determining substitute cycles of load

Calculations of fatigue life under operational loads require replacing a random cycles load (Fig1.a) with sinusoidal cycles. To determine load equivalent cycles, one of the cycles counting method is used: peak counting method, simple-range counting method, full cycles counting method, range-pair counting method and rainflow counting method [4].

By means of one of the cycles counting method, sinusoidal cycles are obtained. The recived cycles vary in the stress amplitude values S_{ai} and mean values S_{mi} (Fig.1b). Sets of cycles with different amplitude values S_{ai} and mean value S_{mi} are replaced by equivalent cycles with a mean value S_{mi}=0 and equivalent amplitude S_{azi} on the basis of two-parameter characteristics.

a) b)

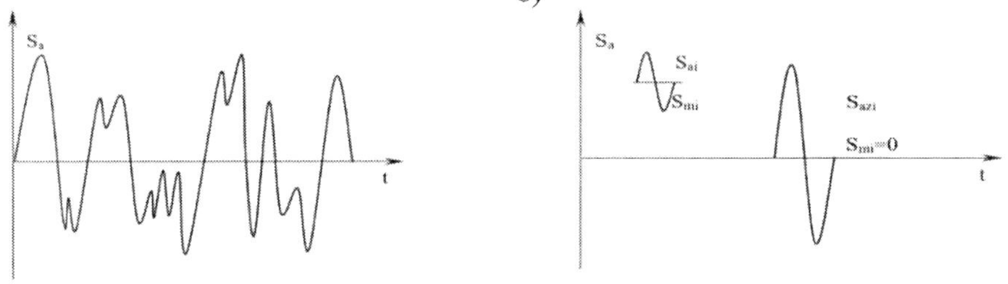

Fig. 1. Schematic presentation of loads: a) random course of load, b) selected load cycles.

Two-parameter fatigue characteristics are used to determine equivalent S_{azi} amplitudes. In this paper, the IM model (Fig.2a) and II model (Fig. 2b) were used to determine the load substitute spectra.

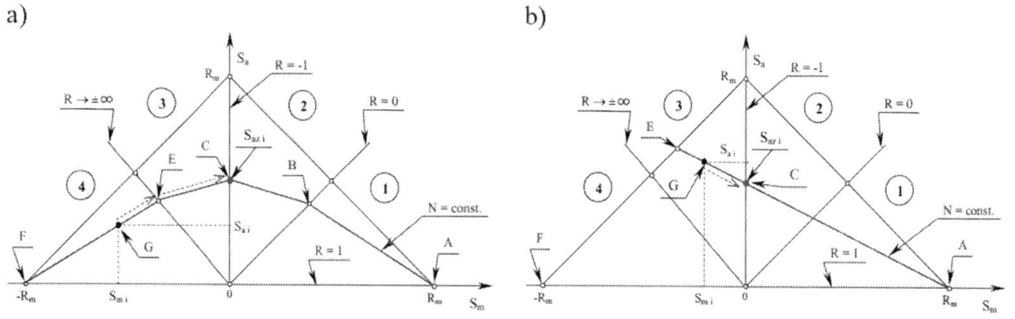

Fig. 2. Schematic presentation of the two-parameter characteristics models: a) Model IM, b) Model II.[2]

Model IM (Fig. 2a) was created on the basis of the model I presented in paper [4]. This model generalizes the Serensen's stress limit plot on the range of high cycle and low cycle fatigue.

Model IM (Fig. 2a) is divided into four zones, which are determined by required values of stress ratio R. A line of constant fatigue life (N = const.) is plotted taking into account the stress amplitude Sai and the average value S_{mi} for sinusoidal cycle, and on this basis the determination of the equivalent amplitude S_{azi} is performed. Selection of stress amplitude value S_{ai} on the ordinate axis, and mean value S_{mi} on the abscissa axis determines the G point. This point is interpolated to point C, in which the line of constant fatigue life intersects with the ordinate axis, determining the load equivalent cycle with the average value $S_m = 0$. Several load cycles, which are located on the same line of constant fatigue life, have the same equivalent stress amplitude S_{azi} and are characterized by the stress ratio R = -1 regardless of in which zone of two-parameter characteristic they are located.

In the model IM the line of constant fatigue life is described by the following formulas for four zones denoted from 1 to 4 (Fig.2b), which characterized by stress ratio variability. For zone 1 described by the variability of range of the stress ratio $0 < R < 1$ the line of constant fatigue life is represented by the equation:

$$N = N_0 \left[\frac{S_{f(-1)} \cdot \left(S_u + S_{a(R)}^{(T)} - S_{m(R)}^{(T)} \right)}{S_{a(R)}^{(T)} \cdot S_u \cdot (1 + \Psi_N)} \right]^{m(-1)} \tag{1}$$

The line of constant fatigue equation for zone 2, in which the stress ratio R takes values $-1 \leq R \leq 0$ is given in the form:

$$N = \frac{N_0 \cdot \left(S_{f(-1)} \right)^{m(-1)}}{\left(S_{a(R)}^{(T)} + \Psi_N \cdot S_{m(R)}^{(T)} \right)^{m(-1)}} \tag{2}$$

The line of constant fatigue equation for zone 3 characterized by the variability of range of the stress ratio $-\infty < R < -1$ is described by equation as follow:

$$N = \frac{N_0 \cdot \left(S_{f(-1)} \right)^{m(-1)}}{\left(S_{a(R)}^{(T)} - \Psi_N \cdot S_{m(R)}^{(T)} \right)^{m(-1)}} \tag{3}$$

Zone 4 contains variability of cycle stress ratio $1 < R < +\infty$. In this zone the line of constant fatigue is determine by the equation:

$$N = N_0 \left[\frac{S_{f(-1)} \cdot \left(S_u + S_{a(R)}^{(T)} + S_{m(R)}^{(T)} \right)}{S_{a(R)}^{(T)} \cdot S_u \cdot (1 + \Psi_N)} \right]^{m(-1)} \tag{4}$$

Model II (Fig. 2b) for determining the equivalent load amplitude is a generalization of the Goodman's formula over the limiting fatigue life ranges low cycle and high cycle fatigue. This model is based on the Wöhler fatigue diagram for the stress ratio R = -1. Under load conditions characterized by stress ratio R = - 1, point A corresponds to the tensile strength of the material S_u, while point C corresponds to the fatigue limit $S_{f(-1)}$. The ACE line is the constant durability line (N = const.) For the sinusoidal load for the range of the cycle asymmetry coefficient -∞ <R <1 with variable stress amplitudes Sai and mean values S_{mi} is described by the following equation:

$$N = N_0 \left[\frac{S_{f(-1)}}{S_{a(R)}^{(T)}} \left(1 - \frac{S_{m(R)}^{(T)}}{S_u} \right) \right]^{m(-1)} \tag{5}$$

On the basis of the described methods above of determination equivalent stress amplitudes, the block spectra of load for the applied load run were created. By means of the obtained sinusoidal cycles sets using the presented cycles counting methods, load spectra were developed.

3 Results of fatigue life calculations (using a two-parameter fatigue characteristic)

3.1 Operational load

In Figure 3 is presented run of stress changes in a structural element, which were recorded in operational load conditions. Due to use in calculations of fatigue life, run of stress changes is presented in the relative form S_i/S_{max}, where S_{max} is the maximum value occurring in the load run.

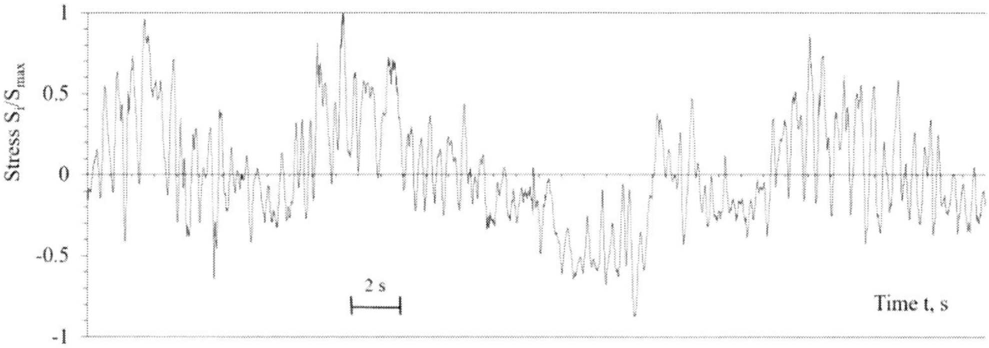

Fig. 3. Fragment of stress changes run in the form of the relative values S_i/S_{max}.[2]

The load run presented in Figure 3 provide to determination of sinusoidal cycles set by use of the full cycle counting method. This method uses the assumption, that all ranges occurring in pairs within a definitive scope form cycles, that form closed hysteresis loops. This method schematization is based on progressive filtering of cycles with increasing amplitude value [5].

The obtained data sets contained sinusoidal cycles with S_{ai}/ S_{max} and S_{mi} /S_{max} parameters from the variability of the stress ratio -∞ < R <1, which is characteristic for broad spectrum loads. By means of determined data were appointed cycle set characterized by mean value Sm = 0 and equivalent amplitude S_{az} received by the use of two-parameter fatigue characteristic. Method of substitute amplitude S_{az} determination, described in point 2, were applied. For assumed operational run determined block load spectra. The basis for the spectra development was the set of sinusoidal cycles determined by the adopted cycles counting methods. The S355J0 steel was used to perform calculations.

a) b)

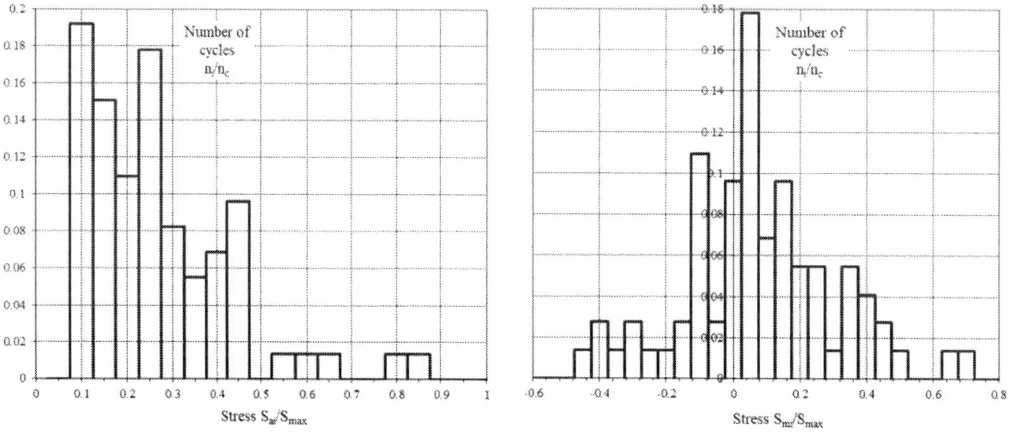

Fig. 4. Histograms of load cycles: a) for stress amplitude, b) for mean values.

The block load spectra (Fig. 5), used in fatigue life calculations, were determined for S_{max} values = 500 MPa for stress changes, taking into account model IM and model II of two-parameter fatigue characteristics. Figure 5a presents a block spectrum allowing the determined substitute amplitudes S_{az} based on the IM model of a two-parameter fatigue characteristic. Whereas, Figure 5b illustrates a block spectrum determined by means on the II model of two-parameter fatigue characteristics.

a) b)

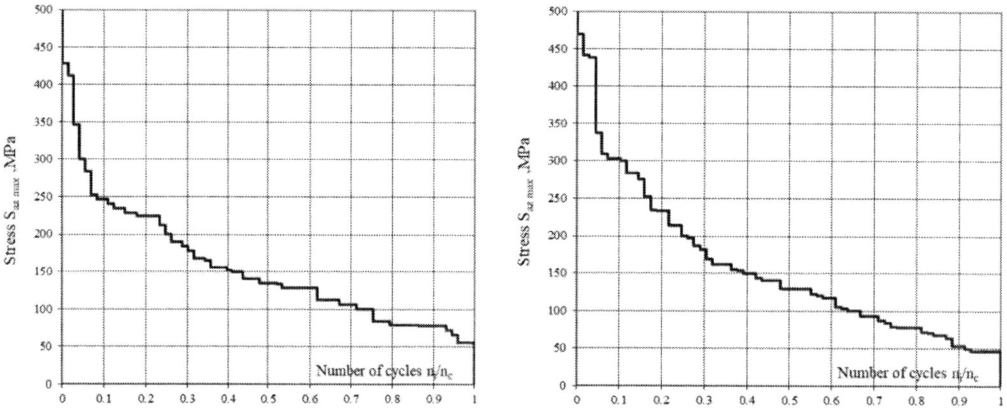

Fig. 5. Substitute block load spectra for S_{max} = 500 MPa, determined for two-parameter fatigue characteristic: a) model IM, b) model II.

3.2 Material properties

The presented two-parameter fatigue characteristics were evolved for S355J0 steel. This material is a low-alloy steel with increased strength. Structure of the steel is composed of ferrite and pearlite grains. The S355J0 steel has the following static properties: tensile strength Su = 678 MPa and yield point Sy = 499 MPa, and Young's modulus E = 208159 MPa. Cyclic properties of S355J0 steel on the basis of the calculations were made: $m_{(-1)}$ = 12,33, $C_{(-1)} = 1.145 \cdot 10^{36}$, $S_{f(-1)}$ = 243.1 MPa, k = 0.1386, ψ_N = 0,1205 [5].

3.3 Calculations terms

Fatigue life calculations was carried by means on the Palmgren-Miner's hypothesis on linear fatigue damage summation expressed as follows:

$$D = \sum_{i=1}^{k} \frac{n_i}{N_i} = 1.0 \tag{6}$$

The fatigue life results were presented in the form of a number of accomplished load blocks, computed from the formula bellow:

$$\lambda = \frac{1}{D} \tag{7}$$

By used of equivalent load spectra carried out the calculations of fatigue life. This calculations allowed to determine fatigue life diagram Sa (N). The fatigue life diagram compares the results of calculations for the model IM, model II of two-parameter fatigue characteristics and fatigue life experimentally determined. To calculate fatigue life, the Wöhler curve was determined experimentally under load conditions characterized by stress ratio R = -1, described by the following formula:

$$\log S_a = -\frac{1}{14{,}81}\log N + 3{,}0177 \tag{8}$$

On the basis of fatigue life calculation results for the IM model, a fatigue life chart is described by the equation:

$$\log S_a = -\frac{1}{12{,}33}\log N + 3{,}1240 \tag{9}$$

For II model of the two-parameter fatigue characteristic, the fatigue life diagram is featured by the equation bellow:

$$\log S_a = -\frac{1}{14{,}11}\log N + 3{,}0283 \tag{10}$$

The results of fatigue life calculations complied in the form of the diagrams (Fig. 6). The chart presents calculations carrying out on the basis of substitute load spectra for the two-parameter fatigue characteristics with the results of experimental tests.

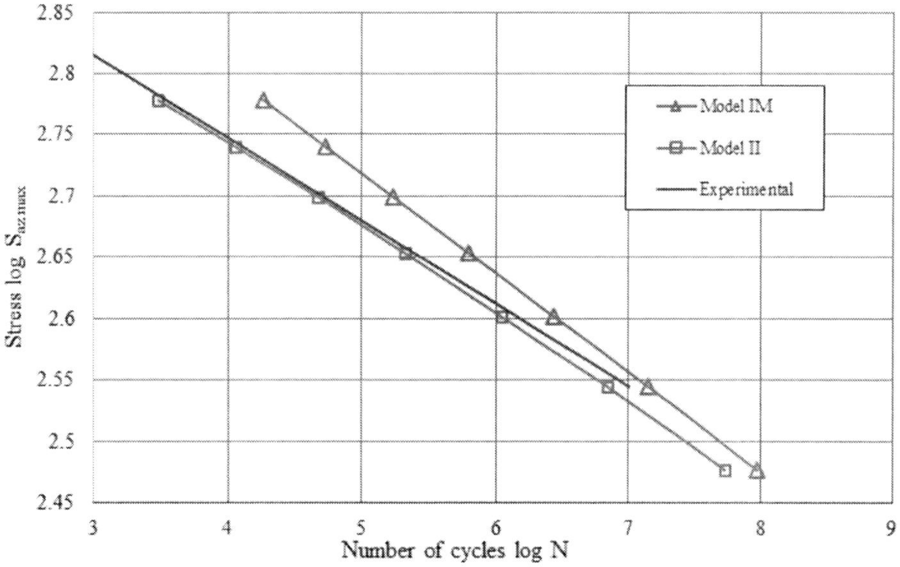

Fig. 6. Fatigue life diagram determined experimentally and for load spectra allowed two-parameter fatigue characteristics: model IM and model II.

4 Analysis of research results and summary

The results of fatigue life calculations based on the adopted cycle counting method and selected two-parameter fatigue characteristics (model IM and model II) are characterized by discrepancy.

Due to calculations using two-parameter fatigue characteristics, the stress average value Sm is taken into account. As opposed to fatigue life calculations by means on Wohler diagram, which ignored the impact of mean value Sm. This has a significant impact to assess the fatigue life of structural element. In particular, element that is exposed to board spectrum of operational loads.

On the fatigue life diagrams is presented that the highest durability is found in the results obtained on the basis of the substitute load spectra for the IM model of the two-parameter fatigue characteristic. The lowest fatigue life was obtained for calculations based on the model II of the two-parameter fatigue characteristic. Fatigue life for model II is similar to the durability determined during experimental tests. Meaning differences in the results of fatigue life calculations between model IM and model II of two-parameter characteristics may be affected the range of using chosen characteristics [3].

Model II is used for the stress ratio $-\infty < R < -1$, ignored zone 4 (Fig.2) with range of stress ratio $1 < R < +\infty$. It can be said that the selection of the appropriate model of the two-parameter fatigue characteristic has a significant impact on the fatigue life calculations results. The choice of the proper model is related to the cyclic properties of researched material. .

In addition to the fatigue characteristics models, the cycle counting method also affects the fatigue life results. In cause that different cycle counting method is used, the calculated durability values would be different from each other.

References

1. Kocańda S., Szala J., PWN (2004)
2. Ligaj B., Sołtysiak R., Polish Meritime Research, no. 1 (2016)
3. Ligaj B., Szala G., Polish Meritime Research, no. 1 (2010)
4. Szala G., Ligaj B., Materials Science Forum, Fatigue Failure and Fracture Mechanics (2012)
5. Szala G., Ligaj B., Instytut Technologii Eksploatacji – Pańswtowego Instytutu Badawczego (2011)

Evaluation of the technical condition of road transport means

Henryk Tylicki[1,*]

[1]University of Applied Sciences, ul. Podchorążych 10, 64-920, Piła, Poland

Abstract. The paper presents the methodology for determining the procedures of the condition and fault location test, using logical relations (0,1) between diagnostic parameters and technical conditions of the means of transport in the form of a binary observation matrix, determination of the status control and fault location test and use of diagnostic information received during use of means of transport. Implementation of the proposals included in the study should increase the efficiency of diagnosing means of transport, and thus contribute to activities that rationalize their exploitation.

1 Introduction

Physical processes occurring in systems and sets of means of transport have a significant impact on the change of their state and are related to their use and handling. These processes are the source of emission of diagnostic signals, the analysis of which allows to generate information on the technical condition of means of transport.

The automation of the system of monitoring the condition of means of transport, in which it stands out, the assessment of their technical condition requires, among other things, the determination of the condition and fault location test, while the problems associated with it come down to determining the relationship between diagnostic parameters and technical conditions of the means of transport, determining the test the condition and location of damage and the use of diagnostic information received during the use and operation of means of transport.

The term "best" used above means the adoption of appropriate criteria and consideration of these problems in terms of seeking an optimal solution to the above problems. This requires the development of a methodology and algorithm for determining the status and fault location test, and formulating assumptions for the implementation of the algorithm. Implementation of the proposals included in the study should increase the efficiency of diagnosing means of transport, and thus contribute to activities that rationalize their exploitation.

2 Characteristics of the problem

Determination of the technical condition of means of transport is connected with the examination of relations between diagnostic parameters - the condition of the means of transport. Based on the findings made in [1,2,4,8,9], it is believed that the use of appropriate procedures should take into account the appropriate level of decomposition (means of transport, arrangement of the means of transport, assembly of arrangements, element of the assembly) and the appropriate scope of the condition assessment (condition/state control , location of damage, condition control and location of damage to the means of transport). The complexity of transport systems as well as the complexity of processes occurring in them cause that the existing relationships between diagnostic parameters and states are usually stochastic relationships. These dependencies in the practice of diagnosing means of transport, according to the theory of technical diagnostics, depending on the degree of disturbance of

* Corresponding author: htylicki@pwsz.pila.pl

diagnostic signals can be determined by a boolean observation matrix - with a small degree of disturbance of diagnostic parameter values or probability matrix with a high degree of disturbance of diagnostic parameters values. It is believed, however, that due to limitations resulting from the passive or passive experimental mode, it is sufficient to analyze the boolean matrix and determine the condition control and fault location test. In order to distinguish possible solutions for the construction of tests, the criterion of state discrimination and the procedure for building the test using the boolean matrix method should be used. As a result of this method, a T_{KS} status control test and a T_{LU} damage location test are obtained. In the first case, based on the binary matrix M_b^d (condition/state table), a boolean matrix M_b^{KS} should be created for the fitness control, in which a subset of pairs of distinguishable S_o, S_i, $i=\overline{1,k}$ states are introduced in place of the states.

$$M_{bij}^{KS}=\begin{cases}0 \text{ when } \left(1-M_{bij}^{d}\right)=0\\ 1 \text{ when } \left(1-M_{bij}^{d}\right)=1\end{cases} \tag{1}$$

Ones which occur in the elements of matrix $M_{bij}^{KS}\in M_b^{KS}$ stand for state discrimination $S_i \in S$ using the parameter $y_j \in Y$, and zero - indistinguishable. Analyzing the matrix M_b^{KS}, for the test T_{KS} parameter $y_j \in Y$ is chosen, which has a maximum number of ones in the column. In the case when the jth column does not contain only ones, the missing ones in the nth column are searched for, and if they occur, then the nth parameter where $y_n \in Y$ is appended to the T_{KS} test. Then the T_{KS} test takes the form:

$$T_{KS}=\{y_j, y_n\} \tag{2}$$

When specifying T_{LU} test elements, we create a boolean matrix M_b^{LU} to damage location, in which a subset of state pairs is introduced instead of states S_l, S_i; $l=\overline{1,k}$; $i=\overline{1,k}$; $i\neq L$.

$$M_{blij}^{LU}=\begin{cases}0 \text{ when } \left|M_{blj}^{d}-M_{bij}^{d}\right|=0\\ 1 \text{ when } \left|M_{blj}^{d}-M_{bij}^{d}\right|=1\end{cases} \tag{3}$$

Ones which occur in the elements of matrix $M_{blij}^{LU}\in M_b^{LU}$ stand for state discrimination between $S_l \in S$ and $S_i \in S$ using the parameter $y_j \in Y$, while zero stands for indistinguishable. Then analyzing the matrices M_b^{LU} for the test T_{LU} parameter $y_j \in Y$, which has a maximum number of ones in the column, is beign chosen. In the case when the jth column does not contain only ones, the missing ones in the nth or nth+1 column are searched for. In the event of their occurrence, *n* and *n + 1* parameters are included in the test T_{LU}. Then the test T_{LU} takes the form:

$$T_{LU}=\{y_j, y_n, y_{n+1}\} \tag{4}$$

In the case of the T_{KS} state control test for using the presented algorithm, a binary matrix M_d^b should be constructed , that one of its rows correspond to the condition of fitness $S^0 \in S$. The matrix elements in this row should have a value of "0", which means that none of the checks $d_j \in T_D$ give any information about the object's damage.

In the case of a fault location test T_{LU} for using the presented algorithm matrix M_d^b it already contains rows characterizing states from a subset of unfitness:

$$S^1=\{\ S_i\ \} \tag{5}$$

3 Methodology for determining the status and fault location test

As shown above, the boolean matrix is determined by means of a set of relations between a set of diagnostic parameters $Y = \{y_n\};\ n = \overline{1, N}$ and a set of states $S=\{s_i\}$. The boolean matrix method thus provides for the determination of the state control test T_{KS} and fault location test T_{LU}. In the first case based on a binary matrix M_b^d create a boolean matrix M_b^{KS} to control the state in which a subset of pairs of distinguishability states S_o and incapacity states S_i, $i = \overline{1, k}$ are introduced instead of states. Ones which occur in the elements of matrix $M_{bij}^{KS} \in M_b^{KS}$ stand for state discrimination $s_i \in S$ using the parameter $y_j \in Y$, while zero stands for indistinguishable. Then analyzing the matrices M_b^{KS}, for the test T_{KS} parameter $y_j \in Y$, , which has a maximum number of ones in the column, is being chosen. In the case when the jth column does not contain only ones, the missing ones in the nth or nth+1 column are searched for. In the event of their occurrence, *nth* and *nth + 1* parameters are included in the test T_{KS}, then the test takes form:

$$T_{KS} = \{y_j, y_n, y_{n+1}\},\ T_{KS} = \{d_j, d_n, d_{n+1}\} \tag{6}$$

where:
d_j - means checking the jth diagnostic parameter.

In the case determining the elements of the T_{LU} test create boolean matrix, M_b^{LU} to locate faults in which a subset of state pairs S_k, S_i; $k = \overline{1, K};\ i = \overline{1, I}$; $i \neq k$ is introduced instead of states. Then analyzing the matrix M_b^{LU}, for the test T_{LU} parameter $y_j \in Y$, which has a maximum number of ones in the column, is being chosen. In the case when the jth column does not contain only ones, the missing ones in the nth or nth+1 column are searched for. In the event of their occurrence, *nth* and *nth + 1* parameters are included in the test T_{LU}.

In order to determine the value of the binary elements of the observation matrix, the significance of changes in the values of diagnostic parameters should be performed $\{y_j\}$ depending on the elements of the set of states $\{s_i\}$. Among many possibilities [1,3,4,5,6] the chosen ones that deserve attention are:

a) method of testing the average value;
b) the method of testing the distance of confidence intervals of the mean value of the diagnostic parameter;

4 Testing the relationship diagnostic parameter - state using the medium test

The relation of the diagnostic parameter - state test is carried out by means of the procedure of testing the significance of changes in the values of the output parameters with the parametric significance test of the two averages for two sets of parameter values $\{\{y_{jl}\}(\Theta_1)\}$ i $\{\{y_{jl}\}(\Theta_k)$;. To do this [4,7,8,9]:

1. Calculate the average value:

$$\overline{y} = \frac{1}{N}\sum_{i=1}^{N} y_i \tag{8}$$

2. Calculate the variance of the diagnostic parameter values:

$$\hat{s}^2 = \frac{1}{N-1}\sum_{i=1}^{N}\left(y_i - \overline{y}\right)^2 \tag{9}$$

3. Testing the hypothesis about equality of variances with the F-Snedecor test:

$$H_o : \sigma_2^o = \sigma_i^2 \; ; i = \overline{1,k}$$

$$H_1 : \sigma_0^2 > \sigma_i^2 \quad \text{or} \; \sigma_0^2 \neq \sigma_i^2 \tag{10}$$

a) the value of F-Snedecor's statistics: $F = \hat{s}_o^2 \,/\, \hat{s}_i^2$ gdy $\hat{s}_o^2 > \hat{s}_i^2$; $F = \hat{s}_i^2 \,/\, \hat{s}_o^2$ when $\hat{s}_i^2 \leq \hat{s}_o^2$;

b) critical value of the distribution $F_\alpha = F(\alpha, N_o\text{-}1, N_i\text{-}1)$
where: α - the level of significance of the test; α=0,05
N_o - number of measurements for the state S_o
N_i - number of measurements for the state S_i

If:

$F > F_\alpha$ - the hypothesis H_o is rejected

$F < F_\alpha$ - there is no reason to reject the hypothesis H_o

4. Testing the hypothesis about the equality of mean values using the variance analysis test

$H_o : m_o = m_i$
$H_o : \neg H_o$

a) in case of $\sigma_0^2 > \sigma_i^2$ the value of the *u* statistics is determined:

$$u = \frac{\left|\bar{y}_{j,o} - \bar{y}_{j,a}\right|}{\sqrt{\dfrac{\hat{s}^z_{j,o}}{N_o} + \dfrac{\hat{s}^z_{j,a}}{N_a}}} \tag{11}$$

and is compared to the critical value of the distribution u_α (distribution N(0,1)): $t_\alpha = t(\alpha, N_o+N_i - 2)$; α=0,05.

when: $u > u_\alpha$ - H_o hypothesis is rejected,
$u < u_\alpha$ - there is no reason to reject the hypothesis H_o.

b) in case of $\sigma_2^o = \sigma_i^2$ the value of the *t* statistics is determined:

$$t = \frac{\left|\bar{y}_{j,o} - \bar{y}_{j,a}\right|}{\sqrt{\dfrac{N_o \cdot \hat{s}_{j,o} + N_a \cdot \hat{s}_{j,a}}{N_o + N_a - 2} \cdot \left(\dfrac{1}{N_o} + \dfrac{1}{N_a}\right)}} \tag{12}$$

and is compared to the critical value of the distribution t_α (Students distribution): $t_\alpha = t(\alpha, N_o+N_i - 2)$; α= 0,05.
when:
$t > t_\alpha$ H_o hypothesis is rejected,
$t < t_\alpha$ there is no reason to reject the hypothesis H_o.

Due to the incomparable measurement conditions, it can be expected that $\sigma_2^o \neq \sigma_i^2$. This is a premise that in order to determine the significance of changes in diagnostic parameter of the state of transport u statistics should be used and the related application procedure. This means that as a result of this procedure, the values of the binary matrix are appropriately assumed: 1 - when the test of averages shows that the hypothesis about the equality of mean values of parameters is rejected $\{\{y_{j1}\}(\Theta_1)\}$ and $\{\{y_{j1}\}(\Theta_k)\}$ and 0 when the test of averages shows that there is no reason to reject the equality hypothesis average values of parameters $\{\{y_{j1}\}(\Theta_1)\}$ and $\{\{y_{j1}\}(\Theta_k)\}$.

5 Investigation of the distance of confidence intervals of the mean value of the diagnostic parameter

Investigation of the distance of confidence intervals of the mean value of the diagnostic parameter for two states of the means of transport [1,3,5,7] is carried out using the distance test procedure for two sets of parameter values $\{\{y_{jl}(\Theta_1)\}$ and $\{\{y_{jl}(\Theta_k)$. It is assumed that the population has a distribution $N(m, \sigma)$, with both mean value m and standard deviation σ are unknown in the population. Moreover, assuming that a small sample was drawn randomly from the population ($n \geq 10$), the confidence interval for the average population is obtained by dependence [7,9]:

$$r_{\alpha j} = t_{\alpha j} \frac{S_j}{\sqrt{n-1}} \tag{13}$$

where: $S_j = \sqrt{\frac{1}{N}\sum_{l=1}^{N}\left(y_{jl} - \overline{y_j}\right)^2}$ – standard deviation from the sample,

$\overline{y_j} = \frac{1}{N}\sum_{l=1}^{N} y_{jl}$ - the arithmetic mean of the results of the trial,

t_α – Student's quantile distribution for N-1 degrees of freedom and significance level $\alpha = 0,05$.

1- α – coefficient of confidence (probability of coverage by the confidence interval average value m of the population, $1 - \alpha = 0,95$),

N – number of objects of the group of means of transport.

Further examination consists in conducting a test for finding common points for the determined confidence intervals for two sets $\{\{y_{jl}(\Theta_1)\}$ and $\{\{y_{jl}(\Theta_k)\}$ according to relation:

$$\begin{aligned} &\text{a) for } \overline{y}_j(\Theta_1) < \overline{y}_j(\Theta_k): \quad \overline{y}_j(\Theta_k) - r_{\alpha j}(\Theta_k) > \overline{y}_j(\Theta_1) + r_{\alpha j}(\Theta_1) \quad \Rightarrow \text{„1”} \\ &\text{b) for } \overline{y}_j(\Theta_1) > \overline{y}_j(\Theta_k): \quad \overline{y}_j(\Theta_1) - r_{\alpha j}(\Theta_1) > \overline{y}_j(\Theta_k) + r_{\alpha j}(\Theta_k) \quad \Rightarrow \text{„1”} \end{aligned} \tag{14}$$

Correspondingly, the values of the binary matrix are assumed: 1 - when the test shows that the confidence intervals of the mean parameters values $\{\{y_{j1}\}(\Theta_1)\}$ and $\{\{y_{j1}\}(\Theta_k)\}$ they do not have common points (they are significantly dependent on the state s_i) and 0, when the distance test shows that the confidence intervals of the mean parameters $\{\{y_{j1}\}(\Theta_1)\}$ and $\{\{y_{j1}\}(\Theta_k)\}$ they have points in common (they are not significantly dependent on the state s_i).

6 The algorithm for determining the status and fault location test

The algorithm of the methodology for determining the diagnostic test of means of transport includes the following stages [8,9,10]:

1. Data acquisition.
2. Optimization of the set of diagnostic parameter values (only in the case of a large number of elements in Y set, e.g. m>10).
3. Organize a data set by specifying a set $\{s_i(\Theta_k),\ i=1,\ldots,1;\ k=1,\ldots,K\}$.

4. The study of the significance of changes in the values of diagnostic parameters $\{y_j(\Theta_k)\}$ depending on the state $\{s_i = f(\Theta_k); \Theta_k \in (\Theta_1, \Theta_b)\}$, ie which of the diagnostic parameters "best" describes the state s_i:
   a) Using the method of testing the distance between confidence intervals of mean values of diagnostic parameters in the case of testing means of transport (number of objects greater than 10). Correspondingly, the values are: 1 - when the test shows that the confidence intervals of the average parameters $\{\{y_{jl}\}(\Theta_1)\}$ and $\{\{y_{jl}\}(\Theta_k)\}$ they have no common points and 0 if the distance test shows that the confidence intervals of the mean parameters $\{\{y_{jl}\}(\Theta_1)\}$ i $\{\{y_{jl}\}(\Theta_k)\}$ they have common points according to the relations (14);
   b) In case of set $\{s_i = f(\Theta_k)\}$ is a multi-element set of results of studies on the significance of changes in diagnostic parameter values $\{y_j(\Theta_k)\}$ depending on the state $\{s_i = f(\Theta_k); \Theta_k \in (\Theta_1, \Theta_b)\}$ are identical for all states $\{s_i = f(\Theta_k)\}$.
5. Designation of the boolean matrix:
   a) 1 – when change of state s_i causes significant changes in the value of the diagnostic parameter y_j;
   b) 0 – when change of state s_i does not cause significant changes in the value of the diagnostic parameter y_j;
6. Designation of the T_{KS} state control test based on the boolean matrix - T_{KS} test as a vector of logical values (0, 1) diagnostic parameters and identical to vector: $\{(y_1, \ldots, y_n, \ldots, y_N)\} \Rightarrow S^0 \vee S^1 = \{(s_1, \ldots, s_n, \ldots, s_N)$ for fitness S^0 and the unfittnes of the means of transport S^1.
7. Determination of the fault location test T_{LU} based on the boolean matrix - test T_{LU} as a vector of logical values <0, 1> diagnostic parameters and identical to vector: $\{(y_1, \ldots, y_n, \ldots, y_N)\} \Rightarrow S^1 = \{(s_1, \ldots, s_n, \ldots, s_N)$, while:
   a) if the logical value of the diagnostic parameter check vector assumes the value "1" - the parameter value is within the limit value range: $y_j = y_{jg}$;
   b) if the logical value of the diagnostic parameter check vector assumes the value "0" - the parameter value is outside the limit value range: $y_j \neq y_{jg}$.
8. Interpretation of test results:
   a) $T_{KS} = \{y_1, \ldots, y_j, \ldots, y_m\}$:
      - $\{0, \ldots, 0, \ldots, 0\}$ – means of transport usable (fit) ("0" means that value $y_j \neq y_{jg}$),
      - $\{1, \ldots, 0, \ldots, 0\}$ – means of transport unfit („1" means that value $y_1 = y_{1g}$),
      - $\{1, \ldots, 1, \ldots, 0\}$ - means of transport unfit („1" means that value $y_j = y_{jg}$),
      - $\{1, \ldots, 1, \ldots, 1\}$ - means of transport unfit („1" means that value $y_m = y_{mg}$),
   b) $T_{LU} = \{y_1, \ldots, y_j, \ldots, y_m\}$:
      - $\{1, \ldots, 0, \ldots, 0\}$ - means of transport unfit, location of fault – state s_i („1" means that value $y_1 = y_{1g}$),
      - $\{1, \ldots, 1, \ldots, 0\}$ - means of transport unfit, location of fault – state s_{i+1} („1" means that value $y_j = y_{jg}$),
      - $\{1, \ldots, 1, \ldots, 1\}$ - means of transport unfit, location of fault – state s_{i+n} („1" means that value $y_m = y_{mg}$),

7 Implementation of the algorithm for determining the status and fault location test

The scope of the algorithm implementation was formulated on the basis of functional requirements related to the implementation of the methodology of optimization of a set of diagnostic parameters to assess the state of the machine and to determine the technical condition and fault location test. On the basis of the above scope of implementation, tasks formulated as components of the assessment process of means of transport were formulated. These are [9,10]:

1. Acquisition of measurement and simulation data:
   a) entering data by the end user;
   b) data transfer using the "copy and paste" mechanisms implemented in the Windows system;
   c) data import from other database systems or from text files;
   d) data edition;
   e) saving the entered data to the database;
   f) reduction of the set of diagnostic parameters.
2. Machine condition assesment:
   a) determination of the relation matrix: technical condition - value of the diagnostic parameter;
   b) entering the relation matrix: technical condition - operation time - value of the diagnostic parameter);
   c) to designate a diagnostic test to check the condition and location of damage to the machine.
3. Data reporting and visualization:
   a) visualization of selected series in the form of line and point charts;
   b) defining the parameters of the visualized series (eg color, thickness or type of line);
   c) visualization of the form of the condition and fault location test;
   d) display in a tabular form the results of simulations.

8 Conclusions

To sum up the issues discussed above concerning the methodology of building diagnostic tests of means of transport, the following conclusions can be made:

1. All presented procedures for determining state and location control tests allow to determine optimal diagnostic tests due to the adopted criteria.
2. Due to the preference in the selection of diagnostic parameters of the similarity method and the method of testing the dependence of the diagnostic - state parameter, in order to build a binary diagnostic matrix in the field of logical relations (0, 1), the method of testing the confidence interval distance of the mean value of the diagnostic parameter should be selected.
3. In the case of building a TD diagnostic test, based on the binary analysis of the Boolean matrix, the method of the check vector should be used $T_D = < T_{KS}, T_{LU} >$:

a) T_{KS} test as a vector of logical values (0, 1) of diagnostic parameters and a state vector of states: $\{(y_1, \ldots, y_n, \ldots, y_N)\} \Rightarrow S^0 \vee S^1 = \{(s_1, \ldots, s_n, \ldots, s_N)$ for the state S^0 and state of incapacity of transport S^1.

b) T_{LU} test as a vector of logical values <0, 1> of diagnostic parameters and a state vector of states: $\{(y_1, \ldots, y_n, \ldots, y_N)\} \Rightarrow S^1 = \{(s_1, \ldots, s_n, \ldots, s_N)$.

References

1. J. S. Bendat, A.G Piersol.: *Metody analizy i pomiarów sygnałów losowych,* PWN, Warszawa (1976)
2. L. Będkowski *Elementy diagnostyki technicznej*, WAT, Warszawa (1991)
3. D.C. Betz: *Application of optical fibre sensors for structural health and usage monitoring*. Dynamics Research Group, Department of Mechanical Engineering, The University of Sheffield. Sheffield (2004)

4. W. Cholewa, J. Kaźmierczak: *Data processing and reasoning in technical diagnostics*. WNT, Warszawa (1995)
5. S. Niziński: *Eksploatacja obiektów technicznych*. ITE, Radom (2002)
6. W.J. Staszewski, C. Boller, G.R. Tomlinson: *Health Monitoring of Aerospace Structures*. John Wiley & Sons, Ltd. Munich, Germany (2004)
7. F. Tomaszewski: *Redukcja informacji diagnostycznej w rozpoznawaniu stanu maszyn*. Diagnostyka. Vol. 26, PTDT, Olsztyn, (2002)
8. H. Tylicki: *Redukcja informacji diagnostycznej w rozpoznawaniu stanu maszyn*. Diagnostyka, vol. 26, Olsztyn, (2002)
9. H. Tylicki: *Wykorzystanie dedykowanego systemu diagnostycznego w rozpoznawaniu stanu maszyn. Sprawozdanie z realizacji prac badawczych „Techniki wirtualne w badaniach stanu, zagrożeń bezpieczeństwa i środowiska eksploatowanych maszyn”*. Bydgoszcz (2011)
10. B. Żółtowski: *Diagnostic system for the metro train*. ICME, Science Press, Chengdu, China, s.337-344. (2006)

Application of harmonic balance method for solving non-linear equations of motion

Robert Kostek[1,*]

[1] University of Science and Technology, Al. Prof. S. Kaliskiego 7, 85-796 Bydgoszcz, Poland

Abstract. This article presents the advantages and limitations of a harmonic balance method applied for solving non-linear equations of monition. This method provides an opportunity to find stable and unstable periodic solutions, which was demonstrated for a few equations. An error of solution decreases rapidly with increase of number of harmonics for smooth time history of acceleration, which shows convergence; whereas, for discontinuous time histories, this method is not effective.

1 Introduction

Non-linear equations of motion are usually solved with implicit and explicit numerical methods, which integrate time histories. Typically the Runge–Kutta, polynomial or direct methods are used to solve the equations [1]. These numerical methods provide an opportunity to simulate chaotic and transient vibration; whereas stable periodic vibrations, are obtained after certain time, when transient component decreases to negligible value. Thus there appears to be some problems with simulation undamped and unstable vibrations, because it is known that numerical methods change the energy of simulated system, thus sometimes vibration amplitude rises and transient component remains large. These phenomena are coupled with stability of numerical methods [2]. Undamped vibrations can be simulated with numerical methods, which integrate time histories; nevertheless, additional procedures and skills are required. Whereas the Harmonic Balance Method (HBM) can be used to find stable and unstable cycles and their spectrums without these problems [3].

2 Convergence of HBM

The Harmonic Balance Method employs a Fourier series to estimate time histories of displacement, velocity and acceleration. These series are presented below:

$$x_h = A_0+\Sigma A_n \sin(\omega_h t+\phi_h), \quad (1)$$

$$x_h' = \Sigma A_n \omega_h \sin(\omega_h t+\phi_h+\pi/2), \quad (2)$$

$$x_h'' = \Sigma A_n \omega_h^2 \sin(\omega_h t+\phi_h+\pi), \quad (3)$$

where x_h - is displacement m, x_h' - velocity m/s, x_h'' - acceleration m/s^2, A_n - amplitude of nth harmonic m, $\omega_h=\omega_1 n$ – frequency of nth harmonic rad/s, ϕ_h – phase angle of nth harmonic rad, and t – time in s. These series are substituted in to the following equation of motion:

$$x'' = f(x, x', t)/m, \quad (4)$$

where m - is mass kg. Acceleration can be computed both from equation (3) and (4), thus an error can be calculated from the following expression:

$$E_{x''(t)} = x_h''-f(x_h', x_h, t)/m, \quad (5)$$

where $E_{x''}(t)$ - is an error. The objective function is the root mean square of this error RMS($E_{x''}(t)$). This leads to optimisation problem, where amplitudes $A_0,\ldots,A_i$, phase angles $\phi_1,\ldots,\phi_i$, and frequencies $\omega_1,\ldots,\omega_i$ are variables. Summarising, HBM replace integration by optimisation.

* Corresponding author: robertkostek@o2.pl

Convergence of harmonic balance method should be studied for non-linear equations, which have exact solutions. First, the equation with cubic non-linearity is studied:

$$x'' = (-k_3 x^3)/m \quad (6)$$

where k_3 - is stiffness co-efficient N/m^3. For three harmonics obtained solutions are near a numerical solution, which demonstrates convergence (Fig. 1). Time histories of x_h'' and $f(x_h', x_h, t)$ are very similar, thus value of RMS($E_{x''}(t)$) is small; nevertheless, some error of period is observed. It should be mentioned that, in spite of fact that the studied equation (6) is strongly non-linear, the amplitude of the first harmonic is the largest, whereas further amplitudes are small.

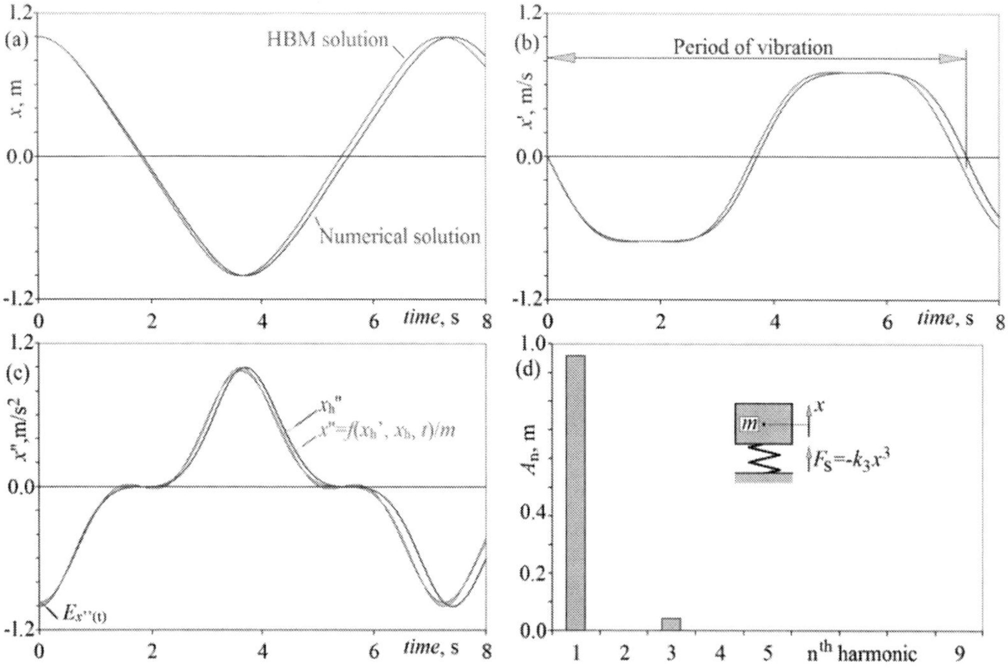

Fig. 1. Time histories of displacement (a), velocity (b), acceleration (c), and spectrum (d) obtained with HBM for the following conditions m=1kg, k_3=1N/m^3, $x_{(t=0)}$=1m and $x'_{(t=0)}$=0m/s, Eq. (6).

Further, two errors can be studied: the error of period E_T and objective function RMS($E_{x''(t)}$). RMS($E_{x''(t)}$) drops rapidly with an increase in the number of harmonic (Fig. 2a), thus for n=7 time histories are very similar to exact solution. The same situation is observed for error of period (Fig. 2b), which is expressed by the following equation:

$$E_T = |(T_{HBM}-T)/T|, \quad (7)$$

where E_T – denotes error of period, T - is exact period and T_{HBM} - represents period obtained with HBM. The same phenomenon is observed for perturbation solution, where the error of period decreases with the increasing order of any solution. The HBM is effective in this case, because time histories are smooth and the amplitudes of higher harmonics are small.

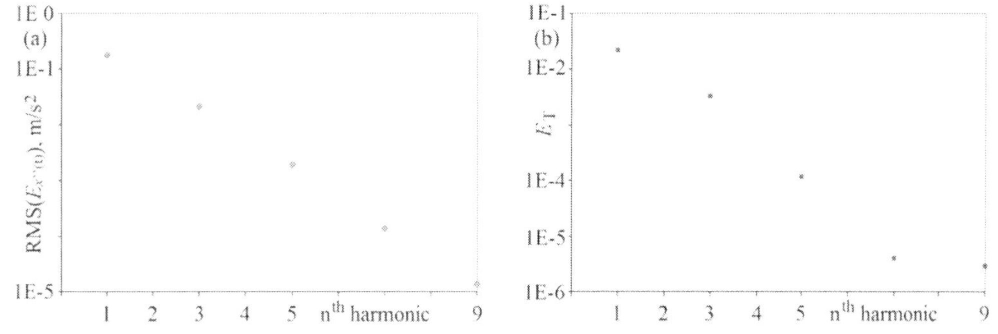

Fig. 2. RMS($E_{x''(t)}$) (a) and error of period E_T (b) against number of harmonic.

Next vibrations described by the following equation are studied:

$$x'' = (-F_0\text{sign}(x))/\text{m}, \quad (8)$$

where F_0 – denotes magnitude of potential force. This force is described by discontinuous function, thus the time history of acceleration is discontinuous, which introduces problems with convergence. The presented solution contains nine harmonics, thus it should be near exact and a numerical solution (Fig.3). For time histories of displacement, good agreement is observed, whereas for time histories of velocity, small differences are noticed. On the other hand differences between time histories of acceleration are obvious. Additional waves are observed, and jumps of acceleration are not well reflected. This shows a slow convergence of the Taylor series to discontinuous functions. Large number of harmonics should be used in this case, which is not effective. The same phenomenon is observed for Coulomb friction. Thus, another solution is applicable - the Walsh function. The spectrum of displacement is depicted in Fig.3d. The amplitude of the first harmonic is the largest, but higher harmonics play a significant role for time history of acceleration.

Next, two errors are studied - RMS($E_{x''(t)}$) and E_T. RMS($E_{x''(t)}$) which drop very slowly with an increase in the number of harmonics (Fig. 4a), thus for n=9 differences between time histories of acceleration are observed. Whereas, E_T drops faster (Fig. 4b), and for nine harmonics period of vibration obtained with HBM is similar to exact period (Fig. 3c).

Summarising this section, HBM is effective for smooth time histories of acceleration, and periodic vibrations. The test problems were without damping force, thus were between stable and unstable regions. Undamped vibrations are difficult to simulation with methods, which integrate time histories.

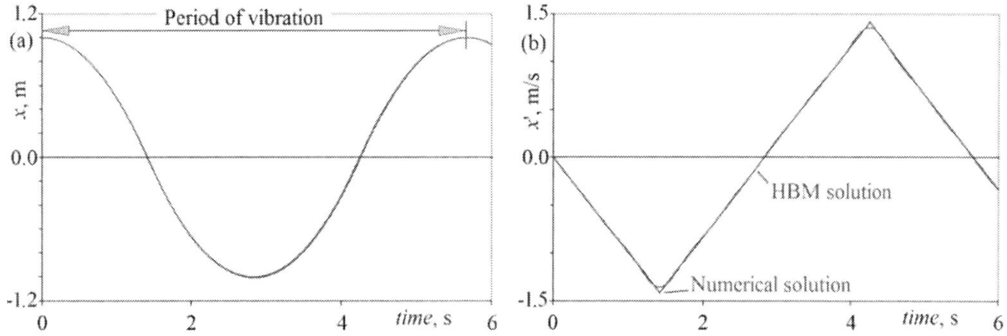

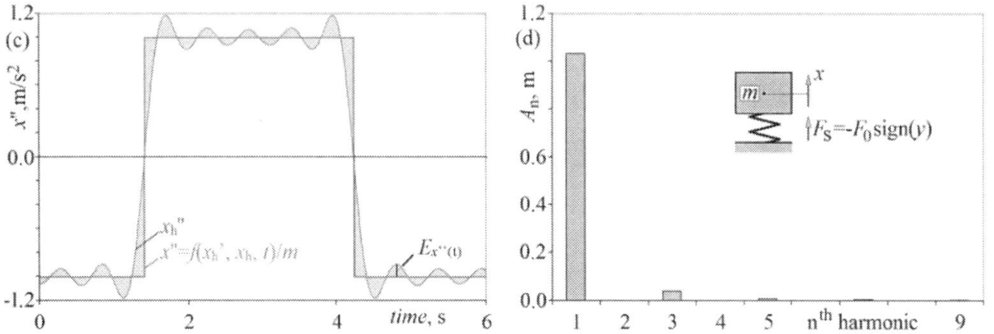

Fig. 3. Time histories of displacement (a), velocity (b), acceleration (c), and spectrum (d) obtained with HBM for the following conditions m=1kg, F_0=1N, $x_{(t=0)}$=1m and $x'_{(t=0)}$=0m/s, Eq. (8).

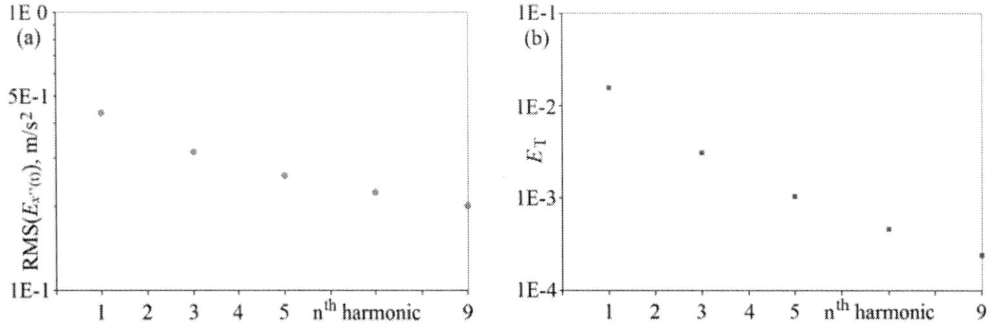

Fig. 4. RMS($E_{x''(t)}$) (a) and error of period E_T (b) against number of harmonic.

3 Application of HBM

Next, the periodic solution of the following equation is studied:

$$x'' = (-k_3x^3 - cx' + F_E\sin(\omega t))/m, \tag{9}$$

where c – denotes damming coefficient N/(m/s), and F_E –amplitude of excitation N. For k_3=1N/m^3, c=0.05 N/(m/s), F_E=7.5N, ω=1rad/s and m=1kg chaotic vibrations are observed, thus HBM can be used to find unstable periodic solutions, which is within any chaotic attractor. Agreement between numerical and HBM solution is excellent for eleven harmonics which shows convergence (Fig. 5). Time histories are smooth but theirs shape is complex, thus eleven harmonics were used. Moreover, amplitudes of higher harmonics are significant. The periodical solution is unstable, because the numerical solution escapes to the chaotic attractor (Fig. 6). This effect is known as the butterfly effect. After a few cycles, a numerical nose is amplified, and the phase point leaves the periodic cycle. Obtained cycle is similar to homoclinic orbit, because of shape of x-x'' trajectory (Figs.6d, 7a).

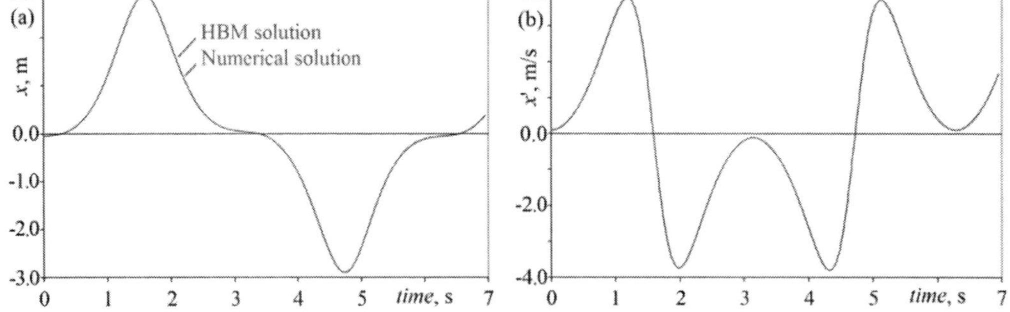

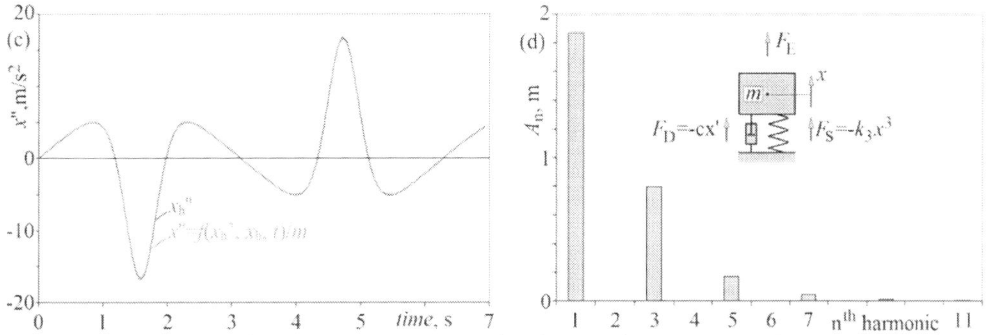

Fig. 5. Time histories of displacement (a), velocity (b), acceleration (c), and spectrum (d) obtained with the HBM and a direct method for k_3=1N/m^3, c=0.05N/(m/s), F_E=7.5N, ω=1rad/s, m=1kg, Eq.(9).

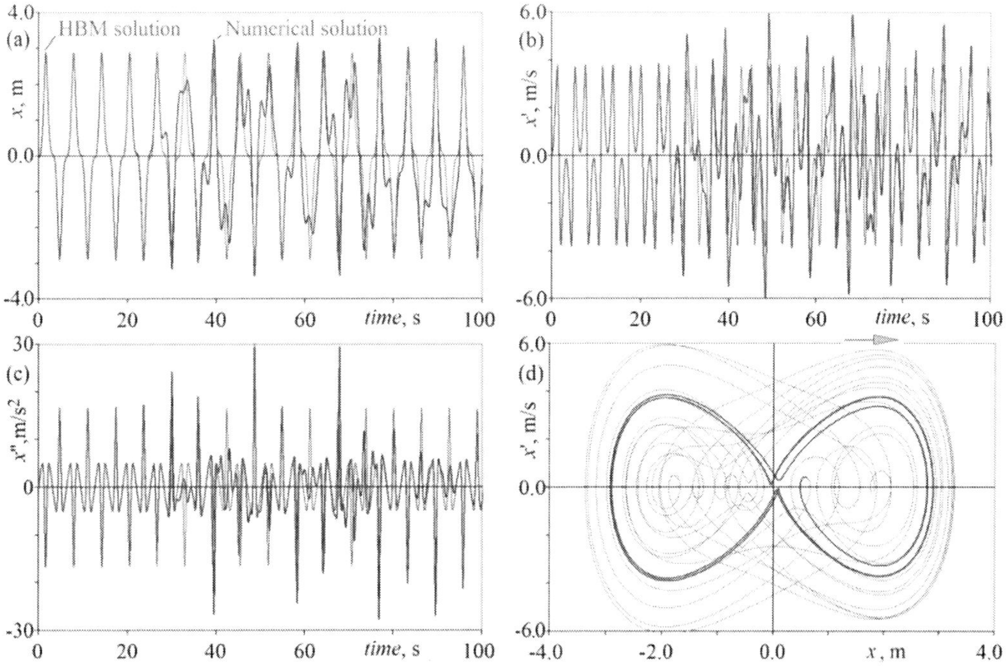

Fig. 6. Time histories of displacement (a), velocity (b), acceleration (c), and phase portrait (d) obtained with the HBM and a direct method for k_3=1N/m^3, c=0.05N/(m/s), F_E=7.5N, ω=1rad/s, m=1kg, Eq. (9).

These vibrations can be studied with x-x'' and x'-x'' trajectories. Trajectory x-x'' of periodic solution is similar to spring characteristic with hysteresis (Fig.7a), whereas shape of chaotic attractor is more complex. For linear vibrations, this trajectory is a straight line. Whereas, trajectory x'-x'' shows complexity of this motion, inner loops are observed. For linear vibrations phase portrait x'-x'' has an elliptical shape. Summarising, shape of chaotic attractor correspond to shape of unstable periodic orbit.

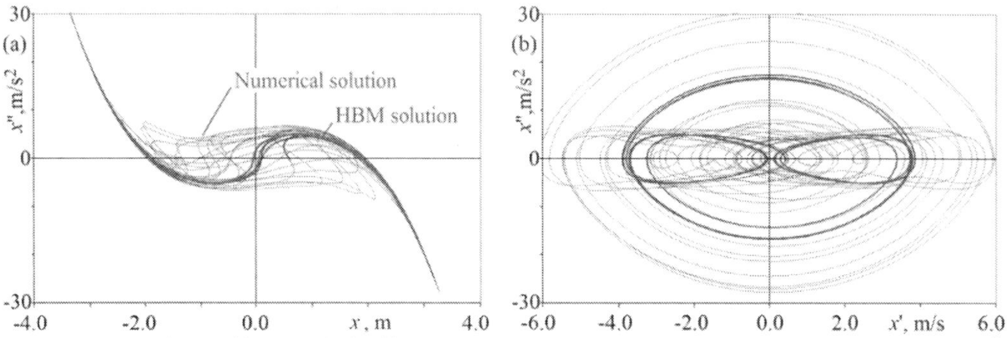

Fig. 7. Trajectories x-x'' (a) and x'-x'' (b), obtained with HBM and a direct method for k_3=1N/m^3, c=0.05 N/(m/s), F_E=7.5N, ω=1rad/s, m=1kg, Eq. (9).

4 Conclusions

The HBM provides an opportunity to find stable and unstable periodic cycles, which play a significant role in the bifurcation theory. Unstable solutions are difficult to obtain with numerical methods, because the numerical solution escapes to an attractor. Obtained results can be compared with a perturbation solution, because perturbation solutions are sums of harmonics. Unfortunately, the HBM is ineffective for discontinues time histories. Summarising, unstable cycles and other phenomena are not observed in linear systems, thus non-linear models should be used in practice, because they better describe real vibrations than linear models.

References

1. R. Kostek, Appl. Math. Comput **219**, 19 (2013).
2. K.J. Bathe, E.L. Wilson, Earthq Eng Struct Dyn, **1** (1972).
3. S.R. Panigrahi, B.F. Feeny, A.R. Diaz, J. Vib. Acoust, **137**, 6 (2015).

AUTHOR INDEX

AUTHOR INDEX